ARTHROPOD BRAIN

ARTHROPOD BRAIN
ITS EVOLUTION, DEVELOPMENT, STRUCTURE, AND FUNCTIONS

Edited by

AYODHYA P. GUPTA
Professor of Entomology
Rutgers University
New Brunswick, New Jersey

A WILEY-INTERSCIENCE PUBLICATION
JOHN WILEY & SONS
New York   Chichester   Brisbane   Toronto   Singapore

Library of Congress Cataloging in Publication Data:

Arthropod brain.
  "A Wiley-Interscience publication."

  Includes bibliographies and index.
    1. Nervous system—Arthropoda. 2. Arthropoda—
Physiology. 3. Brain. I. Gupta, A. P., 1928–
[DNLM: 1. Arthropods—growth. 2. Brain—physiology.
QX 460 A787]

QL434.72.A78 1987   595'.2'04188   87-8253
ISBN 0-471-82811-4

Printed in the United States of America

10 9 8 7 6 5 4 3 2 1

Contributors

Jennifer S. Altman
Institute of Zoology
University of Regensburg
Regensburg, West Germany

Doreen E. Ashhurst
Department of Anatomy
St. George's Hospital Medical
  School
University of London
London, England

K. C. Binnington
Division of Entomology
CSIRO
Canberra City, Australia

W. M. Blaney
Behavioural Entomology Group
Department of Biology
Birkbeck College
University of London
London, England

A. David Blest
Developmental Neurobiology
  Group
Australian National University
Canberra City, Australia

Gary M. Booth
Department of Zoology
Brigham Young University
Provo, Utah

H. Breer
Department of Zoophysiology
University of Osnabrück
Osnabrück, West Germany

Stanley D. Carlson
Department of Entomology
Neurosciences Training Program
University of Wisconsin
Madison, Wisconsin

Steven C. Chamberlain
Institute for Sensory Research
Syracuse University
Syracuse, New York

Jean Chaudonneret
Department of Animal and
  General Biology
University of Dijon
Dijon, France

Thomas A. Christensen
Arizona Research Laboratories
Division of Neurobiology
University of Arizona
Tucson, Arizona

Michel Descamps
Laboratory of Invertebrate
  Endocrinology
University of Science and
  Technology of Lille
Villeneuve d'Ascq, France

Rolf Elofsson
Department of Zoology
University of Lund
Lund, Sweden

Joachim Erber
Institute for Biology
Technical University of Berlin
Berlin, West Germany

Wolf H. Fahrenbach
Laboratory of Electron
  Microscopy
Oregon Regional Primate
  Research Laboratory
Beaverton, Oregon

Wulfila Gronenberg
Institute of Zoology
University of Frankfurt
Frankfurt, West Germany

Ayodhya P. Gupta
Department of Entomology and
  Economic Zoology
Rutgers University
New Brunswick, New Jersey

John G. Hildebrand
Arizona Research Laboratories
Division of Neurobiology
University of Arizona
Tucson, Arizona

Uwe Homberg
Division of Neurobiology
Arizona Research Laboratory
University of Arizona
Tucson, Arizona

Robert Joly
Laboratory of Invertebrate
  Endocrinology,
University of Science and
  Technology of Lille
Villeneuve d'Ascq, France

Haig Keshishian
Department of Biology
Yale University
New Haven, Connecticut

Jenny Kien
Institute of Zoology
University of Regensburg
Regensburg, West Germany

Nikolai Klemm
Department of Zoology
University of Lund
Lund, Sweden

Karl Kral
Department of Zoology
University of Graz
Graz, Austria

Joseph R. Larsen
Department of Entomology
University of Illinois
Urbana, Illinois

Alison R. Mercer
Department of Zoology
University of Otago
Dunedin, New Zealand

Dick R. Nässel
Department of Zoology
University of Lund
Lund, Sweden

Friedrich-Wilhelm Schürmann
Department of Cell Biology
Institute of Zoology
University of Göttingen
Göttingen, West Germany

Monique S. J. Simmonds
Behavioural Entomology Group
Jodrell Laboratory
Royal Botanic Gardens
Kew, Richmond
Surrey, England

Gerhard Wegener
Institute of Zoology
Johannes Gutenberg University
Mainz, West Germany

Preface

The raison d'être of this book is to focus attention on the evolution, development, macro- and microanatomy, and functions of the brain (central nervous system) in the major arthropod groups, as well as in the sister group Onychophora, by bringing together in a single publication most, if not all, significant and up-to-date information. As I began to plan the book some three years ago, it became apparent that the many gaps in our knowledge of the various aspects of the arthropod brain would make the task of synthesizing and systematizing the available information difficult. Thus, perhaps the single most important contribution of this book might be the uncovering of the very gaps in our knowledge of the evolution, development, and functions, including brain metabolism and biochemistry, of the arthropod brain for future research.

Unfortunately, comprehensive comparative studies of the functions of the various brain components in both the developmental stages (immature) and the adults of the representatives of all major arthropod groups are virtually nonexistent; consequently, the lack of such studies presents serious difficulties in not only understanding the evolutionary trends in the arthropod brain, but also in interpreting functional analogies among various groups and correlating newly evolved structural specializations with the concomitant behavioral patterns in various taxa.

The book is organized into three parts: Evolution and Development, Structure and Function, and Techniques; the latter part has been kept small because more comprehensive books on neuroanatomical techniques are presently available. In any multiauthored book some overlap is inevitable, and this book is no exception. However, overlap has been kept to a minimum and retained only where necessary for understanding the discussion at hand. Wherever relevant, overlaps and divergences of opinions in various chapters have been cross-referenced. Most authors have pointed out the dearth of information on their respective topics, and many have suggested areas of further research. Because taxonomic ranking of major arthropod groups is highly controversial, each contributor has used his or her preferred taxonomic categories. Furthermore, each contributor has had complete freedom to develop, interpret, and present his or her view. Each chapter attempts to present an in-depth review of the topics it covers. In the chapters of non-

English-speaking authors, editing has been confined to removing obvious infelicities in order to retain the original style and content.

Organizing a multiauthored book is in many ways comparable to, but more difficult than, conducting an orchestra or directing a complicated movie. While the conductor or the director may, of necessity, compel endless rehearsals and retakes to achieve the desired goal, an editor, alas, does not enjoy such professional luxuries! Thus, an endeavor such as this book could not have been successfully completed without the cooperation and assistance of all the authors who responded to my invitation to contribute and thus made the book possible. To them I am most grateful.

The following individuals, journals, societies, and publishers gave their permissions to reproduce published and/or unpublished materials: J. Boeckh, K. D. Ernst, R. M. Glantz, M. F. Land, K. Schildberger, R. J. Skaer, G. M. Technau, and T. Yamaguchi; *Histochemical Journal* and *Science*; Entomological Society of America, Royal Society of England, and Zoological Society of London; Blackwell, Chapman and Hall, Elsevier, Longman, Oxford University Press, Pergamon Press, Springer-Verlag, John Wiley & Sons, and World Health Organization.

In addition, I am grateful to my wife and children for their understanding, ungrudging support, and enthusiasm during the preparation of the book.

Finally, I hope the book will stimulate further research to fill the gaps that it reveals in our knowledge of the arthropod brain.

AYODHYA P. GUPTA

New Brunswick, New Jersey
August, 1987

Contents

III. TECHNIQUES

ARTHROPOD BRAIN

PART I

EVOLUTION
AND DEVELOPMENT

CHAPTER 1

Evolution of the Insect Brain, With Special Reference to the So-Called Tritocerebrum

Jean Chaudonneret
Department of Animal and General Biology
University of Dijon
Dijon, France

1.1. INTRODUCTION

It is well known that the insect brain comprises three pairs of ganglia, namely the protecerebrum, the deutocerebrum, and the tritocerebrum. In their treatise on the nervous system of invertebrates, Bullock and Horridge (1965) define the tritocerebrum as the "ventral caudal inferior part of the brain which gives rise to nerves to the labrum, the stomatogastric system, and a postoral commissure." They also state that there are two separate tracts in the frontal connective (which I refer to as the ventral root of the frontal ganglion: RGF) (Fig. 1.4) and that the tritocerebrum occasionally has two free postoral commissures, but only one in most insects. However, this commissure seems to be lacking in some instances, especially in higher insects, whereas the double tritocerebral commissure occurs particularly in lower insects. In these insects, the upper commissure (the anteriormost) is generally labeled C1 and the inferior one (the posteriormost) as C2.

Furthermore, when the tritocerebral commissure is double, its relations with the next anatomical elements [ventral dilator muscles of the cibarium: VDMC (dvc in Fig. 1.3B), of the pharynx: VDMP (dvph in Fig. 1.3B), RGFs] are variable.

Thus, the tritocerebral commissural system presents us with a complicated problem, which has given rise to various interpretations (see Section 1.8). It is obvious that this problem is closely related to the morphological significance of the tritocerebral centers (here, the word morphology is used in the sense defined by Snodgrass, 1951). Unfortunately, the few authors who have been interested in this problem have not generally studied the structure of these centers and the intracentric pathways of the constituent fibrillar tracts of the free commissures and those of the nerves in question.

To fill this gap, we studied five insects that belong to very distant groups: the collembolan, *Anurida* (which, according to Denis, 1928, has two free commissures that are very distant from each other); the thysanuran, *Thermobia* (which, according to Chaudonneret, 1950, has two commissures that are separated by the VDMC, dvc in Fig. 1.5); the nymph of the ephemeropteran, *Cloeon* (which, according to Noars, 1962, has only one free commissure); the dermapteran, *Forficula* (in which, according to Moulins, 1969, both the commissures lie above the VDMC); and the larva of the mosquito, *Culex* (which is devoid of free commissures). In this way, I hope to point out the morphological and evolutionary significance of the tritocerebral centers on one hand and of the corresponding commissural system on the other.

1.2. METHODS

The purpose of this morphological study is not to demonstrate the aspect of synaptic branchings or the proper pathways of some nerve fibers, but to establish the architectural pattern of the chief fibrillar tracts, and destination of nerves, and the various interorganic relations. This task requires extremely accurate reconstructions from serial sections of whole heads.

The specific staining methods for the nervous system are not generally suitable for detailed study of the general anatomy of the head, especially because they make tissues very fragile or necessitate the opening of the head capsule, and eventually require the preliminary dissection of the nerve centers. Furthermore, their behavior is capricious (Strausfeld, 1976, for example). It is for these reasons that I used topographical triple stains such as azan, and especially Prenant's. The latter provides pictures whose sharpness is perfectly adequate to indicate the main tracts of intracentric nervous fibers as well as the intracentric roots of nerves.

The methods for obtaining reconstructions are those described in my earlier paper (Chaudonneret, 1967); they were slightly modified, as I used large micrographs instead of drawings.

1.3. *ANURIDA MARITIMA* (COLLEMBOLA)

In insects, the classical tritocerebrum is ventrally and posteriorly followed by the subesophageal ganglion (SEG), which results from the coalescence of the three pairs of gnathal ganglia. This is obvious on a sagittal section of a typical SEG. As one can see in Figures 1.3B, 1.4, 1.7, and 1.8, the three neuromeric neuropil masses (NMs) (CnM, CnMx, CnLb) and the corresponding commissures (CM, CMx, CLb) are clearly set apart from each other by two ventral cortical crests (coarsely dotted) in the interneuromeric planes. The presence of such interneuromeric (or, better, bineuromeric) cortical crests is the very consequence of the coalescence of successive ganglia as explained in Figure 1.2A.

In collembolans, the appearances are somewhat puzzling in the sense that there are four NMs, instead of three, and three cortical crests, instead of two. This can be seen on a sagittal section of the SEG of *Anurida*, for example (Fig. 1.1). Each of these four NMs, has its own commissure inside the SEG. The same situation is found in other collembolans that I have studied: *Tomocerus vulgaris* (Chaudonneret, 1950) and *Isotomurus maculatus* (Fig. 1.3A).

One can find the same pattern in embryos. Figure 1.2B (redrawn from a photograph by Tyszkiewicz, 1976) clearly shows four NMs protruding into the ventral cellular rind in the embryo of *Tetrodontophora*. The author indicates the maxillary and labial neuropil centers (NCs) by numerals 3 and 4; numeral 2 indicates ambiguously the two anteriormost ventral NMs, for, according to the classical opinion of embryologists, they both belong to the mandibular ganglion. Nothing, however, is said about this appearance in the author's text. Numeral 1 indicates the tritocerebral NM, because the section is slightly parasagittal.

Figure 1.1 shows the right half of the cephalic nerve centers of an adult *Anurida*, reconstructed from serial sections. It clearly shows that the second

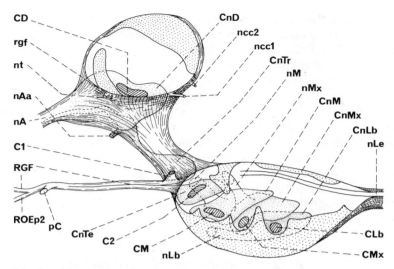

Figure 1.1. *Anurida maritima*: reconstruction of the cephalic nerve centers, mesal view of the right half (ca × 350). *Hatched areas* = sections of commissures; *finely dotted areas* = neuropil centers; *coarsely dotted areas* = section of the cellular rind.

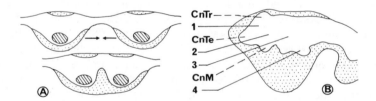

Figure 1.2. (A) Drawing showing how an interneuromeric (bineuromeric) cortical crest appears when two successive ganglia of a nervous chain come to join each other. *Dotted areas* = cellular rind; *hatched areas* = sections of commissures. (B) *Tetrodontophora bielanensis*: longitudinal section of the embryo; simplified drawing from a light microscope photograph by Tyszkiewicz, 1976. The numerals are Tyszkiewicz's. According to him, 1 = neuropil of the tritocerebrum (CnTr); 2 = neuropil of mandibles; 3 = neuropil of maxillae; 4 = neuropil of labium. In my opinion, numeral 2 points out both the tetrocerebral (CnTe) and the mandibular (CnM) neuropils.

NM (CnM) of the SEG is actually mandibular in nature, for it gives rise to the mandibular nerve (nM); it has its own commissure (CM). On the contrary, the first NM (CnTe) has nothing to do with mandibular innervation and has its own commissure (C2). It does, however, give rise to the inferior half of the RGF, which Denis (1928) named "nerf hypopharyngien." This root is classically considered as belonging to the tritocerebrum; in fact, the true tritocerebral NM (CnTr) lies in the inferior part of the peristomodeal connective (PCO), it is obviously set apart from the anteriormost NM of the SEG, and it is a good distance from the deutocerebral one (CnD). The tri-

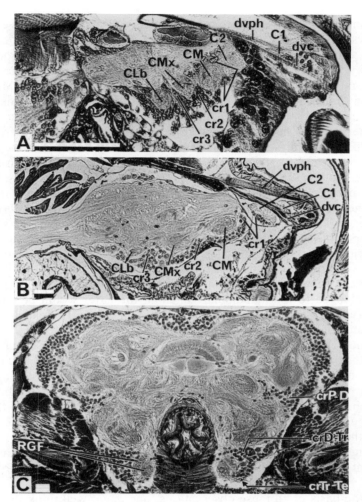

Figure 1.3. (A) *Isotomurus maculatus*: light microscope photograph of a portion of a sagittal section of the head showing the subesophageal ganglion with its four commissures and its three cortical crests. (B) *Thermobia domestica*: light microscope photograph of a portion of a sagittal section of the head showing the subesophageal ganglion with its three commissures and its two cortical crests. The two neuromeric components of the first cortical crest of collembolans are not fused (cr1). Both of the commissures C1 and C2 are free. (C) *Thermobia domestica*: light microscope photograph of a portion of a transverse section of the head showing the brain and the emerging points of the ventral roots of the frontal ganglion (RGF) and their two component fibrillar tracts. Scale = 50 μm.

tocerebral NM has its own commissure (C1 = Denis' b) and gives rise to the dorsal component of the RGF.

This root is very long in *Anurida*, which has a relatively long snout (that is to say the precerebral part of the head is very elongated). The true tritocerebral commissure (C1) lies between the VDMC and the VDMP, but it does not innervate them. Owing to the great distance between the mouth, on the one hand, and the SEG and the PCO, on the other, all these muscles are innervated by nerves (not shown in the drawing) originating from what looks like a commissure (pC = Denis' c), which is situated at the anterior end of the straight part of the RGFs. This transverse link (which also lies between the two sets of VDMC and VDMP) is only an anastomosis, because it does not connect two nerve centers, but rather only two nerves.

In the shortsnouted forms, such as *Tomocerus catalanus, Onychiurus fimetarius* (Denis, 1928), and even *Isotomurus maculatus* (Hutasse, 1972 and personal observations), the RGFs are relatively short; in such cases, these cibariopharyngeal muscles are innervated by the tritocerebral commissure itself (C1 = Denis' K: Fig. 1.3A).

One can see that if the two anterior NMs of the SEG (CnTe and CnM) both belonged to the mandibular ganglion, the latter would be enormous, compared with the maxillary (CnMx) or the labial (CnLb) ganglia. Such a situation would be very difficult to explain.

In short, I can say that the anteriormost NM of the SEG (CnTe) of *Anurida* (and other collembolans) does not belong to the mandibular ganglion.

One might think that the NM, which lies in the PCO (CnTr), and the anteriormost NM of the SEG (CnTe) both belong to the tritocerebrum, because they both take part in the building of the RGF. If such were the case, how could we explain that the volume of these two NC as a whole is much larger than that of any of the gnathal centers, although there are no corresponding functional appendages? Furthermore, how could one explain the fact that the two components of a single ganglion are so distant from each other?

On the basis of all these observations, one might well think that the anteriormost NM of the SEG (CnTe) belongs to a supernumerary neuromere, the tetrocerebrum, because it would be the fourth one in the sequence of the cephalic nerve centers. The preceding NM would then be called the tritocerebrum *s.str.*

1.4. *THERMOBIA DOMESTICA* (THYSANURA: LEPISMATIDAE)

I have described the neuroanatomical pattern of this insect previously (Chaudonneret, 1950) and will briefly outline it and add some new details. From Figures 1.4 and 1.3B, we can see that the SEG contains only the three gnathal NC (CnM, CnMx, CnLb) and their commissures (CM, CMx, CLb), set apart from each other by two well-developed cortical crests. The lower part of

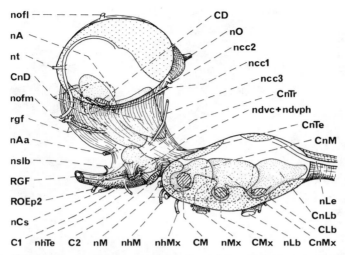

Figure 1.4. *Thermobia domestica*: reconstruction of the cephalic nerve centers, mesal view of the right half (ca × 80).

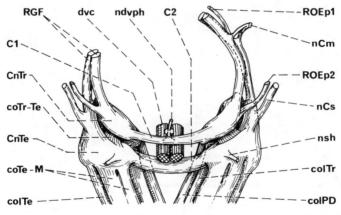

Figure 1.5. *Thermobia domestica*: reconstruction of the dense neuropil masses of the so-called tritocerebrum and main corresponding fibrillar tracts; anteroventral view (ca × 150).

the PCO encloses two distinct NMs (CnTr and CnTe: Figs. 1.4, 1.5, and 1.6A), but they are very close to each other; laterally, they are set apart by a cortical crest, which is perfectly comparable to those that are situated in the interneuromeric planes of the SEG (or of the brain proper: Fig. 1.3C). However, it is less developed. Both of these NMs form the two halves of the RGF. In their proximal part, the two tracts of this root are separated by the continuation of the abovementioned cortical crest (Figs. 1.3C and 1.6C:

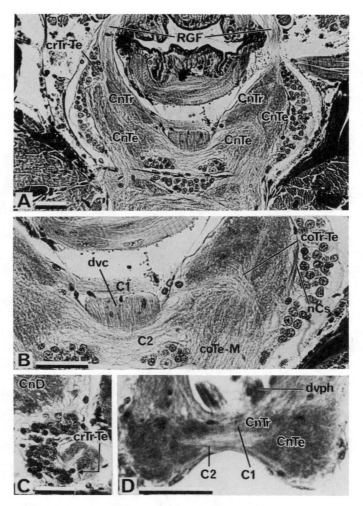

Figure 1.6. (A) *Thermobia domestica*: light microscope photograph of a portion of a transverse section (ventrally and posteriorly inclined) of the head showing the double structure of the so-called tritocerebrum and the connective tract uniting its two ipsilateral components. (B) Same, more enlarged. (C) *Thermobia domestica*: light microscope photograph of a portion of a transverse section of the brain at the level of the emerging point of the ventral root of the frontal ganglion showing the cortical crest between the deutocerebrum (CnD) and the dorsal component of the root and also the prolongation of the tritotetrocerebral rind crest (crTr-Te) between the two components of the root. (D) Nymph of *Cloeon dipterum*: light photograph of a portion of a transverse section of the head showing the structure of the "tritocerebral" region with the two commissures wrapped in the same neurilemma sheath. Commissure C2 is tangentially sectioned. Scale = 50 μm.

crTr-Te). The two tracts we are dealing with remain quite distinct for a long distance, occasionally up to the vicinity of the frontal ganglion.

Each of the two NMs of the PCO has its own commissure (C1 for CnTr and C2 for CnTe; the latter is quite near the SEG); the anteriormost of these two free commissures (C1) runs between the VDMC on the one hand, and the VDMP on the other (dvc and dvph: Figs. 1.3B and 1.5) and innervates them (ndvc + dvph: Figs. 1.4 and 1.5). It is obvious that the tetrocerebral NM of the collembolans (CnTe) and its commissure (C2) have shifted ahead, away from the SEG. Thus, the tetrocerebrum has come to join the true tritocerebrum.

Figure 1.5 shows the dense NMs of the tritocerebrum (in the classical sense) and the main corresponding fibrillar tracts, in anteroventral view. The loose neuropil, the cellular rind, and the neurilemma are assumed to be cut away.

In addition to the trito- and tetrocerebral NCs (CnTr and CnTe), their commissures (C1 and C2), and the two constituent fibrillar tracts of the RGFs, this figure reveals a short and stout bundle of fibers (coTr-Te) uniting the ipsilateral trito- and tetrocerebral NCs. This bundle is comparable to those that unite two successive neuromeres. The interneuromeric connectives, uniting the tetrocerebrum with the mandibular ganglion, are also to be seen (coTe-M). The connectives uniting the tritocerebrum to the deutocerebrum cannot be seen because, due to the "endocranial bending" of the nervous system, they are perpendicular to the plane of the drawing.

Dorsal to these interneuromeric connections (that is to say posteriorly in anatomical terms) are transneuromeric connections running toward the SEG and the ventral chain. They come from (or go to) the proto- and deutocerebrum (colPD), from (or to) the tritocerebrum *s.str.* (colTr) and, *independently*, from (or to) the tetrocerebrum (colTe). Thus, the tritocerebron *s.str.* and the tetrocerebrum behave exactly as the true neuromeric NCs do, longitudinally as well as transversely.

It may also be remarked that both of these neuromeres are concerned with sensory nerves: the roots of the epipharyngeal organs (ROEP1, ROEP2), the clypeal sensory nerve (nCs), the hypopharyngeal sensory nerve (nsh), and some motor nerves (the clypeal motor nerve (nCm), which is related with both the trito- and tetrocerebrum, via the two bundles of the RGF), as I have had the opportunity to observe in some favorably oriented sections.

1.5. *CLOEON DIPTERUM* NYMPH (EPHEMEROPTERA)

In this insect, the so-called tritocerebral lobes are united by only one commissure (Fig. 1.7), as already described by Vassal (1938), Hanström (1940), and Noars (1962). It is the same in other genera: *Silphurus* (Drenkelfort,

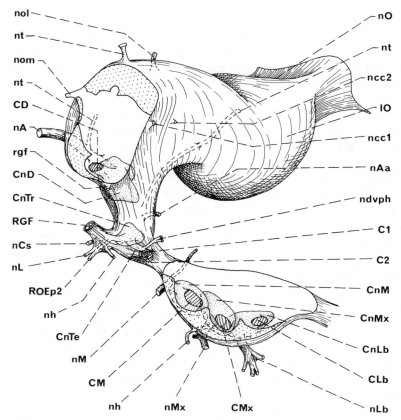

nol
nt
nom
nt
CD
nA
rgf
CnD
CnTr
RGF
nCs
nL
ROEp2
nh
CnTe
nM
CM
nh   nMx   CMx

nO
nt
ncc2
IO
ncc1
nAa
ndvph
C1
C2
CnM
CnMx
CnLb
CLb
nLb

Figure 1.7. Nymph of *Cloeon dipterum*: reconstruction of the cephalic nerve centers, mesal view of the right half (ca × 240).

1910), and *Colobriscus* (Wiseley, 1965). According to Vayssière (1882), this commissure is nearly as long in *Heptagenia* as in the Acrididae.

This commissure, which is clearly distant from the SEG, originates from the PCO in the vicinity of the emerging points of the RGF and gives off the nerves to the VDMP (ndvph). This is the reason why Hanström (1940) thought that this commissure was homologous to C1. Nevertheless, in *Ecdyonurus*, according to Moulins (1971), this musculature is innervated by two nerves directly originating from the PCO, thus leading him to infer that the single "tritocerebral" commissure of the Ephemeroptera corresponds to C2 of the Thysanura, the commissure C1 being absent. Therefore, a more careful study of the tritocerebral region in the Ephemeroptera is needed.

Figure 1.7, which can easily be compared with Figure 1.4, shows that the "tritocerebral lobe" of the nymph of *Cloeon* contains the two NCs: the trito- (CnTr) and the tetrocerebral (CnTe) ones, which I have detected in the other two examples. These two centers lie close together side by side, but remain clearly distinct from each other. They are situated near the middle of the PCO, which is much longer than in *Thermobia* (whose head is globulous). They are distant from both the deutocerebral NM (CnD) and the mandibular ganglion, which lies in the SEG as usual. Each of them constitutes half of the RGF, and gives rise to the single free poststomodeal commissure. The innervation of the VDMP is supplied by the only dorsal component of this commissure. The VDMC are lacking.

In other words, the pattern is the same as in *Thermobia*, with the following differences: first, the trito- and tetrocerebrum have come nearer to each other and have shifted more anteriorly and, second, the two corresponding commissures are imbedded in the same neurilemma sheath. In fact, the two commissural tracts are remarkably distinct from each other in the interior of the only apparent commissure (Fig. 1.6D).

1.6. *FORFICULA AURICULARIA* (DERMAPTERA)

In this insect, we find again two free postoral commissures (Fig. 1.8), as already reported by Kühnle (1913), Lhoste (1957), and Moulins (1969). They are a good distance from the RGF, which seems to emerge from the deutocerebral lobe!

The upper commissure innervates the cibarial (ndvc, nsc) and the pharyngeal musculature (nvph). No nerves arise from the inferior commissure. At first glance, one might think that these two commissures are homologous with those of thysanurans. Nevertheless, the VDMC curiously do not run between these commissures, but under the lower one. Clearly, reinvestigations are needed.

In Figure 1.8, we again find the trito- and tetrocerebral NCs (CnTr and CnTe), which are only much closer to each other than in the previous examples. They have shifted more anteriorly than in *Cloeon* and have come to join the deutocerebral NC (CnD). These trito- and tetrocerebral centers are dorsoventrally elongated and give rise by halves to the RGF, which emerges from their upper parts. These two centers also give rise to the poststomodeal commissures, which emerge from their lower part. This disposition, as well as the great anteroposterior elongation of the brain, seems to be related to prognathism (*cf.* Popham, 1962).

A closer examination shows that both the trito- and tetrocerebral NCs give rise by halves to the inferior commissure (C1 and C2). By contrast, the upper commissure (pC) arises from the tritocerebrum only.

This is better seen on a frontal reconstruction of the so-called tritocerebral lobes (Fig. 1.9). It is obvious that the inferior commissure, which comprises two distinct tracts of nerve fibers (C1 and C2) coming from the tritocerebral neuropil (CnTr) for the uppermost, from the tetrocerebral neuropil (CnTe) for the lower, is homologous with C1 and C2 of the previously studied in-

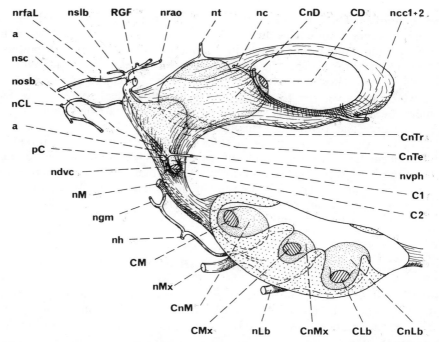

Figure 1.8. *Forficula auricularia*: reconstruction of the cephalic nerve centers, mesal view of the right half (ca × 100).

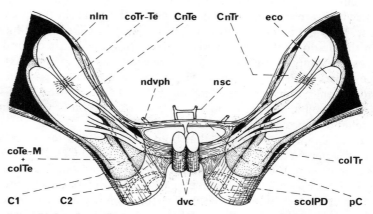

Figure 1.9. *Forficula auricularia*: reconstruction of the dense neuropil masses of the so-called tritocerebrum and main corresponding connective and commissural fibrillar tracts, anterior view (ca × 200).

14

sects, but both of these commissures are wrapped in the same neurilemma sheath, just as in the nymph of *Cloeon*. Here, they are only much closer to each other than in the ephemeropteran nymph.

As for the upper commissure (pC), it serves the ventral cibariopharyngeal muscles and is to be considered as a pseudocommisure, that is to say, a simple anastomosis (probably with chiasmatic fibers), uniting two symmetrical nerves whose pathways converge to the sagittal plane. This pseudocommissure is comparable to the one we found between the anterior parts of the RGFs in *Anurida*, although its relative situation with the PCOs and these roots is very different. This can easily be explained by the large differences between the local topography in these two cases.

Another anastomosis of this kind can be found in *Forficula* in the pattern of innervation of the anterior frontal retractor muscles of the labrum (nrfaL: Fig. 1.8) and between the nerves to the ventral cibarial semicircular muscles (nsc). These anastomoses are marked "a" in Fig. 1.8. Moreover, it must be remembered that, in some insects, the ventral cibariopharyngeal musculature is innervated by nerves coming from the PCOs, or even from the recurrent nerve, without any connection with the commissural system. This has been pointed out by Rousset (1966) in the larva of the neuropteran, *Chrysopa* and by Moulins (1971) in the nymph of the ephemeropteran, *Ecdyonurus*. In these insects, the "tritocerebral" commissure is single, but it is highly probable that it has the same structure as the nymph of *Cloeon* or *Forficula*. Furthermore, we will see later that in the larva of the mosquito, *Culex* (in which the two commissures are present, although they are not free), the innervation of the ventral cibariopharyngeal dilator muscles does not come from C1. It can be thought, thus, that the tritocerebral fibers, which innervate these muscles, are basically independent of the true tritocerebral commissure, although they generally run inside it. The usual fusion of these nerves with the commissure C1, as well as the possible fusion of the commissures C1 and C2 with each other, are the expression of a general tendency shown by nerves with neighboring pathways, whatever their neuromeric belonging may be.

Other points of interest are shown in Figure 1.9: there is an interneuromeric connective (coTr-Te) between the trito- and tetrocerebral NCs, and each component of the inferior commissure is double-rooted. This is reminiscent of the double commissures, which have been described in metameric ganglia of the ventral chain in some insect embryos. This double structure is generally considered as a primitive feature. I was, thus, astonished when I discovered the double roots of the commissures C1 and C2, because the "tritocerebral" region is one of the most modified in the nervous system; these modifications are the consequence of the backward migration of the mouth and the (correlated?) bending of the neural axis, as I have explained in my previous papers (Chaudonneret, 1950, 1966, 1978b).

In short, I can say that the pattern of the so-called tritocerebral lobes of *Forficula* is basically the same as in *Thermobia* and in the nymph of *Cloeon*.

That is to say, they comprise two neuromeres, namely the tritocerebrum *s.str.* and the tetrocerebrum, each possessing its own commissure and both giving rise, by halves, to the RGF. Here, this pattern is somewhat obscured by the presence of a pseudocommissure.

1.7. *CULEX HORTENSIS* LARVA (DIPTERA)

In this insect there are no free poststomodeal commissures (Fig. 1.10). We find anew, however, the trito- and tetrocerebral NCs (CnTr and CnTe), which are very close to each other. They are located in the middle part of the PCO. Both of them give rise, by halves, to the RGF; each has its own commissure (C1 and C2), but both of these commissures have sunk into the ventral part of the PCO and into the dorsal part of the SEG. In this part, they are separated from the mandibular commissure (CM) by a long narrow cortical lamina in the same way that the three gnathal commissures are set apart from one another.

No nerves arise from the tritocerebral commissure *s.str.* (C1) or from the tetrocerebral one (C2). Here, the nerves, which serve the VDMC and VDMP, come from the upper posterior region of the PCO; because I have

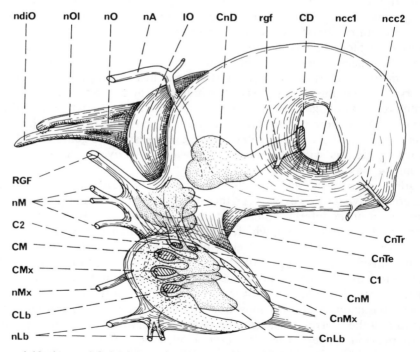

Figure 1.10. *Larva* of *Culex hortensis*: reconstruction of the cephalic nerve centers, mesal view of the right half (ca × 200).

not been able to see their emerging point with sufficient accuracy, they are not represented in my drawing.

Because the commissures C1 and C2 are intracentral, the VDMC (which are present) must run above them. As a consequence, and contrary to the current opinion, the relative situation of the VDMC with the commissural system cannot be a good criterion to establish the morphological significance of the poststomodeal commissure(s): these muscles can run between C1 and C2, as it is the primitive pattern in my opinion (*cf.* the collembolans and thysanurans, for example); above these commissures when they have sunk into the SEG (*cf.* the larva of *Culex*); or below these commissures when they have fused with each other, while remaining independent of the SEG (*cf. Forficula*).

In Figure 1.10, we can see that the mandibular nerves (nM, which are deeply rooted inside the mandibular neuropil mass) leave the central nervous system at the level of the "tritocerebral lobes," as already pointed out by Christophers (1960) in the larva of *Aedes aegypti*. Thus, the emerging point of a nerve does not infallibly show its neuromeric significance.

1.8. DISCUSSION

In each of the five insects studied here, there are two distinct NCs between the deutocerebral and the mandibular ones. They show various features of neuromeric units, as we have seen, especially in *Thermobia* and *Forficula*. Thus, it is reasonable to say that these two NCs belong to two different neuromeres, namely the tritocerebrum *s.str.* and the tetrocerebrum.

Several criticisms have been leveled at the idea of such a supernumerary segment in the insect head. Let us consider the various types of arguments used against this point.

1.8.1. The Double Constitution of the Tritocerebrum *s.l.* Actually Does Not Prove the Existence of Two Distinct Neuromeres

This argument has been used by Bitsch (1963, 1973), Hutasse (1972), and Moulins (1971). It is based on the fact that the deutocerebrum generally comprises two more or less distinct NCs; one of these is considered as motor, the other as sensory. All the authors agree in ascribing these centers to a single neuromere.

In fact, this duality of the deutocerebral center is the consequence of the necessity of integrating the numerous sensory messages coming from the antennary sensilla. This requires a great development of the corresponding synaptic areas and, consequently, the hypertrophy of the sensory NC; this center always appears as a simple lobe of the whole deutocerebral neuropil. As for the tritocerebrum (*auctorum sensu*), the two constituent NCs, which are widely distant from each other in collembolans, are always distinct from each other, even if they often lie closely side by side (as in most other insects). Contrary to what happens with the deutocerebrum, each of the two so-called tritocerebral centers is both, at once, sensory and motor (*cf. Thermobia*) and has its own commissure (C1, C2, which are always discernible, even if they lie closely side by side under the same neurilemma sheath). Moreover, these two centers are united by a nervous tract of connective type (coTr-Te: Figs. 1.5, 1.6B, and 1.9), and each of them receives trans-neuromeric links as every typical neuromere does.

1.8.2. A Double Commissure is a Primitive Feature of Neuromeres

Raised by Bitsch (1963), this argument is based on the fact that both the ganglia of the same pair are united by two commissures in phyllopodous crustaceans as well as in the thoracoabdominal nervous chain in some insect embryos.

In insects, the "tritocerebral" region is deeply altered (*cf. supra*); as a consequence, it is not likely that a primitive feature will be preserved. However (and contrary to all my expectations), *Forficula* shows (Fig. 1.9) that each of the commissures C1 and C2 (which make up the only true visible commissure) has a double root into the corresponding NM. Perhaps this duality of the roots may be a remnant of the presumed primitive duality mentioned.

1.8.3. The Commissure C1 Has Nothing To Do With a True Interganglionic Commissure: It Belongs to the Stomatogastric Sympathetic System; C2 is the True Tritocerebral Commissure

This argument has been put forward in one way or another by numerous authors, more particularly Bitsch (1963, 1973), François (1969), Hutasse (1972), Moulins (1971), Rähle (1970), and Rousset (1966).

The argument is based on the fact that the ventral musculature of the cibarium and of the pharynx (considered as a visceral one) is innervated by the commissure C1. In other words, this commissure would only represent a simple anastomosis between visceral nerves. However, it would link a pair of stomatogastric centers: the "centres rostraux" imagined by Denis (1928). It would be considered to represent the ventral part of a hypothetical "anneau sympathique périoral," whose dorsal part would correspond to the frontal ganglion.

As we have already seen (Section 1.6.), however, the innervation of the ventral cibariopharyngeal musculature is sometimes actually entirely independent of the commissure C1; consequently, this commissure is basically independent of the stomatogastric system.

According to Denis (1928) (and this is more or less tacitly admitted by some other authors), the hypothetical sympathetic center (or rostral center), corresponding to C1, would be situated in collembolans, at the very least, in the anterior region of the SEG; it is precisely in this place that I have pointed out the tetrocerebral center and shown that C2 corresponds to this center, but not C1, as the mentioned authors postulated. It must be added that it would be very difficult to understand in what manner a so-called sympathetic center would be able to slip into the neuromeric sequence of the other centers in such an intimate way as to mimic exactly a true neuromere (cf. especially the case of *Thermobia*).

1.8.4. The Existence of a True Neuromere Implies the Existence of a Corresponding Metamere; No Embryological Study Has Been Able to Find the Least Sign That Points to a Tetrocephalic Segment

This argument is certainly the most widely used, especially by Weber (1952), Bitsch (1963, 1973), Matsuda (1965), Brückmoser (1965), Wada (1966), Malzacher (1968), Rohrschneider (1968), Scholl (1969), Larink (1969, 1970), Tamarelle (1972), Rempel (1975), and Haget (1977).

In fact, no corresponding celomic cavities, or even simple mesodermal masses, have ever been seen by embryologists. But it must be remarked that, in such a matter, appearances of celomic cavities are clear only at the levels of the abdomen, the thorax, and the gnathal region. Already, "chez presque tous les Insectes évolués, on n'observe pas de cavités coelomiques mandibulaires" (Haget, 1977). Their appearances are far from typical in the pregnathal region, whose interpretation is still controversial. Little cavities only rarely appear in the mesoderm of the intercalary segment (= tritocephalic *s.str.*): *Locusta* (Roonwal, 1937), *Anurida* (Garaudy-Tamarelle, 1969). In fact, the ontogenic processes are very deeply altered in this region by the backward migration of the mouth (only the last stages of which are discernible in embryos), by the correlated upward migration of the ventral preoral areas, and by the hypertrophy of the protocephalic and mandibular territories; the model I proposed (Chaudonneret, 1966) accounts for all these phenomena.

In the adult, the tetrocephalic segment is very reduced; very few structures should be considered as originating from the tetrocerebral mesoderm: only a part of the posterior clypeal muscles of the labrum or weak parts of the pseudotentorium of collembolans (Chaudonneret, 1950, 1973). The consequence is that the embryonic mesoderm of this vestigial segment is greatly reduced. In fact, it is practically indistinguishable from the tritocephalic *s.str.* one. In this tetrocephalic segment, the development of celomic vesicles is impossible and it is certainly unrealistic to hope for their discovery, even in lower insects. In his study on the embryogenesis of *Lepisma*, Larink (1970) wrote: "Im Interkalarsegment findet man nur während der kurzen Zeit der Segmentierung getrennte Somite. Sie treten bald mit dem Mandibularmesoderm in Verbindung."

As far as the nervous system is concerned, the trito- and tetrocerebral ganglia are so close to each other in larval or adult insects that they are already fused when they appear in the embryo. The collembolans are a noteworthy exception in the sense that, in the adult as well as in the embryo, the tritocerebrum *s.str.* is located in the PCO, whereas the tetrocerebrum is situated in the SEG. In fact, the presence of an additional neuromere (the tetrocerebral one) in the SEG of a collembolan was pointed out as early as 1900 by Folsom in a paper (very remarkable for its time) about the embryogenesis of *Anurida*. His Figure 28 exactly corresponds to what I found in the adult of this same collembolan. Unfortunately, Folsom considered the first neuromere as the mandibular one; he was, thus, led to conclude that the labium is homologous with the first maxillipede of crustaceans; this view is untenable.

In a paper dealing with the embryology of the collembolan *Isotoma*, Philiptschenko (1912) did not find in his transverse sections the neuromere discovered by Folsom in longitudinal sections and, for that reason, he rejected Folsom's observations.

In a succint account of the embryogenesis of the collembolan *Orchesella*, Brückmoser (1965) also denied Folsom's observations. Nevertheless, he actually represented four NCs in the SEG. He arbitrarily and erroneously ascribed the anteriormost of these centers to the tritocerebrum.

In her study of the embryogenesis of various collembolans, Tamarelle (1972) formally denied the existence of a supernumerary neuromere in the SEG of these insects and considered the anteriormost NM of this nervous center as mandibular in nature. Nevertheless, in her photographs of longitudinal sections of embryos of *Thyphlogastrura* and *Pseudosinella*, this so-called mandibular mass is strongly bilobate; the two lobes are set apart by a cortical crest and each of them is as large as the maxillary or labial NMs. In her text, she said nothing about the bilobate mass, which needs to be explained, especially if one considers that Tamarelle mentioned two commissures corresponding to this so-called mandibular ganglion in the embryo of *Anurida*.

The last study dealing with the embryogenesis of collembolans is Tyszkiewicz's (1976). This author also showed four neuropil centers in the subesophageal ganglion and, without giving the least proof, ascribed the two anteriormost centers to the mandibular neuromere (as we saw above).

In other words, the actual facts about collembolans (reported in illustrations of recent embryologists) strongly agree with Folsom's figure, whatever these embryologists have said (or not said) about them. These facts also perfectly agree with the reconstruction I have presented for *Anurida* and with its interpretation.

1.9. CONCLUSION

The detailed study of the cephalic nervous centers of five very different insects (a collembolan, a lepismid, a nymph of Ephemeroptera, a dermapteran, and a larva of mosquito) shows that the tritocerebrum in the classical sense actually comprises two neuromeres: the tritocerebrum *s.str.* and the tetrocerebrum. Each of these neuromeres is concerned with motor and sensory nerves. They are linked with each other and with the contiguous ones by short connective tracts (interneuromeric tracts) and with the more distant ones by long connective tracts (transneuromeric tracts). Both take part in the building of the ventral roots of the frontal ganglion. Lastly, each of them has its own commissure.

About the evolution of these two neuromeres, I can say that in collembolans, the tetrocerebrum and its commissure lie in the subesophageal ganglion, but the tetrocerebral center is clearly set apart from the mandibular one by a cortical crest, which characterizes the intersegmental levels in the polyneuromeric nerve centers. The tritocerebrum *s.str.* lies in the peristomodeal connectives and is a good distance from the deutocerebrum; its own commissure is free. The ventral cibarial dialator muscles make their way between the tritocerebral commissure and the front of the subesophageal ganglion (which contains the tetrocerebral one).

In lepismids, the tetrocerebrum shifts anteriorly and reaches the tritocerebrum. The commissures corresponding to both of these neuromeres are free and separated by the ventral cibarial dilator muscles.

In Pterygota, these two neuromeres are always very close to each other and, as a whole, they tend to draw nearer to the deutocerebrum. Their commissures are often enclosed under the same neurilemma sheath; then, the ventral cibarial dilator muscles (when present) run below this composite commissure. When the peristomodeal connectives are short, both the trito- and tetrocerebral commissures may fuse with the subesophageal ganglion. Then, the ventral cibarial dilator muscles run above them.

The ventral cibariopharyngeal musculature is always innervated by nervous fibers issued from the tritocerebrum *s.str.* Generally, these fibers run inside the tritocerebral commissure and leave it as short nerves to the muscles. But, in some instances, these fibers are independant of this commissure; they appear as nerves of their own coming from the upper part of the peristomodeal connectives. These symmetrical nerves possibly unite into a pseudocommissure that is only an anastomosis.

Numerous authors, anatomists, and embryologists have criticized the idea of a neuromeric duality of the tritocerebrum in the classical sense. However, their arguments do not stand the scrutiny of facts.

The tetrocerebral neuromere can also be found in other arthropods: in 1959, Legendre discovered a reduced supernumerary segment in the prosoma of spiders. Here, this segment is well defined (celomic vesicles, ganglionic anlagen, appendicular anlagen). This segment very early is annexed by the segment of the pedipalps; it is for this reason that Legendre named it "segment fugace." The ganglion of this segment lies exactly in the same sequential position as the tetrocerebrum does: as a matter of fact, it is situated behind the entire protocerebrum–deutocerebrum (rostral center)–tritocerebrum (cheliceral center) complex, and in front of the nerve centers of the segment of the pedipalps.

The tetrocerebrum is also present in the malacostracan Crustacea (Chaudonneret, 1978a). The Myriapoda show discreet but significant traces of it (Chaudonneret, 1978b, in which I have suggested phylogenetic models for the evolution of the nervous and endocrine systems in the phylum Annelida-Arthropoda).

The tetrocerebral neuromere, thus, belongs to a tetrocephalic metamere, which is always very reduced and located between the tritocephalic (strictly speaking) and the mandibular segments in Mandibulata, between the segments of the chelicerae and of the pedipalps in Chelicerata (they are serially homologous with the previously named ones in Mandibulata).

The tetrocerebrum and the tetrocephalon, therefore, can be considered as integral parts of the cephalic architecture of all arthropods.

1.10. SUMMARY

The subesophageal ganglion of collembolans contains four neuropil masses, instead of three. The anteriormost of these masses (whose commissure lies inside the subesophageal ganglion) has nothing to do with mandibular innervation, but rather with the tritocerebrum (which lies in the peristomodeal connective and has a free commissure): it takes part in the building of the ventral root of the frontal ganglion. Apparently, the neuropil mass in question belongs to a supernumerary neuromere that exists between the tritocerebral and the mandibular neuromeres, and is named "tetrocerebrum."

In the other four insects studied, the tetrocerebrum such as is found in collembolans joins more or less closely the true tritocerebrum (tritocerebrum *s.str.*). Each of these two centers always gives rise by halves to the ventral root of the frontal ganglion and has its own commissure. Both of these commissures may be free, enclosed in the same neurilemma sheath (there is, thus, only one apparent commissure), or sink into the subesophaeal ganglion. A pseudocommissure may occur in the pattern of innervation of the ventral cibariopharyngeal musculature. Furthermore, the tetrocerebrum, as well as the tritocerebrum *s.str.* show various features that are typical of neuromeric units.

The various anatomical or embryological criticisms that have been leveled at the idea of a supernumerary neuromere (and a supernumerary segment) in the insect head are carefully examined.

ABBREVIATIONS USED IN FIGURES

a = anastomosis.
C1 = tritocerebral commissure; C2 = tetrocerebral commissure; CD =

deutocerebral commissure; CLb = labial commissure; CM = mandibular commissure; CMx = maxillary commissure; CnD = deutocerebral neuropil center; CnLb = labial neuropil center; CnM = mandibular neuropil center; CnMx = maxillary neuropil center; CnTe = tetrocerebral neuropil center; CnTr = tritocerebral neuropil center; colPD = transneuromeric connective tract coming from the proto- and deutocerebrum; colTe = transneuromeric connective tract coming from the tetrocerebrum; colTr = transneuromeric connective tract coming from the tritocerebrum; coTe-M = tetrocerebro-mandibular connective tract; coTr-Te = tritotetrocerebral connective tract; crD-Tr = deutotritocerebral cortical crest; crP-D - protodeutocerebral cortical crest; crTr-Te = tritotetrocerebral cortical crest; cr1 = tetromandibular cortical crest; cr2 = mandibulomaxillary cortical crest; cr3 = maxillolabial cortical crest.

dvc = ventral cibarial dilator muscle; dvph = ventral pharyngeal dilator muscle.

eco = situation of the cellular rind.

l0 = optic lobe.

nA = deutocerebral antennary nerve; nAa = tritocerebral accessory antennary nerve; nc = nervus connectivus; ncc1 = medial nerve of corpus cardiacum; ncc2 = lateral nerve of corpus cardiacum; ncc3 = tritocerebral nerve of corpus cardiacum; nCL = nerve to clypeolabrum; nCm = motor clypeal nerve; nCs = sensory clypeal nerve; ndi0 = nerve to the imaginal disc of the eye; ndvc = nerve to the ventral dilator muscles of the cibarium; ndvph = nerve to the ventral dilator muscles of the pharynx; ngm = nerve to the mandibular gland; nh = hypopharyngeal nerve; nhM = mandibular nerve to the hypopharynx; nhMx = maxillary nerve to the hypopharynx; nhTe = tetrocerebral nerve to the hypopharynx; nL = nerve to the labrum; nLb = labial nerve; nLe = Leydig's nerve; nlm = neurilemma; nM = mandibular nerve; nmst = nerve to the parietal musculature of the stomodeum; nMx = maxillary nerve; n0 = optic nerve; nofl = nerve to the lateral frontal organ; nofm = nerve to the median frontal organ; n0l = nerve to the larval eye; no1 = nerve to the lateral ocellus; nom = nerve to the median ocellus; nosb = nerve to the laterobuccal sensory organ; nrao = nerve to the retractor muscle of the mouth angle; nrfaL = nerve to the anterior frontal retractor muscle of the labrum; nsc = nerve to the ventral semicircular muscles of the cibarium; nsh = hypopharyngeal sensory nerve; nslb = nerve to the lateral suspensorial muscle of the mouth; nt = tegumentary nerve; nvph = nerve to the ventral dilator muscles of the pharynx.

pC = pseudocommissure.

RGF = ventral root of the frontal ganglion; rgf = dorsal root of the frontal ganglion; ROEp1 = medial root of the epipharyngeal organ; ROEp2 = lateral root of the epipharyngeal organ.

scolPD = section of the connective tract from proto- and deutocerebrum (lying in the posterior region of the peristomodeal connective).

REFERENCES

Bitsch, J. 1963. Morphologie céphalique des Machilides (Insecta Thysanura). *Ann. Sci. Nat. Zool.* (12° série) **5**: 501–706.

Bitsch, J. 1973. Morphologie de la tête des Insectes. Partie générale, pp. 42–100. In P. P. Grassé (éd.), *Traité de Zoologie*, vol. 8(1). Masson, Paris.

Brückmoser, P. 1965. Embryologische Untersuchungen über den Kopfbau der Collembole *Orchesella villosa*. L. *Zool. Jahrb. Anat.* **82**: 299–364.

Bullock, T. H. and G. A. Horridge. 1965. *Structure and Function in the Nervous Systems of Invertebrates,* vol. 2, pp. 801–964. Freeman, San Francisco and London.

Chaudonneret, J. 1950. La morphologie céphalique de *Thermobia domestica* (Packard) (Insecte Aptérygote Thysanoure). *Ann. Sci. Nat. Zool.* (11° série) **12**(10): 145–302.

Chaudonneret, J. 1966. La construction phylogénétique de la tête des Insectes. I- Le squelette. *Bull. Sci. Bourgogne* **24**: 241–263.

Chaudonneret, J. 1967. Méthodes d'anatomie morphologique. *Ann. Soc. Ent. Fr. (N.S.)* **3**(3): 757–765.

Chaudonneret, J. 1973. Interprétation des formations dites endosquelettiques du Collembole *Isotomurus* (Insecte Aptérygote Entotrophe). II. Le vincularium. *Bull. Soc. Zool. Fr.* **98**(4): 499–513.

Chaudonneret, J. 1978a. Evolution de la chaîne nerveuse latérale chez quelques Malacostracés. *Arch. Zool. Exp. Gén.* **119**(1): 163–184.

Chaudonneret, J. 1978b. La phylogenèse du système nerveux annélido-arthropodien. *Bull. soc. Zool. Fr.* **103**(1): 69–95.

Christophers, R. 1960. *Aedes aegypti, the Yellow Fever Mosquito: Its Life History, Bionomics and Structure.* Cambridge University Press, Cambridge.

Denis, J. R. 1928. Etudes sur l'anatomie de la tête de quelques Collemboles suivies de considérations sur la morphologie de la tête des insectes. *Arch. Zool. Exp. Gén.* **68**: 1–291.

Drenkelfort, H. 1910. Neue Beiträge zur Kenntnis der Biologie und Anatomie von *Silphurus lacustris* Eaton. *Zool. Jahrb. Anat.* **29**: 527–617.

Folsom J. W. 1900. The development of the mouth parts of *Anurida maritima*. *Bull. Mus. Comp. Zool. Harvard* **36**: 87–157.

François, J. 1969. Anatomie et morphologie céphalique des Protoures (Insecta Apterygota). *Mém. Mus. Nat. Hist. Nat. Paris (A)* **59**(1): 1–144.

Garaudy-Tamarelle, M. 1969. Les vésicules coelomiques du segment intercalaire (= prémandibulaire) chez les embryons du Collembole *Anurida maritima* Guér. *C. R. Acad. Sci. Paris* **269D**(2): 198–200.

Haget A. 1977. L'embryologie des Insectes, pp. 1–387. In P. P. Grassé (éd.), *Traité de Zoologie*, vol. 8 (5B). Masson, Paris.

Hanström, B. 1940. Inkretorische Organe, Sinnesorgane und Nervensystem des Kopfes einiger niederer Insektenordnungen. *Kungl. Svensk. Vetensk. Akad. Handl.* **18**: 1–265.

Hutasse, F. 1972. Anatomie et morphologie céphaliques d'*Isotomurus maculatus*

(Agren) 1903. Interprétations morphologiques de la tête des Collemboles (Insecta. Apterygota). Thèse d'Etat, Dijon; A.O. du C.N.R.S., n° 3667, Paris.

Kühnle, K. F. 1913. Vergleichende Untersuchungen über das Gehirn, die Kopfnerven und die Kopfdrüsen des gemeinen Ohrwurms (*Forficula auricularia* L.). *Jena. Z. Naturwiss.* **50:** 147–276.

Larink, O. 1969. Zur Entwicklungsgeschichte von *Petrobius brevistylis* (Thysanura, Insecta). *Helgol. Wiss. Meeresunters.* **19:** 111–155.

Larink, O. 1970. Die Kopfentwicklung von *Lepisma saccharina* L. (Insecta, Thysanura). *Z. Morphol. Tiere.* **67:** 1–15.

Legendre, R. 1959. Contribution à l'étude du système nerveux des Aranéides. *Ann. Sci. Nat. Zool.* (12° série) **1:** 339–473.

Lhoste, J. 1957. Données anatomiques et histophysiologiques sur *Forficula auricularia* L. (Dermaptère). *Arch. Zool. Exp. Gén.* **95**(2): 75–252.

Malzacher, P. 1968. Die Embryogenese des Gehirns paurometaboler Insekten. Untersuchungen an *Carausius morosus* und *Periplaneta americana.* *Z. Morphol. Tiere.* **62:** 103–161.

Matsuda, R. 1965. Morphology and evolution of the insect head. *Mem. Am. Entomol. Inst.* **4:** 1–334.

Moulins, M. 1969. Etude anatomique de l'hypopharynx de *Forficula auricularia* L. (Insecte, Dermaptère): Téguments, musculature, organes sensoriels et innervations. Interprétation morphologique. *Zool. Jahrb. Anat.* **86:** 1–27.

Moulins, M. 1971. La cavité préorale de *Blabera craniifer* Burm. (Insecte, Dictyoptère) et son équipement sensoriel: Étude anatomique, ultrastructurale et physiologique de l'épipharynx et l'hypopharynx. *Zool. Jahrb. Anat.* **88**(4): 527–586.

Noars, R. 1962. Nouvelles données sur l'anatomie céphalique des larves de *Cloeon* (Ephéméroptères). *Bull. Soc. Linn. Lyon* **31**(2): 38–43.

Philiptschenko, J. 1912. Beiträge zur Kenntnis der Apterygoten. III. Die Embryonalentwicklung von *Isotoma cinerea.* Nic. *Z. Wiss. Zool.* **103:** 519–560.

Popham, E. J. 1962. Prognathism and brain structure in *Forficula auricularia* L. (Dermaptera) and *Dysdercus intermedius* Dist. (Hemiptera-Homoptera). *Zool. Anz.* **168:** 36–43.

Rähle, W. 1970. Untersuchungen an Kopf und Prothorax von *Embia ramburi* Rimsky-Korsakof 1906 (Embioptera, Embiidae). *Zool. Jahrb. Anat.* **87:** 248–330.

Rempel, J. G. 1975. The evolution of insect head: The endless dispute. *Quaest. Entomol.* **11**(1): 7–24.

Rohrschneider, I. 1968. Beiträge zur Entwicklung des Vorderkopfes und der Mundregion von *Periplaneta americana.* *Zool. Jahrb. Anat.* **85:** 537–578.

Roonwal, M. L. 1937. Studies on the embryology of the African Migratory Locust, *Locusta migratoria migratorioides* Reiche and Frm. (Orthoptera, Acrididae). II. Organogeny. *Philos. Trans. R. Soc. Lond.* **B227:** 175–244.

Rousset, A. 1966. Morphologie céphalique des larves de Planipennes (Insectes Névroptéroides). *Mém. Mus. Nat. Hist. Nat. Paris (A)* **42:** 1–199.

Scholl, G. 1969. Die Embryonalentwicklung des Kopfes und Prothorax von *Carausius morosus* Br. (Insecta, Phasmida). *Z. Morphol. Tiere* **65:** 1–142.

Snodgrass, R. E. 1951. Anatomy and morphology. *J. N.Y. Entomol. Soc.* **59**(2): 71–73.

Strausfeld, N. J. 1976. *Atlas of an Insect Brain.* Springer, Berlin-Heidelberg-New York.

Tamarelle, M. 1972. Contribution à l'embryologie des Collemboles Arthropléones. Thèse d'Etat, Bordeaux, n° 80; A.O. du C.N.R.S., n° 7391, Paris.

Tyszkiewicz, K. 1976. The embryogenesis of the central nervous system of *Tetrodontophora bielanensis* (Waga) (Collembola). *Acta Biol. Cracov. (ser. Zool.)* **19:** 1–21.

Vassal, M. A. 1938. Recherches sur l'hypopharynx des Ephémères. *Bull. Sci. Bourgogne* **8:** 133–140.

Vayssière, A. 1882. Recherches sur l'organisation des larves d'Ephémères. *Ann. Sci. Nat. Zool.* **13:** 1–37.

Wada, S. 1966. Analyse der Kopf-Hals-Region von *Tachycines* (Saltatoria) in morphogenetische Einheiten. *Zool. Jahrb. Anat.* **83:** 185–326.

Weber, H. 1952. Morphologie, Histologie und Entwicklungsgeschichte der Articulaten. *Fortschr. Zool.* **9:** 18–231.

Wiseley, B. 1965. Studies on Ephemeroptera. III. *Colobriscus humeralis* (Walker); morphology and anatomy of the winged stages. *N. Z. J. Sci.* **8:** 398–415.

CHAPTER 2

Evolutionary Trends in the Central and Mushroom Bodies of the Arthropod Brain
A Dilemma

Ayodhya P. Gupta
Department of Entomology and Economic Zoology
New Jersey Agricultural Experiment Station
Cook College, Rutgers University
New Brunswick, New Jersey

Any study of evolution necessarily involves theories, but if we stopped short with known facts our phylogenetic trees would wither at the roots. Particularly when we attempt to reconstruct events that took place in remote precambrian times we can have recourse only to our imagination.

R. E. Snodgrass, 1958

2.1. INTRODUCTION

To discuss the evolutionary trends in the central and mushroom bodies of the arthropod brain in general, it will be useful to comment on the evolution of the various arthropod groups themselves. It is generally accepted that arthropods as well as the Onychophora originated from a lobopod annelid ancestor. It is no coincidence then that the arthropod brain resembles the annelid brain in its basic plan, especially the ventral nerve cord (Horridge, 1965). The extant arthropods can be divided into aquatic (Aquatic Chelicerata, and Crustacea) and terrestrial (Terrestrial Chelicerata, Myriapoda, and Insecta or Hexapoda; some decapods and isopod crustaceans) groups.

Figure 2.1 shows a monophyletic arrangement of the various groups of arthropods (for a comprehensive discussion of both the mono- and polyphyletic origin of arthropods, see Gupta, 1979a). Note that according to this arrangement, the lobopod ancestor gave rise to both the Myriapoda and Trilobitomorpha; the latter were the progenitors of both the Crustacea and the Aquatic Chelicerata. Exactly when the Aquatic Chelicerata and the ancestors of Myriapoda colonized land is controversial and mysterious. According to Størmer (1977), the arthropod invasion of land occurred during the late Silurian and Devonian times, some 400 million years ago. It is also believed that aquatic arthropods colonized land independently several times.

The question of whether or not the exoskeleton in arthropods developed before or after land colonization is controversial. Contrary to Manton's (1973) view that uniramians (Onychophora, Myriapoda, and Hexapoda) colonized land in soft-bodied condition, there is paleontological evidence that some myriapod-like creatures, with well-sclerotized exoskeletons and articulated uniramous legs, were already present in the Lower and Middle Cambrian (500 million years) (Bergström, 1979). Finally, I (Gupta, 1979c) have suggested that both apterygote and pterygote insects originated from myriapodan ancestors (Fig. 2.2). Their origin from Crustacea and Trilobitomorpha also has been suggested. Note that four major evolutionary lines are recognized among pterygotes: Paleopteroid, consisting of the two primitive orders, Odonata and Ephemeroptera; the Orthopteroid line; the Hemipteroid line; and the most advanced Neuropteroid line, consisting of some of the most highly evolved orders. All the controversies notwithstanding, land colonization had profound effects in bringing about important anatomical and physiological changes, some of which may have had important bearing on the evolution of the arthropod brain. I have suggested elsewhere

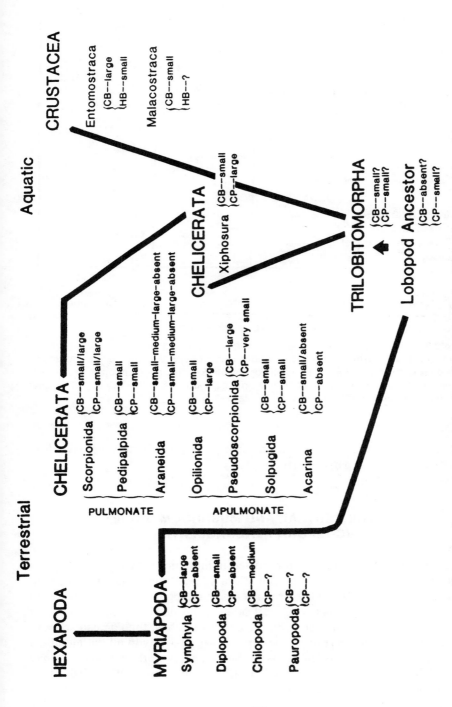

Figure 2.1. Occurrences and relative sizes of the central bodies (CBs) and corpora pedunculata (CP) or mushroom bodies in the major arthropod groups, other than insects. (Based on information primarily from Bullock and Horridge, 1965; phylogenetic arrangement of taxa based on Gupta, 1983.)

29

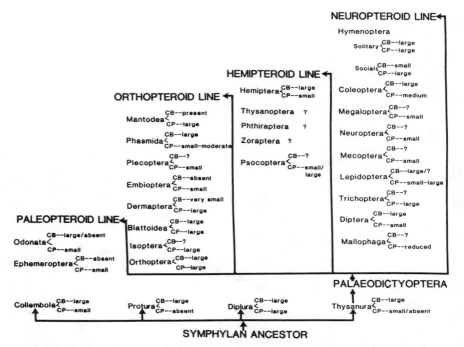

Figure 2.2. Occurrences and relative sizes of the central bodies (CBs) and corpora pedunculata (CP) or mushroom bodies in various insect orders, arranged under four major evolutionay groups. Small/absent = small or absent; Small-large = variously developed; ? = unknown. (Based on information primarily from Bullock and Horridge, 1965; phylogenetic arrangement of taxa based on Gupta, 1979b.)

(Gupta, 1983) that land colonization most probably affected the evolution of the neurohemal organs of arthropods.

2.2. THE CENTRAL AND MUSHROOM BODIES (CORPORA PENDUNCULATA)

The central body (CB) has been reported in all major groups of arthropods (Figs. 2.1, 2.2) and if we accept Hanström's (1925) view that the so-called hemiellipsoid body (HB) of the Crustacea is homologous with the mushroom body or corpus pedunculatus (CP) of other arthropods, on the basis that a tract connects these neuropils with the olfactory lobes, the CP also seem to be present in all major arthropod groups, not necessarily in all the members belonging to these groups. In terms of their sizes, both the CB and CP may be small, large, or equally developed in various subgroups. Although drawing any evolutionary conclusions on the basis of size alone may seem unwarranted and even misleading, it is hard to imagine that a structure will attain a well-developed stage without having any selective advantage, which apparently we do not presently know; the neuronal diversity of the CB even in the closely related species suggests that it is not a static component of the brain, and indeed dramatic increase in the volume of CP in newly emerged adults of the rove beetle, *Aleochora* (Staphylinidae) during the first 20 days of its adult life occurs (Bieber and Fuldner, 1979). A comparative study of the size of the CB in all major arthropod groups suggests that a large CB is probably a primitive or ancestral stage, whereas a smaller one represents a more advanced condition. Conversely, a small CP appears to be an ancestral condition and an enlarged one an advanced stage (see Section 2.2.2.3).

2.2.1. The Central Body

Apparently, the CB is a new structure that evolved for the first time in arthropods, and was supposedly absent in the lobopod annelid ancestor of the arthropods. I hazard this guess because the CB is apparently absent in polychaet annelids (see Fig. 16.6 in Bullock and Horridge, 1965).

2.2.1.1. Structure and Functions. According to Horridge (1965), the CB is present in all arthropods as a median neuropil mass. In insects, it is generally composed of two major subdivisions: the central body (located dorsolaterally) and the ellipsoid body (Power, 1943). The structure of the CB has been described in several insects (Jawlowski, 1936; Satija and Das, 1963; Klemm, 1968; Strausfeld, 1970, 1976; Satija et al., 1975; Williams, 1975; Larsen et al., 1979; Ghaffar et al., 1984; Homberg, 1985; see also Chapter 15 by Homberg, in this volume).

Packard (1886) apparently found no CB in the cave beetle, *Anophthalamus tellkampfii*, although according to Ghaffar et al. (1984), the CB is the largest component of the central complex (CB, ellipsoid body, and the ventral tubercle) in the cave beetle, *Neaphaenops tellkampfii* (Carabidae). In another cave beetle, *Glacicavicola bathyscioides* (Leiodidae), Larsen et al. (1979) found only the CB and no ellipspoid body or the ventral tubercle in the central complex (see also Section 2.2.1.2).

Several functions have been proposed for the CB. According to Huber (1965), the CBs are excitation centers for both the subesophageal and thoracic ganglia (and thus affect locomotory and acoustic behavior). And both Drescher (1960) and Huber (1965) considered the CP as inhibitory centers for the CBs. F. W. Schürmann (personal communication, 1986) considers the CB (as well as CP) as centers in long loops with delayed input and output, in contrast to certain giant fiber systems that function with short delay. Ghaffar et al. (1984) reported that in the cave beetle, *N. tellkampfii*, there are very few fiber tracts that connect the CB with other neuropils of the brain; however, they described a few fibers coming from the olfactorio globularis tract of the antennal lobe and a few others that connect it to the pons cerebralis (PC). Although unconfirmed, vision seems to have played no role in the evolution of large CB, because although several cave beetles, the dipluran, *Japyx,* all known proturans, and some ants are blind, they possess CBs. Could it be that in these arthropods, large CBs compensate for loss of vision in some manner?

Although Chapman (1979) regarded the CB as the center for premotor outflow from the brain to the ventral nerve cord, as far as is known, no experimental proof exists that CBs (or CP) represent premotor areas. Both the CB and CP are "far away from descending brain neurons, linked by long chains of interneurons (in the brain) of unknown integrative properties to efferent and most afferent brain neurons" (F. W. Schürmann, personal communication, 1986). Howse (1975) commented that the central complex may modulate total responsiveness and supposedly controls circadian activity rhythms (see also Chapter 15 by Homberg in this volume).

2.2.1.2. Presence in Various Arthropod Groups.

In Insecta or Hexapoda. Note (Fig. 2.2) that, except in the known case of social Hymenoptera and some Dermaptera, the CB is relatively large or well developed in most insect orders in which it is known. According to Hanström (1928, 1930, 1940), Howse (1974), and Howse and Williams (1969), the CB is relatively constant in different orders. The latter authors established a direct correlation between the head width and the volume of the CB. Groth (1971) observed a similar correlation in Diptera, although a linear one. According to Groth, the CP:CB ratio is a meaningful index, and he found that this ratio is very high in social and subsocial Hymenoptera and Isoptera, but small in all nonsocial insects (including *Pieris, Blaberus, Tenthredo, Mantis,* and *Chrysops*). Note that, based on their respective sizes, this CP:CB ratio is *small* for most of the apterygote orders and Odonata, Hemiptera, and even in the highly evolved orders of Lepidoptera and Diptera.

It is uncertain whether or not the CB has been secondarily lost or simply unreported in some Odonata, Ephemeroptera, Plecoptera, Embioptera, Isoptera, Megaloptera, Neuroptera, Mecoptera, Trichoptera, Mallophaga, and Psocoptera; clearly, the CBs in these orders need to be studied.

In Other Arthropod Groups. If we look at all the groups as a whole, we would have to conclude that each major group shows both the primitive (large CB) and the advanced (small CB) conditions (Figs. 2.1, 2.2). Among aquatic arthropods (Fig. 2.1), interestingly, the most primitive arthropod, *Limulus polyphemus* has both an advanced CB (small) and CP (large) (Fahrenbach, 1976, 1977), which means that the CP:CB ratio is high, suggesting the presence of complex behavior in this animal, if indeed a high CP:CB ratio is indicative of the evolution of complex behavior, such as in social Hymenoptera. However, *Limulus* is not considered to possess complex behavior patterns, but is known to show well-developed adaptive behavioral changes required by its migrations from the high supralittoral through the intertidal and subtidal environments (Rudloe, 1979).

Among the Crustacea, the more primitive lower Crustacea (Entomostraca) possess primitive, well-developed CBs, but their more highly evolved cousins, the terrestrial Decapoda (Malacostraca) possess the apomorphic small CB.

Among the terrestrial arthropods, the Myriapoda show the same diversity as the aquatic and terrestrial Crustacea. The more primitive Symphyla have well-developed CB, but the more highly evolved Diplopoda possess small CB, with the Chilopoda falling in between the two, with moderately developed CB. The CB has not yet been reported in Pauropoda.

Among the Chelicerata, all, except the Pseudoscorpionida, seem to possess the advanced (small) CB, including the most primitive scorpions. The diversity in the size of the CB is most pronounced in the terrestrial Chelicerata, where many Araneida have well-developed CB, most others small to moderately developed CB, and some Acarina (e.g., *Bdella*) with no CB at all (Bullock and Horridge, 1965); apparently these latter animals secondarily lost this brain component. It may be that somehow a *small* CB (not a large one) is associated with some specific behavioral pattern, regardless of whether or not the animal is terrestrial (e.g., social Hymenoptera) or aquatic (e.g., *Limulus*).

I believe it is safe to predict that similar diversity will be evident in Hexapoda also, when more information about these structures will be available in future.

2.2.1.3. Plesiomorphic and Apomorphic Trends. I can tentatively suggest that a relatively large or well-developed CB probably represents a primitive or ancestral situation, most commonly shared by members of the primitive Crustacea, all apterygote and several pterygote insect orders, the symphylan Myriapoda (e.g., *Hanseniella*; Tiegs, 1940), and many terrestrial Chelicerata. In other words, it has remained unchanged in many of the present-day arthropods, without having undergone any evolutionary changes since its appearance in the early arthropods. F. W. Schürmann (personal communication, 1986) also considers the CB as the most stable brain component. By the same token, any modification (e.g., smaller size or absence) of the plesiomorphic situation in some arthropods may be considered an advanced or apomorphic situation, as is probably the case in social Hymenoptera, in which the CB are reduced in size; the apparent absence (?) of CB in some Isoptera represents a secondary loss.

If we take the opposite position that a *small* CB is the primitive condition, and, as the arthropods evolved toward a more active mode of life and terrestriality, the CB concomitantly evolved toward *greater size*, it would be difficult to support this thesis with respect to the most highly evolved social Hymenoptera among insects and the terrestrial decapod Crustacea, diplopod Myriapoda, and most certainly the Chelicerata (Arachnida), many of which have a rather active mode of life, but have *small* CBs. If we argue further that the primitive small CB remained unchanged since its appearance in the early arthropods, it would be difficult indeed to explain the presence of a *large* CB in most insect orders! Furthermore, it would also be hard to explain the large size of CB in the blind dipluran, *Japyx* and in the primitive symphylan Myriapoda (e.g., *Hanseniella*).

It should be noted here that the *large* size of the CB does not seem to be necessarily associated with 1) active mode of life and terrestriality, nor 2) behavioral complexity (Horridge, 1965). For example, the CB is large in the dipluran *Japyx*, which is primitive and blind, but small in social Hymenoptera, which are highly evolved and display very complex behavior. The only exception to this seems to be the web-spinning spiders that have well-developed CB.

Then why did the CB evolve toward smaller size in those arthropods that display complex behavior? Could it be that the functions of the CB shifted from these structures to some other structure and they lost the selective advantage needed to remain large? This evolutionary dilemma is hard to explain in the light of our present knowledge of the functions of the CB and would have to await more detailed works on the functional roles of this brain component in many more arthropods, especially comparative studies of volume changes during various developmental stages (immature) and in the adults, concomitant with progressive functional and behavioral changes during the life cycle.

2.2.2. The Corpora Pedunculata or Mushroom Bodies

Dujardin (1850) noted that the calyxes of the corpora pedunculata (CP) were especially large in "intelligent" insects, such as social Hymenoptera (bees and ants), and thought that they were comparable to the cerebral cortex in vertebrates and hence the seat of intelligence.

2.2.2.1. Homology With the Hemiellipsoid Body of Crustacea. If we accept that the so-called hemiellipsoid body (HB) of the Crustacea is homologous with the CP of other arthropods, as suggested by Hanström (1925), on the basis that a tract connects these structures with the olfactory lobes, the CP also seem to be present in all arthropods. The homology of the CP in all major arthropod groups has not yet been established, however. The HB in Crustacea is not considered to be a part of the main brain mass, but rather as migrated equivalents of CP of insects, especially in the stalked-eyed crustaceans (Horridge, 1965).

2.2.2.2. Structure and Functions. Typically, a CP consists of a cup-shaped calyx, often composed of several neuropil zones with intrinsic and extrinsic fibers, and a stalk or pedunculus, which extends through the calyx and whose fibers may be arranged into external and internal layers, producing a double-layered cylindrical structure in some arthropods, and which gives rise to the alpha- and beta-lobes posteroventrally.

The CP have been described in several insects (von Alten, 1910; Thompson, 1913; Hanström, 1925, 1928, 1930, 1940; Jawlowski, 1936; Goosen, 1949; Ratzendorfer, 1952; Satija and Das, 1963; Goll, 1967; Strausfeld, 1970, 1976; Pearson, 1971; Schürmann, 1973; Howse, 1974; Weiss, 1974, 1981; Satija et al., 1975; Larsen et al., 1979; Mobbs, 1982; Ghaffar et al., 1984; see also Chapter 11 by Schürmann in this volume).

Several functions have been attributed to the CP. They are considered to be involved in the regulation of complex behavior and serve as summation centers for excitation from different sources (Chapman, 1979). Strausfeld (1976) suggested that the CP were primarily olfactory–taste centers. But he is now convinced (personal communication, 1984) that this view is wrong, because the CP receive many inputs from other sensory centers. According to him, intracellular recordings and Lucifer yellow or Cobalt filling by Erber and his associates (references, Chapter 21 by Erber et al. in this volume) strongly suggest that the CP are a "multimodal centre with complex integrative properties which can be modified by conditioning. In the fly, the CP also receive many different inputs and some of their output neurons terminate amongst outgoing axons of descending neurons." According to him, enormous structural variation in the CP suggests that they are designed differently in each species with regard to the numbers of K-cells and their connections in the alpha-, beta-, and pedunculus lobes.

Schürmann proposed (1974) that the CP function as a comparaing and weighting system for different modalities and qualities (stimuli from different sensory systems). According to him (personal communication, 1986), the CP are evolutionarily unstable. They may have evolved as "higher order 'antennal'" neuropils. This is supported by his work on Onychophora (Chapter 8 in this volume). This view is consistent with those of Hanström (1930) and Holmgren (1916).

Some attempts have been made to correlate the CP with optic activity. According to F. W. Schürmann (personal communication, 1986), in fast-flying insects with large eyes, the central glomerular neuropils did not develop or modify, but the nonglomerular neuropil, with its sets of large and fast neurons continued to be dominant or became so. According to him, during their evolutionary diversification, the CP have received input from the visual system, as shown by Mobbs (1982) in the honey bee and by Honnegger and Schürmann (1975) in crickets. Intracellular recordings show multimodal input processing and output in the CP. However, both Forel (1874) and Rabl-Ruckhard (1875) disputed the role of CP in vision, because blind ants have well-developed CP. Similarly, Satija and Das (1963), Larsen et al. (1979), and Ghaffar et al. (1984) have reported well-developed CP in blind cave beetles and concluded that there is no direct relationship between the size of the CP and optic activity. This is supported also by the presence of large CP in the dipluran, *Japyx*, which is blind.

As a matter of fact, there appears to be an inverse correlation between well-developed CP and optic activity. For example, Flögel (1878), Newton (1879), and Power (1943) reported rudimentary development of CP in Diptera that generally show high optic activity. Note that CP are small in Odonata, which also possess large eyes. Jonescu (1909) found that the drones of the honey bee (*Apis mellifera*) had the smallest CP (compared with the largest in workers), but their optic lobes were twice as large as in workers and queens. Weiss (1974) concluded that in *Periplaneta americana*, it is the antennal input, not the optic, to the CP that is most significant. Indeed, Menzel et al. (1974) and Erber et al. (1980) also have demonstrated the possible association of olfactory memory in Hymenoptera with the CP (Chapter 21 by Erber et al. in this volume). According to Mobbs (1982), the CP in the honey bee process second-order antennal and fourth- and higher order visual information. He goes on to say that, on the basis of the nature of the connections of the CP, these structures might play an olfactory role in some insects and as visually mediated memory centers in others. According to him, in most insects the CP are involved in analyzing olfactory and mechanosensory information from the antennal lobes, but in the Hymenoptera an additional visual input is found. If indeed this is so in general, it would not only neatly explain the absence of the CP in the Protura (Horridge, 1965), which have no eyes and antennae, but would also provide a more satisfactory explanation for the large size of the CP in those insects that have well-differentiated antennae and a diversity of antennal sensilla types.

As a matter of fact, Callahan (1979) has suggested that among the lower orders of insects, the antennal sensilla types are fewer. For example, Collembola show only one type: thick- and thin-walled trichodea. Thysanura show three types: thick- and thin-walled trichodea; short, thin-walled basiconica; and coeloconica. He has suggested that these basic types of sensilla became further modified in higher orders. For example, our work on the German cockroach, *Blattella germanica* (Dictyoptera), revealed five types: trichodea, two types of chaetica, basiconica, and campaniformia (Wheeler and Gupta, 1986), Lepidoptera showed seven types, and some Hymenoptera 12 types (Callahan, 1975). According to Callahan (1979), the surface texture of these sensilla also became much more complex in higher orders. For example, in Coleoptera, Lepidoptera, and Hymenoptera one would find corrugated, helical, terraced, and equiangular surfaces.

Callahan (1979) also suggested that antennal movement also shows a gradual complexity from the lower to the higher orders. For example, among the Collembola, the antennal movement is very restricted, and may explain why the CP are small in these insects. As a matter of fact, he suggests that "waving" would be a better term for antennal movements among orders that are less highly evolved than Dermaptera. In higher orders, according to him, the slower antennal waving has been superimposed with high frequency antennal vibrations.

And finally, what about the role of predatory, defensive (aggressive), and fast locomotory behaviors on the evolution of larger CP? For example, Diplura are predatory and fast moving animals and possess large CP, as do Odonata, and the soldiers and workers in the orders Isoptera and Hymenoptera, respectively; this remains to be demonstrated, however.

2.2.2.3. Presence in Various Arthropod Groups.

In Insecta or Hexapoda. Note that unlike the CB, the CP show great diversity in terms of their presence or absence and size in Insecta (Fig. 2.2). Among the Apterygota, the CP are small in Collembola and Thysanura, large in Diplura (perhaps owing to the predatory and fast locomotory habits), and absent in Protura, probably owing to the absence of antennae (Horridge, 1965). According to Rensch (1959), volume differences in closely related species are due to an increase in the number of intrinsic K-cells. Moreover, sexually dimorphic and caste-related differences in the CP volume are known to occur (Chapter 11 by Schürmann in this volume).

In the two palaeopteroid orders Odonata and Ephemeroptera also, they are small. Beyond the palaeopteroid line, however, the CP appear to have evolved toward greater structural complexity and size, probably in conjunction with the evolution of complex behavior patterns, although this does not seem to be true in ants (Pandazis, 1930). For example, the CP are well-developed in workers and soldiers of Isoptera (Horridge, 1965), some Blattoidea, Mantodea, Dermaptera, and Cursoria (grasshoppers) among the orthopteroid group; in both the social and solitary Hymenoptera (von Alten, 1910); and in some Lepidoptera and Trichoptera among the most highly evolved neuropteroid group. According to Horridge (1965), the CP size is not correlated with complex behavior. Bernstein and Bernstein (1969), however, tentatively suggested that in the workers of the ant, *Formica rufa*, the foraging efficiency is correlated to the increase in the number of K-cells in the CP (hence, to the size of the CP).

By and large, however, the CP are *small* in many insect orders. Furthermore, within the same orders, the more highly evolved members have smaller CP than the less evolved ones. For example, among termites the CP in the higher termites (*Macrotermes, Apicotermes*) are relatively smaller, with long thin lobes, than those in lower termites (*Zootermite*) (Howse and Williams, 1969). Ratzendorfer (1952) has reported that advanced CP may be found in both the hemi- and holometabolous insects, thus suggesting the diversity in the relative sizes of the CP in various orders.

In Other Arthropod Groups. Among the aquatic arthropods, the situation in the most primitive Xiphosura seems to be anomalous, in that the horseshoe crab, *Limulus polyphemus* has large CP, but their functional significance is unknown (W. Fahrenbach, personal communication, 1986). According to him, CP in *L. polyphemus* are examples of convergent evolution, not derived from the ancestral form. The lower Crustacea have reduced HB and the higher ones possess well-developed HB or none at all, perhaps lost secondarily; the latter possibility is not unreasonable to suggest, because there are numerous instances of this in vertebrates. For example, many fishes have secondarily lost their swim bladders and have become bottom dwellers.

Among the terrestrial arthropods, the CP are unknown (absent?) in the Myriapoda and show great diversity among the Chelicerata. For example, in Gonoleptidae, Phalangida (Opilionida), some Scorpionida, and Araneida, these structures are well developed, but very small (rudimentary?) in Pseudoscorpionida, and absent in some Araneida and Acarina, perhaps secondarily. The CP are small in many Araneida, some Scorpionida, Pedipalpida, and Solpugida (Solifugae) (Horridge, 1965).

Note that among the terrestrial Chelicerata, the most primitive Scorpionida show both the primitive and advanced conditions of the CP. When more information is available, Insecta will most probably reveal similar diversity.

2.2.2.4. Plesiomorphic and Apomorphic Trends. It appears that small CP represent the ancestral or plesiomorphic situation and enlarged ones the most advanced or apomorphic condition. It should be noted, however, that the CP in various arthropod groups are "not necessarily homologous" (Horridge, 1965). Until we understand more fully the functions of the CP, the evolutionary dilemma in terms of the prevalent size diversity and their presence in some arthropods and absence in others will remain unresolved.

2.3. SUMMARY

The evolution of the central and mushroom bodies in various major arthropod groups, considered monophyletic in this chapter, remains enigmatic. This is largely owing to the facts that (1) both these brain components show both ancestral (plesiomorphic) and advanced (apomorphic) stages in many arthropod groups and their subgroups, regardless of the taxonomic and phylogenetic positions of these taxa; and (2) the presently known functions of these brain components are inadequate to explain their evolutionary diversities in size in various arthropod taxa.

On the basis of the information gathered, it appears that a relatively large central body (CB) represents the ancestral stage, whereas a smaller one is a more advanced condition. Conversely, small mushroom bodies or corpora pedunculata (CP) appear to be a primitive condition and enlarged ones an advanced stage.

The explanation for the evolutionary diversity of these brain components would most likely come from comparative studies of volume changes during various developmental stages (immature) and in the adults, concomitant with the progressive functional and behavioral changes during the life cycle. Such studies are long overdue.

ACKNOWLEDGMENTS

I thank Wolf Fahrenbach and Friederich Schürmann for reading the initial draft of this manuscript and their comments. New Jersey Agricultural Experiment Station Publication No. F-08112-02-86, supported by state funds, U.S. Hatch funds, and Rutgers Research Council.

REFERENCES

von Alten, H. 1910. Zur Phylogenie des Hymenopterengehirns. *Jena Z. Naturwiss.* **46:** 511–590.

Bergström, J. 1979. Morphology of fossil arthropods as a guide to phylogenetic relationships, pp. 3–56. In A. P. Gupta (ed.), *Arthropod Phylogeny*. Van Nostrand-Reinhold, New York.

Bernstein, S. and R. A. Bernstein. 1969. Relationships between foraging efficiency and the size of the head and component brain and sensory structures in the red wood ant. *Brain Res.* **16:** 85–104.

Bieber, M. and D. Fuldner. 1979. Brain growth during the adult stage of a holometabolous insect. *Naturwissenschaften* **66:** 426.

Bullock, T. H. and G. A. Horridge. 1965. *Structure and Function in the Nervous Systems of Invertebrates*, vol. 2. Freeman, San Francisco.

Callahan, P. S. 1975. Insect antennae with special reference to the mechanism of scent detection and the evolution of the sensilla. *Int. J. Insect Morphol. Embryol.* **4**(5): 381–430.

Callahan, P. S. 1979. Evolution of antennae, their sensilla and the mechanism of scent detection in Arthropoda, pp. 259–298. In A. P. Gupta (ed.), *Arthropod Phylogeny*. Van Nostrand-Reinhold, New York.

Chapman, R. F. 1979. *The Insects: Structure and Function*. Elsevier, New York.

Drescher, W. 1960. Regenerationsversuche am Gehirne von *Periplaneta americana* unter Berücksichigung von Verhaltensänderung und Neurosekretion. *Z. Morphol. Ökol. Tiere* **48:** 576–649.

Dujardin, F. 1850. Mémoire sur le système nerveux des insects. *Ann. Sci. Natur. Zool.* **14:** 195–206.

Erber, J., T. Masuhr, and R. Menzel. 1980. Localization of short-term memory in the brain of the bee, *Apis mellifica*. *Physiol. Entomol.* **5:** 343–358.

Fahrenbach, W. H. 1976. The brain of the horseshoe crab (*Limulus polyphemus*.). I. Neuroglia. *Tissue Cell* **8:** 395–410.

Fahrenbach, W. H. 1977. The brain of the horseshoe crab (*Limulus polyphemus*). II. Archtecture of the corpora pedunculata. *Tissue Cell* **9**(1): 157–166.

Flögel, J. H. L. 1878. Über den einheitlichen Bau des Gehirn in den verschiedenen Insekten-Ordnungen. *Z. Wiss. Zool. Suppl.* **30:** 556–592.

Forel, A. 1874. Les fourmis de la Suisse. *Nouv. Mem. Soc. Helv. Sci. Natur.* **26:** 1–200.

Ghaffar, H., J. R. Larsen, G. M. Booth, and R. Perkes. 1984. General morphology of the brain of the blind cave beetle, *Neaphaenops tellkampfii* Erichson (Coleoptera: Carabidae). *Int. J. Insect Morphol. Embryol.* **13**(5/6): 357–371.

Goll, G. 1967. Strukturuntersuchungen am Gehirn von *Formica. Z. Morphol. Ökol. Tiere* **59:** 143–210.

Goosen, H. 1949. Untersuchungen an Gehirnen Verschieder grosser, jeweils verwandter Coleopteren- und Hymenopteren *Arten. Zool. Jahrb. Abt. Allg. Zool. Physiol.* **62:** 1–63.

Groth, U. 1971. Vergleichende Untersuchungen über die Topographie und Histologie des Gehirns der Dipteren. *Zool. Jahrb. Anat.* **88:** 203–313.

Gupta, A. P. (editor) 1979a. *Arthropod Phylogeny.* Van Nostrand-Reinhold, New York.

Gupta, A. P. 1979b. Arthropd hemocytes and phylogeny, pp. 669–735. In A. P. Gupta (ed.), *Arthropod Phylogeny.* Van Nostrand-Reinhold, New York.

Gupta, A. P. 1979c. Origin and affinities of Myriapoda, pp. 373–390. In M. Camatini (ed.), *Myriapod Biology.* Academic Press, New York.

Gupta, A. P. 1983. Neurohemal and neurohemal-endocrine organs and their evolution in arthropods, pp. 17–50. In A. P. Gupta (ed.), *Neurohemal Organs of Arthropods.* Charles C. Thomas, Springfield.

Hanström, B. 1925. The olfactory centres in crustaceans. *J. Comp. Neurol.* **38:** 221–250.

Hanström, B. 1928. *Vergleichende Anatomie des Nervensystems der wirbellosen Tiere unter Berücksichtigung seiner Funktion.* Springer-Verlag, Berlin.

Hanström, B. 1930. Über das Gehirn von *Termopsis nevadensis* u. *Phyllium pulchrifolium* nebst Beiträgen zur Phylogenie der Corpora Pedunculata der Arthropoden. *Z. Morphol. Ökol. Tiere* **19:** 732–773.

Hanström, B. 1940. Inkretorische Organe, Sinnesorgane und Nervensystem des Kopfes einiger niederer Insekternordnungen. *Kgl. Svensk. Vetenskaps. Akad. Handl. (B)* **18:** 4–266.

Holmgren, N. 1916. Zur vergleichenden Anatomie des Gehirns von Plychaeten, Onychophoren, Xiphosuren, Arachniden, Crustaceen, Myriapoden und Insekten. *Kgl. Svensk. Vetenskaps. Akad. Handl.* **56:** 1–303.

Homberg, U. 1985. Interneurons of the central complex in the bee brain (*Apis mellifica,* L.). *J. Insect Physiol.* **31:** 251–264.

Honnegger, H. W. and F. W. Schürmann. 1975. Cobalt sulphide staining of optic fibers in the brain of the cricket, *Gryllus campestris. Cell Tissue Res.* **159:** 213–225.

Horridge, G. A. 1965. ArthropodA: General anatomy, pp. 801–964. In T. H. Bullock and G. A. Horridge (eds.), *Structure and Function in the Nervous Systems of Invertebrates,* vol. 2. W. H. Freeman, San Francisco.

Howse, P. E. 1974. Design and function in the insect brain, pp. 180–194. In L.

Barton-Browne (ed.), *Experimental Analysis of Insect Behaviour.* Springer-Verlag, Berlin.

Howse, P. E. 1975. Brain structure and behavior in insects. *Annu. Rev. Entomol.* **20**: 359–379

Howse, P. E. and J. L. D. Williams. 1969. The brains of social insects in relation to behavior. *Proc. 6th Congr. IUSSI* **1969**: 59–64.

Huber, F. 1965. Neural integration (central nervous system), pp. 333–406. In M. Rockstein (ed.), *The Physiology of Insecta* vol. 2. Academic Press, New York.

Jawlowski, H. 1936. Über den Gehirnbau der Käfer. *Z. Morphol. Ökol. Tiere* **32**: 67–91.

Jonescu, C. N. 1909. Vergleichende Untersuchungen über das Gehirn der Honigbiene. *Jena Z. Med. Naturwiss.* **45**: 11–180.

Klemm, N. 1968. Monoaminhaltige Strukturen im Zentralnervensystem der Trichoptera (Insecta). *Teil I. Z. Zellforsch. Mikrosk. Anat.* **92**: 487–502.

Larsen, J. R., G. Booth, R. Perks, and R. Gundersen. 1979. Optic neuropiles absent in cave beetle, *Glacicavicola bathyscioides* (Coleoptera: Leiodidae). *Trans. Am. Microsc. Soc.* **98**(3): 461–464.

Manton, S. M. 1973. Arthropod phylogeny—A modern synthesis. *J. Zool. Soc.* **171**: 111–130.

Menzel, R., J. Erber, and T. Masuhr. 1974. Learning and memory in the honeybee, pp. 195–217. In L. Barton-Browne (ed.), *Experimental Analysis of Insect Behaviour.* Springer-Verlag, Berlin.

Mobbs, P. G. 1982. The brain of the honeybee *Apis mellifera.* I. The connections and spatial organisation of the mushroom bodies. *Philos. Trans. R. Soc. (B)* **298**: 309–354.

Newton, E. T. 1879. On the brain of the cockroach *Blatta orientalis. Q. J. Microsc. Sci.* **19**: 340–356.

Packard, A. S. 1886. The cave fauna of North America, with remarks on the anatomy of the brain and origin of the blind species. *Nat. Acad. Sci. Mem.* **4**: 3–156.

Pandazis, G. 1930. Über die relative Ausbildung der Gehirnzentren bei biologisch verschiedenen Ameisenarten. *Z. Morphol. Ökol. Tiere* **18**: 114–169.

Pearson, L. 1971. The corpora pedunculata of *Sphinx ligustri* L. and other Lepidoptera: An anatomical study. *Philos. Trans. R. Soc. (B)* **259**: 477–516.

Power, M. E. 1943. The brain of *Drosophila melanogaster. J. Morphol.* **72**: 517–559.

Rabl-Ruckhard, T. 1875. Studien über Insektengehirne. I. Das Gehirn der Ameise. *Arch. Anat. Physiol. Wiss. Med.* **42**: 480–499.

Ratzendorfer, C. 1952. Volumetric indices for the parts of the insect brain. A comparative study in cerebralization of insects. *J. N.Y. Entomol. Soc.* **60**: 129–152.

Rensch, B. 1959. Trends towards progress of brain and sense organs. *Cold Spring Harbor Quart. Biol.* **24**: 291–303.

Rudloe, A. 1979. *Limulus polyphemus*: A review of the ecologically significant literature, pp. 27–35. In E. Cohen (ed.), *Biomedical Applications of the Horseshoe Crab (Limulidae).* Alan R. Liss, New York.

Satija, R. C. and B. Das. 1963. Histological studies on the brain of *Orthophagus catta* (F.) (Coleoptera). *Res. Bull. Panjab Univ. Sci.* **14**: 231–241.

Satija, R. C., K. Sumal, and R. Sumeet. 1975. Morphogenetic studies on the brain of *Callosobruchus analis* (Coleoptera) *Res. Bull. Panjab Univ. Sci.* **26:** 61–68.

Schürmann, F. W. 1973. Über die Struktur der Pilzkorper des Insektengehirns. III. Die Anatomie der Nervenfarsen in der Corpora Pedunculata bei *Acheta domesticus* L. (Orthoptera): Eine Golgi-Studie. *Z. Zellforsch. Mikrosk. Anat.* **145:** 247–285.

Schürmann, F. W. 1974. Bemerkungen zur Funktion der Corpora Pedunculata im Gehirn der Insekten aus morphologischer Sicht. *Exp. Brain Res.* **20:** 406–432.

Snodgrass, R. E. 1958. Evolution of arthopod mechanisms. *Smithson. Misc. Collect.* **138**(2): 1–77.

Størmer, L. 1977. Arthropod invasion of land during Silurian and Devonian times. *Science* (Washington, D. C.) **197:** 1362–1364.

Strausfeld, N. J. 1970. Variations and invariants of cell arrangements in the nervous system of insects (A review of neuronal arrangements in the visual system and corpora pedunculata). *Verh. Zool. Ges.* **64:** 97–108.

Strausfeld, N. J. 1976. *Atlas of An Insect Brain.* Springer-Verlag, Berlin.

Thompson, C. B. 1913. A comparative study of the brain of three genera of ants with special reference to the mushroom bodies. *J. Comp. Neurol.* **23:** 515–572.

Tiegs, O. W. 1940. The embryology and affinities of the Symphyla, based on a study of *Hanseniella agilis*. *Q. J. Microsc. Sci.* **82:** 1–225.

Weiss, M. J. 1974. Neuronal connections and the function of the corpora pedunculata in the brain of the American cockroach, *Periplaneta americana*. *J. Morphol.* **142:** 21–69.

Weiss, M. J. 1981. Structural patterns in the corpora pedunculata of Orthoptera: A reduced silver analysis. *J. Comp. Neurol.* **203:** 515–553.

Wheeler, C. M. and A. P. Gupta. 1986. Effects of exogenous juvenile hormone I on the numbers and distribution of antennal and maxillary and labial palp sensilla of male *Blattella germanica* (L.) (Dictyoptera: Blattelliadae). *Experientia* **42:** 57–58.

Williams, J. D. L. 1975. Anatomical studies of the insect nervous system: A ground plan of the midbrain and an introduction to the central complex in the locust *Schistocerca gregaria* (Orthoptera). *J. Zool. Lond.* **167:** 67–86.

CHAPTER 3

The Cellular Mechanisms That Shape Neuronal Diversity and Form
Insights from the Embryonic Grasshopper

Haig Keshishian

Department of Biology
Yale University
New Haven, Connecticut

3.1. INTRODUCTION

Neuronal morphogenesis is one of the striking events of insect development. Early in embryogenesis hundreds of seemingly indistinguishable ganglion cells are generated from a sheet of dividing neuroblasts. Within hours after birth, each initially spheroidal ganglion cell differentiates into a stereotyped and uniquely specified neuron. This remarkable transformation, of an epithelium into a nervous system, has remained one of the more elusive problems of development.

This review examines the cellular mechanisms that shape neuronal form and cellular diversity in the insect. The focus will be on recent efforts using the grasshopper embryo. The grasshopper embryo, perhaps more than any other insect preparation, has proved a suitable model for analyzing neural development with single cell resolution. In the grasshopper, it is possible to follow the entire sequence of neuronal development in situ, from the differentiation of the neuroblast to the morphogenesis of the neuron. Precise developmental manipulations of single neurons or their precursors are now routinely feasible.

In most cases, the three-dimensional structure of an embryonic grasshopper neuron is predictable and uniquely specified. This simplifies the description of normal branch development. It also facilitates experimental tests directed at specific arborizations. Another advantageous feature is that the embryo generates a miniature version of the adult nervous system, so that morphological, physiological, and neurochemical properties are remarkably complete by late embryogenesis.

This review will approach the problem of neuronal morphogenesis from three standpoints. First, the cellular mechanisms that control neuronal diversity will be addressed. The determinative events that control each neuronal phenotype operate early in development, so that soon after cell birth a neuron's fate is sealed. The role of cell lineage and position in dictating the morphological and functional fate of a neuron will be assessed.

Secondly, the review will consider how neuronal processes grow and navigate the embryonic environment. Most axons and dendrites extend behind exploring growth cones. An important, and yet unresolved problem in this area is to identify the local and general features that are recognized by a growth cone. Progress has come from looking at how neurons pioneer the epithelial expanses of the periphery, and also from how neuritic processes choose fascicles in the embryonic central nervous system (CNS).

Finally, the review will turn to specific examples of neuronal morphogenesis. These were selected to reveal a few of the neuronal strategies used to generate characteristic branch structures.

3.2. NEUROGENESIS AND DETERMINATION

3.2.1. The Neurogenic Cells

Insect central nervous systems are generated by two kinds of neurogenic cell. Most insect neurons are the progeny of the bilaterally symmetrical, segmental *neuroblasts* (NBs) (Wheeler, 1893; Bate, 1976b; Hartenstein and Campos-Ortega, 1984; Doe and Goodman, 1985a). The remaining neurons are derived from the *midline precursor* cells (MPs) (Bate and Grunewald, 1981; Thomas et al., 1984).

Each grasshopper segment possesses a stereotyped number of NBs. These are large, slightly flattened cells, measuring up to 30 μm in diameter (Bate, 1976b). The number and pattern of the NBs are segment specific. In *Schistocerca americana,* the basic pattern in a hemisegment consists of seven rows containing from two to five cells each, for a total of 29 cells. The opposite hemisegment has an exact mirror image of NBs. In addition, one unpaired NB, distinct from the MPs, lies on the posterior midline (Bate, 1976b; Goodman and Spitzer, 1979; Doe and Goodman, 1985a).

Each NB is individually identifiable, by both its position and its progeny. The segmental differences seen in the ventral nerve cord arise from the addition or deletion of one or more NBs to a basic pattern of 29 per side. Thus, the three thoracic segments each have one extra NB on each side, and the prothoracic segment has an additional anterior unpaired NB. Some segments may have partial patterns, such as the most posterior 11th abdominal segment (Doe and Goodman, 1985a).

The MP cells are arranged in longitudinal arrays of seven cells, aligned on the midline (Bate and Grunewald, 1981; Goodman et al., 1981). This pattern is repeated segmentally, although there are also segment-specific variations in number (Doe and Goodman, 1985a). The MPs differ from the NBs principally in two ways. Unlike NBs the MPs are not organized into bilaterally symmetrical pairs. The MPs also produce neurons by a single symmetrical division, rather than the budding and division behavior typical of NBs (Bate and Grunewald, 1981).

The properties of NBs and MPs are in all likelihood broadly conserved among insect embryos. Several homologies to the pattern of grasshopper neuroblasts have been recently described for *Drosophila melanogaster* NBs (Hartenstein and Campos-Ortega, 1984). A close homology is also found in the pattern and the progeny of the MP cells of the embryos of flies and grasshoppers (Thomas et al., 1984).

3.2.2. Neurogenesis and Cellular Diversity

Each NB successively buds as many as 50 daughter cells, through a fixed number of cell divisions. These *ganglion mother cells* (GMCs) then each divide once to generate sibling pairs of neurons. All NB cell lineages end in the death of the neuroblast (Bate, 1976b). The midline precursors each generate only two daughters by symmetrical division, circumventing the GMC cell stage (Bate and Grunewald, 1981).

It is likely that the roster of neurons generated by a given neuroblast or midline precursor is invariant. The clone often consists of uniquely identifiable cells, where phenotypes are preordained by parentage and birth order (Goodman and Spitzer, 1979; Taghert and Goodman, 1984). The fixity of the lineages permits the characterization of conserved phenotypes within a clone. A good example of this is found with the lineage derived from the unpaired NB.

Each segment's single posterior unpaired neuroblast buds up to 50 GMCs, producing a clone of approximately 100 dorsal unpaired median (DUM) neurons (Goodman and Spitzer, 1979, 1981a, b). Several neuronal phenotypes are conserved among the DUM neurons in each segment of the ventral nerve cord. The DUM neurons probably share the same neurotransmitter, octopamine (Goodman et al., 1979). They also have similar dendritic anatomy, consisting of bilaterally symmetrical dendrites and axons.

Goodman and Spitzer (1979, 1981b) noted systematic trends in the phenotypes of neurons as related to birth order. The largest efferent neurons are generated early in the lineages. The early cells also have active spiking cell body membranes. The later arising neurons are often small, local interneurons. These cells often have passive, nonspiking cell body membranes, or are completely nonspiking.

Goodman and Spitzer (1979) also noted that the birth order of the first few identified neurons was stereotyped. The large number of DUM cells made it impractical to determine whether or not this applied to all the progeny of the NB. To test this hypothesis, Taghert and Goodman (1984) characterized the entire clone generated by NB 7-3. This is the last NB to develop segmentally (Doe and Goodman, 1985a), producing fewer neurons than most NBs. Examining both morphology and transmitter phenotypes, they found that the total number and birth order of the neurons within the clone was invariant.

However, although clonally related neurons often share similar or identical phenotypes, phenotypes unique to a single cell within a clone are also found. For example, Taghert and Goodman (1984), and Keshishian and O'-Shea (1985a, b) have both demonstrated neurotransmitter phenotypes in which only one or two cells in a hemisegment express the molecule.

Midline precursor cells also produce seemingly invariant lineages (Goodman et al., 1981). Homologous MP progeny have strongly conserved phenotypes in each segment. Nevertheless, the intersegmental homologies can vary systematically. For example, Bate et al., (1981) have found an anterior-to-posterior gradient in the programmed cell death for the MP3 progeny, with increasing probability of cell death among the more posterior MP3 daughters. Also, the size of dendritic arbors and neuritic projections declined in the more posterior daughters.

In conclusion, it is likely that each NB and MP produces a highly stereotyped clone of daughter neurons. The neurons produced by segmentally homologous precursor cells often share similar phenotypes, and may produce homologous neurons through similar birth orders. It is likely that these homologous cell lineages are probably responsible for the intersegmental homology of neurons found in many insect central nervous systems.

3.2.3. The Determination and Differentiation of Neurogenic Cells

From the preceding account it follows that neuronal identity is largely predictable by cell lineages from the NB or MP cell. However, how a neuronal phenotype is encoded in a lineage is not mechanistically known. As will be shown here, the identity of a NB is established during its differentiation through intercellular or positional mechanisms acting on an epithelium of uncommited ectodermal cells. Although the phenotypes of nearly a thousand neurons are fixed by lineage within an array of about 30 hemisegmental NBs, the NBs seem to acquire their individual identities via positional signals.

Neuroblasts and midline precursors are derived from a portion of segmental ectoderm fate mapped by Poulson (1950). The neurogenic ectoderm (NE; sometimes called the neuroepithelium) in *Drosophila melanogaster* extends laterally from the ventral midline to the tracheal placodes (Hartenstein and Campos-Ortega, 1984). In the grasshopper, it has been possible to directly test the competency of ectodermal cells to become neurogenic. Doe and Goodman (1985b) have mapped the neurogenic ectoderm to a paramedial strip, consisting of 150 ectodermal cells in each hemisegment. Within the NE the cells are apparently multipotent (Taghert et al., 1984; Doe and Goodman, 1985b; Doe et al., 1985). In the grasshopper, each of the 150 hemisegmental NE cells has the option to differentiate into neurogenic cells (the NBs and MPs), gliogenic cells, or nonneuronal support cells, that may contribute to body wall epithelia (Doe and Goodman, 1985a).

Each neuroblast originates as a NE cell that enlarges and delaminates from the ectodermal sheet (Doe and Goodman, 1985a). Delamination places the NB dorsally sandwiched between the remaining NE cells and the mesoderm. Associated with each NB are two classes of cells, also arising from the neurogenic ectoderm. A single *cap* cell lies tightly apposed to the ventral side of each neuroblast, while a few elongated *sheath* cells envelope the neuroblast, and later its daughters (Doe and Goodman, 1985a; Doe et al., 1985).

Position, rather than lineage, seems to control the identities of specific neuroblasts (Taghert et al., 1984; Doe and Goodman, 1985b). A lesioned neuroblast will be replaced by nearby NE cells. The NEs can come from diverse locales, yet having delaminated they adopt the lesioned NBs identity, and will generate the correct clone of daughters (Taghert et al., 1984; Doe and Goodman, 1985b; Doe et al., 1985). In other words, the same NE, depending on its position when it delaminates, can differentiate into different NBs. The source of the positional cue is not known. However, GMCs and differentiating neurons probably do not convey positional information to the delaminated neuroblasts; a replacement NB will ignore the partially generated lineages of lesioned NBs to generate duplicate sets of neurons (Doe and Goodman, 1985b).

Thus, a multipotent and seemingly uniform neurepithelial sheet will generate a highly stereotyped array of neurogenic cells. If that array of neurogenic cells is manipulated, the presumably uncommitted ectodermal cells will recreate the neurogenic array using cues that persist through early neurogenesis.

3.2.4. Neuronal Determination and Fate Switching

A rapid narrowing of developmental options occurs in early neurogenesis. The transition from a multipotent NE to highly committed postmitotic neurons is accomplished by a series of determinative events, whose timing can be revealed by cell lesioning experiments. As was illustrated earlier, the neurogenic ectoderm regulates to replace, with correct identity, lesioned neuroblasts. However, there are at present no examples of either GMCs or neurons themselves being replaced, either by the NB making an additional GMC, or by a GMC switching its identity (Doe and Goodman, 1985b; Taghert et al., 1984; Kuwada and Goodman, 1985). From this, it follows that there is no evidence for number regulation of grasshopper neurons.

There are now two examples where one neuron will *switch* its fate to replace a killed cell, provided the lesion is made early enough. Fate-switching has been seen for the GMC-1 progeny of the 1-1 neuroblast (the aCC and pCC neurons; Goodman et al., 1984; Doe and Goodman, 1985b; Doe et al., 1985) and for the MP3 progeny (the H-cell and its sib; Doe et al., 1985; Kuwada and Goodman, 1985).

In both the H-cell and aCC/pCC examples of fate-switching, one of the sib neurons was killed at successively later times. Beyond a critical period after birth, the surviving neuron will express either one of the two possible cell types, on average with equal frequency. This period, about 5 hr for the aCC/pCC switch, probably marks the time when the two neurons become committed to their respective neuronal fates (Goodman et al., 1984; Doe and Goodman, 1985b). When one sib-cell is killed earlier than this, the remaining neuron always develops into the same one of the two possible cell types. If one reasons that the killing experiment is directed randomly, on average in half of the kills one of the sib neurons had switched its fate.

A model to explain this phenomenon is that the two neuronal fates are hierarchically related. When only one cell is present, only the dominant fate is followed. In the presence of both cells, the two fates assort; the mechanism by which this is done is unknown. It is probably dependent, however, on some kind of intercellular interaction between the sib neurons.

Although there are as yet only two reported examples of neuronal fate-switching in the grasshopper, it is possible that it is a general property of all GMC or MP progeny. The behavior of neuronal fate-switching is characteristic of cells belonging to lineage *equivalence groups*. This phenomenon was first described in the gonad of the nematode, *Caenorhabditis elegans* by Kimble (1981) and Kimble and White (1981). A similar fate-switching occurs among the progeny of the O/P teloblasts in the embryonic leech, *Helobdella triserialis* (Shankland and Weisblat, 1984).

In summary, neuronal fate may be modeled as being shaped initially by positional mechanisms, that function to define and fix cell lineages in neuroblasts. Each NB arises from a multipotent cellular epithelium, the neurogenic ectoderm. The NE cells can differentiate into one of several cell types to create widely differing progeny. However, once delaminated a neuroblast adopts a specific identity, defined by a characteristic clone of daughter GMCs. Nevertheless, the sibling ganglion cells derived from each GMC are born equivalent. Initially, they are capable of pursuing either of two cell fates to yield two uniquely specified neurons. The final determination is apparently made within hours of birth by undefined positional or intercellular mechanisms.

3.3. MECHANISMS OF NEURITE GROWTH AND GUIDANCE

Soon after its birth, the anatomical fate of a neuron is largely determined. Nonetheless, the cell still faces the task of differentiating into its mature, branched form. That transformation is achieved through both passive and active cellular mechanisms; the most prominent of the latter being the active growth cone-mediated exploration of the embryonic environment.

The mechanical characteristics of neuronal growth cones have been studied intensively in tissue culture (and reviewed by Johnston and Wessells, 1980; Letourneau, 1982; Wessells, 1982). In general, growth cones advance across a substrate by means of a cyclical propulsive mechanism. This is often modeled as achieved through (1) the extrusion from the body of the growth cone of filopodia and lamellipodia, to encompass large areas of the substrate (e.g., Tosney and Wessells, 1983); (2) The adhesion of these fine processes onto the substrate, with apparent selectivity for specific kinds of substrate (e.g., Letourneau, 1979); and (3) that the body of the growth cone is drawn forward by contracting the adherent filopodial processes, possibly through an actin–myosin interaction (modeled by Bray, 1973).

3.3.1. Neuritic Exploration and Axon Guidance

An advancing growth cone responds to its substrate in a predictable fashion, and in insects this is vividly demonstrated by the stereotyped trajectories taken by axons in the CNS and periphery (Bentley and Keshishian, 1982). Embryonic limb-buds are relatively simple and sparse epithelial structures. The behavior of pioneering axons in these relatively simple environments has helped to illuminate important features of process guidance in general.

The thoracic leg limb-buds have highly stereotyped peripheral nerves in common with most of the peripheral nervous system. These are founded in early limb-buds by small numbers of uniquely specified pioneer neurons (Bate, 1976a; Keshishian, 1980). Collectively, the pioneers establish each branch of a peripheral nerve (Bentley and Keshishian, 1982; Ho and Goodman, 1982; and Keshishian and Bentley, 1983a-c). The pioneers are not endowed with unique guidance capacities, and follower neurons can pioneer if need be (Keshishian and Bentley, 1983c). The remarkably simple behavior of these neurons permits several generalizations about growth cone guidance (Bentley and Caudy, 1983b):

1. The orientation of an emerging neurite from the soma of a neuron is highly stereotyped. As yet unidentified orienting cues, probably on a segmental or the appendage level, stereotype the direction of neurite emergence and its initial trajectory (Bentley and Caudy, 1983b; Caudy and Bentley, 1986). However, different axons respond to these "gradients" in differing ways; some pioneering axons that are near neighbors can emerge with widely differing angles. Furthermore, different axons can grow in opposite directions within the same segment, as is seen when afferent and efferent growth cones collide in the coxa and femur (Keshishian and Bentley, 1983a, b).

2. Growth cone filopodia are capable of sampling large areas of the embryonic landscape (Keshishian and Bentley, 1983a). The growth cone follows the apparent "choices" made by filopodia for specific cell surfaces or epithelial regions (Bentley and Caudy, 1983b).

3. Growth cones and their filopodia are sensitive to segmental borders, and can turn or align on them. The significance of segmental borders is particularly revealed where guidepost neurons have been lesioned (Bentley and Caudy, 1983a), or where the pioneering axons arise precociously before guideposts have differentiated (Caudy and Bentley, 1986).

4. Perhaps the most precise way of providing a local guidance cue is to uniquely label the surface of a single cell, to serve as a guidepost (Bentley and Keshishian, 1982). In the leg these surfaces are often the newly delaminated somata of afferent neurons, that arise at given loci with high reliability (Bentley and Keshishian, 1982; Ho and Goodman, 1982; Keshishian and Bentley, 1983a, b). In the leg, the trajectory of a pioneer neurite, especially where it makes abrupt turns, is shaped by the stereotyped disposition and the timed appearance of other neurons in the pathway. It is possible to disrupt the normal guidance of

a pioneering neurite, where it makes a sharp turn, by selectively lesioning the guidepost (Bentley and Caudy, 1983a).

The trajectory taken by a neurite across an epithelium, therefore, is not exclusively programmed as a sequence of elongations and timed left or right turns. Rather, major directional changes result from neuron-specific selective affinities for local surface features, such as guideposts. These cells are influencing guidance in concert with other effects, such as general epithelial polarities or segment borders. However in some cases, especially where neurites grow in generally linear trajectories, guidance may be exclusively achieved through apparent epithelial polarizations.

The latter point is illustrated by the generally linear trajectories taken by afferents in the imaginal wing discs of *Drosophila melanogaster*. Putative polarizations of the disc epithelium seem sufficient to project axons in the appropriate proximal directions. Single-cell cues, such as more proximal neurons, play no necessary role for guidance (Blair and Palka, 1985; Blair et al., 1985; Schubiger and Palka, 1985). A similar phenomenon occurs in the antennae of the grasshopper. In those buds, axons grow in linear trajectories down a tube (Bate, 1976a; Ho and Goodman, 1982; Berlot and Goodman, 1984). In the antenna, however, a selective neuritic contact occurs at the first sharp turn at the base (Berlot and Goodman, 1984).

3.3.2. Labeled Pathways and Surface Cues

Although there are examples of neuritic guidance (in the periphery) that is achieved in the absence of neuronal cell recognition, such mechanisms could not explain the complex trajectories that occur in the CNS. Most neuronal growth occurs in environs more complex than the peripheral limb-buds. In a developing ganglion the surfaces of other neurons will serve as substrates. How growth cones select and fasciculate upon axonal processes has been examined intensively in the thoracic ganglia of the grasshopper embryo (Goodman et al., 1984). The "labeled pathways" hypothesis proposes that neuronal surfaces are characteristically labeled. Growth cones from other cells can recognize and preferentially respond to the specific labels. The path taken by a neurite within the neuropil results from a series of highly selective filopodial contacts on respective fascicles.

Surface sampling and selectivity is illustrated in the behavior of the G neuron, as its growth cone explores the neuropil (Raper et al., 1983 a, b; 1984; Bastiani et al., 1984). Although the filopodia of the G growth cone have broad access to dozens of nearby neurites, they preferentially accumulate on a single fascicle consisting of the three P and two A axons. Lesioning one component of the fascicle, the P axons (Raper et al., 1984) disrupt normal G cell behavior, so that its growth cone loses fascicle preference.

A similar and unique preference by the aCC neuron for its target, the U axons, has also been tested (Goodman et al., 1984). Normally, the aCC neurite sprouts pointing anteriorly. It reverses its direction to grow posteriorly after fasciculating on the posteriorly advancing U axons. Lesioning the U axons blocks the direction change, despite the availability of other nearby posteriorly directed neurites.

Cells homologous to the aCC/pCC system are found in *Drosophila melanogaster* (Thomas et al., 1984). Monoclonal antibodies directed against neuritic tissue in the fly will distinguish small neuritic subsets, including those involved in establishing early pathways. In one case (the Mab SOX2), a monoclonal antibody will selectively label the *Drosophila* aCC and none of the other local neuritic surfaces, including the aCC sib. Using the methods of molecular biology it may prove feasible to identify the molecules used to label pathways in fruitfly embryos (Goodman et al., 1984).

3.3.3. Passive Mechanisms

Finally, the growth of neurites can occur through growth cone-independent, passive mechanisms. Most often, the passive elongation of a neurite occurs in response to stretch. An example of this has been found by Keshishian and Bentley (1983b), where the growth of the leg results in a massive elongation of afferent axons. It is likely most neurites insert new membrane and elongate over their entire length, and this probably is involved in the uniform growth exhibited by most neurons postembryonically. Bray (1984) has elegantly shown passive growth in vitro. Neurites readily elongate in response to pulling or stretching forces, even when their growth cones have been severed.

3.4. MORPHOGENESIS

3.4.1. The Timecourse of Morphogenesis

How the mechanisms of growth cone guidance are harnessed to generate a neuron of characteristic form remains largely unknown. It is not known how a filopodium can recognize a given surface. Nor do we know how a single neuron can generate several growth cones, each founding a different neuronal branch, with its own characteristic guidance behavior and affinities. Nevertheless, the ability to follow the development of uniquely specified neurons in situ has permitted detailed chronologies of events occurring during differentiation. A surprising result from the developmental timecourses of diverse kinds of neurons is that the cells follow similar and perhaps common programs for differentiation (Goodman and Spitzer, 1979, 1981b; Taghert and Goodman, 1984; Keshishian and O'Shea, 1985b). Morphogenesis, the acquisition of membrane properties, sensitivity to neurotransmitters, synapse formation, and the expression of neurotransmitters all arise in a generally obeyed sequence.

Most grasshopper neurons develop their anatomy in an orderly and step-wise fashion as well. A typical example of the acquisition of adult morphology in an embryonic neuron is found for the brain arborizations of the descending contralateral movement detector (DCMD) (Bentley and Toroian-Raymond; 1981). A quantitative study of branch structure and dimensions was made from the 40% stage of embryogenesis to the adult. The mature dendritic morphology is largely complete by as soon as the 60% stage, with the hatchling neuron structurally similar to the adult cell. The principal difference between the embryonic and adult neurons was one of size, not form. Bentley and Toroian-Raymond (1981) noted a stereotyped program of process sprouting and elongation, with little evidence of the pruning back of central branches. Early processes featured short lateral filopodial sprouts, but the origin of each dendritic branch was through the stereotyped emergence of a process at a characteristic time and place.

Although most of the major branches of a cell are usually founded in an orderly and error-free fashion in early neuronal morphogenesis, there are several examples of remodeling.

3.4.2. Neuronal Remodeling

An attractive system for studying both morphogenesis and synaptogenesis is the wing stretch receptor (SR). In a morphogenetic study of the embryonic CNS arborizations of SR, Heathcote (1981) found that the developing central processes have both appropriate (i.e., adult corresponding) and inappropriate branches. The inappropriate branches are trimmed off by the 70% stage. The embryonic branches are characteristically less ramified and have fewer sprouts than their adult counterparts. The role of synaptic activity on the development of these processes is unclear. The wing does not develop fully until the adult stage. Nevertheless, it is likely that the SR is making functional connections onto CNS targets soon after hatching, and possibly during embryogenesis. Heathcote (1980) found that the synaptic strength and conduction properties of the SR synapses reach adult proportions during early nymphal development.

Another example of remodeling is found for the axonal branching of DUM cells (Goodman and Spitzer, 1979). Here the pruning of supernumerary processes of DUM cells is a mechanism for refining unique efferent branch anatomies. The neuron DUMETi normally projects symmetrically out of nerve 5 of the ganglion to innervate on both sides the extensor tibiae muscle of the leg. At the time it first reaches that muscle (approximately the 60% stage), it sports numerous collateral branches extending into nerve 5. During the next 10% of development, these branches are subsequently lost, leaving only the appropriate ETi trajectory.

To test whether or not the branch loss was influenced by the peripheral target, Whitington et al. (1982) severed the thoracic leg on one side. The bilateral structure of DUMETi permits a comparison of control and experimental branches in the same cell. The central anatomy of DUMETi was not altered by denying peripheral targets. Other motoneurons were similarly unaffected (Whitington and Seifert; 1982). The control side of DUMETi reduced its peripheral branch collaterals normally. On the experimental side the neuron failed to reduce its peripheral branching, and retained its immature multibranched appearance. These results suggest that local branch remodeling may be influenced by peripheral targets, but normal ganglionic differentiation and survival is largely autonomous.

Perhaps the most dramatic example of neuronal remodeling in grasshopper occurs in H cells (Goodman et al., 1981). In *Schistocerca nitens*, the two MP3 progeny cells initially develop indistinguishable and symmetrical projections, with phenotypes not unlike several other MP progeny. Some 15–20% of development later, with equal frequency one of the two neurons transforms into an elaborately branched H-shaped neuron, acquiring by the way other new phenotypes, such as overshooting action potentials. The sib neuron retains its simpler morphology.

A morphological transformation is not obligatory for normal H-cell development in other species. In the closely related grasshopper *S. americana*, the H-cell and its sib develop without transformation, and the cells will show early fate-switching (Kuwada and Goodman, 1985). The reason for this discrepancy is not known.

3.4.3. The Role of Innervation

Usually, the major branches of insect neurons arise with predictable order and behavior during morphogenesis, yielding highly stereotyped cells. This might imply that insect neurons are restricted to invariable developmental programs, with limited morphological plasticity. In fact, there is good evidence that some interneurons can significantly adjust the growth and relative sizes of their individual arbors in response to the level of embryonic synaptic inputs. An excellent example of this kind of plasticity is found in the development of the medial giant interneuron (MGI) of the terminal ganglion.

Shankland and Goodman (1982) made a detailed morphogenetic analysis of the normal development of the grasshopper MGI. Each MGI receives as its major synaptic input a massive convergent innervation from a subset of the mechanosensory neurons of the contralateral cercal appendage. The sensory arbors converge to form a cup-shaped glomerulus, into which is nestled the egg-shaped contralateral arbor of the MGI.

During normal morphogenesis, the MGI adds branches in an orderly fashion, with each one navigating the local neuropil behind growth cones. Many of the major branches of the MGI are established before the arrival of the first wave of sensory axons. By the 60% stage, the neuron has established a good part of its mature form (Shankland and Goodman, 1982). Nevertheless, luxuriant sprouting and branching occurs soon after the afferents arrive on the scene.

Postembryonic deafferenting of the MGI in crickets (Murphey et al., 1975) revealed a moderate loss of branch structure. Embryonic deafferentation during morphogenesis has a much more significant effect (Shankland and Goodman, 1982; Bentley and Keshishian, 1982). Removing cerci before the arrival of the afferents severely retards (by up to 40%) the overall branching on MGI. Bilaterally denervated MGIs can only develop up to the level found normally at the 70% stage in controls.

Unilateral cercal ablation leaves the innervated cell relatively unharmed, permitting it to arborize normally. The deafferented MGI fails to sprout to a normal level in those arbors experimentally denied innervation. Of equal importance, the cell sprouts supernormally to invade new neuropilar regions. In conclusion, the neuron develops within a fixed number of "arborization domains," where branch density is promoted by synaptic input. However, each domain does not develop autonomously. There is an interdependence of the regions, so that retardation of one domain permits proliferation elsewhere.

3.5. SUMMARY

The analysis of the relatively simple and well-defined nervous systems of grasshopper embryos has provided numerous insights into the determination and differentiation of highly specialized identifiable neurons. The insect CNS is derived from a multipotent neurogenic ectoderm, whose cells delaminate to establish a stereotyped array of neurogenic cells. The unique identities of each neurogenic cell is determined positionally, whereas the characteristic phenotypes of the progeny neurons are preordained by their cell lineages. Nevertheless, neurons are capable of fate-switching, based upon local intercellular interactions. Morphogenesis itself is rooted in the growth and guidance of individual processes, achieved by diverse extrinsic cues in the microenvironment of the elongating neurite. These include both regional cues, such as putative epithelial polarities, as well as surface cues found on single cells, as seen in peripheral guideposts and CNS labeled pathways.

Neurons differentiate in very similar ways, following what may be a commonly obeyed sequence. Morphogenesis usually occurs through the relatively error-free and orderly establishment of each branch, so that a structure resembling the adult form is evident by about 20% of embryogenesis after sprouting. There are several cases, however, in which neurons modify their form through branch loss. Finally, the density of processes within different branches may be adjusted to match the levels of embryonic synaptic innervation.

ACKNOWLEDGMENTS

I thank R. Wyman, E. Aceves-Pina, M. Anderson, and S. Waddell for comments on the manuscript. I also thank Drs. C. Doe, J. Kuwada, and C.

Goodman for making available prepublication manuscripts of their work. The author is supported by the McKnight Foundation Scholars award.

REFERENCES

Bastiani, M. J., J. Raper, and C. S. Goodman. 1984. Pathfinding by neuronal growth cones in grasshopper embryos III. Selective affinity of the G growth cone for the P cells within the A/P fascicle. *J. Neurosci.* **4:** 2311–2328.

Bate, C. M. 1976a. Pioneer neurons in an insect embryo. *Nature (London)* **260:** 54–56.

Bate, C. M. 1976b. Embryogenesis of an insect nervous system. I. A map of the thoracic and abdominal neuroblasts in *Locusta migratoria*. *J. Embryol. Exp. Morphol.* **35:** 107–123.

Bate, C. M. and E. B. Grunewald. 1981. Embryogenesis of an insect nervous system II. A second class of neuron precursor cells and the origin of the intersegmental connectives. *J. Embryol. Exp. Morphol.* **61:** 317–330.

Bate, C. M., C. S. Goodman, and N. C. Spitzer. 1981. Embryonic development of identified neurons: Segment-specific differences in the H cell homologues. *J. Neurosci.* **1:** 103–106.

Bentley, D. and M. Caudy. 1983a. Pioneer axons lose directed growth after selective killing of guidepost cells. *Nature (London)* **304:** 62–64.

Bentley, D. and M. Caudy. 1983b. Navigational substrates for peripheral pioneer growth cones: Limb-axis polarity cues, limb-segment boundaries and guidepost neurons. *Cold Spring Harbor Symp. Quant. Biol.* **48:** 573–585.

Bentley, D. and H. Keshishian. 1982. Pathfinding by peripheral pioneer neurons in the grasshopper. *Science (Washington, D.C.)* **218:** 1082–1088.

Bentley, D. and A. Toroian-Raymond. 1981. Embryonic and postembryonic morphogenesis of a grasshopper interneuron. *J. Comp. Neurol.* **201:** 507–518.

Berlot, J. and C. S. Goodman. 1984. Guidance of peripheral pioneers in the grasshopper: Adhesive hierarchy of epithelial and neuronal surfaces. *Science (Washington, D.C.)* **223:** 493–496.

Blair, S. S. and J. Palka. 1985. Axon guidance in cultured wing discs and disc fragments of *Drosophila*. *Dev. Biol.* **108:** 411–419.

Blair, S. S., M. A. Murray, and J. Palka. 1985. Axon guidance in cultured epithelial fragments of *Drosophila* wing. *Nature (London)* **315:** 406–409.

Bray, D. 1973. Model for membrane movements in the neuronal growth cone. *Nature (London)* **244:** 93–96.

Bray, D. 1984. Axonal growth in response to experimentally applied mechanical tension. *Dev. Biol.* **102:** 379–389.

Caudy, M. and D. R. Bentley. 1986. Pioneer growth cone morphologies reveal proximal increases in substrate affinity within leg segments of grasshopper embryos. *J. Neurosci.* **6:** 364–397.

Doe, C. Q. and C. S. Goodman. 1985a. Early events in insect neurogenesis. I. Development and segmental differences in the pattern of neuronal precursor cells. *Dev. Biol.* **111:** 193–205.

Doe, C. Q. and C. S. Goodman. 1985b. Early events in insect neurogenesis. II. The role of cell interactions and cell lineage in the determination of neuronal precursor cells. *Dev. Biol.* **111:** 206–219.

Doe, C. Q., J. Y. Kuwada, and C. S. Goodman. 1985. From epithelium to neuroblasts to neurons: The role of cell interactions and cell lineage during insect neurogenesis. *Proc. Roy. Soc. London* (*B*) **312:** 67–81.

Goodman, C. S. and N. C. Spitzer. 1979. Embryonic development of identified neurones: Differentiation from neuroblast to neurone. *Nature* (*London*) **280:** 208–214.

Goodman, C. S. and N. C. Spitzer. 1981a. The mature electrical properties of identified neurones in grasshopper embryos. *J. Physiol.* **313:** 369–384.

Goodman, C. S. and N. C. Spitzer. 1981b. The development of electrical properties of identified neurones in grasshopper embryos. *J. Physiol.* **313:** 385–413.

Goodman, C. S., M. Bate, and N. C. Spitzer. 1981. Embryonic development of identified neurons: Origin and transformation of the H cell. *J. Neurosci.* **1:** 94–102.

Goodman, C. S., M. O'Shea, R. E. McCaman, and N. C. Spitzer. 1979. Embryonic development of identified neurons: Temporal pattern of morphological and biochemical differentiation. *Science* (*Washington, D.C.*) **204:** 1219–1222.

Goodman, C. S., M. J. Bastiani, C. Q. Doe, S. duLac, S. L. Helfand, J. Y. Kuwada, and J. B. Thomas. 1984. Cell recognition during neuronal development. *Science* (*Washington, D.C.*) **225:** 1271–1279.

Hartenstein, V. and J. A. Campos-Ortega. 1984. Early neurogenesis in wild type *Drosophila melanogaster*. *Wilh. Roux Arch.* **193:** 308–325.

Heathcote, R. D. 1980. Physiological development of a monosynaptic connection involved in an adult insect behavior. *J. Comp. Neurol.* **191:** 155–166.

Heathcote, R. D. 1981. Differentiation of an identified sensory neuron (SR) and associated structures (CTO) in grasshopper embryos. *J. Comp. Neurol.* **202:** 1–18.

Ho, R. K. and C. S. Goodman. 1982. Peripheral pathways are pioneered by an array of central and peripheral pioneer neurones. *Nature* (*London*) **297:** 404–406.

Johnston, R. N. and N. Wessells. 1980. Regulation of the elongating nerve fiber. *Curr. Topics Dev. Biol.* **16:** 165–206.

Keshishian, H. 1980. Origin and morphogenesis of pioneer neurons in the grasshopper metathoracic leg. *Dev. Biol.* **80:** 388–397.

Keshishian, H. and D. Bentley. 1983a. Embryogenesis of peripheral nerve pathways in grasshopper legs I. The initial nerve pathway to the CNS. *Dev. Biol.* **96:** 89–102.

Keshishian, H. and D. Bentley. 1983b. Embryogenesis of peripheral nerve pathways in grasshopper legs II. The major nerve routes. *Dev. Biol.* **96:** 103–115.

Keshishian, H. and D. Bentley. 1983c. Embryogenesis of peripheral nerve pathways in grasshopper legs III. Development without pioneer neurons. *Dev. Biol.* **96:** 116–124.

Keshishian, H. and M. O'Shea. 1985a. The distribution of a peptide neurotransmitter in the postembryonic grasshopper central nervous system. *J. Neurosci.* **5:** 992–1004.

Keshishian, H. and M. O'Shea. 1985b. The acquisition and expression of a peptidergic phenotype in the grasshopper embryo. *J. Neurosci.* **5:** 1005–1015.

Kimble, J. 1981. Alterations in cell lineage following laser ablations of cells in the somatic gonad of *Caenorhabditis elegans*. *Dev. Biol.* **87:** 286–300.

Kimble, J. and J. G. White. 1981. On the control of germ cell development in *Caenorhabditis elegans*. *Dev. Biol.* **81:** 208–219.

Kuwada, J. Y. and C. S. Goodman. 1985. Neuronal determination during embryonic development of the grasshopper nervous system. *Dev. Biol.* **110:** 114–126.

Letourneau, P. C. 1979. Cell substratum adhesion of neurite growth cones, and its role in neurite elongation. *Exp. Cell Res.* **124:** 127–138.

Letourneau, P. C. 1982. Nerve fiber growth and its regulation by extrinsic factors, pp. 213–254. In N. Spitzer (ed.), *Neuronal Development*. Plenum, New York.

Murphey, R. K., B. Mendenhall, J. Palka, and J. S. Edwards. 1975. Deafferentation slows the growth of specific dendrites of identified giant interneurons. *J. Comp. Neurol.* **158:** 407–418.

Poulson, D. F. 1950. Histogenesis, organogenesis and differentiation in the embryo of *Drosophila melanogaster*, pp. 168–274. In M. Demerec (ed.), *Biology of* Drosophila. Wiley, New York.

Raper, J., M. J. Bastiani, and C. S. Goodman. 1983a. Pathfinding by neuronal growth cones in grasshopper embryos I. Divergent choices made by growth cones in sibling neurons. *J. Neurosci.* **3:** 20–30.

Raper, J., M. J. Bastiani, and C. S. Goodman. 1983b. Pathfinding by neuronal growth cones in grasshopper embryos II. Selective fasciculation onto specific neuronal pathways. *J. Neurosci.* **3:** 31–41.

Raper, J., M. J. Bastiani, and C. S. Goodman. 1984. Pathfinding by neuronal growth cones in grasshopper embryos IV. The effects of ablating the A and P axons upon the behavior of the G growth cone. *J. Neurosci.* **4:** 2239–2345.

Schubiger, M. and J. Palka. 1985. Genetic suppression of putative guidepost cells: Effect on establishment of nerve pathways in *Drosophila* wings. *Dev. Biol.* **108:** 399–410.

Shankland, M. and C. S. Goodman. 1982. Development of the dendritic branching pattern of the medial giant interneuron in the grasshopper embryo. *Dev. Biol.* **92:** 489–506.

Shankland, M. and D. Weisblat. 1984. Stepwise commitment of blast cell fates during the positional specification of the O and P cell lines in the leech embryo. *Dev. Biol.* **106:** 326–342.

Taghert, P. H. and C. S. Goodman. 1984. Cell determination and differentiation of identified serotonin-immunoreactive neurons in the grasshopper embryo. *J. Neurosci.* **4:** 989–1000.

Taghert, P. H., C. Q. Doe, and C. S. Goodman. 1984. Cell determination and regulation of neuroblasts and neurones in grasshopper embryos. *Nature (London)* **307:** 163–165.

Thomas, J. B., M. J. Bastiani, M. Bate, and C. S. Goodman. 1984. From grasshopper to *Drosophila*: A common plan for neuronal development. *Nature (London)* **310:** 203–207.

Tosney, K. W. and N. K. Wessells. 1983. Neuronal motility: The ultrastructure of veils and microspikes correlates with their motile activity. *J. Cell. Sci.* **61:** 389–411.

Wessells, N. K. 1982. Axon elongation: A special case of cell locomotion, pp. 225–246. In R. Bellairs, A. Curtis, and G. Dunn (eds.), *Cell Behaviour.* Cambridge University Press, Cambridge.

Wheeler, W. M. 1893. A contribution to insect embryology. *J. Morphol.* **8:** 1–160.

Whitington, P. M. and E. Seifert. 1982. Axon growth from limb motoneurons in the locust embryo: The effect of target limb removal on the path taken out of the central nervous system. *Dev. Biol.* **93:** 206–215.

Whitington, P. M., M. Bate, E. Seifert, K. Ridge, and C. S. Goodman. 1982. Survival and differentiation of identified embryonic neurons in the absence of their target muscles. *Science (Washington, D.C.)* **215:** 973–975.

PART II

STRUCTURE
AND FUNCTION

CHAPTER 4

The Brain of the
Horseshoe Crab,
Limulus polyphemus

Wolf H. Fahrenbach
Laboratory of Electron Microscopy
Oregon Regional Primate Research Center
Beaverton, Oregon

Steven C. Chamberlain
Institute for Sensory Research
Syracuse University
Syracuse, New York

4.1. INTRODUCTION

The acknowledged antiquity of the horseshoe crab, its evolutionary conservatism, and its long-disputed systematic position all contributed to generate a wealth of anatomical studies, imaginative phylogenetic speculations, and spirited polemical exchanges. Much of this attention has been directed toward the structure of the brain, undoubtedly the largest among arthropods. All the studies of the late 19th Century (e.g., Packard, 1893; Viallanes, 1893) have been thoroughly discussed by the major "classical" authors on the *Limulus* brain, specifically Patten and Redenbaugh (1900), Patten (1912), Holmgren (1916), Hanström (1926a), and Johansson (1933), who also did extensive work on numerous reconstructions of the brain. Recent work, which will be covered in the pertinent sections, has dealt mainly with the optic ganglia and the corpora pedunculata. During some of these studies, it became apparent to us that an overall anatomical plan of the brain was not available. Extracting analogies between various older nomenclatures for different neuronal groupings has not been invariably successful. Hence, our aim in this article is to provide readers with a descriptive and annotated atlas of the brain of sufficient detail so that future investigators will not have to redescribe the gross anatomy of the brain to define the location of specific neurons. We have largely adopted and expanded the terminology of Chamberlain and Wyse (1986). That system incorporates terms in common use for well-known structures, and assigns geographical terms to previously unnamed or multiply named anatomical features. It should be observed that we are using the term ganglion in the broad sense of a neuronal mass that may or may not contain neuropil. Future research may permit functional designations to be superimposed on the present terminology when such information becomes available. Our terminology and its correspondence with that of four earlier workers is shown in Table 4.1.

Beyond the patently protocerebral, that is, visual regions of the brain, the segmentation of the remainder is steeped in controversy, especially with regard to the absence of the usual arthropodan first antennae and the corresponding deutocerebrum. We will refrain from speculation, but primarily point out which connections can be discerned between ganglia. We arbitrarily define the boundary of the brain as the posterior border of the posterior lateral neuropil (the "antennal glomerulus" of Holmgren, 1916, and Johansson, 1933), approximately level with the start of the circumesophageal connectives.

TABLE 4.1. LOCATION, NUMBER, SIZE, AND NOMENCLATURE OF *LIMULUS* BRAIN STRUCTURES

	Depth From Dorsal Surface	# Elements (Unilateral)	Patten (1912)	Holmgren (1916)	Hanström (1926)	Johansson (1933)
Peripheral nerves						
LON—lateral optic nerve	0–400 μm	9200–10,500 fibers[a]	Lateral eye nerve	Facettenaugennerv	Komplexaugennerv	Komplexaugennerv
MON—median optic nerve	400–500 μm	231 fibers[b]	Parietal eye nerve	Ozellarnerv	Linsenaugennerv	Linsenaugennerv
VON—ventral optic nerve	800–900 μm	370–500 fibers[c]	Lateral olfactory nerve	Nervus olfactorius lateralis	Ventralaugennerv	Lateraler Riechnerv
Corpus pedunculatum						
CP—corpus pedunculatum	300–3000 μm	NA	Cerebral hemisphere	Globuli	Corpus pedunculatum	Globulimasse
CPA—anterior lobe of the corpus pedunculatum	600–3000 μm	NA	Anterior lobe of the cerebral hemisphere	Globulus I (C)		Globulimasse I (vorderer Stiel)
CPDA—dorsal anterior lobe of the corpus pedunculatum	1100–1700 μm	NA	Median internal (gustatory) lobe	Globulus III (B)		Globulimasse III (oberer Stiel)
CPGN—glomerular neuropil of the corpus pedunculatum	NA	NA	Subcortical neuropil	Glomerulischicht	Neuropil des corpus pedunculatum	Stielglomerulischicht
CPK—Kenyon cells of the corpus pedunculatum	300–3000 μm	5.0×10^7 somata[d]	Cells of the cerebral cortex	Globulizellen	Ganglienzellenschicht des corpus pedunculatum	Globulizellen
CPL—lateral lobe of the corpus pedunculatum	300–3000 μm	NA	Lateral lobe of the cerebral hemisphere	Globulus II (C)		Globulimasse II (lateraler Stiel)
CPP—posterior lobe of the corpus pedunculatum	1900–3000 μm	NA	Posterior lobe of the cerebral hemisphere	Globulus I (C)		Globulimasse I (hinterer Stiel)
CPPN—peduncular neuropil of the corpus pedunculatum	NA	NA	Subcortical neuropil	Stielfasern	Neuropil des corpus pedunculatum	Faserschicht
CPS—stalk of the corpus pedunculatum	1300–1900 μm	5000 fibers[e]	Cerebral peduncle	Stiel des Globulus	Pedunculus	
Neuropils						
CBN—central body neuropil	600–1400 μm	NA	Olfactory neuropil	Zentralkörper Gestreifter Körper	Zentralkörper	Zentralkörper
CEC—circumesophageal connective	800–2000 μm	NA	Midbrain	SchlundKommissur	Schlundkonnective	Schlundring
CN—central neuropil of the brain	800–2100 μm	NA	Central neuropil & commissural masses	Frontallobe (anterior)	Lobus frontalis des Vorderhirns	Zerebrallobus (anterior)
LN—laminar neuropil	0–700 μm	NA	First optic ganglion	Äussere Sehmasse	Lamina des Komplexauges	Äussere Sehmasse
MN—medullar neuropil	400–1000 μm	NA	Second optic ganglion	Innere Sehmasse	Medulla des Komplexauges	Innere Sehmasse

65

TABLE 4.1. LOCATION, NUMBER, SIZE, AND NOMENCLATURE OF *LIMULUS* BRAIN STRUCTURES (*Continued*)

	Depth From Dorsal Surface	# Elements (Unilateral)	Patten (1912)	Holmgren (1916)	Hanström (1926)	Johansson (1933)
OGN—ocellar ganglion neuropil	600–700 μm	NA	Parietal eye ganglion	Ozellarganglion	Sehmasse des Linsenauges	Sehmasse des Linsenauges
PLN—posterior lateral neuropil	1400–1900 μm	NA	Cheliceral lobe	Antennalglomerulus		Antennalglomerulus
AC—anterior commissure (protocerebral bridge)	800 μm	NA		Brücke	Stielkommissur (Brückenkommissur)	Stielkommissur
OT—optic tract	800 μm	NA	Ocellar tract			Sehtraktus
PC—posterior commissure	2000 μm	NA	Commissure	Antennalkommissur		Antennalkommissur (Zerebralkommissur)
Ganglion cell groups						
CBG—ganglion cells of the central body	600–1400 μm	15,000 somata	Olfactory lobe	Globulizellen	Ganglienzellenschicht des Zentralkörpers	Globulizellen
DLPG1—dorsal lateral posterior ganglion #1	700 μm	500 somata				Zellengruppe F
DLPG2—dorsal lateral posterior ganglion #2	1100–1200 μm	30 somata				Zellengruppe E
DMG—dorsal medial ganglion	600–1500 μm	1000 somata	Olfactory cells		Grosszelliges Gebiet der Ganglienzellenschicht des Protocerebrums	Zellengruppen I (anterior) H (posterior)
LG—laminar ganglion cells	0–700 μm	5000–6000 somata'	First optic ganglion		Kleinzellige Ganglienzellenschicht der Lamina	

	Size	Somata	English name	German names		
MG—medullar ganglion cells	300–1000 μm	9500–11,000 somata[f]	Second, third, fourth optic ganglia		Ganglienzellenschicht der Medulla	
OG—ocellar ganglion cells	600–700 μm	100–150 somata[f]	Parietal eye ganglion	Ozellarganglion	Ganglienzellenschicht der Sehmasse des Linsenauges	Linsenaugenganglion
RG—raphe ganglion	1800–1900 μm	50 somata[g]	Hemisphere association cells			
VLPG1—ventral lateral posterior ganglion #1	1400–2000 μm	500 somata				
VLPG2—ventral lateral posterior ganglion #2	1400–1700 μm	400 somata				Zellgruppen C, D (Antennalganglion)
VMG—ventral medial ganglion	1400–2000 μm	600 somata[g]	Chelicero-hemisphere cells			Zellgruppen A (anterior) B (posterior)
VMPG—ventral medial posterior ganglion	1700–1800 μm	2–3 somata				
VP—ventral photoreceptors	600–900 μm	Variable, <20 somata[h]	Ganglion cells of the lateral olfactory nerve		Rudimente der Sehzellen des Ventralauges	

[a] Based on counts of ommatidia in adult animals (Chamberlain, 1978), counts of efferent fibers (Evans et al., 1983), and counts of rudimentary photoreceptor axons (Fahrenbach, 1975a). The lower totals given by Fahrenbach (1975) apply to younger (smaller) animals.

[b] Sum of retinular, arhabdomeric, and rudimentary cell axons from Fahrenbach (1975a,b). The number of efferent fibers in the MON is unknown.

[c] Sum of afferent (Fahrenbach, 1975; Evans et al., 1983) and efferent (Evans et al., 1983) fibers. There are about 300 afferent fibers and between 70 and 200 efferent fibers.

[d] Number given by Fahrenbach (1977) for a 25-cm adult.

[e] Number given by Fahrenbach (1979) for a 5-cm animal.

[f] Ranges given by Chamberlain (1978).

[g] Numbers given by Chamberlain and Engbretson (1982).

[h] The total number of ventral photoreceptors is about 300; the number which lie along the portion of the nerve within the brain is variable.

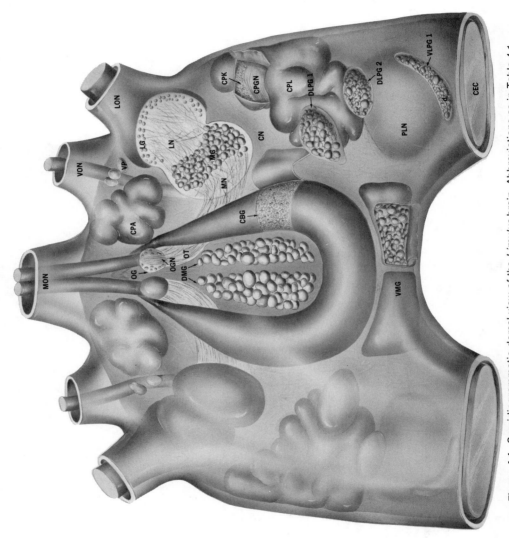

Figure 4.1. Semidiagrammatic dorsal view of the *Limulus* brain. Abbreviations as in Table 4.1.

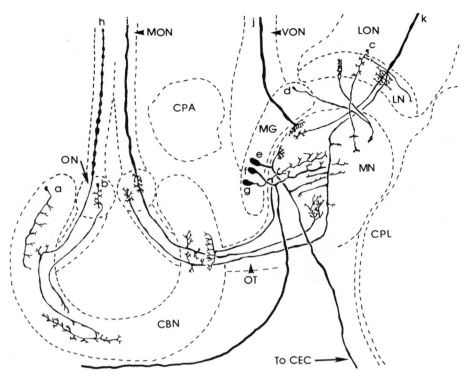

Figure 4.2. Composite diagram in a horizontal plane showing some neurons and axons in the dorsal region of the brain. The profiles are based on single neurons stained with methylene blue, Golgi, or cobalt impregnation procedures. (a) Intrinsic neurons in the central body ganglion cell layer send a branched process into the central body neuropil. (b) Intrinsic neurons in the ocellar ganglion give rise to a process that bears branched collaterals in the ocellar ganglion neuropil and passes through the optic tract to make connection with attenuated branches in the central body neuropil. (c) Intrinsic neurons of the lamina send a process with short collaterals through the laminar neuropil and across the chiasma to the medullar neuropil where it ends in several layers of short collaterals. (d) Neurons in the medullar ganglion cell layer next to the chiasma bear a branched process which makes a small tufted arborization in the medullar neuropil with one branch and sends the other across the chiasma to the laminar neuropil where it arborizes in an elongated region that nearly spans the neuropil. (e) Some large neurons in the postero-medial part of the medullar ganglion cell layer send a short process into the medullar neuropil and a long one into the ipsilateral circumesophageal connective. (f) Some large neurons in the posteromedial parts of the medullar ganglion cell layer send a larger arborization into the adjacent medullar neuropil and a long axon across the anterior commissure to the contralateral brain. (g) Some large neurons in the same region send an arborized branch into the medullar neuropil and a second branch along the optic tract to a branched termination in the neuropil of the central body. (h) Beaded axons in the median optic nerve pass through the ocellar ganglion without giving off collaterals and ramify extensively in the neuropil of the central body. (i) Smooth axons in the median optic nerve arborize in the ocellar ganglion, pass into the optic tract, give off collaterals under the central body, and terminate in an arborization in the posterior medullar neuropil. (j) Axons of ventral photoreceptors terminate in a small arborization along the

69

4.2. OVERVIEW

The entire nervous system in *Limulus* resides, in a manner of speaking, inside the circulatory system. Thus, all the nerves and the brain are encapsulated in sturdy connective tissue sheaths laced with scattered striated muscle cells. This covering serves simultaneously as the wall of blood vessels (Dumont et al., 1965).

The brain (Figs. 4.1–4.6) is a large, nearly spherical body up to 7 mm in diameter in an adult, with a flat or slightly concave dorsal surface. On the anterior surface, the optic nerves from the lateral, median, and ventral eyes enter dorsally (Figs. 4.7–4.9). On the posterior surface, the massive circu-mesophageal connectives exit dorsolaterally. Except for the optic ganglia and the central body (Fig. 4.10) on the dorsal surface, the entire surface of the brain is formed by the bulging lobes of the corpora pedunculata (Figs. 4.11–4.13). During development and into adult life, these lobes increase in size, both absolutely and relative to the remainder of the brain (Figs. 4.14–4.15), with the result that adult horseshoe crabs have the largest brains of any arthropod (Hanström, 1926b; Fahrenbach, 1977).

The central neuropil of the brain with its adherent specialized neurons, neuropil, and commissures has a broad, H-shaped form, whose anterior arms were frequently referred to in the older literature as frontal lobes. In the adult, this central mass is cradled in the great ventral expanse of the corpora pedunculata, which bulge upward laterally and anteriorly to give the brain its spherical shape.

4.2.1. Neuroglia

Arthropod neuroglia is being treated in a separate chapter in this volume (Chapter 14 by Carlson), and *Limulus* neuroglia has been explored in some detail already (Fahrenbach, 1976). Three types of neuroglial cells can be discerned in the horseshoe crab brain. Vascular neuroglia (Fig. 4.7A,B) fills all circulatory, i.e., hemocoelic, spaces of the brain with an open meshwork of cells. Aside from supporting neurosecretory fibers and their terminals, vascular glia is laden with glycogen and mitochondria and appears to serve more metabolic than supportive functions. Stellate astrocytes (Fig. 4.15A), the second cell type, are abundant and function as multiple, perineuronal

---

surface of the medullar neuropil next to the medullar ganglion cell layer. (k) Axons of eccentric cells in the lateral optic nerve give rise to both long and short collaterals in the lamina and bifurcate there. One branch traverses the chiasms and terminates near the surface of the medullar neuropil. The other branch sends out collaterals as it passes through the medullar neuropil, enters the optic tract, gives off collaterals under the central body, and terminates in the ocellar ganglion. Abbreviations are defined in Table 4.1. Based on Patten (1912), Hanström (1926a), Chamberlain and Barlow (1980), and Batra and Chamberlain (1985).

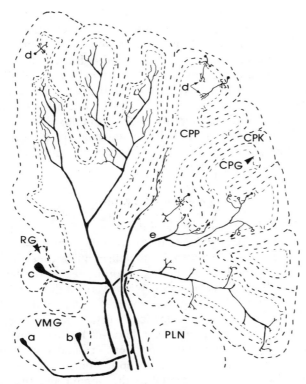

Figure 4.3. Composite diagram in a horizontal plane showing some neurons in the ventral regions of the brain. The neuronal profiles are more schematic than in Fig. 4.2 because single neurons have not been stained in isolation. (a) Neurons in the medial portion of the ventral medial ganglion send their processes to the peduncular neuropil and Kenyon cell somata of the ipsilateral corpus pedunculatum. The fibers pass between the peduncular neuropil and the Kenyon cell layer along the tracts of Kenyon cell neurites. These neurons are immunoreactive to substance P. (b) Neurons in the lateral portion of the ventral medial ganglion have branched processes. One branch leaves the brain by way of the ipsilateral circumesophageal connective. The other enters the ipsilateral corpus pedunculatum and ramifies in the glomerular neuropil. (c) Neurons in the raphe ganglion bear branched processes. One branch ramifies in the peduncular neuropil of the ipsilateral corpus pecunculatum. The other leaves the brain via the ipsilateral circumesophageal connective. (d) The intrinsic neurons, or Kenyon cells, of the corpus pedunculatum send branching neurites both to the adjacent glomerular neuropil and to the deeper peduncular neuropil. (e) Axons of unknown origin terminate in the glomerular neuropil with club-shaped and annular endings. Abbreviations as in Table 4.1. Based on Fahrenbach (1977, 1979) and Chamberlain and Engbretson (1982).

sheaths and as the principal sustentacular cells in neuropil and trabecular partitions of the brain. Velate astrocytes (Fig. 4.7C), the third type, are restricted to the tightly packed masses of association (Kenyon) neurons in the corpora pedunculata and the central body. Each cell has a complex, honeycombed structure and envelops up to 150 adjacent small neurons.

In contrast to that of most terrestrial arthropods, the *Limulus* central nervous system is devoid of a blood–brain barrier, a feature apparent not only in the sporadic distribution of junctional complexes between glial cells, but also in the accessibility of the circumneuronal space to tracer ions and molecules, including, on occasion, its own hemocyanin (Willmer and Harrison, 1979; Harrison and Lane, 1981).

4.2.2. Optic Lamina

The lateral optic nerves from the compound eyes enter the brain laterally at its dorsal anterior aspect and spread out over the convex face of the lamina (Figs. 4.2, 4.8). This mass of neurons and neuropil is geometrically complex. When viewed down the axis of the lateral optic nerve, the lamina forms a sprung horseshoe that opens laterally. When viewed in horizontal sections, the profile of the lamina is always a blunt crescent with its convex surface facing roughly anteriorly and its concave surface facing posteriorly. The convex, anterior surface bears a broken layer of laminar ganglion cells distributed in small groups between the fascicles of retinal axons. Fibers exiting the concave side of the lamina form an optic chiasma before entering the medulla. For details of the geometry of the lamina and the chiasma, see Chamberlain and Barlow (1982).

The spatial array of ommatidia in the compound eye is represented in the lamina by a first-order, point-to-point mapping of the visual afferents (Chamberlain, 1978; Chamberlain and Barlow, 1982). The characteristic subunit structure of the lateral optic nerve, which corresponds to horizontal strips of ommatidia in the lateral eye (Snodderly and Barlow, 1970), is not strongly expressed in the lamina, but reappears in the retinotopic organization of the medulla. Individual eccentric cell axons decrease their diameter as they enter the laminar neuropil, then give rise to numerous short small-diameter varicose collaterals, which innervate the surrounding neuropil, and several longer collaterals, which penetrate to the edges of the laminar neuropil. Finally, each axon bifurcates and sends two processes across the chiasma to the medulla (Chamberlain and Barlow, 1980).

The remainder of the wiring pattern of the lamina has not been worked out. Besides processes from the laminar neurons (Fig. 4.2C) and the axons of retinal eccentric cells, the lamina also receives axons of retinal rudimentary photoreceptors and retinular cells, in addition to retrograde processes from neurons in the medulla (Patten, 1912; Hanström, 1926a). Unlike that of many other arthropods, the lamina of *Limulus* does not seem to be organized into optic cartridges. The intrinsic neurons of the lamina appear to

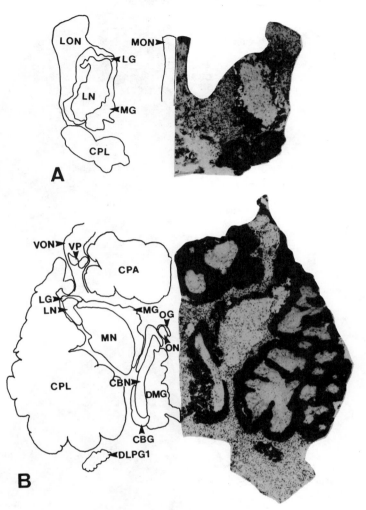

Figure 4.4. Frontal series of the brain and orientation diagrams. (A) Level of the lateral optic nerve and lamina. (B) Level of the central body. Abbreviations as in Table 4.1.

respond transiently to the cessation of illumination (Wilska and Hartline, 1941; Oomura and Kuriyama, 1953; Snodderly, 1971).

4.2.3. Optic Medulla

The medulla (Figs. 4.2, 4.8) consists of a substantial ovoid mass cradled in the concavity of the lamina. Posteroventrally, it gives rise to a bundle of fibers, the optic tract, that passes just ventrally to the central body and

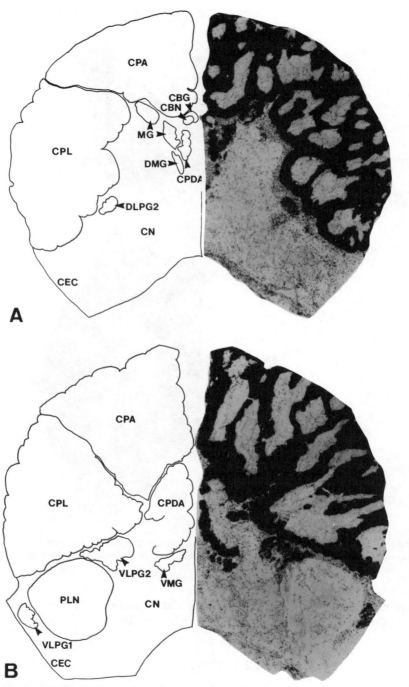

Figure 4.5. Frontal series of the brain and orientation diagrams. (A) Level just ventrad to the anterior commissure. (B) Level of dorsal anterior lobe of the corpus pedunculatum and posterior lateral neuropil. Abbreviations as in Table 4.1.

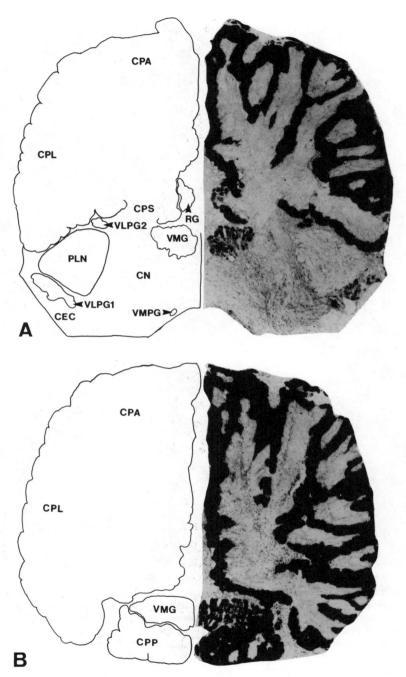

Figure 4.6. Frontal series of the brain and orientation diagrams. (A) Level of the raphe ganglion and posterior lateral neuropil. (B) Level of the most ventral part of the ventral medial ganglion. Abbreviations as in Table 4.1.

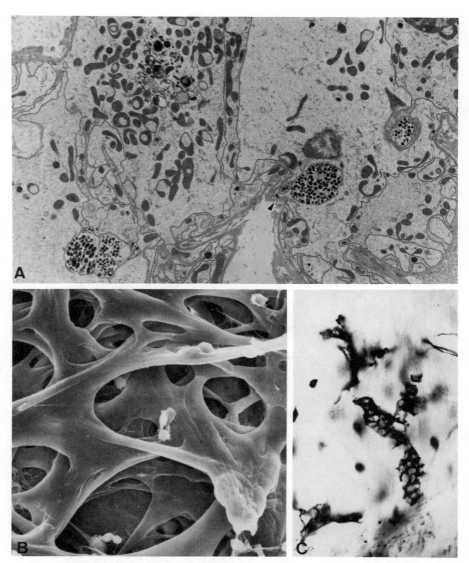

Figure 4.7. Neuroglia of the brain. (A) Electron micrograph of vascular neuroglia in the vicinity of the dorsal medial ganglion. The cells are heavily interdigitated, replete with mitochondria, and convey at least two types of neurosecretory fibers, of which one axon is approaching a neurohaemal terminal (*arrowhead*; 8000×). (B) Scanning electron micrograph of vascular glia, illustrating the cancellous nature of this tissue. Isolated hemocytes are scattered about (800×). (C) Golgi-impregnated velate astrocytes, the characteristic neuroglial cells of the Kenyon cell layer in the corpora pedunculata. Each honeycomb recess harbors one neuron (400×).

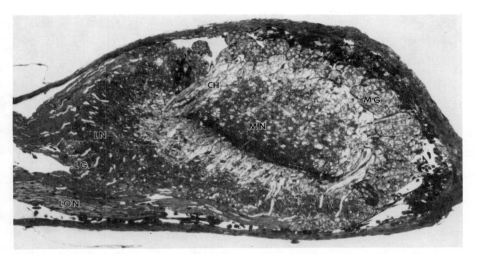

Figure 4.8. Horizontal section of the lateral optic nerve, its entrance into the lamina, and chiasmatic fibers entering the medulla. The midline of the brain lies above the picture (CH = chiasma). Other Abbreviations as in Table 4.1 (190×).

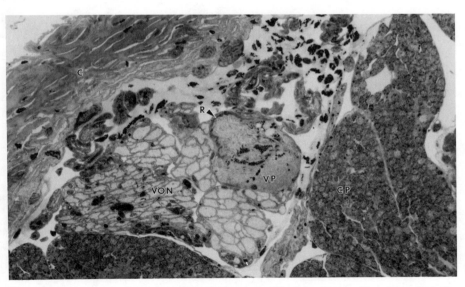

Figure 4.9. Ventral optic nerve inside the capsule (C) of the brain with adjacent photo-receptor cells. The rhabdomere (R) is visible in one of the cells. Other Abbreviations as in Table 4.1 (470×).

77

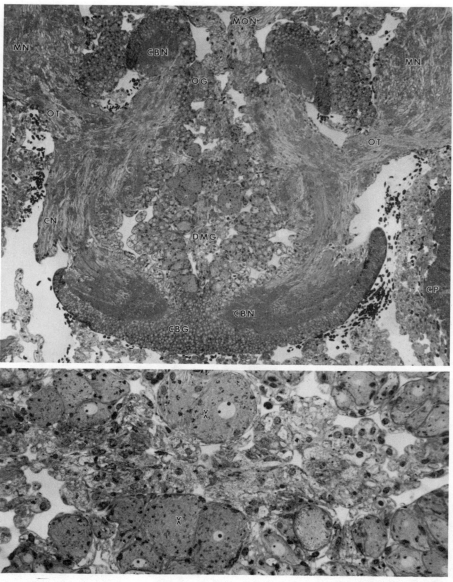

Figure 4.10. (A) Horizontal section through the central body. Several layers of its neuropil are evident. Open spaces are hemocoelic in nature and contain darkly stained hemocytes. Abbreviations as in Table 4.1 (150×). (B) Dorsal medial ganglion (enlargement of A turned 90°). The largest neurons (x) stain intensely with paraldehyde fuchsin for neurosecretory material and are immunoreactive for bombesin. The longitudinal partition consists of vascular glial cells (420×).

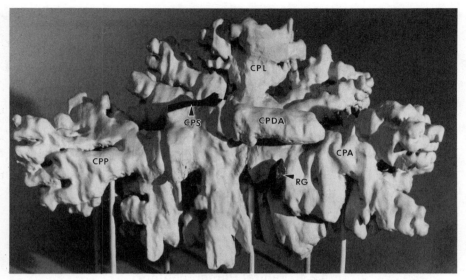

Figure 4.11. Wax plate reconstruction of the neuropil of one corpus pedunculatum, seen across the midline from slightly in front. Abbreviations as in Table 4.1.

connects the medullar neuropil with the ocellar ganglion. Posteromedially, the medullar neuropil is continuous with the central neuropil of the brain; anteriorly, it receives the fibers of the optic chiasma. The dorsal, medial, and ventral surfaces of the medullar neuropil are covered by the somata of the medulla. These neurons are a heterogeneous assemblage with cell bodies ranging in size from 10–70 μm in diameter (Fahrenbach, 1975). The majority, however, are larger than those of the laminar group, which uniformly are about 13 μm. The medullar neuropil is stratified into a number of more or less compacted layers (Hanström, 1926a).

The medulla appears to have an extremely complex organization. It receives afferent fibers from the median, ventral, and lateral eyes (Chamberlain and Barlow, 1980; Batra and Chamberlain, 1985), as well as from mechanoreceptors on the opisthosoma (Chamberlain, 1978). The retinotopic map of the afferents from the lateral compound eye, owing to the interaction of the crossing at the chiasma and the expression of the subunit structure of the optic nerve, has numerous cuts (Chamberlain and Barlow, 1982). The medullar group of neurons probably contains a diverse array of functional subsets (Fig. 4.2D–G). Patten (1912) reported three different types of intrinsic medullar neurons on the basis of connections revealed by vital staining with methylene blue. Hanström (1926a), using the Golgi technique, found four different kinds of medullar neurons, including two types that did not overlap with Patten's categories. Backfilling with cobalt ions revealed that about 150 medullar neurons send fiber collaterals to the ipsilateral circumesophageal connective, and a smaller number send collaterals to the contralateral connective (Chamberlain, 1978). Recently, O'Connor et al. (1982) demonstrated that a small cluster of medullar neurons gives rise to a circumesophageal network of dopaminergic fibers.

The response characteristics of medullar neurons are complex and varied. Snodderly (1971) noted four types of medullar cell responses: "ON" cells had large receptive fields and responded both at light onset and cessation; "ON-OFF" units had large receptive fields and responded both at the onset of light and at its cessation; "OFF" units had small receptive fields and responded only to light cessation; "Delayed OFF" responded several seconds after the cessation of light.

A conspicuous posterolateral protrusion of fine-textured neuropil, which was occasionally mentioned in the old literature as an additional optic mass, does not receive visual afferents from any of the optic nerves (Chamberlain, 1978), and is probably a bulge of the central neuropil rather than part of the medullar neuropil proper.

4.2.4. Ventral Photoreceptors

Numerous single and clustered ventral photoreceptor cells are always found along the portion of the ventral optic nerve that runs within the brain (Fig. 4.9). They have the same structure as those scattered along the ventral optic nerve outside the brain. The lucent rhabdomeral lobe bears the light-sensitive rhabdom and the arhabdomeral lobe contains the nucleus (Calman and Chamberlain, 1982). Clusters of these photoreceptors may reach 250 μm in diameter. Within each cluster, the lobes of the cells can be distinguished. Considering the long space constants measured for ventral photoreceptor axons by Behrens and Fahy (1982), these receptors were probably the source of the photosensitivity of the brain reported by Snodderly (1969).

4.2.5. Ocellar Ganglion

The ocellar ganglion (Figs. 4.2, 4.10A) is a small spherical mass of neuropil and neurons situated medially to the anterior arms of the central body. It receives the median optic nerve on its anterior surface and the optic tract on its posterior surface. The intrinsic neurons are about 15 μm in diameter and give rise to a process that arborizes in the ocellar neuropil and continues to branch out in the neuropil of the central body (Fig. 4.2B) (Hanström, 1926a). Extrinsic input to the ocellar neuropil consists of fibers from the median ocellus and from the compound lateral eye by way of the optic tract (Chamberlain and Barlow, 1980).

4.2.6. Central Body

The central body is a horseshoe-shaped mass of neuropil and neurons (Fig. 4.10A) that sits superficially on the dorsal surface of the central neuropil of the brain. Over most of its curvature, it is confluent with the underlying neuropil, but anteriorly the tips of the horseshoe plunge ventrally into the raphe between the hemispheres of the corpora pedunculata and terminate along the midline far ventrad to the anterior parts of the central neuropil. Although numerous previous reports have stated the contrary, the central body is a bilaterally symmetrical structure and only the anterior tips drift out of their side-by-side position in the constrained space between the corpora pedunculata.

The central body neuropil (Fig. 4.15B) forms the core of the structure and is covered along its curve on the inner, dorsal, and outer surfaces by a deep layer of small neurons that resemble, but are slightly larger than, the Kenyon cells of the corpora pedunculata. Axons of these intrinsic neurons arborize within the central body neuropil (Fig. 4.2A) (Hanström, 1926a). The neuropil itself has distinct subdivisions along a radial gradient, but its structure and connectivity remain to be investigated.

The optic tract connects the ocellar ganglion and the medulla just ventral to the central body, and carries inputs to the central body from nerve fibers originating in the median ocellus, lateral compound eye (Chamberlain and Barlow, 1980), and the medullar neurons (Hanström, 1926a). Fibers from some neurons in the dorsal medial ganglion also innervate the central body neuropil (Chamberlain and Engbretson, 1982; Chamberlain et al., 1983).

4.2.7. Dorsal Medial Ganglion

This large cluster of somata fills the concavity of the central body and extends ventrally between the two anterior arms of the central neuropil (Fig. 4.10A,B). The limited information available suggests that this group contains functional and structural subdivisions that remain to be elucidated. The neurons range from less than 10 μm to 100 μm in diameter. Many send an axon to the ipsilateral circumesophageal connective, but a few send one to the controlateral connective (Chamberlain, 1978). Cytochemical procedures yield diverse subsets of neurons, including a paired cluster that stains with paraldehyde fuchsin (Fahrenbach, 1973; probably also detected by Scharrer, 1941) and is immunoreactive for bombesin (Wyse, 1983), and disjointed sets of neurons immunoreactive to antibodies against substance P (Chamberlain and Engbretson, 1982), serotonin (Chamberlain et al., 1983), and proctolin (Wyse, 1983). Most of the immunoreactive neurons send processes into the neuropil of the central body. The diversity of axonal inclusions in the adjacent central body neuropil (Fig. 4.15B) mirrors the variety of neuronal types. The brain participates at least to some degree in the production of several neurohumoral substances, specifically a chromatophorotropic factor (Brown and Cunningham, 1941; Dores and Herman, 1980), a hyperglycemic factor (Dores and Herman, 1980), and various cardioregulatory transmitters (Corning et al., 1971; see also Pax and Sanborn, 1967a, b; Benson et al., 1981; Augustine et al., 1982).

4.2.8. Corpora Pedunculata

The corpora pedunculata, which form the ventral hemisphere of the *Limulus* brain, can best be visualized as two cauliflower-like masses, each attached to the central neuropil through a stalk (Fig. 4.11). Developmentally, each corpus starts as a main stalk that subsequently gives rise to the multiply branched anterior, lateral, and posterior lobes, and a more anterior stalk that ultimately fuses with the main stalk and connects to the unbranched dorsal anterior lobe. The extensive arborization of the main lobes is covered uniformly by a deep layer of interneurons (Kenyon cells, formerly called globuli cells). The neuropil is differentiated into a peripheral zone of glomerular neuropil adjacent to the Kenyon cell layer and a central zone of peduncular neuropil (Fig. 4.12A-C).

Input to each corpus consists of approximately 10,000 fibers (Fahrenbach, 1977, 1979) that can be segregated into five categories (Fig. 4.4). Type A is responsible for the organization of the glomeruli of the glomerular neuropil and terminates in both annular and club-shaped endings (Fig. 4.12B,D). Type B has bushy endings in the glomeruli, but contains round synaptic vesicles instead of the flat ones found in type A fibers. Fiber types C, D, and E have opaque vesicles of distinctive morphologies and diverse targets of innervation, which include the somata of the Kenyon cells (type D). The somata of the type D fibers are probably those in the medial portion of the ventral medial ganglion that possess immunoreactivity for substance P (Chamberlain and Engbretson, 1982). The glomerular neuropil is also reactive for FMRFamide-like peptides (Watson et al., 1984) and catecholamines (O'Connor et al., 1982).

The Kenyon cell count for the corpora pedunculata amounts to about 100 million, a value far above that recorded for the brain of any other arthropod. These neurons (Fig. 4.13A), reminiscent of vertebrate cerebellar granule cells, are intrinsic to the corpora pedunculata (Fig. 4.3D). Their input is received from collateral arborizations that enter the glomeruli, and their output impinges in highly convergent fashion on dendrites of efferent neurons in the peduncular neuropil. They receive additional input from all the other afferents, distributed variously between the glomeruli, the outflow synaptic regions, and, most rarely, the somata.

The output fibers, onto which the Kenyon cells converge, are confined to the peduncular neuropil and are distinguished by particularly empty-appearing cytoplasm (Figs. 4.12A,B; 4.13B). At least some of these are fibers from the neurons of the raphe ganglion (Fig. 4.3) (Chamberlain, 1978; Fahrenbach, 1979). In a 50-mm animal, approximately 300 of these efferent fibers leave through the main stalk of each corpus, where they constitute the population with the largest neurite diameters.

The functional attributes of the corpora pedunculata are largely conjectural. An apparent pathway exists from the primary chemosensory input of the legs and opisthosoma to the posterior lateral neuropil, and from there

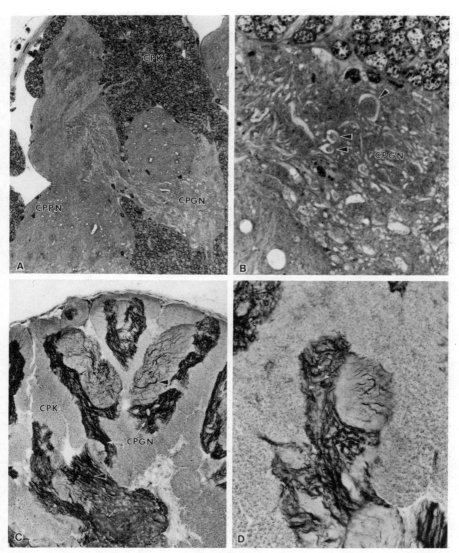

Figure 4.12. Histology of the corpora pedunculata. Abbreviations as in Table 4.1. (A) Low power view illustrating the Kenyon cell layer and subdivisions of the neuropil. Large, lightly stained fibers in the peduncular neuropil are efferent dendrites (400 ×). (B) The glomerular neuropil shows several annular afferent terminals (*arrowheads*). Two efferent dendrites at the lower left lie in the peduncular neuropil (1600 ×). (C) Reduced silver impregnation in which the Kenyon cells and their branches remain unstained. Afferent axons form a dense tangle in the glomerular neuropil, whereas efferent fibers are the coarse and scarcer neurites in the peduncular neuropil (*arrowheads*; 140 ×). (D) High power view of a preparation similar to C, illustrating the annular terminals of the glomeruli (*arrowhead*; 380 ×).

83

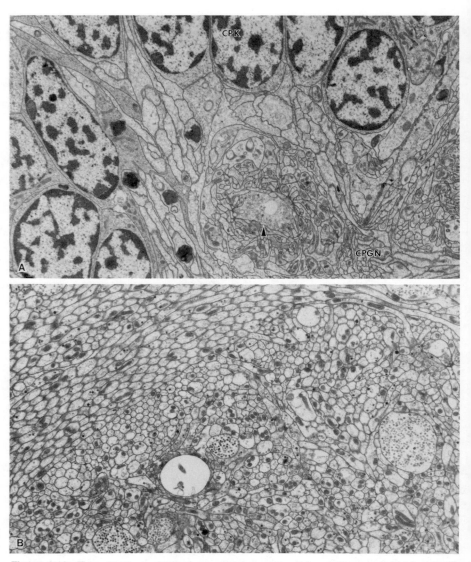

Figure 4.13. Fine structure of the corpora pedunculata. Abbreviations as in Table 4.1. (A) Transition zone between the Kenyon cell layer and the glomerular neuropil. One small glomerulus with its central, club-shaped afferent (*arrowhead*) is shown. Kenyon cell axons enter the neuropil in small fascicles (7800×). (B) Peduncular layer of neuropil. All small packed fibers without intervening glia are Kenyon cell axons. The prominent pale fiber is an efferent dendrite surrounded by Kenyon cell synapses. Granule-filled afferents of at least two types are also shown (10,000×).

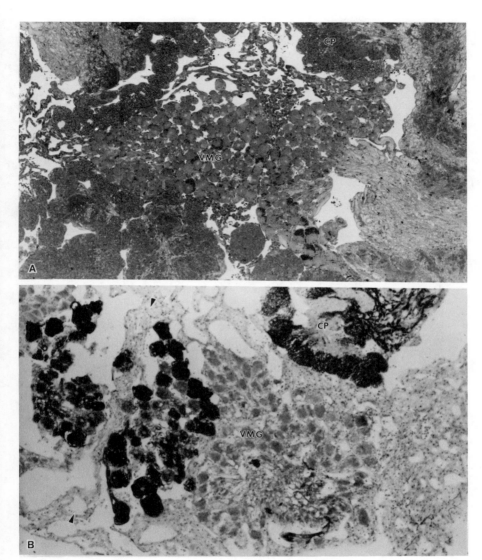

Figure 4.14. Horizontal section of the ventral medial ganglion (*anterior up*). Abbreviations as in Table 4.1. (A) The midline of the ganglion is approximately in the middle of the picture. A massive fiber tract emerging from the ganglion at the right branches and connects to the stalk of the corpus pedunculatum, the posterior lateral neuropil, and the circumesophageal connective (90×). (B) The same plane of section (the midline indicated by *opposed arrowheads*). The section has been immunocytochemically stained for substance P; only the large medial neurons of the ganglion are reactive as are their terminations in the corpora pedunculata (upper right: 110×).

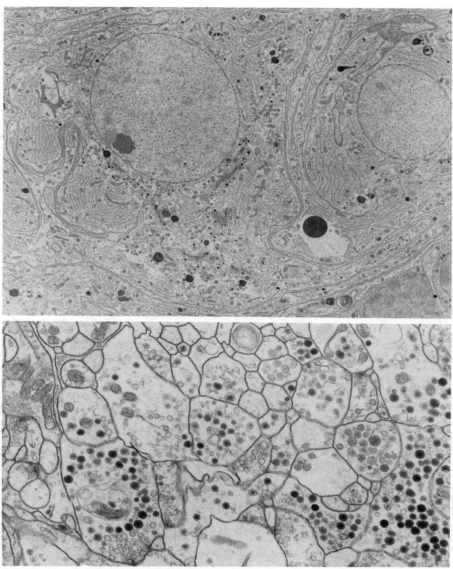

Figure 4.15. (A) Representative neurons of central ganglia, in this case the ventral medial ganglion. Stellate glial cells typically invade the somata of neurons as single or multiple contorted sheets. The neurons are active in secreting material ranging from dense-cored vesicles near the scattered Golgi bodies to large, seemingly colloidal droplets (6500 ×). (B) Representative neuropil of the central nervous system, here in the central body. Typical for *Limulus* is the bewildering variety of vesicle and granule types, found frequently in combination in any one axonal profile (27,000 ×).

to the corpora pedunculata. Secondary input from the visual system and from peripheral mechanoreceptors is also possible (Chamberlain, 1978). The vast literature on insect corpora pedunculata has only marginal significance concerning the structure and function of these bodies in the horseshoe crab.

4.2.9. Raphe Ganglion

This small grouping of neurons (Patten, 1912; Fahrenbach, 1979) is located in the median seam between the hemispheres of the corpora pedunculata (Figs. 4.3, 4.6A, 4.11). Neurites from these cells enter the neuropil of the corpora pedunculata through the superficial layer of Kenyon cells rather than through the stalk. In the peduncular neuropil, the dendritic branch divides in an open, arborescent fashion, whereas the axonal branch leaves via the stalk and descends through the ipsilateral circumesophageal connective (Fig. 4.3C) (Patten, 1912; Chamberlain and Engbretson, 1982). Kenyon cell terminals converge in enormous numbers onto the raphe neuron dendrites, which are readily identified by their characteristically empty-appearing cytoplasm (Fig. 4.12A,B). Hence, it appears that the axons of the raphe neurons constitute a numerically small, but important, route for outflow of highly processed information from the corpora pedunculata.

4.2.10. Ventral Medial Ganglion

This massive collection of predominantly large neurons (Fig. 4.14A) contributes a substantial tract of fibers to the posterior commissure and, although the region has not been studied in detail, appears to send fibers to the ipsilateral posterior lateral neuropil, circumesophageal connective, and corpus pedunculatum. The results of Chamberlain (1978) and Chamberlain and Engbretson (1982) suggest that this group contains more than one functional subdivision. Neurons in the medial portion of the group possess immunoreactivity for substance P (Figs. 4.3A; 4.14B) and send their fibers to the peduncular neuropil and the Kenyon cells of the ipsilateral corpus pedunculatum. Neurons in the lateral portion of the group send out a branched fiber, which joins the ipsilateral circumesophageal connective posteriorly, and anteriorly innervates the glomerular neuropil of the ipsilateral corpus pedunculatum (Fig. 4.3B). These data suggest that many of the neurons in the ventral medial ganglion are involved in the processing of chemosensory information.

4.2.11. Posterior Lateral Neuropil

These distinctive spherical, fine-textured masses of neuropil (Figs. 4.5B, 4.6A), located at the ventrolateral corners of the brain, have elicited comments from virtually all students of the *Limulus* brain. Most investigators have treated them as glomeruli of the hypothetical first antennae (Holmgren,

1916; Johansson, 1933) or assigned them mistakenly to the chelicerae (Patten, 1912). By all appearances, they are the termini of an enormous number of very fine sensory fibers from the posterior region of the animal. Coarse neurite branches from the ventral medial ganglion enter the posterior lateral neuropil and the stalk of the corpus pedunculatum, possibly both ipsi- and contralaterally, and thereby provide the structural basis for relaying integrated sensory messages to the corpora pedunculata.

4.2.12. Ventral Lateral Posterior Ganglion #1

This set of neurons (Figs. 4.5B, 4.6A) is composed of a heterogeneous band of cells wrapped around the posterior surface of the posterior lateral neuropil. The somata range from 20–90 μm in diameter and are clustered in disjointed groups.

4.2.13. Ventral Lateral Posterior Ganglion #2

The position of this ganglion (Figs. 4.5B, 4.6A) lies between the corpus pedunculatum and the central neuropil of the brain. The somata range from 15–95 μm in diameter, and evidently more than one type is present. The fibers enter the central neuropil posteriorly.

4.2.14. Dorsal Lateral Posterior Ganglion #1

This group of neurons (Fig. 4.4B) is located between the corpus pedunculatum and the central neuropil of the brain somewhat dorsally and laterally to the dorsal posterior lateral ganglion #2. The somata are relatively uniform in appearance and range from 65–90 μm in diameter. The axons enter the central neuropil posteromedially.

4.2.15. Dorsal Lateral Posterior Ganglion #2

This small mass of somata (Fig. 4.5A) lies between the corpus pedunculatum and the central neuropil of the brain just dorsad to the ventral lateral posterior ganglion #2. The cell bodies are all large, 40–90 μm in diameter, with axons that enter the central neuropil posteromedially.

4.2.16. Ventral Medial Posterior Ganglion

This predictably located set of only two or three cells (Fig. 4.6A) lies between the central neuropil and the vascular sheath of the brain. Nothing is known of the structure or function of the preceding five groups of neuronal cell bodies.

4.2.17. Peripheral Nerves

The large lateral optic nerve (Figs. 4.4A, 4.8) enters the brain at its dorsal anterior aspect and spreads out over the surface of the lamina. The nerve trunk is organized into subunits that reflect horizontal strips of ommatidia in the retina (Snodderly and Barlow, 1970). The lateral optic nerve contains axons from eccentric and retinular cells in the compound eye, axons from the rudimentary photoreceptors behind the compound eye, and efferent octopaminergic axons from neurons in the brain that terminate in the retina and partly govern the circadian rhythm of the eyes (Waterman and Wiersma, 1954; Nunnemacher and Davis, 1968; Fahrenbach, 1970, 1971, 1973, 1981; Barlow and Chamberlain, 1980; Evans et al., 1983). Of these, only the central connections of the eccentric cell axons are known (Fig. 4.2K). These fibers innervate the lamina and medulla, pass through the optic tract, give off collaterals to the central body neuropil, and ultimately terminate in the ocellar neuropil (Chamberlain and Barlow, 1980).

The ventral optic nerve (Fig. 4.9) enters the anterior surface of the brain ventrad and medially to the lateral optic nerves. Once inside the sheath, it turns dorsally, then medially, and ends in the medulla. Individual photoreceptor axons terminate in a small planar array on the surface of the medullar neuropil (Fig. 4.2J) (Batra and Chamberlain, 1985). The ventral optic nerve also contains efferent fibers that innervate the photoreceptors (Battelle et al., 1982; Evans et al., 1983), but the loci of the somata are unknown.

Both median optic nerves run in a single sheath that contacts the anterior surface of the brain on the midline. After entering the brain, the two nerves separate and run posterolaterally to the paired ocellar ganglia. The median optic nerve contains axons of ocellar retinular and arhabdomeric cells, axons of the fused endoparietal eye, and efferent fibers that terminate in the periphery. Centrally, some of these fibers pass through the ocellar ganglion and arborize in the neuropil of the central body (Fig. 5.2H); others ramify in the ocellar neuropil, continue through the optic tract to ramify beneath the central body, and finally terminate in the ipsilateral medulla (Fig. 4.2I) (Chamberlain and Barlow, 1980). No correspondence between axon types and innervation patterns has yet been found, nor are the loci of the somata of efferent fibers known.

Other peripheral nerves that appear to enter the brain, in fact, run along its surface in or under the sheath and enter the central nervous system posteriorly to the posterior lateral neuropil, that is, they enter into the circumesophageal ring.

4.2.18. Tracts, Commissures, and Connectives

Although the brain abounds in well-defined tracts, none has been specifically traced. The optic tract (Fig. 4.10A), a sizeable fiber bundle connecting the ocellar ganglion and posterior medulla, passes just ventrad to the anterior arm of the central body. Those fibers known to pass along the optic tract have been mentioned above. The remaining tracts of the central neuropil of the brain, some of which are quite conspicuous, remain to be investigated.

All of the commissural tracts lie in the posterior half of the brain. Along its posterior curve, the neuropil of the central body crosses the midline. The anterior commissure, presumably the protocerebral bridge of other arthropods, lies parallel and just ventrad to the posterior curve of the central body. The structure of the anterior commissure suggests that it connects the stalks of the corpora pedunculata. The posterior commissure, its coarse fibers running next to the esophageal margin of the brain, appears to originate, in part, in the ventral medial ganglion, and is directed into the posterior lateral neuropil. Fibers of opisthosomal mechanoreceptors also appear to cross the midline in the posterior commissure (Chamberlain, 1978).

Posteriorly, the central neuropil of the brain grades imperceptibly into the broad and complex circumesophageal connectives. Our division between central neuropil and circumesophageal connectives at the level of the posterior lateral neuropil and ventral lateral posterior ganglion #1 is arbitrary.

At the present time, no behavioral pattern or control of physiological function can be reliably assigned to any identifiable neuron or specific circuitry. Hence, we will withhold speculative comments and expect future elucidation of such connections.

4.3. SUMMARY

The brain of the horseshoe crab, *Limulus polyphemus*, has been described and its partly new topographical nomenclature collated with that of classical authors. In the most general terms, 12 sets of neuronal groups and seven specific regions of neuropil, either paired or unpaired, have been identified anterior to the posterior lateral neuropil, formerly the antennal glomerulus. The structural organization of the brain has been illustrated in the form of an atlas, which includes current data on the numerical composition of tracts and neuronal groupings. Neuronal interconnections between all the optic ganglia—lamina, medulla, and ocellar ganglion—and the central body have been described. Broad coverage of the histology and ultrastructure of the corpora pedunculata, associated groups of neurons, and their interconnections has been included. The details on peripheral nerves, tracts, commissures, and groups of neurons should provide a wealth of landmarks for future work on the brain.

ACKNOWLEDGMENT

Parts of the studies performed by the authors and quoted in this article were supported by grants RR00163, RR05694, and EY00392 to WHF and by EY03446 to SCC from the National Institutes of Health.

REFERENCES

Augustine, G. J., Jr., R. H. Fetterer, and W. H. Watson, III. 1982. Amine modulation of the neurogenic *Limulus* heart. *J. Neurobiol.* **13:** 61–74.

Barlow, R. B., Jr. and S. C. Chamberlain. 1980. Light and a circadian clock modulate structure and function in *Limulus* photoreceptors, pp. 247–269. In T. P. Williams and B. N. Baker (eds.), *The Effects of Constant Light on the Visual Process.* Plenum, New York.

Batra, R. and S. C. Chamberlain. 1985. Central connections of *Limulus* ventral photoreceptors revealed by intracellular staining. *J. Neurobiol.* **16:** 435–441.

Battelle, B.-A., J. A. Evans, and S. C. Chamberlain. 1982. Efferent fibers of *Limulus* eyes synthesize and release octopamine. *Science (Washington, D.C.)* **216:** 1250–1252.

Behrens, M. E. and J. L. Fahy. 1981. Slow potentials in nonspiking optic nerve fibers in the peripheral visual system of *Limulus*. *J. Comp. Physiol.* **141:** 239–247.

Benson, J. A., R. E. Sullivan, W. H. Watson, III, and G. J. Augustine, Jr. 1981. The neuropeptide proctolin acts directly on *Limulus* cardiac muscle to increase the amplitude of contraction. *Brain Res.* **213:** 445–454.

Brown, F. A. and O. Cunningham. 1941. Upon the presence and distribution of a chromatophorotropic principle in the central nervous system of *Limulus*. *Biol. Bull. (Woods Hole)* **81:** 80–95.

Calman, B. G. and S. C. Chamberlain. 1982. Distinct lobes of *Limulus* ventral photoreceptors. II. Structure and ultrastructure. *J. Gen. Physiol.* **80:** 839–862.

Chamberlain, S. C. 1978. *Neuroanatomy of the visual afferents in* Limulus polyphemus. Institute for Sensory Research, Syracuse, New York, Special Report ISR-S-17.

Chamberlain, S. C. and R. B. Barlow, Jr. 1980. Neuroanatomy of the visual afferents in the horseshoe crab (*Limulus polyphemus*). *J. Comp. Neurol.* **192:** 387–400.

Chamberlain, S. C. and R. B. Barlow, Jr. 1982. Retinotopic organization of lateral eye input to *Limulus* brain. *J. Neurophysiol.* **48:** 505–520.

Chamberlain, S. C. and G. A. Engbretson. 1982. Neuropeptide immunoreactivity in *Limulus* I. Substance P-like immunoreactivity in the lateral eye and protocerebrum. *J. Comp. Neurol.* **208:** 304–315.

Chamberlain, S. C. and G. A. Wyse. 1986. An atlas of the brain of the horseshoe crab, *Limulus polyphemus*. *J. Morphol.* **187:** 363–386.

Chamberlain, S. C., B.-A. Battelle, and G. A. Wyse. 1983. Localization of serotonin-like immunoreactivity in the *Limulus* protocerebrum. *Soc. Neurosci. Abstr.* **9:** 76.

Corning, W. C., R. Lahue, and R. Von Burg. 1971. Supraesophageal ganglion influences on *Limulus* heart rhythm: Confirmatory evidence. *Can. J. Physiol. Pharmacol.* **49:** 387–393.

Dores, R. M. and W. S. Herman. 1980. The localization of two putative neurohormones in the central nervous system of *Limulus polyphemus*. *Comp. Biochem. Physiol.* **67A:** 459–463.

Dumont, J. N., E. Anderson, and E. Chomyn. 1965. The anatomy of the peripheral nerve and its ensheathing artery in the horseshoe crab, *Xiphosura* (*Limulus*)*polyphemus*. *J. Ultrastruct. Res.* **13**: 38–46.

Evans, J. A., S. C. Chamberlain, and B.-A. Battelle. 1983. Autoradiographic localization of newly synthesized octopamine to retinal efferents in the *Limulus* visual system. *J. Comp. Neurol.* **219**: 369–383.

Fahrenbach, W. H. 1970. The morphology of the *Limulus* visual system. III. The lateral rudimentary eye. *Z. Zellforsch.* **105**: 303–316.

Fahrenbach, W. H. 1971. The morphology of the *Limulus* visual system. IV. The lateral optic nerve. *Z. Zellforsch.* **114**: 532–545.

Fahrenbach, W. H. 1973. The morphology of the *Limulus* visual system. V. Protocerebral neurosecretion and ocular innervation. *Z. Zellforsch.* **144**: 153–166.

Fahrenbach, W. H. 1975. The visual system of the horseshoe crab, *Limulus polyphemus*. *Int. Rev. Cytol.* **41**: 285–349.

Fahrenbach, W. H. 1976. The brain of the horseshoe crab (*Limulus polyphemus*) I. Neuroglia. *Tissue Cell* **8**: 395–410.

Fahrenbach, W. H. 1977. The brain of the horseshoe crab (*Limulus polyphemus*) II. Architecture of the corpora pedunculata. *Tissue Cell* **9**: 157–166.

Fahrenbach, W. H. 1979. The brain of the horseshoe crab (*Limulus polyphemus*) III. Cellular and synaptic organization of the corpora pedunculata. *Tissue Cell* **11**: 163–200.

Fahrenbach, W. H. 1981. The morphology of the *Limulus* visual system. VII. Innervation of photoreceptor neurons by neurosecretory efferents. *Cell Tissue Res.* **216**: 655–659.

Hanström, B. 1926a. Das Nervensystem und die Sinnesorgane von *Limulus polyphemus*. *Kungl. Fysiografiska Sällskapets Handlingar. N.F.* **37**(5): 1–79.

Hanström, B. 1926b. Untersuchungen über die relative Grösse der Gehirnzentren verschiedener Arthropoden unter Berücksichtigung der Lebensweise. *Z. Mikrosk. Anat. Forsch.* **7**: 135–190.

Harrison, J. B. and N. J. Lane. 1981. Lack of restriction at the blood-brain interface in *Limulus* despite atypical junctional arrangements. *J. Neurocytol.* **10**: 233–250.

Holmgren, N. 1916. Zur vergleichenden Anatomie des Gehirns von Polychaeten, Onychophoren, Xiphosuren, Arachniden, Crustaceen, Myriapoden und Insekten. *Kungl. Svenska Vetenskapsakademiens Handlingar.* **56**: 1–303.

Johansson, G. 1933. Beiträge zur Kenntniss der Morphologie und Entwicklung des Gehirns von *Limulus polyphemus*. *Acta Zool.* **14**: 1–100.

Nunnemacher, R. F. and P. P. Davis. 1968. The fine structure of the *Limulus* optic nerve. *J. Morphol.* **125**: 61–70.

O'Connor, E., W. Watson, III., and G. A. Wyse. 1982. Identification and localization of catecholamines in the nervous system of *Limulus polyphemus*. *J. Neurobiol.* **13**: 49–60.

Oomura, Y. and H. A. Kuriyama. 1953. On the action of the optic lobe of *Limulus longispina*. *Jpn. J. Physiol.* **3**: 165–169.

Packard, A. S. 1893. Further studies on the brain of *Limulus polyphemus* with notes on its embryology. *Mem. Nat. Acad. Sci.* **6**: 289–331.

Patten, W. 1912. *The Evolution of the Vertebrates and their Kin.* Blakiston's, Philadelphia, Pennsylvania.

Patten, W. and W. A. Redenbaugh. 1900. Studies on *Limulus* II. The nervous system of *Limulus polyphemus*, with observations upon the general anatomy. *J. Morphol.* **16**: 91–200.

Pax, R. A. and R. C. Sanborn. 1967a. Cardioregulation in *Limulus* I. Gamma aminobutyric acid, antagonists and inhibitor nerves. *Biol. Bull.* (*Woods Hole*) **132**: 381–391.

Pax, R. A. and R. C. Sanborn. 1967b. Cardioregulation in *Limulus* II. Inhibition by 5-hydroxytryptamine and antagonism by bromlysergic acid diethylamide and picrotoxin. *Biol. Bull.* (*Woods Hole*) **132**: 392–403.

Scharrer, B. 1941. Neurosecretion. IV. Localization of neurosecretory cells in the central nervous system of *Limulus*. *Biol. Bull.* (*Woods Hole*) **81**: 96–104.

Snodderly, D. M., Jr. 1969. Processing of visual inputs by the ancient brain of *Limulus*. Doctoral Dissertation, The Rockefeller University, New York.

Snodderly, D. M., Jr. 1971. Processing of visual inputs by the brain of *Limulus*. *J. Neurophysiol.* **34**: 588–611.

Snodderly, D. M., Jr. and R. B. Barlow, Jr. 1970. Projection of the lateral eye of *Limulus* to the brain. *Nature* (*London*) **227**: 284–286.

Viallanes, M. H. 1893. Études histologiques et organologiques sur les centres nerveux et les organes des sens des animaux articulés. I. Le cerveau de la Limule (*Limulus polyphemus*). *Ann. Sci. Nat. Zool. Palaeontol.* (ser. 7)**14**: 405–456.

Waterman, T. H. and C. A. G. Wiersma. 1954. The functional relation between retinal cells and optic nerve cells in *Limulus*. *J. Exp. Zool.* **126**: 59–85.

Watson, W. H., III, J. R. Groome, B. M. Chronwall, J. Bishop, and T. L. O'Donohue. 1984. Presence and distribution of immunoreactive and bioactive FMRFamide-like peptides in the nervous system of the horseshoe crab, *Limulus polyphemus*. *Peptides* **5**: 585–592.

Willmer, P. G. and J. B. Harrison. 1979. Cation accessibility of the peripheral nervous system in *Limulus polyphemus*—An electrophysiological study. *J. Exp. Biol.* **82**: 373.

Wilska, A. and H. K. Hartline. 1941. The origin of "off-responses" in the optic pathway. *Am. J. Physiol.* **133**: 491–492.

Wyse, G. A. 1983. Serotonin-, proctolin-, and bombesin-like immunoreactivity: Histochemical localization in *Limulus* nervous system. *Soc. Neurosci. Abstr.* **9**: 75.

CHAPTER 5

Histology and Ultrastructure of the Acarine Synganglion

K. C. Binnington
Division of Entomology
CSIRO
Canberra City, Australia

5.1. INTRODUCTION

Acarines (mites and ticks) are unsegmented and generally small arthropods that have a highly condensed central nervous system. These attributes do not encourage their use as experimental animals, and consequently the understanding of acarine neurobiology is more rudimentary than for some other arthropods. Many members of the acarine order demand attention, however, because of their tremendous economic and medical importance. Therefore, the physiology of some species, including aspects of their neurobiology, has received considerable attention. A general overview of the tick nervous system and its relationship to the circulatory and neuroendocrine systems was given by Binnington and Obenchain (1982) and recent studies on the ultra-structure of the tick neuroendocrine system have been summarized by Binnington (1986).

5.2. GENERAL ANATOMY AND HISTOLOGY

The acarine nervous system, with its fusion of all central ganglia into a periesophageal synganglion provides an extreme example of the condensation typical of most chelicerates (Horridge, 1965). This concentration and forward migration of chelicerate ganglia is considered an evolutionary progressive feature (Hanström, 1928; Ioffe, 1963). In ticks, the synganglion is surrounded by a sheath of the circulatory system that extends dorsally to form an aorta and encloses major anterior and lateral nerve trunks (Binnington and Obenchain, 1982). Mites differ from ticks, and most other chelicerates, in not having circulatory vessels (Mitchell, 1964).

The preesophageal part of the tick synganglion includes a protocerebrum (Ioffe, 1963) as well as optic, cheliceral, and palpal ganglia (Fig. 5.1A). Although some ticks are regarded as eyeless, a study of some of these species revealed optic ganglia that receive nerves from simple photoreceptors (Binnington, 1972). It is not known if all mites have optic ganglia and optic nerves, but Moss (1962) stated that all tromidiforme mites studied to that time contained ocellar nerves. Certainly, it should not be assumed that the so-called

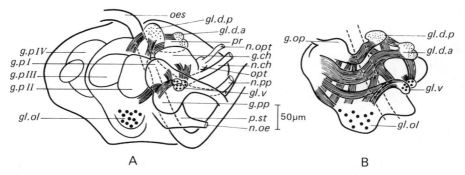

A   B

Figure 5.1. Diagrams of tracts demonstrated by histological staining in the synganglion of *Dermacentor pictus*. A) Lateral view. B) Intracerebral tracts. Anterior to right of the page for A and B. g.ch. = cheliceral ganglion; g.op. = opisthosomal ganglion; g.pp. = palpal ganglion; g.p.I–g.p.IV = pedal ganglia I–IV; gl.d.a. = anterior dorsal glomeruli; gl.d.p. = posterior dorsal glomeruli; gl.ol. = olfactory glomeruli; gl.v. = ventral glomeruli; n.ch. = cheliceral nerve; n.opt. = optic nerve; n.pp. = palpal nerve; n.e. = esophageal nerve; es. = esophagus; p.st. = stomadeal pons; pr. = protocerebrum. (Redrawn from Ioffe, 1963.)

eyeless acarines do not contain optic nerves (e.g., Robinson and Davidson, 1914; Eichenberger, 1970) and some form of dermal photoreceptors.

The arthropodan tritocerebrum is thought to be represented in ticks by a postoral commissure and stomodeal bridge (Obenchain, 1974a), but an equivalent to the deutocerebrum of arthropods is absent because ticks, like all chelicerates, lack antennae (Horridge, 1965). Prominent lobes on the ventral surface of the synganglion may receive axons (Section 5.2.3) from the tick olfactory organ situated on the front legs (Foelix and Axtell, 1972).

The postesophageal part of the synganglion is composed of pedal ganglia II–IV and posterior opisthosomal ganglia (Fig. 5.1).

5.2.1. Embryology and Morphogenesis

Observations on the embryology of the acarine nervous system have been made by Wagner (1894), Aeschlimann (1958), Ignatowicz (1974), and Ioffe (1984). A basic metameric arthropod pattern with some reduction in the number of abdominal ganglia can be seen in early embryogenesis before all ganglia become fused into a synganglion in the fully developed embryo. Aeschlimann (1958) found that in the argasid, *Ornithodoros moubata*, after metamerization of the germ band, rudiments of the nervous system can be detected; this precedes slightly the growth of cephalothoracic appendages. A wide and deep neural groove separates a bilaterally symmetrical chain of 11 paired ganglia made up of a cerebral ganglion, six thoracic ganglia, and four abdominal ganglia. Aeschlimann (1958) supports Wagner's (1894) view that the posterior abdominal ganglion probably represents a fusion of several ganglia. Longitudinal connectives are present before transverse commissures develop. As the germ band shrinks, there is a forward migration of ganglia around the esophagus, which results in the dorsocaudal displacement of the protocerebrum.

An interesting morphogenetic feature of acarine embryology is the development of six legs in the larva, compared with eight in subsequent instars. Ioffe (1984) found that in the unfed larvae of three ixodid ticks, the fourth pedal ganglia (corresponding to the absent legs) were smaller than the other ones and that they grew during and after larval feeding to reach a size comparable to the other pedal ganglia in the pharate nymph. Ioffe (1984) also shows that in unfed larvae of *Ixodes persulcatus* the ratio of the volume of the synganglion to the volume of the whole tick is much greater than in the case of the unfed nymph (about four times greater) and the unfed female (about 10 times), Ioffe postulates that there may have been a reduction in the size of the pedal ganglia and the number of legs in the larva in order to accommodate the relatively large ventral lobes assumed to receive fibers from the olfactory organs present on the front legs. In support of this hypothesis, Ioffe points out that the olfactory Haller's organs are of considerable functional significance in ixodid ticks, in which they are the major host-seeking receptors (Ioffe, 1984).

5.2.2. Types of Neurons

No information appears to be available on the light microscopy of nerve cells in mites, but histological techniques, including neurosecretory staining and vital staining by methylene blue, have provided some basic information on neurons of the tick synganglion. Obenchain (1974a) concluded that all of the perikarya constituting the cortex of the synganglion of *Dermacentor variabilis* were unipolar with one possible exception; a group of cells lying above the esophagus in the ventral part of the preesophageal synganglion seemed to be stellate and possibly multipolar (Section 5.3.2.). Globuli cells, small and with scant cytoplasm, are aggregated into a distinct mass adjacent to the putative olfactory lobes in the ventral preesophageal part of the synganglion (Ioffe, 1963; Obenchain, 1974a). The synganglion of ticks is rich in cells considered to be "neurosecretory," and a detailed analysis of neurosecretory cell bodies (Obenchain, 1974b) and axonal pathways (Obenchain and Oliver, 1975) has been made in the tick, *D. variabilis*. Other authors have provided further information particularly on the distribution of neurosecretory cells within the synganglion (review by Binnington and Obenchain, 1982). Obenchain (1974b) was able to classify most of the neurosecretory cells of *D. variabilis* according to the four types described in insects (Nayer, 1955; Delphin, 1965), but found some cells that were difficult to classify as "neurosecretory" or "nonneurosecretory." This problem is discussed in greater detail in Section 5.3.2. Also present in the tick synganglion are a considerable number of cell bodies that contain catecholamines (Binnington and Stone, 1977); the relationship between these cells and those described by conventional neurosecretory stains is not known.

Tsvileneva (1964) provided an extensive description of neurons in three species of ixodid ticks, using methylene blue staining (Figs. 5.2, 5.3). An interesting conclusion was that true association neurons were present only in the preesophageal part of the synganglion with association-motor neurons in the postesophageal part. This finding was not confirmed by Eichenberger (1970) for an argasid tick, and Obenchain (1974a) suggested that the methylene blue technique may have failed to stain all types of neurons in the study by Tsvileneva. Nevertheless, there seems little doubt that association-motor neurons are an important part of the tick integration system. Local interneurons or amacrine cells were found by Tsvileneva in the stomodeal bridge of a species of *Boophilus* (Fig. 5.4); it would be interesting to know if these neurons were of the nonspiking type, which is often the case for local interneurons of other arthropods (review by Seigler, 1984).

5.2.3. Neuropilar Tracts

Bundles of axons forming distinctive tracts are a striking feature of sections through the neuropil of the tick synganglion and in *Dermacentor pictus*. Ioffe

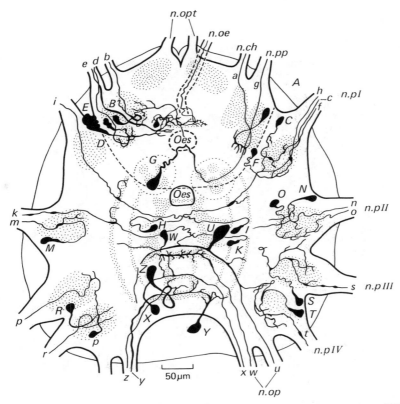

Figure 5.2. Axons and collaterals of association motor neurons in the synganglion of *Boophilus*. gl.ol. = olfactory glomerulus; c.n. = central neuropil; n.op. = opisthosomal nerve; n.opt. = optic nerve; n.p.I–n.p.IV = pedal nerves I–IV; n.pp. = palpal nerve; O.n I–II = opisthosomal neuropile I and II, es = esophagus. (Redrawn from Tsvileneva, 1964.)

(1963) distinguished five layers of commissures and connectives linking various glomeruli of the synganglion (Fig. 5.1); these tracts are similar to those described in Araneae (Hanström, 1928). Tsvileneva (1964) depicts the layers of neuropil described by Ioffe (1963) as being composed of axons and terminals of individual sensory and motor-association neurons (Figs. 5.2, 5.3) rather than of true commissures and connectives. Neuropil, organized into discrete dense areas of terminals (glomeruli) form putative olfactory lobes in the ventral synganglion (Section 5.2.2). Other glomeruli in the preesophageal part of the synganglion include anterior and posterior dorsal glomeruli and a more ventral glomerulus (Fig. 5.1). Ioffe (1963) concluded that *D. pictus* did not have a central body and that this may be a consequence of a reduction in eyes and optic ganglia of ticks. She proposed that the dorsal glomeruli may correspond to the corpora pedunculata of other arthropods.

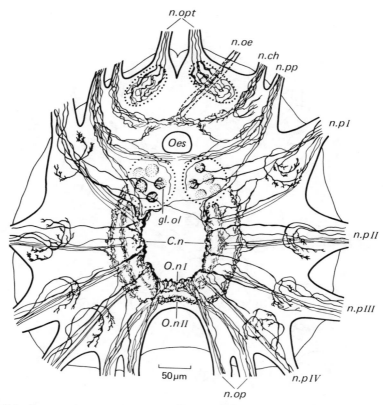

Figure 5.3. Sensory axons and terminals in the synganglion of *Boophilus*. Abbreviations as in Figure 5.2. (Redrawn from Tsvileneva, 1964.)

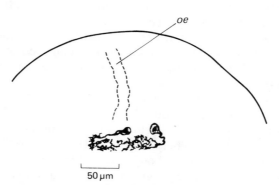

Figure 5.4. Two amacrine neurons stained by methylene blue in the stomodeal bridge of *Boophilus*. e = esophagus. (Redrawn from Tsvileneva, 1964.)

5.3. ULTRASTRUCTURE

5.3.1. Neural Lamella, Perineurium, and Glial Cells

In common with ganglia of other arthropods, the acarine synganglion is ensheathed by an acellular neural lamella and by perineurial glial cells (Fig. 5.5A). The neural lamella of two ixodid and one argasid tick species was found to be from 2–5 μm thick and to consist of fibers embedded in an amorphous layer.

The fibers are banded (Binnington and Lane, 1980) with a periodicity similar to that of insect collagen (Ashhurst, 1968). In insects, the barrier properties of the sheath are enhanced by tight junctions between perineurial cells. In *Boophilus microplus,* there are extensive areas of gap junctions between glial cell membranes (Fig. 5.5A, inset) but in both this tick (Binnington and Lane, 1980) and in *Amblyomma variegatum* (Hart et al., 1980) tight junctions are not seen and lanthanum ions are able to penetrate the sheath. Mites have an even less-developed sheath. Coons and Axtell (1971) reported that in *Macrocheles muscaedomesticae*, there was only a very thin structureless neural lamella and no junctions between perineurial and glial cells. Another mite, *Amystis baccarum,* also has a very thin neural lamella without collagen fibers, but in this case, there are gap junctions between glial cells (Binnington, unpublished data). The absence of tight junctions in acarines may reflect differences in ionic composition of the hemolymph between them and insects. The blood–brain barrier of insects may be necessary to cope with low sodium/potassium ratios in the hemolymph, which have not been found in ticks (discussed in greater detail by Binnington and Obenchain, 1982). As in insects (Sohal et al., 1972), ultrastructural differences, particularly in electron density between subperineurial glial cells and glial cells within or closely apposed to the neuropil of acarines (Coons and Axtell, 1971; Binnington and Lane, 1980), may indicate functional differences (Fig. 5.5B,C). The structure of the tick perineurial cells changes during development; in recently moulted females of *B. microplus,* the cells contain distended cisternae of rough endoplasmic reticulum, which may contain collagen precursors (Ashhurst and Costin, 1976; Binnington and Lane, 1980), whereas in fed females this material is no longer present but cytoplasm is beginning to accumulate glycogen. In postengorgement females, the perineurial cells are packed with glycogen, which is presumably used during the postfeeding reproductive stage of the tick's life cycle.

5.3.2. Neuronal Cell Bodies

Ultrastructural examination is useful in showing the diversity of cell types and can indicate whether or not cells may be neurosecretory or nonneurosecretory. In *B. microplus,* electron microscopy distinguished eight cell types, but of these only the globuli cells could be classified as being definitely

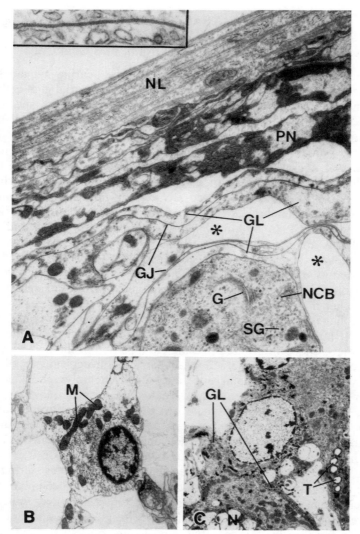

Figure 5.5. Thin sections through the synganglion of *B. microplus*. (A) Neural lamella (NL) overlying the perineurium (PN). There are extensive gap junctions (GJ and see inset) between glial processes. Beneath lie glial cell processes (GL) which ensheath a nerve cell body (NCB) and enclose extracellular spaces (*). The nerve cell body contains Golgi bodies (G) and a few secretory granules (SG) (12,800 ×). (B) A subperineurial glial cell, relatively nonelectron dense. M = mitochondria (5200 ×). (C) The more electron-dense type of glial cell (GL) closely opposed to the neuropil (N), T = trachea (5200 ×).

not neurosecretory; electron microscopy confirms that the globuli cells have a high nucleus/cytoplasm ratio and shows that the cytoplasm contains a poorly developed rough endoplasmic reticulum/Golgi system and no dense granules (Binnington, 1981a). As found in another chelicerate, *Limulus polyphemus* (Fahrenbach, 1979) (see also Chapter 4 by Fahrenbach and Chamberlain in this volume), the globuli cells of *B. microplus* are generally separated from each other only by narrow extracellular spaces and not by glial processes of the type that ensheath most other tick neurons (Binnington, 1981a). There are problems in defining other cells of *B. microplus* as "neurosecretory" or "nonneurosecretory." The difficulty of determining whether a cell contains clumped secretory granules from a single or small number of sections can be overcome by serial sectioning, but there is still the problem of knowing how many secretory granules defines a cell as neurosecretory. Furthermore, it is unlikely that all granular neurons are neurosecretory in the sense of releasing their products into the hemolymph (i.e., neuroendocrine). In *B. microplus,* many central and peripheral terminals as well as neuroendocrine ones contain dense granules. Only those cell bodies that contain dense granules large enough to be seen only in neurohemal structures can be considered as being definitely neuroendocrine. There are at least two such cell types in *B. microplus* (Binnington, 1986). Other cells rich in the smaller granules, which occur in terminals of both neuropil and neurohemal structures, are not easily classified. It is possible that some of these cells may be neuroendocrine and also have peptidergic terminals within the neuropil, especially because central neuropilar terminals of neuroendocrine cells occur in other invertebrates (Golding and Whittle, 1977). There is a reasonably good correlation between the presence of small (< 100 nm) diameter granules in neurons of both vertebrates and invertebrates and their probable adrenergic nature. The synganglion of *B. microplus* contains an abundance of adrenergic cells, and dopamine has been demonstrated biochemically in the synganglion (Megaw and Robertson, 1974). Furthermore, there are many fluorescent varicosities in the neuropil of *B. microplus* after the Falck-Hillarp reaction for catecholamines, suggesting that catecholamines may be widely used by the tick for central neurotransmission. Fluorescent axons from central terminals are also seen in many peripheral nerves, including terminals in the salivary glands (Binnington and Stone, 1977).

5.3.3. Neuropil

The only detailed ultrastructural study on acarine neuropil has been on *B. microplus* (Binnington, 1981a). Its neuropil can be broadly divided into three ultrastructural types. The first, (Fig. 5.6A) corresponding to the tracts seen by light microscopy, is composed of axons or processes that contain few or no synaptic specializations and vesicles. In some cases, electron-dense bodies, which may be very large (up to 350 × 500 nm in diameter), are present

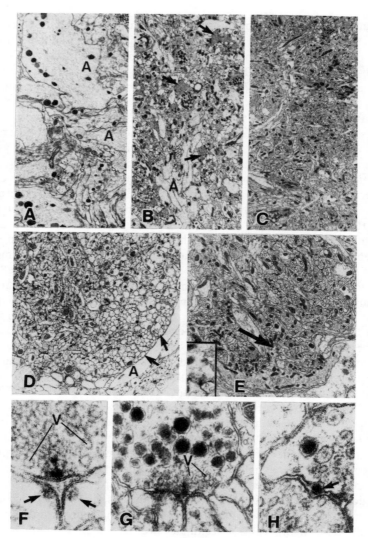

Figure 5.6. Thin sections through the synaganglion of *B. microplus*. (A) Tract-type neuropil with many axons (A) and few synaptic areas. Large granules are present in many of the axons (6200 ×). (B) Neuropil of a pedal ganglion containing relatively electron-lucent axons (A) and more electron-dense, vesicle-rich terminals (*arrows*; 3300 ×). (C) Neuropil of an olfactory lobe; this glomerular neuropil has a greater density than that of the pedal ganglia (3300 ×). (D) A single glomerulus of the olfactory lobe bordered by a crescent-shaped axon (A) and containing bundles of small diameter processes (*arrows*) which may be from the associated globuli cells (4600 ×). (E) Crescent-shaped process forming synapses (*arrow* and *inset*) with small diameter processes which may belong to globuli neurons (5300 ×; *inset* 20,000 ×). (F) Transverse section through a dyad-type synapse; the electron-lucent vesicles (V) clustered around the presynaptic density are similar to those more distal to the "T" bar. The postsynaptic process contains electron-dense material, which is situated in two regions (*arrows*) opposite the presynaptic density (120,000 ×).

in the processes. These bodies may represent aggregations of smaller neurosecretory granules that are also present, or alternatively they could be lysosomal. The neuropil of the pedal ganglia provides an example of the second and predominant form of neuropil (Fig. 5.6B). These ganglia contain an apparently unorganized mixture of, on the one hand, processes that are largely free of vesicular or granular inclusions and, on the other, terminals rich in electron-lucent vesicles, granules, and synaptic densities. The third type of neuropil is that associated with glomeruli, such as those found in the olfactory lobes (Fig. 5.6C). Vesicle-free processes of the globuli cells, which are present in the cortex at this site, are arranged in bundles that enter the glomerular neuropil of the olfactory lobes (Fig. 5.6D). Individual glomeruli often have a crescent-shaped border formed by an axon, whose branches contain presynaptic densities and clustered vesicles (Fig. 5.6E); these branches form synapses (Fig. 5.6E, inset) with processes that probably belong to globuli cells (Fahrenbach, 1979).

As in insects (review by Osborne, 1980), putative active zones of neuropilar terminals can be identified through the presence of presynaptic densities, which, when cut transversly, appear as a double ''T'' (Fig. 5.6F) and longitudinally as two parallel bars of electron-dense material. Terminals of the neuropil of *B. microplus* can be broadly divided into two types, those containing mainly electron-lucent vesicles (Fig. 5.6F) and those with a large number of dense granules (in addition to the smaller lucent vesicles clustered at, or near, the active zones; Fig. 5.6G). Opposite the presynaptic density is a postsynaptic specialization in the form of a dense bar or spot depending on the plane of section (Fig. 5.6F). Processes containing presynaptic densities are often in juxtaposition with two postsynaptic processes (Fig. 5.6F). A series of such arrangements (sometimes called ''dyads'') may be present in the one terminating process. The convergence of more than one presynaptic process onto a single postsynaptic fiber is also seen, although much less frequently. Reciprocal synapses, described for example in *L. polyphemus* (Fahrenbach, 1979), were not observed in the sections examined (Binnington, 1981a). The release of neurotransmitter or neuromodulatory chemicals into extracellular space not associated with a synapse is indicated by section profiles such as the one shown in Figure 5.6H.

5.3.4. Neurohemal Structures

A retrocerebral organ complex described by light and electron microscopy in *Argas (P.) arboreus* (Roshdy et al., 1973; Roshdy and Marzouk, 1982)

(G) In this terminating process, the vesicles (V) aggregated around the ''T'' bar are quite different from the electron-dense granules occurring more distally (59,000 ×). (H) A granule (*arrow*) is present in the extracellular space between two terminating processes. This indicates that exocytotic release of dense cores from granules, not associated with the ''T'' bars, may occur in the neuropil (130,000 ×).

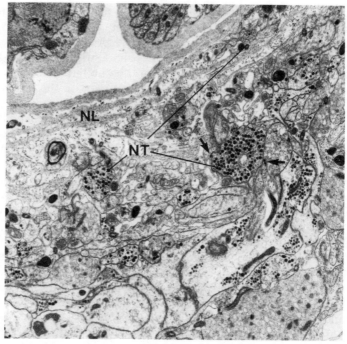

Figure 5.7. A superficial region of the tick synganglion showing neurohemal terminals (NT). Arrows indicate "T" bar configurations. NL = neural lamella (9000 ×).

and by light microscopy in *D. variabilis* (Obenchain and Oliver, 1975) and *Hyalomma dromedarii* (Marzouk et al., 1985) has been postulated to be analogous to the insect corpus cardiacum. However, a detailed ultrastructural study of the complex in *B. microplus* (Binnington, 1983) did not show aggregations of neuroendocrine terminals typical of neurohemal organs. In *B. microplus*, the complex appears to function as an endocrine gland and is composed of modified cells of the periganglionic sheath of the circulatory system; also present is a group of neuronal perikarya probably analogous to the hyperesophageal ganglion observed in *D. variabilis* (Obenchain and Oliver, 1975). Small spherical bodies termed lateral organs also have been postulated to be neurohemal (Chow et al., 1974; Obenchain and Oliver, 1975; Panfilova, 1978), but in *B. microplus* they are endocrine glands and have an ultrastructure compatible with a role in secreting a lipidic hormone (Binnington, 1981b). The bulk of neuroendocrine terminals seen in *B. microplus* are in contact with the extracellular space of the neural lamella/perineurial sheath of the synganglion (Fig. 5.7); thus *B. microplus* appears to release neurohormones in a diffuse manner from the surface of the synganglion without using a discrete peripheral neurohemal structure such as the corpus cardiacum of insects (Binnington, 1983).

5.4. SUMMARY

Most information on the acarine synganglion is from ticks. Early in embryogenesis, the basic arthropodan metameric chain of ganglia fuses into a mass of ganglia called the synganglion. Tick larvae when hatched lack a fourth pair of legs, and the corresponding pedal ganglia, though present, are relatively small. The glial sheath of the synganglion differs from that of insect ganglia in not having tight junctions and in being more permeable to lanthanum. In common with other arthropods, the synganglion includes globuli cells and distinctive neurosecretory cells. However, most cell bodies contain some dense granules and it is often difficult to make a distinction between neurosecretory and nonneurosecretory cells. Tracts through the neuropil include axons of association-motor neurons, but the contribution made to these tracts by true association neurons, especially in the postesophageal part of the synganglion, requires some clarification. Dense neuropil of glomeruli is readily distinguished from that containing axonal tracts and ganglionic neuropil. The surface of the tick synganglion is rich in neuroendocrine terminals, and it is thought that ticks lack a discrete peripheral neuroendocrine organ analogous to the corpus cardiacum of insects.

REFERENCES

Aeschlimann, A. 1958. Développement embryonnaire d'*Ornithodoros moubata* (Murray) et transmission transovarienne de *Borrelia duttoni*. *Acta Trop.* **15**(1): 15–62.

Ashhurst, D. E. 1968. The connective tissue of insects. *Annu. Rev. Entomol.* **13**: 45–74.

Ashhurst, D. E. and N. M. Costin. 1976. The secretion of collagen by insects: Uptake of [³H] Proline by collagen-synthesizing cells in *Locusta migratoira* and *Galleria mellonella*. *J. Cell Sci.* **20**: 377–403.

Binnington, K. C. 1972. The distribution and morphology of probable photoreceptors in eight species of ticks (Ixodoidea). *Z. Parasitkd.* **40**(4): 321–332.

Binnington, K. C. 1981a. *Nervous, neuroendocrine and endocrine systems of ticks.* Doctoral thesis, University of Cambridge, Cambridge.

Binnington, K. C. 1981b. Ultrastructural evidence for the endocrine nature of the lateral organs of the cattle tick *Boophilus microplus*. *Tissue Cell* **13**(3): 475–490.

Binningtonn, K. C. 1983. Ultrastructural identification of neurohaemal sites in a tick: Evidence that the dorsal complex may be a true endocrine gland. *Tissue Cell* **15**(2): 317–327.

Binnington, K. C. 1986. Ultrastructure of the tick neuroendocrine system, pp. 152–164. In J. R. Sauer and J. A. Hair (eds.), *Morphology, Physiology and Behavioural Biology of Ticks*. Ellis Horwood, Chichester.

Binnington, K. C. and N. J. Lane. 1980. Perineurial and glial cells in the tick *Boophilus microplus* (Acarina: Ixodidiae): Freeze-fracture and tracer studies. *J. Neurocytol.* **9**: 343–362.

Binnington, K. C. and F. D. Obenchain. 1982. Structure and function of the circulatory, nervous and neuroendocrine systems of ticks, pp. 351–398. In F. D. Obenchain and R. Galun (eds.), *Physiology of Ticks.* Pergamon Press, Oxford.

Binnington, K. C. and B. F. Stone. 1977. Distribution of catecholamines in the nervous system of the cattle tick *Boophilus microplus* Canestrini. *Comp. Biochem. Physiol.* **58c:** 21–28.

Coons, L. B. and R. C. Axtell. 1971. Cellular organization in the synganglion of the mite *Macrocheles muscaedomesticae* (Acarina: Macrochelidae). An electron microscopic study. *Z. Zellforsch. Mikrosk. Anat.* **119:** 309–320.

Chow, Y. S., S. H. Lin, and C. H. Wang. 1974. An ultrastructural and electrophysiological study of the brain of the brown dog tick *Rhipicephalus sanguineus* (Latreille). *Chinese Bio. Sci.* **1**(4): 83–92.

Delphin, F. 1965. The histology and possible functions of neurosecretory cells in the ventral ganglion of *Shistocerca gregaria Forskal* (Orthoptera: Acrididae). *Trans. R. Entomol. Soc. London* **117:** 167–214.

Eichenberger, G. 1970. Das Zentralnervensystem von *Ornithodorus moubata* (Murray), Ixodoidea: Argasidae, und seine postembryonale Entwicklung. *Acta Trop.* **27**(1): 15–53.

Fahrenbach, W. H. 1979. The brain of the horseshoe crab (*Limulus polyphemus*). III. Cellular and synaptic organization of the corpora pendiculata. *Tissue Cell* **11**(1): 163–200.

Foelix, R. G. and R. C. Axtell. 1972. Ultrastructure of Haller's organ in the tick *Amblyomma americanum* (L.). *Z. Zellforsch. Microsk. Anat.* **124:** 275–292.

Golding, D. W. and A. C. Whittle. 1977. Neurosecretion and related phenomena in annelids. *Int. Rev. Cytol.* **5**(suppl): 189–302.

Hanström, B. 1928. *Vergleichende Anatomie des Nervensystems der wirbellowen Tiere unter Berücksichtigung seiner Funktion.* Springer, Berlin.

Hart, R. J., D. J. Beadle, and R. P. Botham. 1980. The penetration of ionic lanthanum into the central nervous system of the tick *Amblyomma variegatum. Physiol. Entomol.* **5:** 401–405.

Horridge, G. A. 1965. Arthropoda: Details of the group, pp. 1165–1270. In T. H. Bullock and G. A. Horridge (eds.), *Structure and Function of the Nervous Systems of Invertebrates*, vol. 2. Freeman, San Francisco.

Ignatowicz, S. 1974. Observations on the biology and development of *Hypoaspis aculeifer* Canestrini, 1885 (Acarina: Gamasides). *Zool. Pol.* **24**(1): 41–59.

Ioffe, I. D. 1963. Structure of the brain of *Dermacentor pictus* Herm. (Chelicerata, Acarina.) *Zool. Zh.* **42**(10): 1472–1484.

Ioffe, I. D. 1984. Possible correlation between the reduction of legs IV in the larvae and the development of the nervous apparatus in the parasitiformes (Acarina), pp. 286–294. In D. A. Griffiths and C. E. Bowman (eds.), *Acarology VI*, vol. 1. Ellis Horwood Ltd, Chichester.

Marzouk, A. S., F. S. A. Mohamed, and G. M. Khalil, 1985. Neurohemal-endocrine organs in the camel tick *Hyalomma dromedarii* (Acari: Ixodoidea: Ixodidae). *J. Med. Entomol.* **22:** 385–391.

Megaw, M. W. and H. A. Robertson. 1974. Dopamine and noradrenaline in the salivary gland and brain of the tick *Boophilus microplus*: Effect of reserpine. *Experientia* **30**: 1261–1262.

Mitchell, R. 1964. The anatomy of an adult chigger mite *Blankaartia acuscutellaris* (Walch). *J. Morphol.* **114**: 373–392.

Moss, W. W. 1962. Studies on the morphology of the trombidiid mite *Allothrombium lerouxi* Moss (Acari). *Acarologia* **4**(3): 313–345.

Nayer, K. K. 1955. Studies on the neurosecretory system of *Iphita limbata* Stal. I. Distribution and structure of the neurosecretory cells of the nerve ring. *Biol. Bull.* (*Woods Hole*) **108**: 206–307.

Obenchain, F. D. 1974a. Structure and anatomical relationships of the synganglion in the American dog tick *Dermacentor variabilis* (Acari: Ixodidae). *J. Morphol.* **142**(2): 205–223.

Obenchain, F. D. 1974b. Neurosecretory system of the American dog tick *Dermacentor variabilis* (Acari: Ixodidea). I. Diversity of cell types. *J. Morphol.* **142**(4): 433–446.

Obenchain, F. D. and J. H. Oliver. 1975. Neurosecretory system of the American dog tick *Dermacentor variabilis* (Acari: Ixodidea). II. Distribution of secretory cell types, axonal pathways and putative neurohemal-neuroendocrine associations; comparative histological and anatomical implications. *J. Morphol.* **145**(3): 269–294.

Osborne, M. P. 1980. The insect synapse: Structural functional aspects in relation to insectididal action, pp. 29–40. In *Insect Neurobiology and Pesticide Action* (*Neurotox 79*) [Proc. Soc. Chem. Ind. Symp., Uni. York, Sept. 1979.] Society of Chemical Ind., London.

Panfilova, I. M. 1978. Lateral organs of *Ixodes persulcatus* (Parasitiformes, Ixodidae). *Zool. Zh.* **57**(2): 190–196.

Robinson, L. E. and J. Davidson. 1914. The anatomy of *Argas persicus* (Oken, 1818). Part 3. *Parasitology* **6**: 382–424.

Roshdy, M. A. and A. S. Marzouk. 1982. The subgenus *Persicargas* (Ixoidea: Argasidae: Argas). 35. The lateral segmental organs and peritracheal gland in immature and adult *A.* (*P.*) *arboreus*. *Z. Parasitkd.* **66**: 335–343.

Roshdy, M. A., N. M. Shoukrey, and L. B. Coons. 1973. The subgenus *Persicargas* (Ixodoidea: Argasidae: Argas). 17. A neurohemal organ in *A.* (*P.*) *arboreus* Kaiser, Hoogstraal and Kohls. *J. Parasitol.* **59**: 540–544.

Seigler, M. V. S. 1984. Local interneurons and local interactions in arthropods. *J. Exp. Biol.* **112**: 253–281.

Sohal, R. S., S. P. Sharma, and E. F. Couch. 1972. Fine structure of the neural sheath, glia and neurons in the brain of the housefly *Musca domestica*. *Z. Zellforsch. Mikrosk. Anat.* **135**: 449–460.

Tsvileneva, V. A. 1964. The nervous system of the ixodid ganglion. *Zool. Jahrb. Abt. Anat. Ontog. Tiere* **81**: 576–602.

Wagner, Y. N. 1894. Die embrional Entwicklung von *Ixodes calcaratus* Bir. *Trudy Imperatorskago S.-Peterburgskago Obshchestra Estestvoispytatelei* Vypusk 2, Otd. *Zool. Fiziol.* **24**: 1–246.

CHAPTER 6

Comparative Anatomy
of the Crustacean Brain

Dick R. Nässel
Rolf Elofsson
Department of Zoology
University of Lund
Lund, Sweden

6.1. INTRODUCTION

The brain is a complex aggregation of nerve cells that is rather poorly understood as a whole in arthropods. In crustaceans many of the centers and tracts have anatomical names, but often no functions are assigned to them. Others can at least be assigned modalities, such as visual, olfactory,

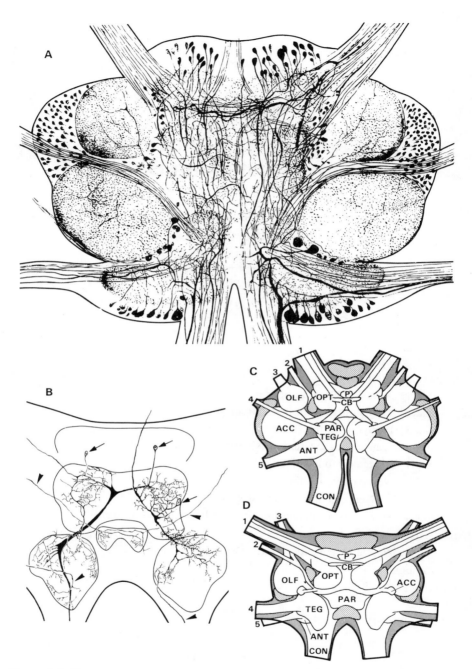

Figure 6.1. The decapod brain seen in horizontal sections. Abbreviations: in protocerebrum: CB = central body; OPT = optic neuropil; P = protocerebral bridge. In deutocerebrum: ACC = accessory lobe; OLF = olfactory lobe; PAR = parolfactory lobe In tritocerebrum: ANT = antennary neuropil; CON = esophageal connective; TEG = teg-

112

or mechanosensory. Until recently, many of these centers were classified solely on which sensory (or motor) nerves connected them to the periphery. Although these centers may primarily subserve the proposed modalities, a large number of central neurons in the brain that have processes in these centers are multimodal (Glantz et al., 1981). Hence, a functional analysis of the brain seems to be possible if one focuses on single pathways. Such studies have been done and will be presented.

Detailed anatomical and physiological studies of the crustacean central nervous system (CNS) have mostly been conducted on decapod crustaceans, such as crabs, crayfishes, and lobsters. Hence, a complete comparative overview of crustacea is not possible to present. Only a sketchy account of the comparative anatomy of the brain of some studied crustaceans can be given in a short review such as this. For details, we refer to earlier comprehensive reviews by Bullock and Horridge (1965) and Sandeman (1982) and concentrate on some selected topics and more recent work. The systematic division of the Crustacea proposed by Bowman and Abele (1982) has been followed.

6.2. ANATOMY OF THE CRUSTACEAN BRAIN

6.2.1. A Basic Plan of the Brain

The bulk of the crustacean brain (supraesophageal ganglia) develops from three paired embryonic ganglia. The relatively large anteriormost of these anlagen gives rise to the protocerebrum in the adult brain and to the centers of the compound and frontal eyes. The following paired ganglion, belonging to the antennulary segment, become the deutocerebrum, and the ganglionic anlagen of the antennal segment becomes the tritocerebrum (Fig. 6.1). In addition, a paired preantennulary ganglion has been identified in some crustacean groups, although no appendages exist for this segment and their significance in the contribution to the brain is a matter of dispute (Anderson, 1973). During development, the stomodeum, which initially separates the antennulary and antennal ganglionic anlagen, is, in malacostracan crustaceans, displaced caudally, and the brain fuses into one unit retaining the tritocerebral commissure behind the esophagus. This fusion is, however, not complete in nonmalacostracans. In these crustaceans, the tritocerebrum is

---

umentary neuropil. The nerves: 1 = optic nerve with olfactory-globular tract; 2 = oculomotor nerve; 3 = antennular nerve; 4 = tegumentary nerve; 5 = antennary nerve. (A) The crayfish brain after methylene blue vital staining. (After Retzius, 1890.) (B) The brain of the crab *Carcinus* after methylene blue staining. Three large (probably multimodal) interneurons are traced here. Their cell bodies are marked with *arrows*. In addition four presumed sensory neurons are shown at *arrowheads*. (After Bethe, 1897.) (C) Schematic diagram of the crayfish brain. Clusters of large cell bodies are shown with dots. (Redrawn from Helm, 1928; Sandeman, 1982.) (D) Diagram of the *Carcinus* brain. (Redrawn from Sandeman, 1982.)

usually situated behind or on both sides of the esophagus on the connectives leading to the ventral nerve cord.

The shape and internal structure of the brain is governed by the input channels from different sensory organs and the output to appendages, head muscles, neurohemal organs, and so on. The degree of complexity of the brain also correlates with the presence and form of associative centers. We will therefore, before discussing the variations within crustaceans, give a short review of brain-associated sensory and other organs.

Three sensory organs innervate the protocerebrum: the compound eyes, the frontal eyes (nauplius eye and frontal organs), and the organ of Bellonci. Leaving this neuromere, are various motor nerves such as the nervus oculomotorius to the eyestalk muscles and axons from peptide-containing neurons combining into the neurohemal sinus gland. Other muscles present in the rostrum, head capsule, and the two antennae as well as the green gland are innervated by small nerves, for example, the rostral and accessory nerves as described by Seabrook and Nesbitt (1966) for *Oronectes virilis*. A considerable variation of the nerves to various muscles and organs of the head occurs in different species. The frontal eyes can be subdivided into four groups and the term includes the nauplius eye, which in many species appears as an eye, and the frontal organs, which usually are reduced eyes. They all terminate in a single, paired, or three-lobed glomerular neuropil situated medially on the anterior margin of the protocerebrum. The neuropil is situated outside (non-Malacostraca) or included in the brain neuropil (Malacostraca) (Elofsson, 1966). The organ of Bellonci is today the name of a complex little organ connected to the lateral protocerebrum (classically referred to as medulla terminalis). Its function is unknown, but chemo- and photoreception have been suggested (Chaigneau 1977; Steele 1984). The sinus gland is a neurohemal organ situated lateral to the optic lobe neuropils. It receives axons originating mostly from the lateral protocerebrum, but also from other parts of the brain and optic neuropils.

The deutocerebrum receives sensory input from chemo- and other receptors on the first antenna (antennulae). A statocyst is present in some decapods and its nerve joins the deutocerebrum (Yoshino et al, 1983). Antennulary muscles are innervated from this part of the brain.

The tritocerebrum derives sensory input from and sends motor output to the second antenna (antennae). This caudal portion of the brain connects with the ventral nerve cord via the connectives passing on each side of the esophagus. The brain contribution to the visceral system varies from one to several nerves, but often two unpaired nerves can be found, as in *Oronectes virilis* (Seabrook and Nesbitt, 1966) and other decapods (Hanström, 1947): A neurohemal organ, the postcommissural organ, emerges from the tritocerebrum. Its nerves leave via the postesophageal commissure behind the esophagus (Chaigneau, 1983).

6.2.2. Comparative Aspects of Brain Morphology

With the above general pattern in mind, a short review of the variations within the Crustacea will be given (some features are shown in Fig. 6.3). The most important variation in the morphology of the protocerebrum is caused by the development of the compound eyes. Several, especially non-malacostracan, groups lack compound eyes altogether, such as the class Cephalocarida and the subclasses Mystacocarida, Cirripedia (as adults), and Copepoda. Few species have compound eyes in the class Ostracoda. In these cases, the protocerebrum is small, housing only the centers of the frontal eyes and the organ of Bellonci. An extreme example is offered by the sessile barnacles (Cirripedia). In these animals, the brain appears as a thin semi-circle in front of the esophagus without any visible separation between the proto-, deuto-, and tritocerebrum. Mixed motor and sensory nerves, being the equivalents of the frontal eye and antennulary nerves, are present as well as small nerves to the anterior part of the digestive tract. Compound eyes and the second antenna are absent and the first antenna is reported to

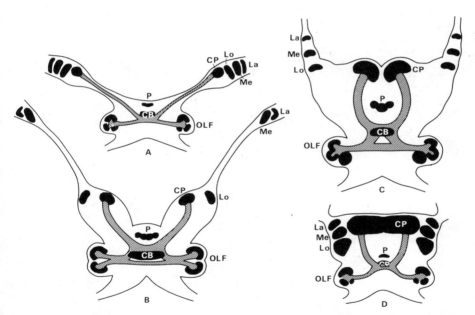

Figure 6.2. Examples of some of the variation in brain morphology within Decapoda. We show the olfactory-globular tracts (*shaded*) connecting olfactory lobes (OLF) and corpora hemiellipsoidale (corpora pedunculata, CP) and the arrangements of the optic lobe neuropils lamina (La), medulla (Me) and lobula (Lo). Note that these neuropils can reside either in the eyestalks or be withdrawn into the cephalic neuropil masses. CB = central body; P = protocerebral bridge. (Redrawn from Hanstöm, 1947.) (A) *Carcinus*-type. (B) *Emerita*-type. (C) *Calocaris*-type. (D) *Athanas*-type.

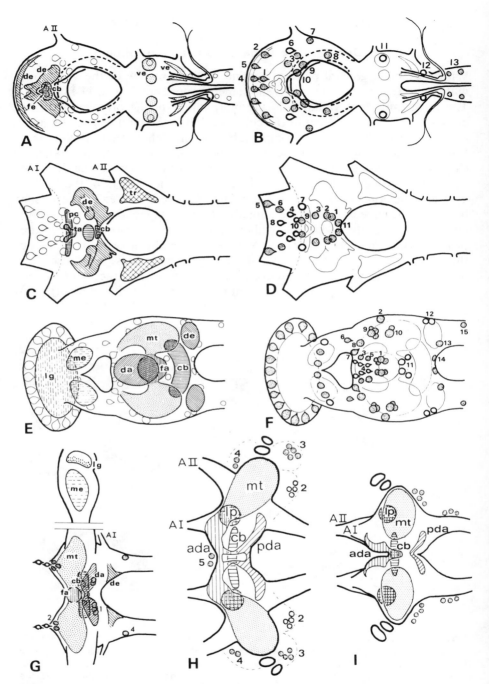

Figure 6.3. The morphology of some crustacean brains and the distribution of monoamines in cell bodies and neuropil after Falck-Hillarp histofluorescence treatment (*horizontal sections*). Catecholaminergic cell bodies are *dotted*. Indolylalkylaminergic cell bod-

lack sensory function (Gwilliam and Cole, 1979). In the remaining species with compound eyes, there is a difference in the number and position of optic centers between the class Malacostraca and the rest of the Crustacea. Malacostracan crustaceans have three lateral centers related to the compound eye (Figs. 6.2, 6.5A) and they are by tradition called lamina ganglionaris, medulla externa, and medulla interna, later, by analogy with insects, termed lamina, medulla, and lobula (Strausfeld and Nässel, 1980). The nonmalacostracan crustaceans have only two lateral optic neuropils; the lamina and the medulla (Fig. 6.3E). The medullae are homologous in the Crustacea, whereas the lobula of malacostracans is a novel structure differentiated from the brain (Elofsson, 1969a). Further details on the optic lobes will be given later. In some malacostracans, the lateral part of the protocerebrum (medulla terminalis) resides in the eyestalk (when present); in nonmalacostracans, this neuropil mass is included in the cephalic part of the protocerebrum.

The center of the frontal eyes is usually situated outside and anterior to the main neuropil of the brain in nonmalacostracans (Elofsson, 1966) and inside in malacostracan crustaceans. In the subclass Copepoda, which lacks compound eyes, the frontal eyes have in some planktonic species developed into large eyes that are very different from compound eyes (Elofsson 1966, 1969b). The organ of Bellonci has no discernible center in the crustacean brain, but its neurons enter the lateral protocerebrum.

The size and shape of the deuto- and tritocerebrum vary within the Crustacea, depending on the size and sensory equipment of the first and second antennae. There is, however, a distinct difference between the malacostracans and the other crustaceans. The former class possesses olfactory lobes in the deutocerebrum. These can, as in many decapods, be very large and subdivided in main and accessory lobes and form two extensive outpocketings on the lateral sides of the brain (Fig. 6.1C,D).

6.2.3. Histology of the Brain

The basic design of the nervous system comprises transverse (commissures) and longitudinal (connectives) connections between the hemiganglia and the metameric ganglia of the animal (Fig. 6.1). With the development of a more

ies are unfilled with *heavy outlines.* Neuropil regions with catecholamine fluorescence are *hatched* or *dotted.* The numbers refer to identifiable cell bodies. Abbreviations: AI = antennular nerve; AII = antennary nerve; ada = anterior dorsal area; cb = central body; da = dorsal area; de = deutocerebrum; fa = frontal area; fe = center of frontal eye complex; lg = lamina; lp = lobus paracentralis; me = medulla; mt = lateral protocerebrum; pc = protocerebral commissure; pda = posterior dorsal area; ta = three-lobed area, tr = tritocerebrum, ve = centers in subesophageal ganglia. (A and B) The ostracod, *Heterocypris incongruens.* (C and D) The copepod, *Cyclops strenuus.* (E and F) The phyllopod, *Daphnia magna* (cf. Fig. 6.5). (G) The anostracan, *Artemia salina.* (H) The isopod, *Asellus aquaticus.* (I) The amphipod, *Gammarus pulex.* H and I are malacostracans, the others nonmalacostracans. (From Aramant and Elofsson, 1976a,b; with permission).

complex behavior, enlarged sensory and associative centers in the brain are superimposed on this simple pattern and defined tracts develop. From a histological view, the neuropil develops from a diffuse unstructured (nonglomerular) to a structured glomerular or stratified neuropil (Maynard, 1962). The more advanced crustacean brains show all three neuropil types as well as specific agglomerations of nerve cell bodies in the rim of the brain. Stratified and columnar neuropil is present in the three outer optic centers of malacostracan species with well-developed compound eyes, but absent or feebly developed in nonmalacostracans. Glomerular neuropil is present especially in the proto- and deutocerebrum of malacostracan species, less so in nonmalacostracans. The hemiellipsoid body is a prominent glomerular neuropil anteriorly in the lateral protocerebrum of many malacostracan species. It is the termination of the largest brain tract, the tractus olfactorioglobularis, starting in the lobus olfactorius and lobus accessorius of the deutocerebrum (Fig. 6.2). In the subclass Hoplocarida, where highly developed compound eyes occur, the lateral protocerebrum has additional glomerular neuropil termed the corpus prolongatum and the corpus reniformae (Hanström, 1947). In the sessile-eyed orders Tanaidacea, Isopoda, and Amphipoda, the corpus hemiellipsoidale is absent. All these glomerular neuropils of the lateral protocerebrum innervated by the tractus olfactorioglobularis, have been homologized with the corpora pedunculata (mushroom bodies) of insects. There are two distinctly ordered neuropil regions in the protocerebrum, namely the central body and the protocerebral bridge (Figs. 6.1C, D, 6.2, 6.3). These are situated medially in the brain, with the protocerebral bridge close to the anterior margin and the central body centrally in the brain. The appearance differs in different species. The central body is present in all crustaceans, whereas the protocerebral bridge cannot be recognized in some nonmalacostracans. Glomerular structured neuropil is further found in the frontal eye centers. A structure with mixed glomerular structure and unstructured diffuse organization, although circumscribed and recognizable, is found in the paracentral lobes behind and lateral to the central body and in the neuropils in front of the central body (Helm, 1928). The tractus olfactorioglobularis bypasses the protocerebrum partly decussating behind the central body (Fig. 6.2). There are a number of small and large commissures in the protocerebrum, which have never been identified properly as to their connections and functions.

The most developed deutocerebrum is found in the orders Stomatopoda and Decapoda. It contains, at most, three well-defined glomerular neuropil regions (Figs. 6.1, 6.2). Most laterally, lobus olfactorius is found, followed by the lobus accessorius and the lobus parolfactorius; the latter is sometimes horseshoe-shaped. The lobus olfactorius can be divided into anterior and posterior sublobes. The lobus olfactorius contains oblong glomeruli arranged around the margin, and the lobus accessorius contains small rounded glomeruli scattered all over the lobe. The tractus olfactorioglobularis is connected both with the lobus olfactorius and the lobus accessorius (Hanström,

1925, 1947; Helm, 1928; Sandeman and Luff, 1973; Tsvileneva and Titova, 1985). Well-developed lobi accessori occur only in decapods apart from prawns and crabs. The development of the glomerular neuropil regions varies considerably within the Malacostraca. The glomerular neuropils are connected to one another by large commissures and to specific branches of the antennulary nerve (Hanström, 1947). In nonmalacostracan crustaceans, the deutocerebrum is much less developed (Fig. 6.3), and only in some rare cases can anything reminiscent of a lobus olfactorius be found. Also in those species with compound eyes, the tractus olfactorioglobularis is missing or feebly developed.

Only in some decapod prawns, the tritocerebrum contains well-developed specific glomerular neuropil, situated at the roof of the antennary nerve in a region called lobus antennarius. A dorsal unstructured diffuse neuropil, the lobus tegumentarius, is connected to the n. tegumentarius (Fig. 6.1) absent in stomatopods. In other malacostracans, weakly developed glomerular neuropil may be found in the tritocerebrum (Hanström, 1947), whereas only unstructured neuropil occurs in nonmalacostracans (Fig. 6.3).

6.3. HISTOCHEMICAL LOCALIZATION OF NEUROACTIVE SUBSTANCES

6.3.1. Biogenic Monoamines

The first demonstration of biogenic amines in the central nervous system of crustaceans with the Falck-Hillarp (Falck, 1962; see also Chapter 23 by Klemm in this volume) method was made on the crayfish brain (Elofsson et al., 1966). Subsequent studies with this method involved representatives of many different crustacean taxa (Goldstone and Cooke, 1971; Elofsson and Klemm, 1972; Aramant and Elofsson, 1976a, b; Elofsson et al., 1977; Myhrberg et al., 1979). In Figure 6.3 the distribution of biogenic monoamines in some crustaceans are shown. The distribution of neurons containing biogenic amines especially catecholamines (and related compounds, see below) can be generalized for the crustacean brain.

The optic neuropils in all species having a compound eye contain catecholaminergic neurons (Elofsson and Klemm, 1972). They vary in position, shape, and possibly also in their circuitry in different species. Some optic pathways terminate in the anterior margin of the protocerebrum in a prominent synaptic field. A more lateral and posterior area of the protocerebrum houses less tightly spaced catecholaminergic terminals from the eye neuropils. Within the protocerebrum, the central body contains catecholamines in all crustacean species so far studied. Catecholaminergic fibers are often found in the protocerebral bridge. In the class Ostracoda and Copepoda, the frontal eye center contains catecholamines. The prominent tract connecting the hemiellipsoidal body of the lateral protocerebrum with the olfactory lobes of the deutocerebrum (tractus olfactorioglobularis) and the marginal zones of the hemiellipsoidal body contains only in *Pacifastacus leniusculus* a few catecholaminergic fibers (Myhrberg et al., 1979). Outside the mentioned catecholaminergic areas of the protocerebrum, a sparse network of reactive fibers occur in the neuropil of the rest of the brain. One case of catecholaminergic sensory cells has been demonstrated with the Falck-Hillarp method. These are strongly fluorescent bipolar cells in the organ of Bellonci of the anostracan species, *Artemia salina* (Aramant and Elofsson, 1976a).

The varied organization of the deuto- and tritocerebrum of the brain in different crustacean groups is reflected in the catecholaminergic fiber pattern (Fig. 6.3). It often appears as a sparse, unstructured diffuse network. In the decapod species, with highly developed deutocerebrum comprising olfactory lobes, the catecholamine-containing neurons are found in the glomerular neuropil. In ostracod and copepod species, the deutocerebrum has catecholamine-containing neurons organized in distinct areas, although different from the decapod pattern.

The unstructured diffuse catecholaminergic network of the tritocerebrum can appear as delineated neuropils in those species where the tritocerebrum is well developed.

The Falck-Hillarp method has the drawback of not discriminating between different catecholamines and the precursor substance dihydroxyphenylalanine (DOPA), which all give fluorescent products upon treatment with gaseous formaldehyde. A biochemical investigation, using the sensitive high-pressure liquid chromatography (HPLC) method with electrochemical detection revealed that the brain of decapod crustaceans contained the catecholamines dopamin (DA) and noradrenaline (NA, norepinephrine) (Elofsson et al., 1982; Laxmyr, 1984). Further the precursor substance DOPA occurred in quantities similar to the biogenic amines. Thus, the catecholamine pattern within the crustacean brain referred to earlier is in fact composed of neurons containing DA and NA. DOPA-containing neurons may also be present or the substance can co-exist with the catecholamines.

The Falck-Hillarp method allows also the visualization of indolamines (indolylalkylamines), such as serotonin (5-hydroxytryptamine, 5-HT). The yellow fluorescence displayed by its fluorophore is, however, often masked by the stronger green fluorescence of the catecholamines. With antibodies raised against 5-HT, the distribution of serotonin immunoreactive neurons could be described for the crayfish (Elofsson, 1983). Generally, the same regions were innervated by serotonin-immunoreactive neurons as those innervated by the two catecholamines and DOPA with some small differences. Whereas in the crayfish indolyl- and catecholaminergic neurons occupy roughly the same areas, this is not always the case. In the cladoceran, *Daphnia magna*, and the copepod, *Cyclops strenuus*, (Fig. 6.3D,F) a separation occurred between yellow and green fluorescent areas, indicating different localization of indolyl- and catecholaminergic neurons (Aramant and Elofsson, 1976a).

6.3.2. Neuroactive Peptides

To date, relatively few investigations on peptide distribution have concerned crustaceans. Two independent groups demonstrated the crustacean hyperglycemic hormone in the eyestalks of *Carcinus moenas* and *Astacus leptodactylus,* respectively (Jaros and Keller, 1979; van Herp and van Buggenum, 1979; Keller, 1983). A number of cell bodies (the so-called medulla terminalis ganglionic X-organ, MTGX) situated on the anterior border of the lateral protocerebrum, their axons, and terminals in the sinus gland were immunoreactive. A diffuse plexus and neurons with terminals in the sinus gland, immunoreactive to substance P, were found in the lateral protocerebrum of *Panulirus interuptus.* In the same investigation, leuenkephalin-like immunoreactivity was reported in the retinular cells and in neurons passing from the lateral protocerebrum to the lobula (Mancillas et al., 1981). In the prawn, *Palaemon serratus* antibodies against neurophysin and (Arg)-vasopressin produced an immunoreaction in the distal lacunae of the organ of Bellonci (no cellular reaction) (van Herp and Bellon-Humbert, 1982). Again, in the same species an antiserum against FRMF-amide gave positive immunoreactions in neurons localized in the optic lobes of the eyestalk (Jacobs and van Herp, 1984).

Recently, 15 antisera were applied to the optic lobes of *Astacus leptodactylus* (van Deijnen et al., 1985); nine of these, anti-FMRF-amide, anti-α-MSH, antivasotocin, antigastrin, anti-CCK, antioxytocin, antisecretin, antiglucagon, and anti-GIP, gave positive immunocytochemical reactions. Different neuronal patterns were obtained, only antioxytocin lacking reaction in neuronal cell bodies. Four antibodies gave a positive reaction also in the sinus gland (α-MSH, vasotocin, CCK, and secretin). Also in the fiddler crab optic lobes some peptides were indicated by immunocytochemistry (Fingerman et al, 1985). Proctolin has been demonstrated biochemically in the brain and eyestalk of *Homarus americanus* (Schwarz et al., 1981) and recently, a detailed immunocytochemical mapping of proctolin-like immunoreactivity in the CNS of lobster and crayfish was published (Siwicki and Bishop, 1986).

6.4. IDENTIFIED NEURONS IN THE CRAYFISH BRAIN

An identified neuron is one that can be recognized by its cell body location, shape of processes, and axonal projection and often its chemistry, inputs, and electrical properties (Sandeman, 1969; Selverston and Kennedy, 1969). Neurons can be displayed by means of intracellular recording and dye injection, backfilling through cut axons or immunocytochemistry. Although the neuronal cell bodies in the CNS of crustacea are not actively involved in synaptic events, they can be found at specific sites with a high constancy (Selverston and Kennedy, 1969; Mellon et al., 1976). Hence, motorneurons and descending neurons, which can be filled from cut peripheral nerves or from the cut esophageal connectives, could serve well in studies of homologous or analogous neurons in different groups of crustaceans. Such studies, however, have not yet been undertaken on a broad comparative basis. Bethe (1897, 1898) and Retzius (1890) (e.g., Fig. 6.1B) have described large interneurons in crab and crayfish brains whose morphology makes them qualify as identifiable neurons. Indeed, some of these neurons have been reinvestigated with modern techniques. Chemically identified neurons, such as neurosecretory cells, although they may be quite useful in a comparative analysis of the brain (see e.g., Bliss et al., 1954; Chiang and Steel, 1985; van Deijnen et al., 1985), will not be discussed here.

6.4.1. Morphologically and Physiologically Identified Neurons

In crayfishes and crabs, several classes of neurons have been identified in the brain. Examples are four large motorneurons for rapid withdrawal of the eyestalks in crabs (Sandeman, 1982) and sets of oculomotor neurons, which are arranged in one group in crabs (Sandeman and Okajima, 1973), and three groups in the crayfish (Mellon et al., 1976). Some interneurons may have very similar morphologies in different species, such as one type of statocyst interneuron investigated in two crab species (Sandeman, 1982). Several central sensory interneurons have been characterized in the crayfish brain. Many of these are present as symmetrical pairs, which may be electrically coupled (Glantz and Kirk, 1980).

There is a large number of neurons with cell bodies in the brain and axonal processes descending down the esophageal connectives to the ventral ganglia. A cross section of a crayfish esophageal connective reveals 2300 axons (Nunnemacher et al., 1974). Most likely, only a fraction of these are descending neurons because backfilling of neurons from one cut esophageal connective in a lobster labeled only 300–350 cell bodies (Notwest and Page, 1981). These cell bodies are located in three clusters (one contralateral and two ipsilateral) in the dorsoanterior median region of the protocerebrum. Of these descending neurons, there are a few large ones (Fig. 6.4): the medial giant, the medial hemigiant, and the lateral hemigiant of the crayfish (Notwest and Page, 1981; Glantz et al., 1981). Also in other crustaceans, there are giant descending neurons (Sandeman, 1982). The medial giant of the crayfish is a multimodal neuron with an enormous axon diameter (ca., 200 μm) involved in the escape reflex (Wiersma, 1958; Glantz and Viancour, 1983). The lateral hemigiant has mechanosensory inputs from antennulary and antennary nerves, which corresponds well with the neuron's distribution of dendritic arbors (Glantz et al., 1981). The medial hemigiant (a defense reflex command neuron) receives bilateral mechanosensory inputs from antennulae, antennae, and maxillopods as well as visual input from motion-sensitive units (Glantz et al., 1981). Hence, this neuron receives monosynaptic bilateral inputs from all three neuromeres of the brain, proto-, deuto-,

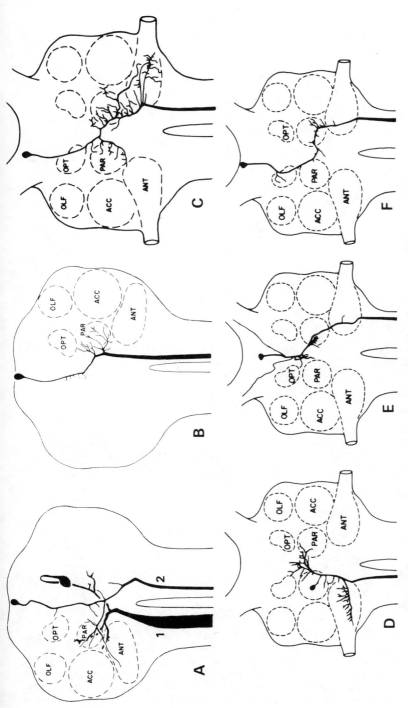

Figure 6.4. Identified descending neurons in the crayfish brain. After physiological recording the cells were intracellularly filled with the dye Lucifer yellow. Abbreviations as in Figure 6.1. (A) The medial giant neuron (1) and a lateral hemigiant neuron (2). (B) The medial hemigiant. (C) Neuron with bilateral monosynaptic inputs from antennular and antennary nerves. (D) Descending varicose neuron. (E) Bilateral optic efferent neuron. (F) Multimodal interneuron—1. (See text for physiology of these neurons. From Glantz et al., 1981, with permission. The median giant in A was drawn after Glantz and Viancour, 1983).

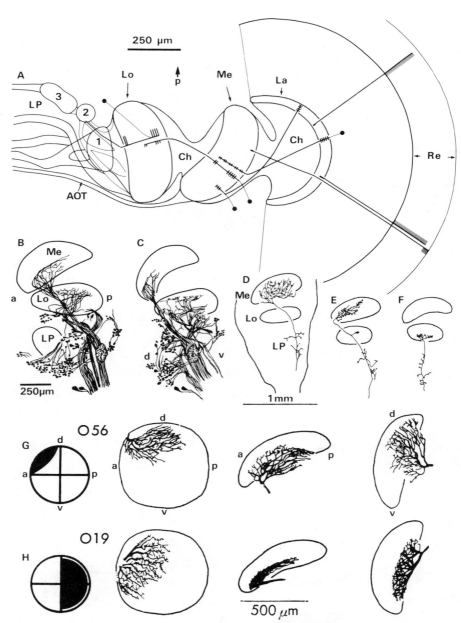

Figure 6.5. The optic lobes of the crayfish. (A) Horizontal view of the optic lobe neuropils in the eyestalk. Some centripetal neurons are depicted schematically. Abbreviations: Ch = chiasma; La = lamina; Lo = lobula; LP = lateral protocerebrum of eyestalk; Me = medulla; Re = receptor layer of compound eye; 1–3 = optic foci receiving inputs from lobula. Several tracts project from the lobula, medulla, and optic foci to other brain centers. One is termed anterior optic tract (AOT) (p = posterior). (B and C) Horizontal (B) and vertical (C) views of optic lobes after cobalt filling of optic nerve. The neurons filled

and tritocerebrum, as well as from the subesophageal ganglia. The distribution of dendrites, however, does not fully correlate with these inputs as far as primary sensory centers goes. Other multimodal descending neurons have been described from the crayfish brain, some of which receive inputs not only from cephalic sensory systems but also from thoracic and abdominal ones (Glantz et al., 1981). Because some of the identified descending neurons are relatively easy to impale and are parts of established reflex pathways, they are favorable probes for sensory integration at the cellular level. From the comparative point of view, little can be said about descending neurons yet. Some neurons of this type have, however, been described in *Daphnia* by Leder (1915).

6.5. THE OPTIC LOBES

No other studied brain neuropil region has an architecture that appears nearly as geometrical as that of the optic lobes. The mosaic pattern of the photoreceptors and optic elements of the compound eye is projected throughout the optic lobes in a columnar arrangement of neurons; the neurons sampling different points in visual space can be monitored at various levels of integration. Thus, the visual pathway offers a unique model system for studies of functional morphology.

Two principal architectures can be resolved in the optic lobes of most crustaceans and insects (Strausfeld and Nässel, 1980): a columnar and a layered organization. The columnar organization is derived from the spatial organization of optic elements and receptor cells in units called ommatidia. For each ommatidium, there is a column of neurons, a cartridge, in the first visual neuropil, the lamina. This columnar organization can be traced also to the second neuropil, the medulla, and (although coarser) in the third, the lobula (Strausfeld and Nässel, 1980). More centrally, the columnar neurons converge onto different sets of higher order neurons. The layered organization results from the different levels of terminations of neurons, from collaterals and dendrites, or from large tangentially arranged processes of neurons. Each layer contains different sets of synaptic or modulatory connections.

6.5.1. Neural Organization of the Optic Lobes

In decapods and other malacostracans the optic lobes consist of three layered and columnar neuropils: the lamina, medulla, and lobula, which are con-

represent connections between brain and optic lobes (a = anterior; p = posterior; d = dorsal; v = ventral). (D–F) Single dye-injected neurons seen in horizontal views. (D) Sustaining fiber (038). (E) Dimming fiber (083). (F) Movement fiber (07). (G and H) Visual fields and dendritic arborization in the medulla of two types of sustaining fibers (056 and 019). To the left are the visual fields. The next row shows the tangential views of the dendrites. The third row is the dorsal view. The fourth row shows anterior views of the dendrites. Similar findings have been presented by Kirk et al. (1982). (B–G are from Yamaguchi et al., 1984, with permission.)

nected by anterioposterior chiasmata (Fig. 6.5A). The optic lobes are often contained within movable eyestalks (Fig. 6.2). The lamina is the only crustacean optic lobe neuropil for which some connections have been analyzed by electron microscopy. With the Golgi method, the neuron types of the lamina have been catalogued in a number of decapods (Hafner, 1973; Nässel, 1975, 1977; Stowe et al., 1977) and some other crustaceans (Nässel et al., 1978; Strausfeld and Nässel, 1980; Elofsson and Hagberg, 1986). In the best known decapod, the crayfish, 10 neuron types can be distinguished in the lamina. Using morphological criteria, these can be grouped into: (1) five types of monopolar neurons, (2) two types of tangential neurons, (3) one type of amacrine neuron, (4) one type of centrifugal neuron, and (5) one type of small field T-neuron. Other decapods have neuron types similar to those in the crayfish, whereas the nondecapods have some types in common and differ in others.

The lamina synaptic neuropil of decapods, mysids, euphausiaceans, and the anostracan, *Artemia*, is bilayered. These layers are defined by the two levels of photoreceptor endings and the dendritic arbors of two types of monopolar neurons (Nässel, 1975, 1977; Nässel and Waterman, 1977). In the crayfish, Golgi electron microscopy has revealed some of the synaptic connections between photoreceptors and five of the monopolar neurons in each cartridge (Nässel and Waterman, 1977). The photoreceptors ending distally in the synaptic layer synapse onto a monopolar (M3) with branches in this layer only, and the proximally terminating receptors synapse onto the monopolar M4, branching proximally. This channeling of synaptic output was suggested to subserve analysis of polarized light within the ommatidium (Nässel and Waterman, 1977). It was later shown that the receptors ending distally in the lamina respond maximally to horizontally polarized light and the proximal ones to vertical e-vector (Sabra and Glantz, 1985), hence supporting the proposed model. Another circuit proposed from synaptic studies is signal amplification by means of convergent input from all seven receptor terminals of one ommatidium onto a single monopolar neuron (M2) (Nässel and Waterman, 1977). Further channeling occurs via the two monopolars M1a and M1b, which receive input from three or four receptors each from both layers of the lamina (Strausfeld and Nässel, 1980). By analogy with insects, some of the lamina circuits may subserve lateral inhibition and feedback loops for light adaptation (Strausfeld and Nässel, 1980). The single violet and seven green receptors of each ommatidium terminate in the medulla and lamina, respectively (Cummins and Goldsmith, 1981), indicating that processing of color information takes place in the medulla.

In other nondecapod crustaceans, there is also a layering of receptor terminals and monopolar dendrites. In the stomatopod, *Squilla mantis,* and the phyllopod, *Daphnia,* the lamina is trilayered. Other studied species have bilayered laminae. The function of the layering of these crustaceans remains obscure. The lamina of studied nonmalacostracans appear simpler in terms of the number of neuron types and their arborization patterns (Nässel et al., 1978).

In decapods, the lamina contain neural processes that may be catecholaminergic, peptidergic, or neurosecretory (Elofsson and Klemm, 1972; Elofsson et al., 1977; van Deijnen et al., 1985; Mancillas et al., 1981). No serotonergic neurons are found in the crayfish lamina (Elofsson, 1983) in contrast to all studied insects (Nässel et al., 1985; unpublished results). Some cells originating in the so-called X-organ appear to send processes to the lamina in the crayfish (Kirk et al., 1983) and isopods (Chiang and Steel, 1985). Some of these aminergic, peptidergic, and neurosecretory fibers may have a modulatory effect on the visual relay neurons of the lamina.

The medulla in decapods and other malacostracans is also clearly columnar and layered. In the crayfish, a large number of neuron types can be distinguished whose processes contribute to the complex architecture with seven main layers (Strausfeld and Nässel, 1980). Some of these neurons remain within a single column, others are columnar but with wide arborizations over several columns, and still others are noncolumnar with wide arborizations over large portions of the projected visual field (Fig. 6.5B,C). Hence, some neurons receive inputs from single lamina cartridges and others may relay larger portions of the visual field or mediate lateral interactions. A large number of medulla neurons connect to lamina and/or lobula neurons, others are amacrines within the medulla and some form direct connections with protocerebral centers or possibly contralateral optic lobes (Fig. 6.5A–E).

Two classes of medulla neurons that relay information directly to protocerebral centers in the brain are the so-called sustaining fibers and dimming fibers (Kirk et al., 1982; Yamaguchi et al., 1984) some of which are shown in Figure 6.5D–H. The sustaining fibers (SFs), originally described by Wiersma and Yamaguchi (1966), are visual neurons with tonic light-on-responses and large defined visual fields and have been found in several decapod species (Wiersma et al., 1982). Fourteen SFs can be distinguished in the crayfish, each receiving inputs from a characteristic portion of the visual field and with a corresponding dendritic field in the columnar array of the medulla (Fig. 6.5G,H) (Kirk et al., 1982; Yamaguchi et al., 1984). Combined, the SFs cover the entire visual field. The SFs project through the optic nerve in the eyestalk and terminate contralaterally in the anterodorsal portion of the medial protocerebrum (Yamaguchi et al., 1984). Here, they arborize close to the axons of the three pairs of rostral giant neurons (neurons with bilateral visual input, possibly from the SFs). The dimming fibers (Fig. 6.5D) have tonic off-responses to light and large spatially characteristic visual fields (Yamaguchi et al., 1984). They also terminate in the anterodorsal protocerebrum like the SFs. Recently, nonspiking amacrine neurons connected to the SFs were found and analyzed in the crayfish medulla (Waldrop and Glantz, 1985). Thus, part of the circuitry of the medulla begins to be unravelled, and it is clear that further anatomical analysis of this neuropil is called for.

The medulla of the studied nonmalacostracans appear simpler in their organization and for instance the pattern of catecholamine fluorescence is less complex (Aramant and Elofsson, 1976a). Only in *Daphnia* and *Artemia* have the types of neurons been analyzed to any extent (Retzius, 1909; Leder, 1915; Nässel et al., 1978).

The lobula has seven main layers in the crayfish (Strausfeld and Nässel, 1980). The neurons of the lobula do not form dendritic arbors restricted to columns the size of single medulla columns. Instead the lobula neurons that are spaced in a columnar way receive inputs from neurons of a larger number of medullary columns (Strausfeld and Nässel, 1980). Some columnar and noncolumnar neurons have been traced to their terminations in optic foci in the lateral protocerebrum of the eyestalk, others continue to protocerebral centers in the brain (Strausfeld and Nässel, 1980). Also, the lobula contains wide field elements (Fig. 6.5B,C,F). Some of these have been identified as motion-sensitive units (MFs) by Kirk et al. (1982) Yamaguchi et al. (1984) (Fig. 6.5F). The MFs are probably presynaptic to premotorneurons in tritocerebrum and are part of the visual pathway underlying the defense reflex (Glantz, 1977; Yamaguchi et al., 1984). It, thus, appears as if one of the roles of the lobula in the crayfish is similar to one proposed for insects: to process and relay motion information (reviewed by Strausfeld and Nässel, 1980). An interesting finding is that the crayfish lobula contains efferents (ascending through the esophageal connectives), which relay mechanosensory information from the body (Wiersma and Yamaguchi, 1966; Wang-Bennett and Glantz, 1985).

Several neuroactive substances have been tentatively identified in different layers of the medulla and lobula of the crayfish: catecholamines, serotonin, substance P, leuenkephalin (in spiny lobsters), FMRF-amide, and vasotocin (Elofsson and Klemm, 1972; Elofsson, 1983; Mancillas et al., 1981; van Deijnen et al., 1985). The medulla and lobula of other decapods and malacostracans have not been studied in any detail, except the distribution of catecholamines (Elofsson and Klemm, 1972).

The neuropil mass, classically termed medulla terminalis in decapods (lateral protocerebrum of the eye stalk), contains optic foci receiving optic lobe efferents and several other neuropil regions where optic lobe neurons send processes on their route to the brain proper (Fig. 6.5A–F). In addition, the corpus hemiellipsoidale (analogue to insect mushroom bodies) is found in the lateral protocerebrum of the eyestalk (Fig. 6.2).

6.6. SUMMARY

The crustacean brain has been best studied in decapod crustaceans, but histological data exist also on other groups, enabling us to discuss some comparative aspects. In general, the brain can be divided into three neuromeres: proto-, deuto-, and tritocerebrum. Each neuromere comprises several sensory-, motor-, and integrative centers supplied by peripheral nerves and central tracts. The presence and organization of these vary among crustaceans of different taxonomic groups. A comparative analysis of brain morphology reveals a fairly homogeneous malacostracan class, distinguished from the rest of the crustaceans, indicating an early separation of evolutionary lines. With histochemical and immunocytochemical techniques, neurons with different neuroactive substances have been mapped especially in decapods. The distribution of indolamine- and catecholamine-containing neurons can be compared in a rather wide range of crustaceans, and a generalized pattern can be described. Within the frame of this pattern, a specific substance can reside in homologous or nonhomologous neurons. In the latter case a similar function may be ascertained by means of analogous neurons. Physiological and morphological identification of neurons has been performed in the brain of crayfishes and crabs. They are components of defined pathways in the CNS, where much of the advances in crustacean brain research has been made. The optic lobes of crustaceans have received attention in studies of visual physiology. These lobes are neuroarchitectonical examples of precise geometries. The columnar and layered organization simplifies studies of single neurons whose visual fields can be related to dendritic arborizations and the retinotopic mosaic.

REFERENCES

Anderson, D. T. 1973. *Embryology and Phylogeny in Annelids and Arthropods.* Pergamon Press, Oxford.

Aramant, R. and R. Elofsson. 1976a. Distribution of monoaminergic neurons in the central nervous system of non-malacostracan crustaceans. *Cell Tissue Res.* **166:** 1–24.

Aramant, R. and R. Elofsson. 1976b. Monoaminergic neurons in the nervous system of crustaceans. *Cell Tissue Res.* **170:** 231–251.

Bethe, A. 1897,1898. Das Nervensystem von *Carcinus maenas.* I. and II. *Arch. Mikrosk. Anat.* **50:** 460–544, 590–640; **51:** 382–452.

Bliss, D. E., J. B. Durand, and J. H. Welsh. 1954. Neurosecretory systems in decapod Crustacea. *Z. Zellforsch.* **30:** 520–536.

Bowman, T. E. and L. G. Abele. 1982. Classification of the recent Crustacea, pp. 1–27. In D. E. Bliss (ed.), *The Biology of Crustacea.* vol. 1. L. G. Abele (ed.), *Systematics, the Fossil Record and Biogeography.* Academic Press, New York, London.

Bullock, T. H. and G. A. Horridge. 1965. *Structure and Function in the Nervous Systems of Invertebrates.* Freeman & Co, San Francisco, London.

Chaigneau, J. 1977. L'organe de Bellonci des Crustaces, mise au point sur l'ultrastructure et sur l'homologie des types avec et sans corps en oignon. *Ann. Sci. Nat. Zool. Paris* (12 Ser.) **19:** 401–438.

Chaigneau, J. 1983. Neurohemal organs in Crustacea, pp. 53–89. In A. P. Gupta (ed.), *Neurohemal Organs of Arthropods*. Charles C. Thomas, Springfield, Illinois.

Chiang, R. G. and C. G. H. Steel. 1984. Structural organization of neurosecretory cells terminating in the sinus gland of the terrestrial isopod, *Oniscus asellus*, revealed by paraldehyde fuchsin and cobalt backfilling. *Can. J. Zool.* **63:** 543–549.

Cummins, D. R. and T. H. Goldsmith. 1981. Cellular identification of the violet receptor in the crayfish eye. *J. Comp. Physiol.* **142:** 199–202.

Elofsson, R. 1966. The nauplius eye and frontal organs of the non-Malacostraca (Crustacea). *Sarsia* **25:** 1–128.

Elofsson, R. 1969a. The development of the compound eyes of *Penaeus duorarum* (Crustacea, Decapoda) with remarks on the nervous system. *Z. Zellforsch.* **97:** 323–350.

Elofsson, R. 1969b. The ultrastructure of the nauplius eye of *Sapphirina* (Crustacea, Copepoda). *Z. Zellforsch.* **100:** 376–401.

Elofsson, R. 1983. 5-HT-like immunoreactivity in the central nervous system of the crayfish *Pacifastacus leniusculus*. *Cell Tissue Res.* **232:** 221–236.

Elofsson, R. and M. Hagberg. 1986. Evolutionary aspects on the construction of the first optic neuropil (lamina) in Crustacea. *Zoomorphology* **106:** 174–178.

Elofsson, R. and N. Klemm. 1972. Monoamine-containing neurons in the optic ganglia of crustaceans and insects. *Z. Zellforsch.* **133:** 475–499.

Elofsson, R. D, Nässel, and H. Myhrberg. 1977. A catecholaminergic neuron connecting the first two optic neuropiles (lamina ganglionaris and medulla externa) of the crayfish *Pacifastacus leniusculus*. *Cell Tissue Res.* **182:** 287–297.

Elofsson, R., T. Kauri, S.-O. Nielsen, and J.-O. Strömberg. 1966. Localization of monoaminergic neurons in the central nervous system of *Astacus astacus* Linne (Crustacea). *Z. Zellforsch.* **74:** 464–473.

Elofsson, R., L. Laxmyr, E. Rosengren, and C. Hansson. 1982. Identification and quantitative measurements of biogenic amines and dopa in the central nervous system and haemolymph of the crayfish *Pacifastacus leniusculus* (Crustacea). *Comp. Biochem. Physiol.* **71C:** 195–201.

Falck, B. 1962. Observations on the possibilities of the cellular localization of monoamines by a fluorescence method. *Acta Physiol. Scand.* **197**(suppl): 1–25.

Fingerman, M., M. M. Hannumate, G. K. Kulkarni, R. Ikeda, and L. L. Vacca. 1985. Localization of substance P-like, leucin-enkephalin-like, methionin-enkephalin-like, and FMRFamide-like, immunoreactivity in the eye stalk of the fiddler crab, *Uca pugilator*. *Cell Tissue Res.* **241:** 473–477.

Glantz, R. M. 1977. Visual input and motor output of command interneurons of the defense reflex pathway in the crayfish, pp. 259–274. In G. Hoyle (ed.), *Identified Neurons and Behaviour of Arthropods*. Plenum, New York.

Glantz, R. M. and M. D. Kirk. 1980. Intercellular dye migration and electrotonic coupling within neuronal networks of the crayfish brain. *J. Comp. Physiol.* **140:** 121–133.

Glantz, R. M. and T. Viancour. 1983. Integrative properties of crayfish medial giant neuron: Steady-state model. *J. Neurophys.* **50**(5): 1122–1142.

Glantz, R. M., M. Kirk, and T. Viancour. 1981. Interneurons of the crayfish brain:

The relationship between dendrite location and afferent input. *J. Neurobiol.* **12**(4): 311–328.

Goldstone, M. W. and I. M. Cooke. 1971. Histochemical localization of monoamines in the crab central nervous system. *Z. Zellforsch.* **116**: 7–19.

Gwilliam, G. F. and E. S. Cole. 1979. The morphology of the central nervous system of the barnacle *Semibalanus cariosus* (Pallas). *J. Morphol.* **159**: 297–310.

Hafner, G. S. 1973. The neural organization of the lamina ganglionaris in the crayfish: A Golgi and E.M. study. *J. Comp. Neurol.* **152**: 255–280.

Hanström, B. 1925. The olfactory centers in crustaceans. *J. Comp. Neurol.* **38**: 221–250.

Hanström, B. 1947. The brain, the sense organs, and the incretory organs of the head in the Crustacea Malacostraca. *Kungl. Fysiogr. Sällsk. Handl.* **58**(9): 1–45.

Helm, F. 1928. Vergleichend-anatomische Untersuchungen uber das Gehirn, insbesondere das "Antennalganglion" der Dekapoden. *Z. Morphol. Ökol. Tiere.* **12**: 70–134.

Jacobs, A. A. C. and F. van Herp. 1984. Immunocytochemical localization of a substance in the eyestalk of the prawn *Palaemon serratus* reactive with an anti-FMRF-amide rabbit serum. *Cell Tissue Res.* **235**: 601–605.

Jaros, P. P. and R. Keller. 1979. Immunocytochemical identification of hyperglycemic hormone-producing cells in the eyestalk of *Carcinus moenas*. *Cell Tissue Res.* **204**: 379–385.

Keller, R. 1983. Biochemistry and specificity of the neurohemal hormones in Crustacea, pp. 118–148. In A. P. Gupta (ed.), *Neurohemal Organs Of Arthropoda*. Charles C. Thomas, Springfield, Illinois.

Kirk, M. D., J. I. Prugh, and R. M. Glantz. 1983. Retinal illumination produces synaptic inhibition of a neurosecretory organ in the crayfish, *Pacifastacus leniusculus* (Dana). *J. Neurobiol.* **14**(6): 473–480.

Kirk, M. D., B. Waldrop, and R. M. Glantz. 1982. The crayfish sustaining fibers. I. Morphological representation of visual receptive fields in the second optic neuropil. *J. Comp. Physiol.* **146**: 175–179.

Laxmyr, L. 1984. Biogenic amines and dopa in the central nervous system of decapod crustaceans. *Comp. Biochem. Physiol.* **77C**: 139–143.

Leder, H. 1915. Untersuchungen über den feineren Bau des Nervensystems der Cladoceren. *Arb. Zool. Inst. Univ. Wien* **20**: 297–392.

Mancillas, J. R., J. F. McGinty, A. J. Selverston, H. Karten, and E. E. Bloom. 1981. Immunocytochemical localization of enkephalin and substance P in retina and eyestalk neurones of lobster. *Nature (London)* **293**: 576–578.

Maynard, D. M. 1962. Organization of neuropil. *Am. Zool.* **2**: 79–96.

Mellon, D., R. H. Tufty, and E. D. Lorton. 1976. Analysis of spatial constancy of oculomotor neurons in the crayfish. *Brain Res.* **109**: 587–596.

Myhrberg, H. E., R. Elofsson, R. Aramant, N. Klemm, and L. Laxmyr. 1979. Selective uptake of exogenous catecholamines into nerve fibres in crustaceans. A fluorescence histochemimcal investigation. *Comp. Biochem. Physiol.* **62C**: 141–150.

Nässel, D. R. 1975. The organisation of the lamina ganglionaris of the prawn, *Pandalus borealis* (Kroyer). *Cell Tissue Res.* **163**: 445–464.

Nässel, D. R. 1976. The retina, and retinal projection on the lamina ganglionaris of the crayfish *Pacifastacus leniusculus* (Dana). *J. Comp. Neurol.* **167**: 341–360.

Nässel, D. R. 1977. Types and arrangements of neurons in the crayfish optic lamina. *Cell Tissue Res.* **179**: 45–75.

Nässel, D. R. and T. H. Waterman. 1977. Golgi EM evidence for visual information channeling in the crayfish lamina ganglionaris. *Brain Res.* **130**: 556–563.

Nässel, D. R., R. Elofsson, and R. Odselius. 1978. Neuronal connectivity patterns in the compound eyes of *Artemia salina* and *Daphnia magna* (Crustacea: Branchiopoda). *Cell Tissue Res.* **190**: 435–457.

Nässel, D. R., E. P. Meyer, and N. Klemm. 1985. Mapping and ultrastructure of serotonin-immunoreactive neurons in the optic lobes of three insect species. *J. Comp. Neurol.* **232**: 190–204.

Notwest, R. R. and C. H. Page. 1981. Anatomical organization of neurons descending from the supraoesophageal ganglion of the lobster. *Brain Res.* **217**: 162–168.

Nunnemacher, R. F., G. Camougis, and J. H. McAlear. 1974. The fine structure of the crayfish nervous system, pp. n11–12. In S. S. Breese (ed.), *5th International Congress on Electron Microscopy* Academic Press, New York.

Retzius, G. 1890. Zur Kenntnis des Nervensystems der Crustaceen. *Biol. Untersuchungen N.F.* **1**: 1–50.

Retzius, G. 1909. Zur Kenntnis des Nervensystems der Daphniden. *Biol. Untersuchungen N. F.* **13**: 107–116.

Sabra, R. and R. M. Glantz. 1985. Polarization sensitivity of crayfish photoreceptors is correlated with their termination sites in the lamina ganglionaris. *J. Comp. Physiol.* **156**: 1–4.

Sandeman, D. C. 1969. Integrative properties of a reflex motoneuron in the brain of the crab *Carcinus maenas*. *Z. Vgl. Physiol.* **64**: 450–464.

Sandeman, D. C. 1982. Organization of the central nervous system, pp. 1–61. In D. E. Bliss (ed.), *The Biology of Crustacea,* vol. 3. H. L. Atwood and D. C. Sandeman (eds.), *Neurobiology: Structure and Function.* Academic Press, New York.

Sandeman, D. C. and S. E. Luff. 1973. The structural organization of glomerular neuropile in the olfactory and accessory lobes of an Australian freshwater crayfish, *Cherax destructor*. *Z. Zellforsch.* **142**: 37–61.

Sandeman, D. C. and A. Okajima. 1973. Statocyst induced eye movements in the crab *Scylla serrata*. III. The anatomical projections of sensory and motor neurons and the responses of the motor neurons. *J. Exp. Biol.* **59**: 17–38.

Schwarz, T. L., G. Lee, and E. A. Krawitz. 1981. Proctocolin-like immunoreactivity in the nervous system of the lobster *Homarus americanus*. *Soc. Neurosci. Abstr.* **7**: 253.

Seabrook, W. D. and H. H. J. Nesbitt. 1966. The morphology and structure of the brain of *Orconectes virilis* (Hagen) (Crustacea, Decapoda). *Can. J. Zool.* **44**: 1–22.

Selverston, A. I. and D. Kennedy. 1969. Structure and function of identified nerve cells in the crayfish. *Endeavour* **28**: 107–113.

Siwicki, K. K. and C. A. Bishop. 1986. Mapping of proctolinlike immunoreactivity in the nervous system of lobster and crayfish. *J. Comp. Neurol.* **243**: 435–453.

Steele, V. J. 1984. Morphology and ultrastructure of the organ of Bellonci in the marine amphipod *Gammarus setosus*. *J. Morphol.* **181:** 97–131.

Stowe, S., W. A. Ribi, and D. C. Sandeman. 1977. The organization of the lamina ganglionaris of the crabs *Scylla serrata* and *Leptograpsus variegatus*. *Cell Tissue Res.* **178:** 517–532.

Strausfeld, N. J. and D. R. Nässel. 1980. Neuroarchitecture of brain regions that subserve the compound eyes of Crustacea and Insects, pp. 1–132. In H. Autrum (ed.), *Handbook of Sensory Physiology*. vol. VII/6B: *Vision in Invertebrates*. Springer-Verlag, Berlin, Heidelberg, New York.

Tsvileneva, V. A. and V. A. Titova. 1985. On the brain structures of decapods. *Zool. Jahrb. Anat.* **113:** 217–266.

van Deijnen, J. E., F. Vek, and F. van Herp. 1985. An immunocytochemical study of the optic ganglia of the crayfish *Astacus leptodactylus* (Nordmann 1842) with antisera against biologically active peptides of vertebrates and invertebrates. *Cell Tissue Res.* **240:** 175–183.

van Herp, F. and C. Bellon-Humbert. 1982. Localization immunocytochimique de substances apparentees a la neurophysine et a la vasopressine dans le pedoncule ocularire de *Palaemon serratus* Pennant (Crustace Decapoda Natantia). *C. R. Acad. Sci. Paris* **295:** 97–102.

van Herp, F. and H. J. M. van Buggenum. 1979. Immunocytochemical localization of hyperglycemic hormone (HGH) in the neurosecretory system of the eyestalk of the crayfish *Astacus leptodactylus*. *Experientia* **35:** 1527–1528.

Waldrop, B. and R. Glantz. 1985. Nonspiking local interneurons mediate surround inhibition of crayfish sustaining fibers. *J. Comp. Physiol.* **156:** 763–774.

Wang-Bennett, L. T. and R. M. Glantz. 1985. Presynaptic inhibition in the crayfish brain. II. Morphology and ultrastructure of the terminal arborization. *J. Comp. Physiol.* **156:** 605–617.

Wiersma, C. A. G. 1958. On the functional connections of single units in the central nervous system of the crayfish *Procambarus clarkii* Girard. *J. Comp. Neurol.* **110:** 421–471.

Wiersma, C. A. G. and T. Yamaguchi. 1966. The neuronal components of the optic nerve of the crayfish as studied by single-unit analysis. *J. Comp. Neurol.* **128:** 333–358.

Wiersma, C. A. G., J. L. M. Roach, and R. M. Glantz. 1982. Neural integration in the optic system. In H. L. Atwood and D. C. Sandeman (eds.), *Biology of Crustacea*. Academic Press, New York.

Yamaguchi, T., Y. Okada, K. Nakatani, and N. Ohta. 1984. Functional morphology of visual interneurons in the crayfish central nervous system, pp. 109–122. In K. Aoki (ed.), *Animal Behaviour: Neurophysiological and Ethological Approaches*. Japan Sci. Soc. Press, Tokyo/Springer-Verlag, Berlin.

Yoshino, M., Y. Kondoh, and M. Hisada. 1983. Projection of sensory neurons associated with crescent hairs in the crayfish *Procambarus clarkii* Girard. *Cell Tissue Res.* **230:** 37–48.

CHAPTER 7

Histology and Ultrastructure of the Myriapod Brain

Robert Joly
Michel Descamps
Laboratory of Invertebrate Endocrinology
University of Science and Technology of Lille
Villeneuve d'Ascq, France

7.1. INTRODUCTION

As in other Arthropoda, the brain of the Myriapoda is composed of three pairs of fused ganglia: the protocerebrum, the deutocerebrum, and the tritocerebrum. Three more or less concentric regions can be recognized in the brain. The internal parts, the cerebral cortex and the medullar area (the neuropil), are wrapped in the neurilemma. We will describe the gross anatomy, the histology, and the fine structure of the brain in the four major groups of Myriapoda: Pauropoda, Symphyla, Diplopoda, and Chilopoda.

7.2. PAUROPODA

All the data concerning the Pauropoda are from Tiegs (1947).

7.2.1. Gross Anatomy

In a dorsal view, the brain of Pauropoda is roughly triangular (Fig. 7.1A). The protocerebrum is relatively large and forms the more posterior part of the brain. Displaced backwards up to the second truncal segment, the protocerebrum is composed of three separate lobes, the *pars intercerebralis* being the more posterior. The deutocerebrum is narrower than the protocerebrum and lies anterior to the latter. A median group of nerve cells constitutes a septum (septal ganglion) between the two ganglia. It is possible that these neurons represent the preantennary ganglion (Tiegs, 1947). The tritocerebrum lies by the side of the esophagus and merges directly into the subesophageal ganglion. So, there are no true circumesophageal connectives (Fig. 7.1B), and the tritocerebrum is in fact composed of two separate lobes, fused only at the anterior part of the brain.

7.2.2. Histology and Fine Structure

Only incomplete observations are available on the internal structure of the brain in Pauropoda. Indeed, these animals are very small (about 2 mm), and

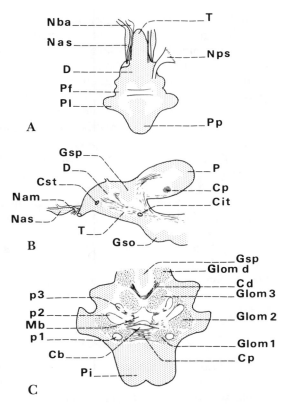

Figure 7.1. The brain of Pauropoda (modified from Tiegs, 1947). (A) Dorsal view. (B) Longitudinal section. (C) Horizontal section. Cb = central body; Cd = deutocerebral commissure; Cit = inferior tritocerebral commissure; Cp = principal commissure; Cst = superior tritocerebral commissure; D = deutocerebrum; Glom d = deutocerebral glomerulus; Glom 1,2,3 = glomeruli 1,2,3; Gso = subesophageal ganglion; Gsp = septal ganglion; Mb = median body; Nam = antennal motor nerve; Nas = antennal sensory nerve; Nba = nerve of the basal antennal sense organ; Nps = pseudocular nerve; p 1,2,3 = peduncles 1,2,3; P = protocerebrum; Pf = protocerebral frontal lobe; Pi = *Pars intercerebralis*; Pl = protocerebral lateral lobe; Pp = protocerebral posterior lobe; T = tritocerebrum. (*Less-densely dotted areas:* areas with cortical nerve cells.)

it is difficult to perform accurate observations. Particularly, there are no data available for the neurilemma.

7.2.2.1. Cerebral Cortex. Much of the dorsal and ventral surface of the protocerebrum is free from nerve cells, the perikarya being located as a thick cortex mainly on the lateral and posterior wall of the protocerebrum. However, dorsally, there is a median cluster of nerve cells (Fig. 7.1A). No protocerebral globuli cells can be observed; thus, the brain of *Pauropus* is similar to those of Symphyla (Holmgren, 1916) and of the chilopod *Geophilus* (Fahlander, 1938) (Sections 7.3.2.3. and 7.5.2.3.).

7.2.2.2. Neuropil. In the protocerebrum of *Pauropus* (Fig. 7.1C), three pairs of compact tracts represent the stalks of the pedunculate bodies, but no precise limits can be determined. These fibrous tracts are associated with a median mass of neuropil, the median body (median glomerulus). Three pairs of masses of intertwined nerve fibers (glomeruli or presumed glomeruli) can also be recognized. Immediately at the rear of the median body, is the central body, which is small and transversaly elongated. There are no ganglion cells associated with it, as in other Myriapoda. Several commissures are distinguishable in the protocerebrum, especially the principal commissure, lying partly behind partly below the median body. The myriapodal commissure (characteristic commissure) was not recognized by Tiegs (1947), but it is possible that it is a part of the principal commissure, although no deutocerebral origin was described. Nevertheless, one of the four other protocerebral commissures is connected with the deutocerebrum and could be the myriapodal commissure.

The most interesting feature concerning other parts of the brain is the presence of two tritocerebral commissures. The inferior commissure is subesophageal, but is not completely free, as in other Myriapoda. The superior tritocerebral commissure is in a preoral position and corresponds to the stomatogastric bridge.

7.2.3. Brain Nerves

The nerves of the stomatogastric system, that is, those issuing from the tritocerebrum and supplying the oral and preoral areas or the stomodeum, as defined by Wigglesworth (1951), are not included here. The following main nerves can be observed (Fig. 7.1B): (1) a pair of large nerves from the pseudoculi, these nerves enter the frontal lobes of the protocerebrum; (2) a pair of large antennary nerves issuing from the deutocerebrum, these nerves, swollen with a thick cortex of nerve cells just before entering the antennae, are mainly, and possibly exclusively, sensory; (3) a pair of presumably motor nerves, arising from the deutocerebrum, just below the antennary nerves; and (4) a pair of thin nerves, ending in the basal antennary sense organ.

7.3. SYMPHYLA

Holmgren (1916) provided the first good description of the myriapod brain in the genus *Scolopendrella* (but some authors think that it was in fact *Scutigerella*). Similar results were obtained in *Hanseniella agilis* (Tiegs, 1940) and in *Scutigerella pagesi* (Juberthie-Jupeau, 1963).

7.3.1. Gross Anatomy

The protocerebrum is the largest part of the brain and covers dorsally the deutocerebrum and the tritocerebrum (Fig. 7.2A). From the front to the rear

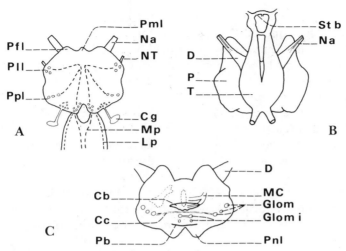

Figure 7.2. The brain of Symphyla. (A) *Scutigerella pagesi:* dorsal view. Cg = cerebral gland; Lp = laterodorsal pathway of deutocerebral neurosecretory products; Mp = median pathway of protocerebral neurosecretory products; Na = antennal nerve; NT = Tömösvary organ nerve; Pfl = protocerebral frontal lobe; Pll = protocerebral lateral lobe; Pml = protocerebral median lobe; Ppl = protocerebral posterior lobe. (B) *Hanseniella agilis:* ventral view. D = deutocerebrum; Na = antennal nerves; P = protocerebrum; St b = stomatogastric bridge; T = tritocerebrum. (C) *Scolopendrella:* protocerebral histology. Cb = central body; Cc = cerebral commissure; Glom = lateral glomeruli; Glom i = internal glomeruli; MC = myriapodal commissure (only drawn in the right side); Pb = protocerebral bridge; Pnl = protocerebral nuchal lobe. Inside the areas defined by the *dashed lines,* the cerebral cortex presents no nerve cells. (A modified from Juberthie-Jupeau, 1963; B modified from Tiegs, 1940; C modified from Holmgren, 1916.)

are first, an unpaired median lobe and then, paired lobes, respectively, the frontal, lateral, posterior, and nuchal lobes; in *H. agilis* (Fig. 7.2B), the nuchal lobes are not well defined. In all Symphyla, the lack of optic lobes must be noted. The deutocerebrum, in ventral position, is composed of two lobes, enlarged laterally and lying by the sides of the esophagus. The tritocerebrum, under the esophagus, is unpaired posteriously and divided anteriorly in two nervous masses that are linked to the deutocerebral lobes. A stomatogastric bridge (without frontal ganglion) joins the anterior parts of the tritocerebrum.

7.3.2. Histology and Fine Structure

7.3.2.1. Neurilemma. It is composed of a thin layer of rather similar small cells, with small lens-shaped nuclei.

7.3.2.2. Cerebral Cortex. Very thick (several cellular layers) in the lateral, posterior, and nuchal lobes of the protocerebrum, the cerebral cortex is thin (only one layer of cells) on the ventral surface. Moreover, some areas of the ventral surface and the dorsal surface of the lateral lobes are free from nerve cells. In the deutocerebrum, a thick cortex is present in the latero-internal-anterior and latero-external-posterior parts. In the tritocerebrum, nerve cells are essentially ventrally located, but some perikarya can also be found in the anterolateral walls.

The nerve cells are generally of the same small type (6–8 μm). In Symphyla, the globuli (clusters of small cortical nerve cells) seem to be absent; nevertheless, in *S. pagesi,* lying by the Tömösvary nerves, two groups of small and dense nucleated cells might represent the globuli (Juberthie-Jupeau, 1963).

The neurosecretory cells (NSCs), studied by Juberthie-Jupeau (1963, 1983), are essentially located in the protocerebrum (Fig. 7.2A) and in the posterodorsal and external regions of the deutocerebral lobes.

7.3.2.3. Neuropil. Some differentiated regions are composed of intertwined axons; they constitute the glomeruli (Fig. 7.2C). In *Scolopendrella* (Holmgren, 1916) and in *Scutigerella pagesi* (Juberthie-Jupeau, 1963) these are: (1) protocerebral lateral glomeruli, extended by a transverse commissure located over the central body, a second commissure, just under the previous one, has uncertain connection with the glomeruli; (2) small juxtaposed glomeruli in the nuchal lobes constitute the protocerebral bridge, under the latter are located two glomeruli; (3) a very large central body (the most developed central body in the Myriapoda) located in front of the protocerebral bridge; and (4) various fibrous tracts, mainly the myriapodal commissure, which is located over the central body, originates from the deutocerebrum, and is connected to two others.

The axonal pathways of NSCs can be observed in the neuropil. Axons issuing from the NSCs of lateral and posterior lobes are gathered in two dorsal and submedian pathways in the ventral nerve cord (Fig.7.2A). The axons of the external NSCs of the nuchal lobe end in the cephalic glands (neurohemal organs). The dorsolateral pathway of the ventral nerve cord is constituted by the axons of the deutocerebral NSCs.

7.3.3. Brain Nerves

Symphyla, being blind are devoid of optic nerves (Fig. 7.2B). Issuing from the protocerebral median lobe is a nerve ending in a small organ composed of about 15 cells. From each lateral lobe arises a thick and short nerve, the Tömösvary nerve (Fig. 7.2A). Two superposed nerves are issued from each antennal lobe. Only few data are available concerning the nerves arising from the tritocerebrum; on the other hand, most of them are stomatogastric nerves.

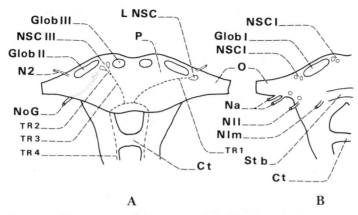

Figure 7.3. The brain of *Tachypodoiulus albipes* (Diplopoda) (modified from Sahli, 1966 in Juberthie-Jupeau, 1983). (A) Posterior view. (B) Anterior view. Ct = tritocerebral commissure; GlobI,II,III = globuli I,II,III; L NSC = lateral neurosecretory cells; N2 = protocerebral nerve no 2; Na = antennal nerves (N3 and N4); NII = *nervus labri lateralis*; NIm = *nervus labri medialis*; NoG = nerve of the Gabe organ; NSC I,III = neurosecretory cells located near the globuli I and III; O = optic lobe; P = protocerebrum; St b = stomatogastric bridge; TR 1,2,3,4 = neurosecretory pathways.

7.4. DIPLOPODA

The first accurate descriptions of the brain of the Julidae, one of the more studied family among the Diplopoda, were presented by Saint-Rémy (1890) and Holmgren (1916). More recently, the study of the Diplopoda brain was completed by, among other authors, Gabe (1954) and Sahli (1966).

7.4.1. Gross Anatomy

In the Julidae, the protocerebrum is divided into two halves by a median groove and shows two frontal lobes extended laterally by well-developed optic lobes (Fig. 7.3A,B). The median posterodorsal region is poorly differentiated and corresponds to the *pars intercerebralis* of insects. The deutocerebrum, composed of two antennal lobes, merges directly in the tritocerebrum. The two tritocerebral lobes are linked by two horizontal commissures: (1) the stomatogastric (stomodeal) bridge, in supraesophageal position; and (2) the transverse (tritocerebral) commissure, in subesophageal position (Fig. 7.3A). It must be noted that the different parts of the brain are not separated by precise limits. The anatomy of the brain varies within the Diplopoda. In the Oniscomorpha (e.g., *Glomeris,* Seifert, 1966), the protocerebral lobes are so much separated that the protocerebral commissure is narrower than the stomatogastric bridge. In the Proterospermophora, numerous species are blind (e.g., Polydesmidae and a Strongylosomidae:

Strongylosoma pallipes, Seifert, 1932); in these species, the protocerebral lobes do not show optical lobes, of course, and so, are not laterally extended. In the Nematophora and in the Penicillata *Polyxenus lagurus* (Nguyen Duy-Jacquemin, 1973), the brain has a structure comparable to that of Julidae.

7.4.2. Histology and Fine Structure

The cerebral structure of Diplopoda, first studied by Saint-Rémy (1890) and Holmgren (1916), was completed, among other authors, by Hörberg (1931) and Sahli (1966; Juberthie-Jupeau, 1983).

7.4.2.1. Neurilemma. The neurilemma is rather thin, particularly in the dorsal region, and shows two parts. The more external part, of noncellular nature, is called neural lamella; the internal part is the perilemma. The latter seems to be constituted of several cellular types, one of which is comparable to neuroglial cells with modified cytoplasmic processes.

7.4.2.2. Cerebral Cortex. Three main cellular types can be found: motor neurons, associative neurons, and neuroglia cells. There are also NSCs in the cerebral cortex. The neurons are unipolar and comparable to those described in the brain of other Arthropoda; however, multipolar neurons are possibly present in the cortex in the posteromedian part of the protocerebrum.

Some cortical regions are differentiated and show clusters of small unipolar chromophilic neurons, the globuli, with reduced cytoplasm. The globuli and the fibers issuing from them constitute the *corpora pedunculata,* but, for convenience, we will call them globuli in this chapter. In Diplopoda, and more generally in Myriapoda, the globuli are located in the protocerebrum. In *Tachypodoiulus albipes* (Julidae), three pairs of globuli are observed: the globuli I, II, III (Fig. 7.3A,B; Fig. 7.4A). The globuli I (*masses ganglionnaires antérieures*; Saint-Rémy, 1890) are located in the anterolateral area of the frontal lobes and are divided into external and internal parts (Fig. 7.4C). The globuli II and III (respectively, *masses ganglionnaires externes* and *masses ganglionnaires internes,* according to Saint-Rémy, 1890) are found in lateral and mediodorsal positions in the posterior area of the frontal lobes (Fig. 7.3A,B). In some cases, the globuli are not easy to demonstrate. In *Polyxenus lagurus* (Penicillata), all the nuclei of the cortical nerve cells are similar, and it is not possible to find the globuli (Nguyen Duy-Jacquemin, 1973). In the Pratinidae (*Jonespeltis splendidus*), the globuli are adjacent to each other; this arrangement and the fact that the axons linking the globuli to the neuropil are very short, lead to a reduced protocerebral volume (Prabhu, 1962).

NSCs are present in the three pairs of brain ganglia. In the Julidae (*T. albipes*) (Sahli, 1966), three groups of protocerebral NSCs can be demonstrated (Fig. 7.3A,B). NSCs seem to be absent in the *pars intercerebralis*

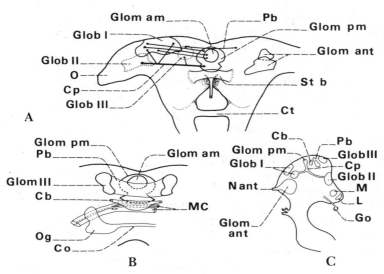

Figure 7.4. Histology of the brain in *Julus* (Diplopoda). (A) Anterior view. (B) Detailed structure of the protocerebral median part. (C) Lateral view. Cb = central body; Co = optical commissure; Cp = principal commissure; Ct = tritocerebral commissure; Glob I,II,III = globuli I,II,III; Glom am = anteromedian glomerulus; Glom ant = antennal glomeruli; Glom pm = posteromedian glomerulus; Glom III = glomerulus III; G o = Gabe organ; L = *lamina*; M = *medulla*; M C = myriapodal commissure; N ant = antennal nerve; O = optic lobe; O g = optic ganglia; P b = protocerebral bridge; St b = stomatogastric bridge. (A and B modified from Holmgren, 1916; C modified from Hörberg, 1931.)

area. Deutocerebral and tritocerebral NSCs are located at the sites of origin of the antennal sensory nerve and the *nervus labri lateralis*. The NSCs of the tritocerebral commissure are located at the junction of this commissure and the periesophageal collar. NSCs, in the same location as in *T. albipes,* are generally described in other Diplopoda. *Pars intercerebralis* NSCs are present in Polydesmidae, Pratinidae, and Penicillata. Nevertheless, in the Pratinidae (*Jonespeltis splendidus*), Prabhu's (1962) account does not agree with Sahli's (1966); according to the latter author, the *pars intercerebralis* NSCs would be in fact migrating blood cells.

In Julidae, Glomeridae, and Penicillata, special structures called neural organs (intracerebral organs) are present in larvae and adults. They are constituted of an external cellular area, incorporated in globulus I, and an internal area. The size (from 10 to 15 μm) and the structure of the internal area vary according to the genus studied. The central area can be occupied by a secretory vacuole (in Julidae; Sahli, 1966) or by concentric chitinous whorls surrounding a central mass (in Glomeridae, Juberthie-Jupeau, 1967; and in *Polyxenus lagurus*, Nguyen Duy-Jacquemin, 1973). The role of these organs remains unknown.

7.4.2.3. Neuropil. The neuropil is constituted both by axons without collateral fibers or axonal endings and by axons with dendrites and/or synapses. The glomeruli are spherical masses of intertwined neurons; their structure is more compact than those of nondifferentiated neuropil. According to Holmgren (1916), the axons issuing from globuli I constitute six or seven tracts ending in a posteromedian or an anteromedian glomerulus (Fig. 7.4A,C). The fibers issuing from globuli II end in one of the last glomeruli; those ending in the posteromedian glomerulus, gathered with some axons issuing from globuli I and III, constitute the principal commissure (Fig. 7.4A,C). In contrast to Chilopoda, in Diplopoda, the glomerular structures are not associated with the principal commissure. Antero- and posteromedian glomeruli are connected with the entire protocerebrum (Fig. 7.4A). Between the median glomeruli is a tract of fibers, the central body; less developed in Diplopoda than in Symphyla, the central body is linked primarily to the *pars intercerebralis* and secondarily to the median glomeruli. Some fibers, coming from globuli III or from the glomeruli associated with these globuli, link the lateral areas of the protocerebrum and constitute the protocerebral bridge. The myriapodal commissure, constituted of fiber tracts issuing from the deutocerebrum, runs partly in front and partly behind the central body (Fig. 7.4B). Generally, the optic lobes show two differentiated areas of neuropil: the *medulla* and the *lamina*; it must be mentioned that in Insecta and in numerous Crustacea, the optic ganglia are constituted by three differentiated areas, the *medulla* being divided into two parts.

The deutocerebrum is characterized by numerous small glomeruli, endings of antennal sensory nerves, and by motor centers linked by a commissure. The deutocerebral lobes are also linked to the protocerebral globuli by a conspicuous fibrous tract; it must be noted that in some Crustacea and Insecta, the protocerebral globuli can also be linked to the antennal ganglia. In contrast to other Arthropoda, there is no decussation in Myriapoda (Diplopoda and others).

The pathways of the neurosecretory cells were studied in *T. albipes* (Sahli, 1966; Sahli and Petit, 1975) (Fig. 7.3A). Axons of globuli I NSCs constitute the nerve of the Gabe organs (cephalic neurohemal organs). Some other pathways can be observed: (1) near the lateral NSCs of the frontal lobes (TR 1); (2) near the globuli III NSCs (TR 2); and (3) in the posterior area of the protocerebrum (TR 3). In the *pars intercerebralis* neuropil, despite the absence of NSCs, secretory pathways can be found. As for the deutocerebral and tritocerebral NSCs, their axons, together with those of the protocerebral NSCs, constitute the pathways called TR4. Such pathways are also present in other Diplopoda.

7.4.3. Brain Nerves

The more detailed studies were performed in Julidae. Saint-Rémy (1890), Holmgren (1916), and Fechter (1961) described six pairs of brain nerves; in

T. albipes (Sahli, 1966) (Fig. 7.3A,B) 10 pairs can be found. Stomatogastric nerves are not studied here.

According to Sahli (1966), three pairs of nerves originate from the protocerebrum. The first two pairs are sensory: N 1 are ocellar nerves and N 2 end laterally in the epidermis; the third pair is constituted by the nerves of Gabe organs. N 3 (antennal motor nerves) and N 4 (antennal sensory nerves) spring from the deutocerebrum; two thin nerves, N 5 and N 6, innervating essentially the antennal muscles, run along the main antennal nerves. N 7 ends in the cephalic anteroventral area. Variations are observed in the same species, in terms of the nerve or the animal studied.

7.5. CHILOPODA

7.5.1. Gross Anatomy

First studied by Saint-Rémy (1890) and Haller (1905), the brain was described by Holmgren (1916) in *Lithobius forficatus*. More recent works have completed this study. The protocerebrum (Fig. 7.5A), laterally extended by the optic lobes, is the more developed and more dorsal part of the brain; the posterodorsal area corresponds to the *pars intercerebralis*. The deutocerebral lobes, elongated in an anteroposterior direction, are prolonged backward by the tritocerebrum; the limits between the deuto- and the tritocerebrum are rather imprecise. Tritocerebral lobes are linked by two commissures: 1) a supraesophageal commissure, the stomatogastric bridge (stomodeal bridge), and 2) a subesophageal commissure, the tritocerebral commissure (Fig. 7.5B); however, the localization of the latter is discussed. The brain is connected with the subesophageal ganglion by two connectives (the periesophageal collar). Some variations can be observed. The ganglia are less distinct in Epimorpha (Geophilomorpha and Scolopendromorpha) than in Anamorpha (Lithobiomorpha and Scutigeromorpha). The protocerebrum is well-developed in Lithobiomorpha, in Scolopendromorpha and, especially, in Scutigeromorpha (Fig. 7.5D). The development of the optic lobes is maximum in Scutigeromorpha, animals with compound eyes; Geophilomorpha (blind animals) have reduced protocerebral lobes (Fig. 7.5C). In contrast to the other Chilopoda, Geophilomorpha show fused deutocerebral lobes. In all Chilopoda, the tritocerebrum is reduced, as in other Tracheata. The stomatogastric bridge is generally wide; nevertheless, in Scutigeromorpha, it is replaced by frontal connectives linking the tritocerebral lobes with the frontal ganglion (Fig. 7.5D), and this arrangement is comparable to that observed in Insecta. The tritocerebral commissure is lacking in *Geophilus*.

7.5.2. Histology and Fine Structure

The first studies (Saint-Rémy, 1890; Holmgren, 1916) were continued, among other authors, by Hanström (1928), Hörberg (1931), and Fahlander (1938)

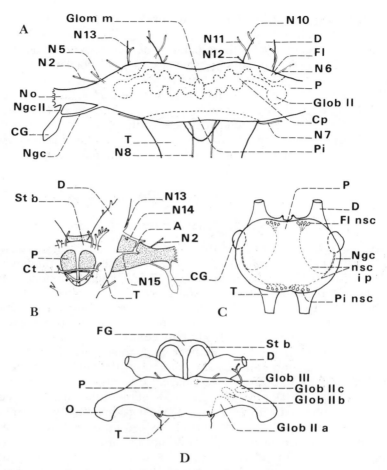

Figure 7.5. The brain of Chilopoda. (A) *Lithobius forficatus* (Lithobiomorpha): dorsal view. (B) *L. forficatus:* ventral view. (C) *Geophilus longicornis* (Geophilomorpha): dorsal view. (D) *Thereuopoda clunifera* (Scutigeromorpha): dorsal view. A = anastomosis; C G = cerebral gland; C p = principal commissure; C t = tritocerebral commissure; D = deutocerebrum; FG = frontal ganglion; Fl = frontal lobe; F l nsc = frontal lobe neurosecretory cells; Glob II, IIa, IIb, IIc, III = globuli II, IIa, IIb, IIc, III; Glom m = median glomerulus; N2 ...... 15 = brain nerves no 2 .... 15; N gc = *nervus glandulae cerebralis*; N gc II = accessory nerve of the cerebral gland; N o = optic nerve; nsc ip = intracerebral pathways of the neurosecretory products; O = optic lobe; P = protocerebrum; P i = *pars intercerebralis*; P i nsc = *pars intercerebralis* neurosecretory cells; St b = stomatogastric bridge; T = tritocerebrum (B modified from Rilling, 1968; C modified from Ernst, 1971; D modified from Fahlander, 1938.)

and, more recently, completed by electron microscopical studies (Jamault-Navarro and Joly, 1977; Jamault-Navarro, 1981).

7.5.2.1. Neurilemma.
The neurilemma is divided into two parts. The more external, called neural lamella, is constituted, in *L. forficatus*, of collagen fibers embedded in an amorphous or granular substance (Fig. 7.7A); brain neural lamella (some few μm) is not so thicker than that around the ventral nerve cord. The more internal part of the neurilemma is called, according to different authors, perilemma or perineurium. In *L. forficatus*, the perilemma is not continuous; so the neurons can be separated from the neural lamella by only thin cytoplasmic processes issuing from cortical neuroglial cells.

7.5.2.2. Cerebral Cortex.
Present all around the protocerebral neuropil, nerve cells are absent, at least partly, at the junction of the proto- and deutocerebrum. The cortical thickness is greater at the rear of the protocerebrum and, dorsally, in the median groove area, where giant cells are found; some other giant neurons are also located near the internal border of the globuli. Glial cells seem to be of various types (three in Geophilomorpha; Lorenzo, 1960). The nucleus of these cells is polymorphous, and the cytoplasm, poor in organelles, is rich in glycogen particles. Glial processes are sinuous and sometimes separated by intercellular spaces filled with electron-dense material. The processes, surrounding the neurons of the cerebral cortex, can indent the cell body, leading to a structure comparable to the trophospongium described in Insecta (Fig. 7.7B).

In *L. forficatus*, motor neurons are unipolar, except in the *pars intercerebralis* area where multipolar neurons could exist. Three types of neurons (at least) can be found: (1) small neurons (10–15 μm) with a round and voluminous nucleus and reduced cytoplasm; (2) greater neurons (about 20 μm) with occasionally a lobated nucleus and voluminous cytoplasm; and (3) neurons of large size and filled with secretory granules: the neurosecretory cells (NSCs). As in Diplopoda, clusters of small nerve cells with reduced and dense cytoplasm can be observed: the globuli (*corpora pedunculata*).

The structure of the protocerebral cortex has been described in *Thereuopoda clunifera* (Scutigeromorpha) (Fahlander, 1938). Four globuli were reported and called, according to topographical data, IIa, IIb, IIc, and III; they correspond, respectively, to globuli Ia, Ib, II, and III in the nomenclature of Holmgren (1916). Globuli II are in dorsolateral position, near the median groove (Figs. 7.5D, 7.6A). In Lithobiomorpha (Fig. 7.5A) and in Scolopendromorpha, only undivided globuli II are present. In Geophilomorpha globuli are absent.

The selected model for the study of cerebral NSCs is *L. forficatus*. The data come mainly from Gabe (1952), Palm (1955), Scheffel (1961), and Joly (1966). In the protocerebrum, NSCs constitute essentially paired symmetrical groups located in the anterolateral area of the frontal lobes. NSCs are

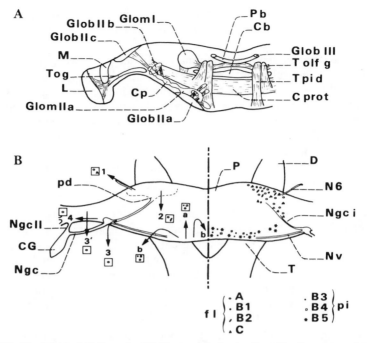

Figure 7.6. The brain of Chilopoda. (A) *Thereuopoda clunifera* (Scutigeromorpha): dorsal view. (B) *Lithobius forficatus* (Lithobiomorpha): dorsal view; right side, NSC localization; left side, protocerebral neurosecretory pathways. Cb = central body; CG = cerebral gland; Cp = principal commissure; C Prot = protocerebral commissure; D = deutocerebrum; Glob IIa, IIb, IIc, III = globuli IIa, IIb, IIc, III; Glom I, IIa, III = glomeruli I, IIa, III; L = *lamina*; M = *medulla*; N gc = *nervus glandulae cerebralis*; N gci = internal pathway of the N gc; N gc II = accessory N gc; N v = *nervus ventralis*; N6 = brain nerve no 6; P = protocerebrum; Pb = protocerebral bridge; pd = "plages dorsales"; T = tritocerebrum; T og = *tractus optico globularis*; T olf g = *tractus olfactorioglobularis*; T pid = tract of fibers linking the *pars intercerebralis* and the deutocerebrum; A, B1, B2, C = frontal lobes (fl) NSC; B3, B4, B5 = *pars intercerebralis* (pi) NSC; a, b = neurosecretory pathways of the *pi* NSC; 1, 2, 3, 3', 4 = other neurosecretory pathways. In the boxes, type of neurosecretory granules present in the axons. (A modified from Fahlander, 1938; B data from Jamault-Navarro, 1981.)

also present in the posterodorsal area (*pars intercerebralis*) (Fig. 7.6B). In the other Chilopoda, NSCs are found in the same location (Gabe, 1966).

Protocerebral NSCs were studied by electron microscopy in *L. forficatus* (Jamault-Navarro and Joly, 1977; Jamault-Navarro, 1981). Three main types of NSCs (A, B, C) can be defined. A and C are present in the frontal lobes. B NSCs are divided into five subtypes: B1 and B2 are found in the frontal lobes and three others (B3, B4, and B5 [Fig. 7.7B]) are located in the *pars intercerebralis* (Table 7.1).

Up to now, serotoninergic and noradrenergic cells have been located by

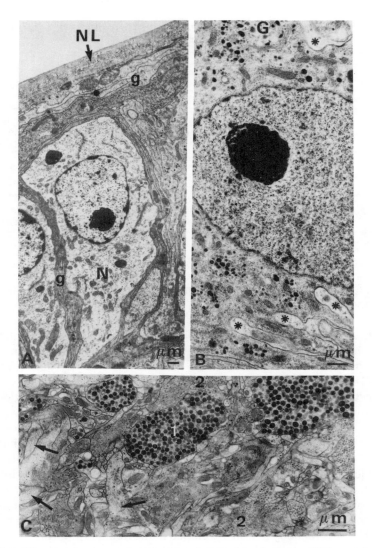

Figure 7.7. Ultrastructural aspects of the brain in *Lithobius forficatus* (Chilopoda). (A) Neurilemma and cerebral cortex. g = glial cell processes; N = neuron; NL = neural lamellae (3300 ×). (B) *Pars intercerebralis* NSC: B5 type. G = Golgi apparatus; * = trophospongium (5500 ×). (C) Portion of neuropil showing axons with neurotubules (*arrows*), high (1) or poor (2) electron density secretory granules (7500 ×). Scale 1 μm. (A and B, courtesy of Dr. Jamault-Navarro, University of Amiens, France.)

TABLE 7.1. PROTOCEREBRAL NEUROSECRETORY CELLS IN *LITHOBIUS FORFICATUS*

| Cell Type | Localization | Cell Size (μm) | Secretory Granules | | | Secretory Pathways |
			Shape	Electron Density	Size (nm)	
A	Frontal	20–40	Globe	+ +	230	Mainly toward
C	lobes	20	Polyedric	+ + +	450	the cerebral
B₁	↓	10–15	Globe	+ +	150	gland
B₂	↓	10–15	Cone or stick	+ +	300 × 150–200	↓
B₃	Pars	20–25	Globe	+ +	200–300	Mainly toward
B₄	intercerebralis	10–15	Globe	Variable	120–200	the neuropil
B₅	↓	30–60	Variable	+ + +	100–300	B₃ nervus ventralis

Data from Jamault-Navarro (1983).

autoradiography in the protocerebrum and in the lateral areas of the deuto- and tritocerebrum (Descamps et al., 1985).

7.5.2.3. Neuropil. In *Thereuopoda clunifera* neuropil (Fahlander, 1938) (Fig. 7.6A), the axons issuing from globuli IIa, corresponding to the *pédoncule antérieur, pédoncule postérieur,* and *tubercule interne,* according to Saint-Rémy (1890), do not reach the median plane of the brain: no median glomeruli are present. Axons of globuli IIc and IIb end in a line of small glomeruli, constituting the principal commissure (*tige externe,* Saint-Rémy, 1890); more axial, according to its topographical position, is the glomerulus IIa. Globuli III are adjacent to reduced glomeruli III; the latter are linked by fibers constituting the protocerebral bridge. Glomeruli I are those of the frontal lobes; they receive fibers issuing not only from the frontal lobes but also from the *pars intercerebralis,* from the periesophageal collar and, probably, from the deutocerebrum. Glomeruli I are linked by two commissures; the upper commissure is related to the protocerebral bridge and the central body. The central body is well developed in the Scutigeromorpha. Numerous protocerebral commissures are observed behind the central body. One of them is wide and joins the optic lobes to each other; a part of this large commissure (C prot; Fig. 7.5A) could be linked to the deutocerebrum and so, corresponds to the myriapodal commissure. As in Diplopoda, the optic ganglia are divided into two differentiated nervous masses: the *medulla* and the *lamina;* they are joined to globuli IIb by a large tract of fibers: the *opticoglobularis* tract. There are numerous glomeruli, endings of sensory nerves, in the deutocerebrum, and motor nerves, linked by transverse commissures; the deutocerebrum is also linked to the protocerebrum, particularly with the *pars intercerebralis* and with the globuli, as well as to the subesophageal ganglion. The cerebral structure of the Scutigeromorpha is comparable to that of Insecta; indeed, the median glomeruli are absent, the glomeruli I are located in the frontal lobes and the globuli are linked to the optic lobes by a large tract of fibers. Nevertheless, a transverse commissure, characteristic of Myriapoda, is linked to the antennal lobes.

In other Chilopoda, some differences can be observed. In Lithobiomorpha, the fibers issuing from globuli II constitute a tract, ending in a reduced median glomerulus (Fig. 7.5A). The tract of fibers is flanked by glomeruli; the more internal (lateral bodies; Holmgren, 1916) are linked by a commissure. The central body is reduced. In Scolopendromorpha, the cerebral structure is comparable to that of Lithobiomorpha; nevertheless, fiber tracts issuing from globuli II end in two different glomeruli, located dorsally over a well-developed central body. In Geophilomorpha, the more important feature is the lack of the central body.

An electron microscope study of the neuropil in *L. forficatus* (Jamault-Navarro, 1979) (Fig. 7.7C) shows, in the axons issuing from the cortical neurons, neurotubules, mitochondria, smooth or rough endoplasmic reticulum, and glycogen particles. Axons are surrounded by glial processes, particularly in the more external part of the neuropil. These processes show numerous glycogen particles and granules of a diameter of more than 300 nm; adjacent axons are sometimes separated by intercellular spaces. If some fibers do not show axonal endings, others show dendrites and synapses; the latter can affect two adjacent axons. Septate junctions are sometimes observed. Tracheae and tracheal cells are also observed in the neuropil. Some neuropilar axons show granules or vesicles, often of the same size in the same axon, but different from one to another (Jamault-Navarro and Joly, 1977; Jamault-Navarro, 1981) (Fig. 7.7C).

There are various pathways for the protocerebral neurosecretory products. Axons issuing from the frontal lobe NSCs (particularly from type A) constitute the main nerve of the cerebral gland. Some axons issuing from B3 NSCs of the *pars intercerebralis* form two symmetrical ventral and intracerebral nerves (*N. ventralis*; Joly and Jamault-Navarro, 1978) joining the nerve of the cerebral gland near its emergence from the protocerebrum (Fig. 7.6B). It must be noted that Ernst (1971) demonstrated, in a light microscopical study in *Geophilus longicornis*, a neurosecretory pathway from the lateral *pars intercerebralis* NSCs to the nerve of the cerebral gland (Fig. 7.5C). In *Lithobius forficatus*, numerous other secretory pathways can be observed (Jamault-Navarro, 1981). Under normal conditions, they are constituted of only few axons with few secretory granules. Pathways issuing from the frontal lobes and the *pars intercerebralis* can be demonstrated (Fig. 7.6B). These neurosecretory pathways seem to be secondary concerning the frontal lobes neurosecretory products, but seem to be principal for the *pars intercerebralis* neurosecretion.

7.5.3. Brain Nerves

The cerebral nerves were first described by Holmgren (1916) and Fahlander (1938). These studies were completed and sometimes discussed by Henry (1948), Applegarth (1952), and Lorenzo (1960). In the more recent works of Rilling (1968) and Joshi et al. (1977), 23 and 12 brain nerves are described, respectively, in *Lithobius forficatus* and in *Scolopendra morsitans*. Brain nerves of *L. forficatus*, excluding the stomatogastric system, are described afterwards (Fig. 7.5A,B).

Eight pairs of nerves arise from the protocerebrum and none is a motor nerve. The first two pairs (N 1 and N 2) end, respectively, in the ocelli and in the Tömösvary organs. N 3 and N 4 are the cerebral gland nerves, and the remaining N 5, N 6, and N 7 end in the anterior area of the epicranium. No data are available concerning the area innervated by N 8; according to Rilling (1968), the pathways of these nerves are comparable to that of the allatocardiac nerves in Insecta. Six pairs of nerves arise from the deutocerebrum. N 9 are thick antennal sensory nerves flanked by three pairs of thin nerves, N 10, N 11, and N 12. N 10 and N 11 originate from antennal muscles and from antennal sensory cells; the area innervated by N 12 is unknown. N 13, N 14, and N 15 are mainly antennal motor nerves; nevertheless, they can also end in some cephalic muscles; N 14 and N 15 are joined by a thick anastomosis. Other nerves originate from the tritocerebrum. N 16, very large nerves, constitute the stomatogastric bridge. Others are stomatogastric nerves, except a pair ending in epicranial muscles. In other Chilopoda, the same nerves can be found. Of course, optic nerves are lacking in blind species (Geophilomorpha and various Scolopendromorpha). N 2 are also lacking in Epimorpha, animals deprived of Tömösvary organs.

7.6. COMPARATIVE ASPECTS OF BRAIN STRUCTURES IN MYRIAPODA

Differences between the various orders of Myriapoda are summarized in Table 7.2.

7.7. SUMMARY

In the Myriapoda (Pauropoda, Symphyla, Diplopoda, Chilopoda), the protocerebrum is the most developed part of the brain. It is generally in anterodorsal position, but in Pauropoda, it lies dorsally and extends backward up to the second truncal segment. The protocerebrum shows laterally the optic lobes, except in blind animals. Generally, the deutocerebrum is well-developed and the tritocerebrum reduced. Two tritocerebral commissures are observed: the stomatogastric bridge, in anterodorsal position, and pos-

TABLE 7.2. THE BRAIN OF MYRIAPODA

	Pauropoda	Symphyla	Diplopoda	Chilopoda
Gross anatomy				
Localization	Protocerebrum: posterodorsal No true subesophageal connectives	Protocerebrum: lobed, anterodorsal	Protocerebrum: anterodorsal No precise limits between the 3 parts of the brain	Protocerebrum: anterodorsal
Optic lobes	Absent	Absent	Present, except in blind species	Present, except in blind species
Tritocerebral commissures				
Anterior = stomatogastric bridge	Present	Present; no ganglion	Present	Present; with frontal ganglion (in Scutigeromorpha)
Posterior = tritocerebral commissure	Not completely free	Present	Present	Present
Histology				
Protocerebral globuli	Absent	Absent; present (?) in *S. pagesi*	3 pairs (I, II, III)	IIa, IIb, IIc in Scutigeromorpha; II undivided in Lithobiormorpha and Scolopendromorpha; absent in Geophilomorpha. III, reduced in Scutigeromorpha; absent in Lithobiomorpha and Scolopendromorpha.

TABLE 7.2. THE BRAIN OF MYRIAPODA (continued)

	Pauropoda	Symphyla	Diplopoda	Chilopoda
Protocerebral glomeruli	3 Pairs (presumed)	3 lateral, 2 internal pairs	Present	Present
Median body (median glomerulus)	1	Absent	Double { anteromedian, posteromedian	Absent in Scutigeromorpha; 2 in Scolopendromorpha; reduced and homologous to posteromedian glomerulus in *L. forficatus*
Protocerebral bridge	Not described	Present: juxtaposed glomeruli	Present: linked to globuli III	Present in Scutigeromorpha; homologous tract in Lithobiomorpha and Scolopendromorpha.
Central body	Small, without ganglia	Very large	Small	Large in Scutigeromorpha and Scolopendromorpha Reduced in Lithobiomorpha Absent in Geophilomorpha
Principal commissure	Present	Absent (?)	Present; not associated with glomerular structures.	Present Line of small glomeruli in Scutigeromorpha Well-developed and associated with glomerular structures in Lithobiomorpha
Myriapodal commissure	Uncertain localization	Well-developed	Well-developed	Present
NSC and neurosecretory pathways	NSC not described	Nuchal lobe NSC → cerebral gland Frontal and lateral protocerebral NSC and deutocerebral NSC → nerve cord	NSC in the 3 pairs of ganglia globuli I NSC → Gabe organs Other NSC → nerve cord	NSC in the 3 pairs of ganglia Protocerebral NSC frontal lobes → mainly cerebral gland *pars intercerebralis* → mainly neuropil and arterial walls

teroventrally, the tritocerebral commissure. The latter is not completely free in Pauropoda and is absent in Geophilomorpha (Chilopoda).

Some differences are observed in the brain internal structure in the various orders of Myriapoda. In Chilopoda, the neurilemma and the cerebral cortex show various glial cells and at least three types of neurons. Globuli (*corpora pedunculata*) are present in Diplopoda (three pairs) and in some Chilopoda (one or two pairs), but absent in Pauropoda and Geophilomorpha (Chilopoda); in Symphyla, they possibly exist, with indistinguishable limits. In the neuropil, numerous tracts of nerve fibers, the glomeruli, join, among other centers, the two protocerebral ganglia; note that (1) the protocerebral bridge (not described in Pauropoda), linking as a rule the glomeruli III, is also present in Lithobiomorpha, Scolopendromorpha, and Symphyla; (2) the principal commissure may be related (numerous Chilopoda) or may not (Diplopoda) to glomerular structures; and (3) the central body, converging area for axons coming from various parts of the brain, is more or less developed; this structure is absent in Geophilomorpha.

The principal differences in the brain structure between Myriapoda and other Arthropoda are: (1) the number of optic ganglia: two (*lamina* and *medulla*), instead of three; (2) the absence of decussation of the fibers linking the proto- and deutocerebrum, and, especially, (3) the existence of a median glomerulus (median body), except in Symphyla and Scutigeromorpha, and of a protocerebral commissure (characteristic of Myriapoda), originating from the deutocerebrum. The brain of the Scutigeromorpha seem to be more closely related to that of Insecta.

The NSCs, not described in Pauropoda, are present in the three other groups of Myriapoda; most of the studies concern the protocerebral NSCs. The neurosecretory pathways run essentially toward the cephalic neuro-hemal organs and the ventral nerve cord. Numerous other pathways are described in Chilopoda.

REFERENCES

Applegarth, A. G. 1952. The anatomy of the cephalic region of a centipede, *Pseudolithobius megaloporus* (Stuxberg) (Chilopoda). *Microentomology* **7**: 127–171.

Descamps, M., R. Joly, and C. Jamault-Navarro. 1985. Autoradiographic localization of 5-hydroxytryptamine and noradrenalin in the central nervous system of *Lithobius forficatus* L. (Myriapoda Chilopoda). *Bijd. Dierk.* **55**: 47–54.

Ernst, A. 1971. Licht—und elektronenmikroskopische Untersuchungen zur Neurosekretion bei *Geophilus longicornis* (Leach), unter besonderer Berücksichtigung der Neurohaemalorgane. *Z. Wiss. Zool.* **182**: 62–130.

Fahlander, K. 1938. Beiträge zur Anatomie und systematischen Einteilung der Chilopoden. *Zool. Bidr. Uppsala* **17**: 1–148.

Fechter, H. 1961. Anatomie und Funktion der Kopfmuskulatur von *Cylindroiulus albipes* (Pocock). *Zool. Jb. Anat.* **79**: 479–528.

156   ROBERT JOLY AND MICHEL DESCAMPS

Gabe, M. 1952. Sur l'emplacement et les connexions des cellules neurosécrétrices dans les ganglions cérébroïdes de quelques Chilopodes. *C.R. Acad. Sci. Paris* **235D**: 1430–1432.

Gabe, M. 1954. Emplacement et connexions des cellules neurosécrétrices chez quelques Diplopodes. *C.R. Acad. Sci. Paris* **239D**: 828–830.

Gabe, M. 1966. *Neurosecretion.* Pergamon Press, Oxford.

Haller, B. 1905. Über den allgemeine Bauplan des Trachaeten—Syncerebrums. *Arch. Mikrosk. Anat.* **65**: 181–279.

Hanström, B. 1928. *Vergleichende Anatomie des Nervensystems der wirbellosen Tiere.* Springer-Verlag, Berlin.

Henry, L. M. 1948. The nervous system and the segmentation of the head in the Annulata. IV Arthropoda. *Microentomology* **13**: 1–26.

Holmgren, N. 1916. Zur vergleichenden Anatomie des Gehirns von Polychaeten, Onychophoren, Xiphosuren, Arachniden, Crustaceen, Myriapoden und Insekten. *K. Sven. Vetenskapsakad. Handl.* **56**: 1–103.

Hörberg, T. 1931. Studien über den komparativen Bau des Gehirns von *Scutigera coleoptrata* (L.). *Lunds Univ. Arsskr. Avd.* **27**(2): 1–24.

Jamault-Navarro, C. 1979. Contribution à l'étude du système endocrinien céphalique chez *Lithobius forficatus* (L.) (Myriapode Chilopode). Thèse 3ème cycle, Amiens, France.

Jamault-Navarro, C. 1981. Cellules neurosécrétrices et trajets axonaux protocérébraux chez *Lithobius forficatus* (L.) (Myriapode Chilopode). Etude ultrastructurale. *Arch. Biol. (Bruxelles)* **92**: 203–218.

Jamault-Navarro, C. and R. Joly. 1977. Localisation et cytologie des cellules neurosécrétrices protocérébrales chez *Lithobius forficatus* (L.) (Myriapode Chilopode). *Gen. Comp. Endocrinol.* **31**: 106–120.

Joly, R. 1966. Contribution à l'étude du cycle de mue et de son déterminisme chez les Myriapodes Chilopodes. *Bull. Biol. Fr. Belg.* **3**: 379–480.

Joly, R. and C. Jamault-Navarro. 1978. Rôle de la *pars intercerebralis* sur l'activité des glandes cérébrales chez *Lithobius forficatus* (L.) (Myriapode Chilopode). Etude ultrastructurale. *Arch. Zool. Exp. Gén.* **119**: 487–496.

Joshi, G. P., P. C. Hurkat, and V. Changulani. 1977. Studies on the morphological aspect of the supraoesophageal and suboesophageal ganglia of *Scolopendra morsitans* (L.) (Myriapoda Chilopoda). *Dtsch. Entomol. Z.* **24**: 175–180.

Juberthie-Jupeau, L. 1963. Recherches sur la reproduction et la mue chez les Symphyles. *Arch. Zool. Exp. Gén.* **102**: 1–172.

Juberthie-Jupeau, L. 1967. Existence d'organes neuraux intracérébraux chez les Glomeridia (Diplopodes) épigés et cavernicoles. *C.R. Acad. Sci. Paris* **264D**: 89–92.

Juberthie-Jupeau, L. 1983. Neurosecretory systems and neurohemal organs of Myriapoda, pp. 204–278. In A. P. Gupta (ed.), *Neurohemal Organs of Arthropods.* Charles C. Thomas, Springfield, Illinois.

Lorenzo, M. A. 1960. The cephalic nervous system of the centipede *Arenophilus bipuncticeps* (Wood) (Chilopoda, Geophilomorpha, Geophilidae). *Smithson. Misc. Collect.* **140**: 1–43.

Nguyen Duy-Jacquemin, M. 1973. Contribution à la connaissance de l'anatomie céphalique, des formations endocrines et du développement postembryonnaire de *Polyxenus lagurus* (Diplopodes Pénicillates). *Thèse Sci. Paris VI*, n° A08186.

Palm, N. B. 1955. Neurosecretory cells and associated structures in *Lithobius forficatus* (L.). *Ark. Zool. K.S. Vetenskap.* **9**: 115–129.

Prabhu, V. K. K. 1962. Neurosecretory system of *Jonespeltis splendidus* (Verhoeff) (Myriapoda, Diplopoda), pp. 417–420. In H. Heller and R. B. Clark (eds.), *Neurosecretion, Mem. Soc. Endocrinol.*, vol. 12. Academic Press, New York.

Rilling, G. 1968. *Lithobius forficatus,* Grosses Zool. Prakt., Helft 13b, G. Fischer-Verlag, Stuttgart.

Sahli, F. 1966. Contribution à l'étude de la périodomorphose et du système neurosécréteur des Diplopodes Iulides. *Thèse Sci. Dijon*, n° 94.

Sahli, F. and J. Petit. 1975. Les plages paracommissurales (formations neurohémales céphaliques) des Diplopodes. *C.R. Acad. Sci. Paris* **280D**: 2001–2004.

Saint-Rémy, G. 1890. Contribution à l'étude du cerveau chez les Arthropodes trachéates. *Arch. Zool. Exp. Gén.* **2**: 1–276.

Scheffel, H. 1961. Untersuchungen zur Neurosekretion bei *Lithobius forficatus* (L.) (Chilopoda). *Zool. Jahrb. Anat.* **79**: 529–556.

Seifert, B. 1932. Anatomie und Biologie des Diplopoden *Strongylosoma pallipes* (Oliv.). *Z. Morphol. Ökol. Tiere* **25**: 362–507.

Seifert, G. 1966. Das stomatogastrische Nervensystem der Diplopoden. *Zool. Jahrb. Anat.* **83**: 448–482.

Tiegs, O. W. 1940. The embryology and affinities of the Symphyla, based on a study of *Hanseniella agilis*. *Q.J. Microsc. Sci.* **82**: 1–225.

Tiegs, O. W. 1947. The development and affinities of the Pauropoda, based on a study of *Pauropus silvaticus*. Part I. *Quart. J. Microsc. Sci.* **88**: 165–267.

Wigglesworth, V. B. 1951. *The Principles of Insect Physiology*. Methuen and Co, London.

CHAPTER 8

Histology and Ultrastructure of the Onychophoran Brain

Friedrich-Wilhelm Schürmann

Department of Cell Biology
Institute of Zoology
University of Göttingen
Göttingen, West Germany

8.1. INTRODUCTION

The systematic position and phylogenetic relationship of the Onychophora is still controversial, as the mixed appearence of "annelid" and "arthropod" features is interpreted differently (Gupta, 1979; Anderson, 1979; Manton, 1979; Ax, 1984; Sawyer, 1984). Comparing brains of various worms and arthropods, Holmgren (1916) lists a number of homologous brain structures for the Onychophora, Annelida, and Arthropoda. Horridge (1965) states in his review that the nervous system of the Onychophora is of little help for the determination of exact affinities. Although some striking similarities of brain compartments in the groups mentioned above should not be over-looked, I do not intend to enter into a broad discussion on homologous structures, but will concentrate on general principles of structural brain organization and on selected brain parts.

Only a few recent investigations on the morphology and/or physiology of onychophoran excitable tissues or organs, other than brain, exist (for muscle and nerve-muscle junctions: Schürmann, 1978a; Camatini et al., 1979; Lanzavecchia and Camatini, 1979; Hoyle and del Castillo, 1979; Hoyle and Williams, 1980; for sense organs: Eakin and Westfall, 1965; Eakin and Brandenburger, 1966; Storch and Ruhberg, 1977). The ultrastructure of ventral cord synapses has been described by Schürmann (1978b). A slow conducting giant fiber system, associated with overall body movements, in the ventral cord lacking cell free connectives, was detected by electrophysiology (Schürmann and Sandeman, 1976). A fast shortening reflex or a quick bending preceded by phasic discharge of two giant fibers can be released after strong touch or pinching of the antennae or tail resembling similar behavior in annelids. Thorough reports on onychophoran behavior are given by Haase (1889), Zacher (1933), Manton (1938, 1950). Responses of muscles to transmitters, such as acetylcholine, dopamin, noradrenalin, and serotonin induced peripherally or within the ventral cord, have been shown in physiological and pharmacological studies (Ewer and van den Berg, 1954; Florey and Florey, 1965; Gardner and Robson, 1978). Monoamines in the ventral nerve cord and in sensory nerves have been demonstrated by the Falck-Hillarp method (Gardner et al., 1978). Neuroactive compounds in the onychophoran brain have not yet been analyzed (for review, see Gardner and Walker, 1982; Klemm, 1985).

The present article focuses on histological and ultrastructural features of the onychophoran brain. Owing to lack of information, a complete survey of all parts of the large brain cannot be given.

8.2. HISTOLOGY

8.2.1. Gross Morphology of the Onychophoran Brain

Onychophora show a well-developped head with a prominent brain, indicating remarkable cephalization resembling Hirudinea and Insecta (Sawyer, 1984). The segmental composition of the adult brain cannot be detected, either from the innervation pattern of the brain or from its compartmentation (Hanström, 1935, but see Feodorov, 1929). There is embryological evidence that the brain is formed from three neuromeres (proto-, deuto-, and tritocerebrum) (Pflugfelder, 1948, 1968; for comparative embryology of Myria-

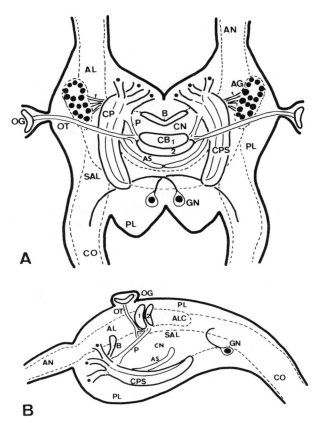

Figure 8.1. Diagram of an onychophoran brain with selected compartments and tracts. (A) Dorsal view: antennal nerve (AN), antennal lobe (AL), antennal glomeruli (AG), subantennal lobe (SAL), bridge (B), corpora pedunculata (CP) with the subcompartmental stalks (CPS); CP proximal fiber bundles stem from frontal globule cell bodies, accessory stalks (AS) are joined to the CP, CP-AG connections marked by *arrowheads*. Central body (CB) with Lamina anterior (1) and posterior (2) linked to the CP by the pedunculus (P) and to the optic ganglion (OG) by a massive optic tract (OT), central neuropil (CN), pericaryal layer (PL) with a pair of giant neurons (GN), commissure (CO) to the ventral cord. (B) Side view, the antennal lobe is continued into the antennal lobe commissure (ALC); other abbreviations as in A.

poda, Hexapoda, and Onychophora, see Anderson, 1966, 1973). Brain nerves are listed by Hanström (1935) and depicted by Henry (1948). As we deduce from electron microscopy, about 15 pairs of nerves, mainly of mixed nature, have to be considered. We could only follow the central projections of the antennal and optic nerves (Fig. 8.1). The eyes have their own peripheral ganglia, and similarly the massive nerve strands of the antennae contain neuropilar parts with synapses and a basal globule cell rind (Fig. 8.8), as already mentioned by Hanström (1935). The massive circumesophageal connectives appear as "Markstrang," with areas of synaptic connectivity as shown for the ventral cord (Schürmann, 1978a). The deep medial invaginations predominantly separate the caudal and rostral pericaryal layer of the hemispheres (Fig. 8.2). This might be considered as a conservation of an onychophoran pecularity so prominent in the ventral nerve cord: the wide separation of the right and left neurosomites.

8.2.2. Pericaryal Layer

The pericaryal rind of the brain is mainly concentrated in its rostro- and caudoventral parts. Most neurons (Figs. 8.2–8.5) belong to the globule cell type (diameter of nuclei 3–8 μm). Grouping of globule cells into clusters—typically present in insect and polychaete brains and in other invertebrates (Hanström, 1928; Clark, 1958; Bullock and Horridge, 1965; Manaranche, 1966)—is not very distinct, though indicated. A clear-cut separation of the mushroom body globule cells from surrounding pericarya is not obvious for the onychophoran brain (Figs. 8.2, 8.4). An appropriate description of neurons other than globule cells and of their position is given by Feodorov (1929). He mentions cell clusters with cells larger than globule cells dorsal to the central body and shows some pairs of big neurons in the caudoventral brain, including a pair of giant descending cells also noted by previous authors (Sänger, 1870; Gaffron, 1884; Holmgren, 1916), which we confirm (Fig. 8.1B,D). Clusters of neurosecretory cells in the laterodorsal and the lateroventral regions of the cerebral ganglion in Onychophora have been histologically identified (Gabe, 1954; Sanchez, 1958). They are apparently not identical to the giant cells we found (for a comparative view of neuroendocrine systems, see Tombes, 1979).

8.2.3. Neuropilar Compartments

Reliable, detailed information on the prominent compartments of the onychophoran brain (Figs. 8.1–8.4) is found in the classical papers (Balfour, 1883; Holmgren, 1916; Feodorov, 1929; Hanström, 1935). The size of the brain and its compartments varies considerably and appears positively correlated with body size. Hanström, comparing brains of different species, emphasizes the uniform expression of compartments within the group. The abundant small-diameter fibers (diameter < 1 μm) stem from globule cells. The thick fibers—so typical of arthropod and annelid relay or plurisegmental neurons—are poorly developed in the onychophoran brain. In his thorough work, Feodorov (1929) describes and depicts a pair of giant fibers forming a chiasma in the central caudal neuropilar masses, before descending into the connectives. We did indeed find a group of thick commissural chiasmatic fibers (diameters, 2–5 μm) in osmicated or Bodian-stained tissue (Fig. 8.3C), probably incorporating Feodorov's giant fibers; but could not trace them completely in serial sections, as the diameter of these fibers diminishes.

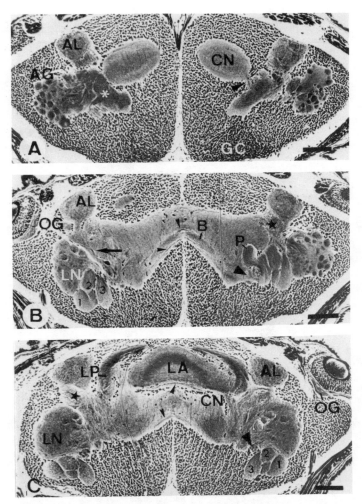

Figure 8.2. Brain of *Peripatoides leuckarti*, light micrographs from selected frontal sections, Bodian stain. Scales 100 μm. (A) Dense package of the uniform globule cells (GC) giving rise to mushroom body fiber bundles, in the proximal stalks (*asterisk*) concentration of connections to other brain compartments, different size and form of antennal glomeruli (AG), median separation of the pericaryal layer. (B) Section posterior to A, AL connected with the central neuropil by a fiber bundle (*arrow*), P closely associated with the proximal corpora pedunculata stalk neuropil, subantennal lobe marked by a star, the antennal glomeruli are followed posteriorly by the postglomerular lateral neuropil (LN), CP stalk is subdivided into three columns (1,2,3) and column 2 further subdivided into a dorsal and ventral part, the accessory stalk (*triangle*) is situated medial to the stalk neuropil, commissural fibers marked by *arrowheads*, the bridge (B) is composed of several fiber bundles. (C) Section posterior to B. Lamina anterior (LA) and posterior (LP) of the bridge appear striated and show fiber continuity in their ventrolateral parts, commissures (*arrowheads*). All abreviations as in Figure 8.1.

163

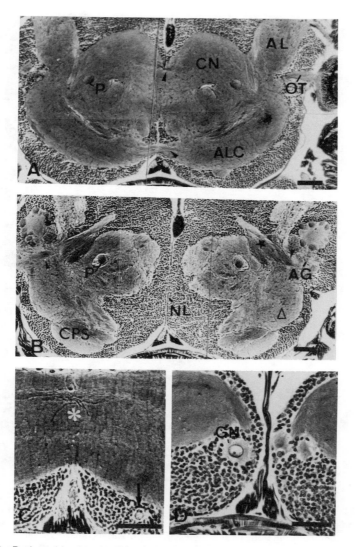

Figure 8.3. *Peripatoides leuckarti*, horizontal brain sections, Bodian stain, scales 100 μm; abbreviations as in Fig. 8.1. (A) Dorsal brain. The antennal lobe is merged with the antennal lobe commissure (ALC) neuropil, connected by a distinct interhemispherical commissure (*arrowhead*). (B) The proximal stalk neuropil is connected with the anterior and posterior group of antennal glomeruli (*arrowhead*), postglomerular lateral neuropil (*triangle*); neural lamella (NL) separates the medial globule cell layers. (C) Ventrocaudal commissure (*asterisk*) from large and giant cell fibers, large neuron (*arrow*). (D) Ventrocaudad neuron (GN).

Contralateral descending giant neurons with and without chiasmatic synapse have been described for insects and many other invertebrate brains (Bullock and Horridge, 1965; Strausfeld and Nässel, 1980). These onychophoran giant fibers are apparently not identical to the ipsilateral medial and lateral giant fiber of the ventral cord, which we traced up to the supraesophageal commissures (diameters, 15–20 μm). The giant descending brain fibers could represent a fast link to the ventral cord giants and mediate antennal receptor information correlated with the fast shortening reflex mentioned earlier. Though the circumscription of compartments and their connectivity is difficult to establish by metal impregnations, the gross compartmentation has been detected. In this paper, we mainly follow the terminology of Horridge (1965) based on German nomenclature.

Compartments with clear-cut borders are the mushroom bodies (corpora pedunculata), the central body, the bridge, and the antennal glomeruli, which together comprise a minor part of the brain neuropil only (Figs. 8.1–8.4). Other neuropilar components with a different packing and orientation of fine fibers, such as the antennal lobes (antennal tracts), subantennal lobes (subantennal tracts), postglomerular lateral neuropil, central commissural mass (Holmgren, 1916), and the antennal commissure, are not clearly separated from the voluminous central neuropil. The paired corpora pedunculata (Figs. 8.2, 8.4A) appear as three associated columns with subdivisions arranged in rostrocaudal direction in the ventral brain. These columns end posterior to the level of the central body. Their form is determined by the parallel orientation of the fine fibers arising from three complexes of globule cells in the anterior brain, which are similar to the Kenyon cell fibers in the mushroom bodies of insects; in fact, these globule cell fibers in Onychophora can be considered as intrinsic elements restricted to the system. As correctly noted by Hanström (1935), the three complexes of globule cells do not send all their fibers in one associated column but in several. The so-called stalks (Holmgren, 1916) represent the proximal bundles of globule cells and lack synapses. Distally, they form the trabeculae (Hanström, 1935), which we have named stalks, analogous to the insect mushroom body stalks. These stalks show a complicated fiber texture in their proximal parts due to a number of fiber bundles connecting with the central neuropil, with the antennal glomeruli, the antennal lobes, and the central body via the so-called pedunculus (tract of tiny fibers) (Figs. 8.1, 8.4B, 8.5H), as described by Hanström (1935). Electron microscopy reveals the proximal stalks as areas with abundant synapses (Fig. 8.10). The mushroom body stalk neuropil is partially separated from surrounding neuropil or from each other by glial envelopes containing collagen extracellularly.

Holmgren (1916) has detected 10 tiny fiber bundles originating from frontal globule cells running medially along the stalks and forming a ventrocaudal commissure. Hanström (1935) suggested that these fiber strands of glomeruli neuropil type might belong to the mushroom body system ("Nebentrabekel"). We found these accessory stalks closely linked with the proximal

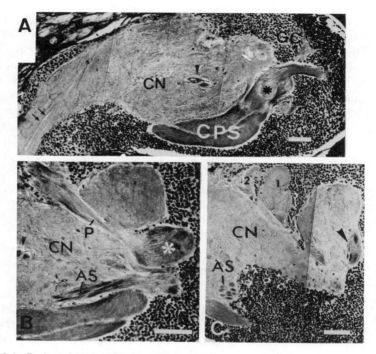

Figure 8.4. *Peripatoides leuckarti*, sagittal brain sections, von Rath osmium impregnation, scales 100 μm; abbreviations as in Figure 8.1. (A) Lateral section through the anterior parts of the corpora pedunculata, light neuropil texture of the proximal stalk (*asterisk*) (site of synaptic interaction with various brain connections), commissure of tiny fibers (*arrowhead*) in the central neuropil (CN), thick fiber (*arrows*) in the connective. (B) Section medial from A, showing the accessory stalk (AS) composed from fiber bundles. The pedunculus (P) has synaptic neuropils in the corpora pedunculata (*asterisk*) and in the dorsal central body commissure (*arrowhead*). (C) Mediosagittal section, lamina anterior (1) and posterior (2); bridge (*arrowhead*) and accessory stalk (AS) show a similar composition.

mushroom body stalks (Fig. 8.4B,C). Because we could not find any other connectivity of this system in silver-stained material, we suggest that the accessory stalks are a peculiar interhemispherical commissure of the proximal mushroom body parts.

The central body (corpus striatum) in the dorsal midbrain (Figs. 8.1, 8.2C, 8.4C) consists of two parts (lamina anterior and posterior), each containing fascicles of different staining intensity. The laminae are connected at their ventrolateral sides by broad fiber bundles. Form and position of the central body in the onychophoran brain appear very dissimilar to the central body in insect brains (Chapter 15 by Homberg in this volume) and the nuchal commissure in polychaete brains considered as homologous by Holmgren (1916). No equivalent for the eightfold subdivision of insect central bodies can be detected. The pericarya of the central body neuropil are found dorsal to and between the lamina anterior and posterior.

The bridge in the onychophoran brain (Fig. 8.2) does not show structural similarities to the bridge in the insect brain, except for its median position in the hemispheres. The antennal glomeruli (approximately 30 on each side, including a pair of large glomeruli) in the ventrolateral, rostral brain parts below the antennal lobes consist of a dense meshwork of tiny fibers resembling very much insect antennal lobe glomeruli (Pareto, 1972; Ernst et al., 1977); but the arrangement of thick and thin fiber elements in the onychophoran antennal glomeruli appears to be inverted, as thicker elements envelope a dense "core" of fine fibers.

Other brain compartments are the antennal lobe (formerly named tract) with the antennal commissure, the subantennal lobe (formerly considered as tract), the lateral neuropil, and the central neuropil (Figs. 8.1–8.4). These brain regions cannot be definitely circumscribed and clearly separated from each other, as glial sheaths are lacking and their fiber masses appear interwoven at many sites. In fact, we found here fibers mainly originating from nearby pericaryal groups, but not forming distinct fiber bundles, as seen in the mushroom and central bodies. These fiber masses represent a common neuropilar pool loosely comparable with the so-called aglomerular neuropil of arthropod brains, although this useful differentiation into glomerular and aglomerular neuropil is artificial. As no Golgi stains of onychophoran brain neurons are available, a more precise description of the neuropil organization cannot be given.

The eye has its own peripheral optic ganglion as depicted by Hanström (1935) and confirmed by Eakin and Westfall (1965). Special central projection areas such as the optic tubercles in insect brains do not exist, in contrast to the huge central areas of the antennal lobes and antennal glomeruli integrating antennal sensory information.

8.2.4. Connections and Commissures

As sharply seen by electron microscopy and in contrast to arthropod brains and many polychete cerebral ganglia, the onychophoran brain does not have a substantial amount of medium-sized and giant fibers serving as relay cells and plurisegmental neurons. Thus, even state-of-the-art methods do not allow elucidation of a comprehensive, dense network of tracts serving for connectivity of brain compartments, known fairly well for both higher invertebrates and vertebrates. Holmgren (1916) and Hanström (1935) considered several large compartment tracts (antennal and subantennal lobes) for which we would like to reserve the term lobe, because these regions must be regarded as areas of synaptic integration more than structures conveying information. In contrast to previous authors, we emphasize that compartmental connectivity in the onychophoran brain is basically achieved by two modes: (1) by bundles of tiny globule cell fibers, and (2) fibers of larger diameters gathered into commissures and in some areas of the central neuropil. We consider as tracts the so-called pedunculus, consisting of tiny fibers of globule cells that link the corpora pedunculata with both parts of the central body. A tract of similar quality, the optic tract (Fig. 8.1), connects the optic ganglion with the central body, which then passes between the antennal lobe and subantennal lobe without ramifications into the latter neuropils (but compare Holmgren, 1916). As mentioned, the accessory stalks might be regarded as a special commissural tract of small fibers between the paired corpora pedunculata. Distinct tracts of large diameter fibers were never detected for the compartments, and no such tract could be traced for a long distance. The commissures of Holmgren and Hanström are of two types: prominent fiber masses as the antennal commissural neuropil and the central neuropil and small fiber bundles to be detected in the mediosagittal plane of the brain. These latter commissures are mainly concentrated in the caudal parts of the neuropil, as stated by Hanström (1935). He describes at least five commissures, including the massive neuropils. We define only the small distinct short fiber bundles crossing the brain midline as commissures (Figs. 8.2B,C, 8.3).

If we consider the brain as a fusion product of segmental neuromeres, then the number of detected commissures does not coincide with the number of 10 separated commissural bundles per segment in the ventral cord (Feodorov, 1926). Thus, commissures in adult brains cannot be homologized with those in the ventral cord and are not indicative of the three brain segments known from embryology. In our opinion, the numbers of tracts and commissures extensively described by Holmgren (1916) do by no means reflect the potential of interaction of compartments in the onychophoran brain.

8.3. ULTRASTRUCTURE

No electron miscroscopic studies so far describe the pericaryal layer in the onychophoran brain and only one investigation gives some basic information on synaptic structure and modes of connectivity in the ventral nerve cord (Schürmann, 1978). We document here only some features of the uniform pericaryal rind (Fig. 8.5) and give some first insight in selected parts of the large neuropil (Figs. 8.6–8.10).

8.3.1. Pericaryal Layer

As in other articulates, the brain is surrounded by an extracellular neurilemm with perpendicularly arranged layers of collagen in a light ground substance (Fig. 8.5A). This lamina is followed by a thin layer of glial cytoplasm with interdigitating membranes. The rare glial cells can hardly be distinguished from neuronal pericarya. No gap and tight junctions typical for insect glia (Lane et al., 1977; also see Chapter 14 by Carlson in this volume) were

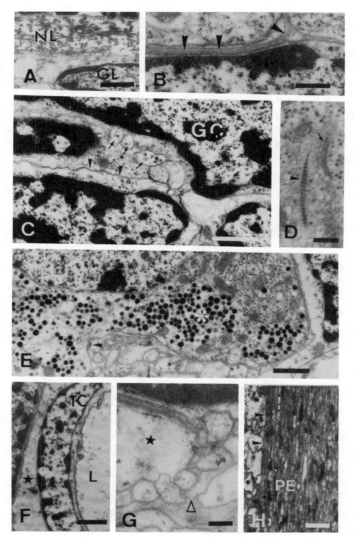

Figure 8.5. *Peripatoides leuckarti*, brain, electron micrographs. (A) Neural lamella (NL) with collagen and a glial cell (GL) of the perilemm, scale 1 μm. (B) Widened extracellular space (*arrowheads*) accompanied by glial and neuronal cytoplasm, scale 0.5 μm. (C) Globule cells (GC) with direct contact of membranes (*arrowheads*), axons with dense core vesicles (*arrows*) surrounded by neuronal cytoplasm, scale 1 μm. (D) Ciliary rootlet in a brain nerve cell (*arrowhead*), Golgi apparatus surrounded by dense core vesicles (*arrow*), scale 0.5 μm. (E). Accumulation of dense core vesicles in a neuron (*asterisk*), desmosomal contact to glial element (*arrowhead*), scale 1 μm. (F) Tracheocyt (TC) with light cytoplasm (*star*), tracheol lumen (L), scale 1 μm. (G) Cytoplasm of the tracheocyt (*star*) is directly adjacent to nerve fibers and glial cytoplasm (*triangle*), scale 0.2 μm. (H) *Pedunculus* (PE) composed of fine fibers is separated by a glial sheath (*arrowheads*) from adjacent neuropil, scale 1 μm.

detected in glial and neuronal cells, but desmosomes and hemidesmosomes at glial membranes border on widened extracellular spaces. Septate junctions of membranes considered as glial elements were occasionally found (Dallai and Giusti, 1979). Widened extracellular space with and without collagenous material is found in the pericaryal layer and the neuropilar rim (Fig. 8.5B). Tracheocytes with large cytoplasm directly border neuronal pericarya and nerve fibers without any membrane specialization (Figs. 8.5F,G).

Neuronal somata primarily of the globule cell type are usually not surrounded by glial membranes, do not have glial invaginations, and are free of synapses (Fig. 8.5C). They greatly resemble insect globule cells (Landolt, 1965). Many somata show accumulation of dense core vesicles at Golgi zones and in their proximal and distal nerve fibers (Figs. 8.5D,E). Some brain neurons show ciliary rootlets (Fig. 8.5D), perhaps indicative for central sensory cells.

8.3.2. Neuropil

8.3.2.1. General Features. Neuropilar compartments and subdivisions show glial sheath often with prominent extracellular walls incorporating collagen (Figs. 8.5H, 8.8A,G). These walls—often arranged in parallel—do not occur as unfenestrated layers. Although distributed in patterns resembling those of annelids (Coggeshall, 1965; Zimmermann, 1968; Baskin, 1971), the glial sheath in the onychophora is structurally different. Thin nerve fibers occur predominantly without glial wrapping and even giant fibers show only poorly developed glial sheaths. Nerve fibers seem to lack neurofilaments known from annelid nervous systems (Günther and Schürmann, 1973).

8.3.2.2. Synapses and Synaptic Connectivity. Synapses in the onychophoran brain neuropil show the features described for the ventral nerve cord contact sites (Schürmann, 1978). They resemble the annelid type synapses (Dhainaut-Courtois and Warembourg, 1969) (Figs. 8.6, 8.7), but show a more distinct presynaptic apposition of paramembranous material with regular patterns canalizing the access of synaptic vesicles to the membrane. Dyad synapses—typical for insects (Schürmann, 1980)—seldom occur in small diameter fibers. As a peculiarity, extensive synaptic areas were detected in the proximal corpora-pedunculata system (Fig. 8.6A). Presynaptic bouton-like endings appear encapsulated by one or several postsynaptic elements. In these presynaptic fibers we often found coated vesicles and coated pits indicating synaptic vesicle endocytosis (Heuser and Reese, 1973; Zimmermann, 1982). Fibers contain clear, round, and dense core vesicles together (Fig. 8.6), but these occur in different ratios which are characteristic for the various brain regions. Concentration of fibers with a predominance of clear vesicles were detected for the antennal glomeruli and some parts of the antennal and subantennal lobes. Fibers with accumulation of dense core vesicles exist in the bridge, the central body, and some foci of the central

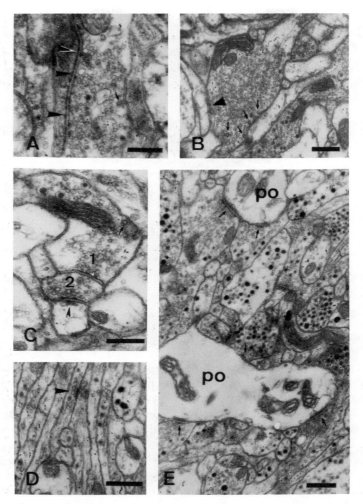

Figure 8.6. *Peripatoides leuckarti,* brain, electron micrographs. (A) Zone of synaptic contact in the proximal corpora pedunculata stalk, subsynaptic paramembranous material (*arrowheads*) and presynaptic figures, note coated pit (*arrow*), scale 0.5 μm. (B) Bouton with clear synaptic vesicles, synaptic site (*triangle*) and coated vesicles (*arrows*), scale 0.5 μm. (C) Serial synapses in the central neuropil, fiber 1 presynaptic to fiber 2, which synapses on another one. Note coated vesicle (*arrow*) and subsynaptic smooth endoplasmic reticulum (*arrowhead*), scale 0.5 μm. (D) Synapse between fine fibers (*arrowhead*), central body, scale 0.5 μm. (E) Convergence of synapses (*arrows*) on postsynaptic elements (po), central body, scale 0.5 μm.

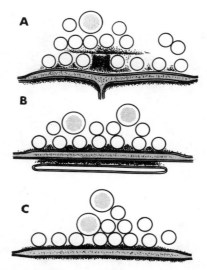

Figure 8.7. (A) Schematic drawing of an insect, (B) onychophoran, and (C) annelid synapse (Schürmann, 1978b).

neuropil. The diameters and distribution of vesicle types are diverse, as shown for *Limulus* synaptic vesicles (Fahrenbach, 1979; see also Chapter 4 by Fahrenbach and Chamberlain in this volume). Subsynaptic densities are often associated with a stack of smooth endoplasmic reticulum (Fig. 8.6C). We frequently found several types of synaptic arrangements: serial synapses, divergent synaptic connectivity, and dendritic-like synaptic convergence, which are nonrandomly distributed (Fig. 8.6). Synapses were found outside the brain in the proximal antennal nerves (Fig. 8.8), which can be considered as an external brain part. "Nerve" synapses occur in spider legs (Foelix, 1975).

To establish some basis for comparison, we took some samples from different brain compartments. The antennal glomeruli represent sites of complex synaptic wiring of possibly fine afferent sensory fibers surrounded and invaginated by light postsynaptic elements (Fig. 8.9). This structuring is more complicated than simple divergence. We confirm the connectivity of the antennal glomeruli with the antennal lobe and the mushroom body system shown by Holmgren (1916). Electron microscopy reveals inhomogeneity of the corpora pedunculata neuropil texture and synaptic equipment (Fig. 8.10). The proximal stalks (formerly named trabeculae) are the initial sites of synaptic interaction and are thickened by fiber bundles connecting with different compartments (bridge, pedunculus, fibers from the anntenal lobe and glomeruli). We tentatively consider these proximal column parts as calyxes, although the synaptic pattern appears more complex than in insect mushroom body calyxes. In addition to the prominent divergent boutons (Fig.

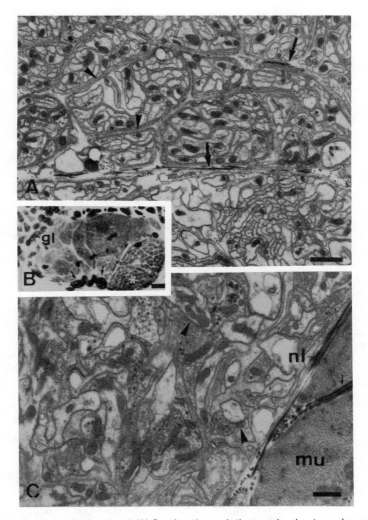

Figure 8.8. *Peripatoides leuckarti.* (A) Section through the proximal antennal nerve. Small fibers are gathered to bundles surrounded by glial sheaths (*arrowheads*) with widened extracellular space, partially filled with collagen (*arrows*), scale 1 μm. (B) Transverse section through the proximal antennal nerve, light micrograph. Note compartments of fiber bundles (*asterisk*), globule cells (gl), glial cells (*arrows*), scale 50 μm. (C) Synaptic neuropil in the proximal antennal nerve (compare A, B), synapses (*arrowheads*), neural lamella with collagen (nl), muscle (mu), scale 0.5 μm.

173

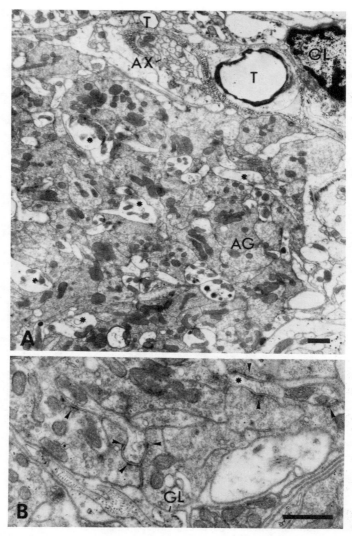

Figure 8.9. *Peripatoides leuckarti.* (A) Antennal glomerulus (AG) with abundant presynaptic profiles (*asterisk*), light elements surround and penetrate the glomerulus, tracheole (T), glial cell (GL), axon bundles (AX), scale 1 μm. (B) Profiles filled with clear vesicles synapse (*arrowheads*) with light postsynaptic elements. Some of these contain vesicles (*asterisk*), glial cytoplasm (GL), scale 0.5 μm.

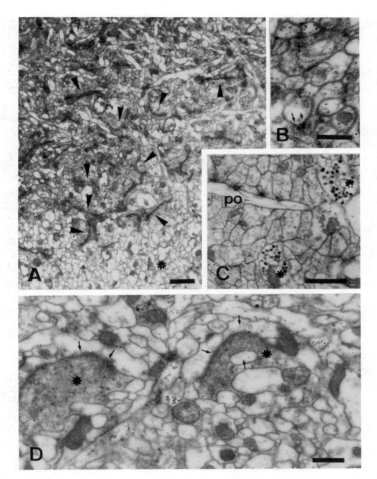

Figure 8.10. *Peripatoides leuckarti,* corpora pedunculata. (A) Proximal stalk with presynaptic boutons (*arrowheads*) and large synaptic contacts, fine stalk fibers (*asterisk*), scale 1 μm. (B) Stalk fibers with small synaptic sites (*arrows*) resembling insect dyad synapses in the mushroom body stalk, scale 0.5 μm. (C) Stalk with postsynaptic profile (po) and elements filled with dense core vesicles (*asterisk*), scale 1 μm. (D) Characteristic presynaptic boutons of the proximal stalk (*asterisk*) with surrounding and invaginating fine fibers (*arrows*), scale 0.5 μm.

8.10) we found a convergent mode of connectivity as well. Synaptic patterns along the stalks are much more mixed than in the insect corpora pedunculata columns. A clear-cut separation of input and output fibers was not detected for the onychophoran mushroom bodies.

The striation of the central body parts is based on package density and vesicle loading of fiber bundles. Although filled with numerous dense core and clear vesicles, the central body does not show as many synapses as other neuropil parts. Pons and pedunculus appear as tracts with scarce synaptic sites.

8.4. GENERAL REMARKS AND CONCLUSIONS

The morphology of the onychophoran brain leads us to consider two aspects: 1) feasibility of experimental approaches, and 2) comparison with annelid and arthropod brains. Neuropilar compartments and connections predominantly comprise tiny globule cell fibers. Large diameter fibers, among them those of descending elements, seem to play a minor role in this type of brain: in this respect, they resemble insect brains of lower order. These morphological facts limit the electrophysiological approach with intracellular recording, a highly successful access to brains and nerve cord of insects, crustaceans and annelids in the past 15 years. Extracellular recordings were tedious to perform for selected nerves and units in the ventral nerve cord of Peripatoides leuckarti (Schürmann and Sandeman, 1976), but might be promising for compartments with obvious geometry, such as the mushroom bodies, if evoked potentials could be obtained.

Compartmentation into cell clusters, so characteristic of many polychetes and leeches, is poorly developed in the onychophoran brain and could be revealed by immunocytochemical or histochemical methods.

When comparing the onychophoran brain with those of annelids and arthropods, we see no convincing need to use this brain for phylogenetic considerations. Homologization by Holmgren (1916) appears artificial on a number of points; the mushroom body system cannot at present be safely attributed to any of the onychophoran brain neuromeres. The T-shaped form of the insect mushroom body Kenyon cells is not present in the onychophoran corpora pedunculata. If the insect brain were to be tilted forward toward the neuraxis, so that the corpora pedunculata pericarya would gain a frontal position, some superficial similarities concerning the position of corpora pedunculata globule cells could be obtained. The fine structural organization (fiber arrangement, synaptic connectivity) shows some striking similarities, but no separation of in- and output to the system, as stated for the insect mushroom bodies (Schürmann, 1974), is obvious. A parallel to the corpora pedunculata of insects and onychophora can be seen in their strong direct relationship to antennal neuropils and the lack or minor development of direct optic connectivity.

The central bodies of insect and onychophoran brains receive different inputs. Whereas the onychophoran central body is closely related to the optic ganglion and to the mushroom body system by the massive pedunculus, such a strong direct connection cannot be stated for insect mushroom bodies.

The important question of the functional necessity of such prominent globule cell masses developed in the onychophoran as well as in other invertebrate brains, cannot yet be answered satisfactorily. Suggestions derived from classical neuroanatomy to understand the brain-specific globule cells await further substantiation by embryological and physiological experiments.

8.5. SUMMARY

The brain of the Onychophora reflects their special evolution by its compartmentation, tracts, and cytological features. Diversification into neuron types and clusters is poorly developed. Compartments are made up of the fine fibers of the abundant uniform globule cells. Giant neurons with giant fibers are scarcely found. Tracts and commissures are mainly formed by bundles of tiny fibers. Only some structural similarities to annelid and arthropod brains were detected concerning position and selected parts of the corpora pedunculata, central body, bridge, and antennal glomeruli. The antennal nerve contains integrative synaptic neuropil. Synapses resembling the annelid type are restricted to the neuropil.

ACKNOWLEDGMENTS

I thank M. Klages and M. Knierim-Grenzebach for technical assistance in histology, electron microscopy, and photography. Parts of the investigation were performed in the Department of Neurobiology, Research School of Biological Science, Canberra, Australian National University, during a visit in 1975–1976, supported by a DFG grant and by a fellowship from the A.N.U. Thanks are due to Prof. Dr. G. A. Horridge of Canberra, and R. Hardie of Armidale, who kindly supplied Onychophora.

Note added in proof: Most recently, a detailed investigation of the ultrastructure of neuron-glia relations has been published: Lane, N. J. and S. S. Campiglia. 1987. The lack of a structured blood-brain barrier in the onychophoran Peripatus acacioi. *J. Neurocytol.* **16,** 93–104.

REFERENCES

Anderson, D. T. 1966. The comparative early embryology of the Oligochaeta, Hirudinea and Onychophora. *Proc. Linn. Soc. New South Wales* **91:** 10–43.

Anderson, D. T. 1979. Embryos, fate maps, and the phylogeny of Arthropods, pp. 59–105. In A. P. Gupta (ed.), *Arthropod Phylogeny,* Van Nostrand-Reinhold, New York.

Ax, P. 1984. *Das Phylogenetische System.* Gustav Fischer-Verlag, Stuttgart, New York.

Balfour, F. M. 1883. The anatomy and development of *Peripatus capensis. Quart. J. Microsc. Sci.* **23:** 213–259.

Baskin, D. G. 1971. The fine structure of neuroglia in the central nervous system of nereid polychaetes. *Z. Zellforsch.* **119**: 295–308.

Bullock, T. H. and G. A. Horridge. 1965. *Structure and Function in the Nervous Systems of Invertebrates.* vol. 1. W. H. Freeman, San Francisco, London.

Camatini, M., E. Franchi, and G. Lanzavecchia. 1979. The body muscles of Onychophora: An atypical contractile system, pp. 419–431. In M. Camatini (ed.), *Myriapod Biology.* Academic Press, London.

Clark, R. B. 1958. The morphology of the supra-oesophageal ganglion of Nephtys. *Zool. Jahrb. (Allg. Zool.)* **68**: 261–296.

Coggeshall, R. E. 1965. A fine structural analysis of the ventral nerve cord and associated sheath of *Lumbricus terrestris* L. *J. Comp. Neurol.* **125**: 393–438.

Dallai, R. and F. Giusti. 1979. The epithelial cell junctions in Onychophora, pp. 433–443. In M. Camatini (ed.), *Myriapod Biology.* Academic Press, London.

Dhainaut-Courtois, N. and M. Warembourg. 1969. Etude ultrastructurale des neurones de la chaine nerveuse de *Nereis pelagica* L. (Annelide Polychete). *Z. Zellforsch.* **97**: 260–273.

Eakin, R. M. and J. L. Brandenburger. 1966. Fine structure of antennal receptors in *Peripatus* (Onychophora). *Am. Zool.* **6**: 614.

Eakin, R. M. and J. A. Westfall. 1965. Fine structure of the eye of *Peripatus* (Onychophora). *Z. Zellforsch.* **68**: 278–300.

Ernst, K.-D., J. Boeckh, and V. Boeckh. 1977. A neuroanatomical study on the organization of the central antennal pathways in insects. *Cell Tissue Res.* **176**: 285–308.

Ewer, D. W. and R. van den Berg. 1954. A note on the pharmacology of the dorsal musculature of *Peripatopsis. J. Exp. Biol.* **31**: 497–500.

Fahrenbach, W. H. 1979. The brain of the horseshoe crab (*Limulus polyphemus*). III. Cellular and synaptic organization of the corpora pedunculata. *Tissue Cell* **11**: 163–200.

Feodorov, B. 1926. Zur Anatomie des Nervensystems von *Peripatus*. I. Das Neurosomit von *Peripatus tholloni* Bouv. *Zool. Jahrb. Anat.* **48**: 273–310.

Feodorov, B. 1929. Zur Anatomie des Nervensystems von *Peripatus*. II. Das Nervensystem des vorderen Körperendes und seine Metamerie. *Zool. Jahrb. Anat.* **50**: 279–332.

Florey, E. and E. Florey. 1965. Cholinergic neurones in the Onychophora: A comparative study. *Comp. Biochem. Physiol.* **15**: 125–136.

Foelix, R. F. 1975. Occurrence of synapses in peripheral sensory nerves of arachnids. *Nature (London)* **254**: 146–148.

Gabe, M. 1954. La neurosecretion chez les invertebres. *Ann. Biol.* **30**: 5–62.

Gaffron, J. 1884. Kurzer Bericht über fortgesetzte *Peripatus*-Studien. *Zool. Anz.* **7**: 336–339.

Gardner, C. R. and E. A. Robson. 1978. A response to monoamines in *Peripatopsis moseleysis* (Onychophora). *Experientia* **34**: 1576.

Gardner, C. R., E. A. Robson and C. Stanford. 1978. The presence of monoamines in the nervous system of *Peripatopsis* (Onychophora). *Experientia* **34**: 1577.

Gardner, C. R. and R. J. Walker. 1982. The roles of putative neurotransmitters and neuromodulators in annelids and related invertebrates. *Prog. Neurobiol.* **18:** 81–120.

Günther, J. and F. W. Schürmann. 1973. Zur Feinstruktur des dorsalen Riesenfasersystems im Bauchmark des Regenwurms. *Zellforsch. Mikrosk. Anat.* **139:** 369–396.

Gupta, A. P. 1979. Origin and affinities of myriapoda, pp. 373–390. In M. Camatini (ed.), *Myriapod Biology*. Academic Press, London.

Haase, E. 1889. *Die Bewegungen von* Peripatus. *Sitzungsberichte Ges. Naturforsch,* pp. 149–151. Freunde, Berlin.

Hanström, B. 1928. *Vergleichende Anatomie des Nervensystems der wirbellosen Tiere.* Springer, Berlin.

Hanström, B. 1935. Bemerkungen über das Gehirn und die Sinnesorgane der Onychophoren. *Lunds Univ. Arskr. N. F.* **31:** 1–37.

Henry, L. M. 1948. The nervous system and the segmentation of the head in the Annulata. V. Onychophora. VI. Chilopoda. VII. Insecta. *Microentomology* **13:** 27–48.

Heuser, J. E. and T. S. Reese. 1973. Evidence for recycling of synaptic vesicle membrane during transmitter release at the frog neuromuscular junction. *J. Cell Biol.* **57:** 315–344.

Holmgren, N. 1916. Zur vergleichenden Anatomie des Gehirns von Polychaeten, Onychophoren, Xiphosuren. *Kungl. Svenska Vet. Akad. Handl.* **56:** 1–299.

Horridge, G. A. 1965. Onychophora, pp. 791–798. In T. H. Bullock and G. A. Horridge (eds.), *Structure and Function in the Nervous Systems of Invertebrates,* vol. 1. Freeman, San Francisco, London.

Hoyle, G. and J. del Castillo. 1979. Neuromuscular transmission in *Peripatus. J. Exp. Biol.* **83:** 13–29.

Hoyle, G. and M. Williams. 1980. The musculature of *Peripatus* and its innervation. *Philos. Trans. R. Soc. Lond. B.* **288:** 481–510.

Klemm, N. 1985. The distribution of biogenic monoamines in invertebrates, pp. 280–296. In R. Gills und J. Balthazar (eds.), *Neurobiology*. Springer, Berlin, Heidelberg.

Landolt, A. M. 1965. Elektronenmikroskopische Untersuchungen an der Perikaryenschicht der Corpora pedunculata der Waldameise (*Formica lugubris* ZETT.) mit besonderer Berücksichtigung der Neuron-Glia-Beziehung. *Z. Zellforsch.* **66:** 701–736.

Lane, N. J., H. le B. Skaer, and L. S. Swales. 1977. Intercellular junctions in the central nervous system of insects. *J. Cell Sci.* **26:** 175–199.

Lanzavecchia, G. and M. Camatini. 1979. Phylogenic problems and muscle cell ultrastructure in Onychophora, pp. 407–417. In M. Camatini (ed.), *Myriapod Biology*. Academic Press, London.

Manaranche, R. 1966. Anatomie du ganglion cerebroide de Glycera convoluta Keferstein (Annelide Polychete), avec quelques remarques sur certains organes prostomiaux. *Cahiers Biol. Marine* **12:** 259–280.

Manton, S. M. 1938. Studies on the Onychophora. VI. The life history of *Peripatopsis. Ann. Mag. Nat. Hist.* **1:** 515–529.

Manton, S. M. 1950. The evolution of Arthropodan locomotory mechanism. I. The locomotion of *Peripatus*. *J. Linn. Soc. Lond.* **41:** 529–570.

Manton, S. M. 1979. Functional morphology and the evolution of the hexapod classes. In A. P. Gupta (ed.), *Arthropod Phylogeny*. Van Nostrand-Reinhold, New York.

Pareto, A. 1972. Die zentrale Verteilung der Fühlerafferenz bei Arbeiterinnen der Honigbienen, *Apis mellifera* L. *Z. Zellforsch.* **131:** 109–140.

Pflugfelder, O. 1948. Entwicklung von *Paraperipatus amboinensis* n. sp. *Zool. Jahrb. Anat.* **69:** 443–492.

Pflugfelder, O. 1968. Onychophora, pp. 42. In *Großes Zoologisches Praktikum*, Part 13a. Fischer, Stuttgart.

Sanchez, S. 1958. Cellules neurosecretrices et organs infracerebraux de *Peripatopsis moseleyi* Wood-Mason (Onychophore) et neurosecretion chez Nymphon gracile Leach (Pycnogonide). *Arch. Zool. Exp. Gen.* **96:** 57–62.

Sänger, N. 1870. *Peripatus capensis* gr. et *Peripatus leucartii* n. sp. In *Trav. 2. Congr. Nat. Russ.,* Moscow 1869.

Sawyer, R. T. 1984. Arthropodization in the Hirudinea: Evidence for a phylogenetic link with insects and other uniramia? *Zool. J. Linn. Soc.* **80:** 303–322.

Schürmann, F. W. 1974. Bemerkungen zur Funktion der Corpora pedunculata im Gehirn der Insekten aus morphologischer Sicht. *Exp. Brain Res.* **19:** 406–432.

Schürmann, F. W. 1978a. On the ultrastructure of nerve-muscle contacts in Onychophora. *Experientia* **34:** 779.

Schürmann, F. W. 1978b. A note on the structure of synapses in the ventral nerve cord of the onychophoran *Peripatoides leuckarti*. *Cell Tissue. Res.* **186:** 527–534.

Schürmann, F. W. 1980. Methods for special staining of synaptic sites, pp. 241–262. In N. J. Strausfeld and T. A. Miller (eds.), *Neuroanatomical Methods. Experimental Entomology,* vol. 1A. Springer-Verlag, New York, Heidelberg, Berlin.

Schürmann, F. W. and D. C. Sandeman. 1976. Giant fibres in the ventral nerve cord of *Peripatoides leuckarti* (Onychophora). *Naturwissenschaften* **63:** 580.

Storch, V. and H. Ruhberg. 1977. Fine structure of the sensilla of *Peripatopsis moseleyi* (Onychophora). *Cell Tissue. Res.* **177:** 539–553.

Strausfeld, N. J. and D. R. Nässel. 1980. Neuroarchitecture of brain regions that subserve the compound eyes of crustacea and insects, pp. 2–133. In H. Autrum (ed.), *Handbook of Sensory Physiology,* vol. 7/6b. Springer, Berlin, New York.

Tombes, A. S. 1979. Comparision of arthropod neuroendocrine structures and their evolutionary significance, pp. 645–667. In A. P. Gupta (ed.), *Arthropod Phylogeny*. Van Nostrand-Reinhold, New York.

Zacher, F. 1933. Onychophora, pp. 79–138. In Kükenthal, W. and T. Krumbach (eds.), *Handbuch der Zoologie,* bd. 3. De Gruyter, Berlin.

Zimmermann, H. 1982. Insights into the functional role of cholinergic vesicles, pp. 305–359. In R. L. Klein, H. Lagercrantz, and H. Zimmermann (eds.), *Neurotransmitter Vesicles*. Academic Press, London.

Zimmermann, P. 1968. Struktur, Verteilung und Funktion der Kontaktzonen im Bauchmark von *Lumbricus terrestris* L. *Z. Zellforsch.* **87:** 137–158.

CHAPTER 9

Organization of the First Optic Neuropil (or Lamina) in Different Insect Species

Karl Kral
Department of Zoology
University of Graz
Graz, Austria

9.1. INTRODUCTION

This chapter reviews the neural organization of the first optic neuropils of the compound eyes in different insect species. It concentrates on studies published during the past 15 years and deals primarily with Golgi techniques and electron microscopy in muscoid flies by Strausfeld, in the honey bee by Ribi, and in the dragonfly, *Sympetrum*, by Meinertzhagen. In addition, it will include works on the lamina of the cockroach, *Periplaneta* (Ribi, 1977); the locust, *Schistocerca* (Nowel and Shelton, 1981); the diurnal butterfly,

Pieris, and nocturnal moth, *Sphinx* (Strausfeld and Blest, 1970); the desert ant, *Cataglyphis* (Meyer, 1979); and "exotic" insects, such as the male firefly, *Phausis* (Ohly, 1975), the mayfly, *Cloeon* (Wolburg-Buchholz, 1977), marchflies (Zeil, 1983a,b), and waterbugs (Wolburg-Buchholz, 1979). The first optic neuropil of mayflies and bibionid flies shows a distinct sexual dimorphism, and that of waterbugs shows a neural superposition that is clearly different from that seen in dipterans.

An overview of the available data reveals that, on the one hand, the first optic neuropil is a highly structured system, which most likely does not function just as a simple relay, but rather as a highly integrative unit. On the other hand, its degree of complexity varies considerably among different insect orders, ranging from the more "primitive" cell arrangement in the cockroach, an ancient insect, to the highly complex arrangement in the honey bee, a most advanced social insect. However, the structure of the neuropil can also vary considerably within an insect order.

9.2. PROJECTION OF THE RETINA INTO THE FIRST OPTIC NEUROPIL

The compound eyes of insects are composed of a number of structural units, the ommatidia. Each ommatidium has its own lens, a fixed group of photoreceptor cells, and a sheath of pigment cells. Centrally, the photoreceptor cells form either a fused rhabdom, in which case all rhabdomeres contributing to the rhabdom share a common field of view, or an open rhabdom, in which case the individual rhabdomeres are separated so that each has its own optical axis. Beneath the retina lies the fenestrated layer, and adjacent to it the first optic neuropil (or lamina) is found in the form of a curved plate. The neuropil shows two main zones: (1) the outer cell body layer containing the somata of the monopolar relay interneurons, and (2) the first synaptic region. The lamina neuropil is in most cases divided into cylindrical synaptic modules, called cartridges. All receptor cells within one ommatidium (as in fused rhabdom apposition and superposition eyes) or all receptors having the same optical axis, but belonging to different ommatidia (as in open rhabdom neural superposition eyes), project to the same cartridge. The number of cartridges is, therefore, always the same as that of ommatidia. Although the majority of the receptor axons terminate in the lamina cartridge, some of them project always to the second visual neuropil (or medulla). In general, one can say that each cartridge represents a retinotopic unit with inputs from one "point" in the visual field. The monopolar cells of the lamina are the first relay interneurons for signals from photoreceptors.

9.3. MONOPOLAR RELAY INTERNEURONS

Concerning their anatomy and physiology, the monopolar cells of the lamina are among the most thoroughly investigated neurons in a great variety of insects: in the large American cockroach (Ribi, 1977), dragonflies (Meinertz-hagen and Armett-Kibel, 1982), grasshoppers (Nowel and Shelton, 1981), Coleoptera (Ohly, 1975), and social Hymenoptera (Ribi, 1975a,b, 1981; Meyer, 1979). The most comprehensive studies, however, have been carried out on muscoid flies (Strausfeld and Nässel, 1981; Shaw, 1984). This abundance of data makes a comparison of the monopolar cells among insects worthwhile and allows an insight into some of the evolutionary trends.

9.3.1. Insects With Apposition Eyes

In the lamina of the fused rhabdom apposition eye, two types of monopolar cells can be distinguished with respect to their spatial extention: small-field cells and wide-field cells. The former have their dendrites exclusively in the parent cartridges and, therefore, receive only direct receptor inputs derived from one ommatidium; the latter extend their dendrites over a wide area of the lamina, corresponding to several cartridges, and therefore receive inputs from many ommatidia. In Figure 9.1A-F the monopolar cells of five different insect orders, with apposition eyes, are illustrated.

In the lamina of the roach, *Periplaneta americana* (Ribi, 1977), only non-stratified wide-field monopolars are described (Fig. 9.1A). Their dendritic branches spread unilaterally or bilaterally from the main fiber and extend upward diffusely over many cartridges. The number of the monopolar cells can vary: usually there are four to six. On the whole, it appears that in the cockroach lamina there is a structural "noise."

Completely different features are observed in the lamina of the dragonfly, *Sympetrum* (Meinertzhagen and Armett-Kibel, 1982; Armett-Kibel and Meinertzhagen, 1985) (Fig. 9.1B). In the lamina of this surviving remnant of a paleopteroid group, there are five types of small-field monopolar cells per cartridge, which are similar to each other. In the area between the distal and proximal lamina, several equally long, unramified branches arise from the main fiber; these branches do not appear to leave the parent cartridge but are reconstructed from serial EM which may underestimate their length. However, although the cells are structurally similar, there is a relatively complex connectivity pattern.

In the locust, *Schistocerca gregaria*, a more advanced insect, small-field monopolars and wide-field monopolars are found (Fig. 9.1C) (Nowel and Shelton, 1981). In contrast to the nonstratified morphology of monopolars in *Sympetrum*, the corresponding cells in *Schistocerca* are bistratified, that is, the cells have a characteristic dendritic pattern in two distinct layers of the neuropil. The monopolars in *Schistocerca* can be classified in six types. Five of the six cells form dendrites in both layers. One cell has branches only in the proximal layer. M1 is the "pipe cleaner" cell. The thick, pipe cleaner-shaped cell is in contact with the receptor axons of one ommatidium. M2, the "side-arm" cell, has bilateral branches that extend to neighboring cartridges. M3 extends its branches over several cartridges. M4, the "uni-

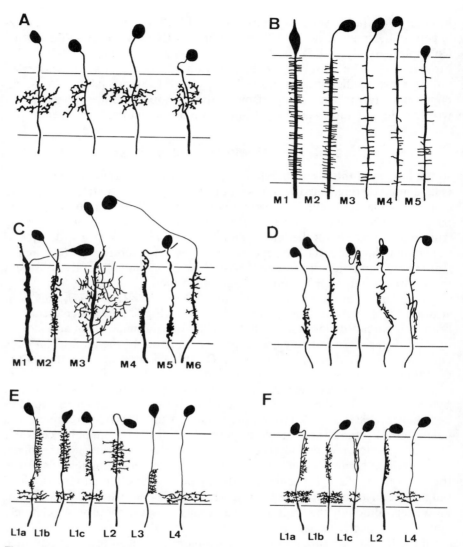

Figure 9.1. Summary diagram of the morphologies of the lamina monopolar cells in different insects with apposition eyes. (A) Wide-field cells in the American roach, *Periplaneta americana*. (B) Five types of small-field cells (M1–5) in the dragonfly, *Sympetrum rubicundulum*. (C) Five major types of the small-field monopolar cells (M1, 4, 5; M2, 6, see text) and the wide-field monopolar cell M3 in the locust, *Schistocerca gregaria*. (D) Small-field monopolars of the diurnal butterfly, *Pieris brassicae*. The small-field cells and the wide-field cells in (E) the honeybee, *Apis mellifera*, and (F) the desert ant, *Cataglyphis bicolor*. (A from Ribi, 1977; B from Meinertzhagen and Armett-Kibel, 1982; C from Nowel and Shelton, 1981; D from Strausfeld and Blest, 1970; E from Ribi, 1974; F from Meyer, 1979.)

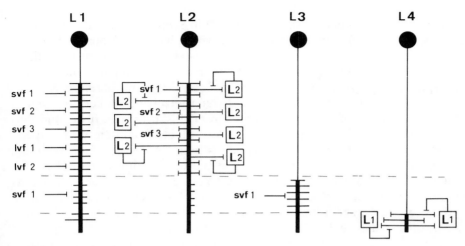

Figure 9.2. Possible connectivity patterns of the four monopolar cells L1-4 in the lamina of the honeybee, *Apis mellifera*: L1 receives synaptic inputs in the distal stratum from short visual fibers svf1-3 and long visual fibers lvf1 and 2, and in the medial stratum from svf1. L2 receives inputs in the distal stratum from all short visual fibers (svf1-3) of the parent cartridge, and inputs from L2 of seven neighboring cartridges. L3 receives inputs from svf1 in the medial stratum, and L4 receives inputs in the proximal stratum from L1 of the six neighboring cartridges. (From Ribi, 1981.)

lateral'' cell, has branches on only one side of the axon. In contrast to all the other cells, M5 has its branches only in the proximal neuropil. Finally M6, the "bean stalk" neuron, lies between the cartridges (and is in respect analogous to the large L5 of *Musca*).

By contrast, in the diurnal cabbage butterfly, *Pieris brassicae* (Strausfeld and Blest, 1970) only small-field monopolars are found (Fig. 9.1D). There are four different types, although a significant stratification cannot be observed. Here and there short branches might reach adjacent cartridges.

In highly advanced social Hymenoptera, such as the honey bee, *Apis mellifera* (Ribi, 1975a,b, 1981), and desert ant, *Cataglyphis bicolor* (Meyer, 1979, 1984), there are both types of cell, small- and wide-field cells, which are, however, significantly more differentiated compared with those of less advanced groups (Fig. 9.1E,F). This differentiation includes (1) multistratification, (2) defined location within a cartridge, and (3) specific direction within the neuropil. Figure 9.2 shows a scheme of the connectivity pattern of the bee monopolars established by Ribi (1981). In the proximal layer, the wide-field neuron L4 possesses bilateral extensions to six more cartridges along the dorsoventral axis of the eye. All short visual fibers (green and blue receptors) of one ommatidium have presynaptic contacts with L1 and 2 in the distal layer. Two short visual fibers are also presynaptic to L1 and 2 in the medial layer. On the other hand, L2 is presynaptic to short visual fibers in the distal layer. This example shows that monopolars can also be presynaptic to visual fibers (also known for M1 in *Sympetrum* and L2 in *Musca*). The long visual fibers (UV receptors), which have their terminals in the medulla, can be in contact with L1 in the distal layer. An analysis of these connectivities shows that L1 receives multiple synaptic inputs from all three receptor types: green, blue, and UV receptors. L1 can therefore be characterized according to Menzel (1977, 1979; Menzel and Blakers, 1976) as a broad-band neuron, L2 receives inputs from all short visual fibers and thus from green and blue receptors, but no UV inputs from long visual fibers. Furthermore, there are feedback synapses from L2 to short visual fibers (in *Sympetrum* and *Musca* such feedback synapses are found only in the proximal lamina). L2 may be, therefore, a color-opponent neuron with inputs either from green or blue receptors. Because L3 receives inputs only from green receptors, according to Menzel (1977) this cell might be a green monochromatic narrow-band neuron. L4 is also connected by collaterals with the broad-band L1 of neighboring cartridges. In the light of these data presented by Ribi (1981), one can see that in the lamina of the bee four physiological types of monopolars may be related to four morphological types. Many characteristics of the physiological cell types in the bee are also true for *Sympetrum* (Fig. 9.3), where M2 receives broad-band inputs; M1 receives input from all short receptor terminals excluding the long visual fibers, and M3 receives selective input from red or from UV receptors and hence is a narrow-band neuron. M5 receives input from the violet-sensitive long visual fiber 7 and its partner 6, and so is probably also a narrow-band neuron (Meinertzhagen et al., 1983; Armett-Kibel and Meinertzhagen, 1985).

In the desert ant, *Cataglyphis bicolor* (Meyer, 1979), the three types of monopolars L1, 2, 4 (L3, described in *Apis*, is not found in *Cataglyphis*) possess dendrites only in the distal and proximal layer, whereas the intermediate layer is absolutely free of dendrites (Fig. 9.1F). L1 has its branches in both layers, whereby the pattern of branches is different in the two layers. L2 is a small-field cell with short branches only in the distal layer. L4 is, like that in *Apis*, a wide-field cell with branches in the proximal layer. The collaterals of L4 are restricted to the dorsoventral axis corresponding to the bisector of the x and y axes in the receptor lattice. L4 could be, therefore, a movement detector. As in *Apis*, nine receptors (green and UV, but no blue as in bee; Mote and Wehner, 1980) of each ommatidium project to the lamina, but in contrast to *Apis*, the six short visual fibers terminate exclusively in the distal layer and only two of the three long visual fibers (UV receptors) terminate finally in the medulla (Meyer, 1984). The ninth cell ends in the most proximal layer of the lamina. All six short visual fibers are in synaptic contact with L1a, 1c, and 2. Synapses also exist between long visual fibers and L1a, 2: (1) Short visual fibers are presynaptic to monopolars and long visual fibers or endings of the ninth cell; (2) monopolars are presynaptic to visual fibers or other monopolars; (3) visual fibers of the ninth cell synapse on long visual fibers and monopolars.

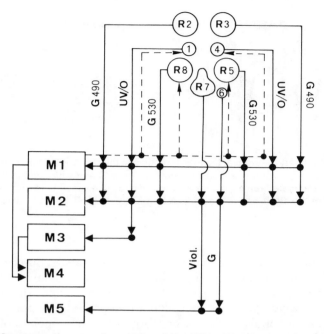

Figure 9.3. Summary diagram of synaptic relationships between the receptor cells of one ommatidium and the monopolar cells of the corresponding cartridge in the lamina of the ventral eye of the dragonfly, *Sympetrum rubicundulum*. The retinular input pathways to monopolars, as well as connections between monopolar cells, are indicated by continuous lines. Broken lines demonstrate reciprocal connections. M = monopolar cell; R = receptor cells; R2/3 and 5/8 = green receptors; R1/4 = either UV or orange receptor; R7 = violet receptor; R6 = green receptor; R1-5 = receptor cells with short visual fibers; R6/7 = receptor cells with long visual fibers. (Redrawn from Meinertzhagen et al., 1983.)

9.3.2. Insects With Superposition Eyes

The monopolars in the lamina of nocturnal insects with superposition eyes mostly belong to the wide-field type. In Figure 9.4A,B, the monopolars from two typical nocturnal insects, which, however, are fairly different in their behavior, are shown: the moth, *Sphinx ligustri* (Strausfeld and Blest, 1970), and the male firefly, *Phausis* (Ohly, 1975). In *Sphinx* (Fig. 9.4B) the organization of the lamina into cartridges is not obvious. The "imprecision" is observed in the entry zone of the receptor axons into the lamina, as they spread in all directions. The wide-field monopolar cells themselves do not show any regular arrangement. Rather, they are similar to the primitive monopolars in *Periplaneta*. The degree of convergence is high. The monopolars receive inputs from as many as 20 or more ommatidia.

In contrast to *Sphinx* and other nocturnal insects, five types of monopolar cells can be clearly observed in *Phausis* (Fig. 9.4A). In addition, besides the wide-field cells, there are also small-field cells. The three wide-field elements

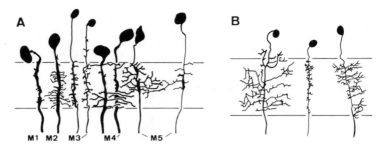

Figure 9.4. Summary diagram of the morphologies of the monopolar cells in two nocturnal insects with superposition eye. (A) Small-field (M1 and 3) and wide-field (M2, 4, 5) monopolar cells in the male firefly, *Phausis splendidula*. (B) One small-field and two wide-field monopolars of the moth *Sphinx ligustri*. (A from Ohly, 1975; B from Strausfeld and Blest, 1970.)

can have contacts with more than 40 retinotopic channels. Some monopolars have large asymmetrically arranged branches, which points to a fairly unorganized wiring of the neuropil, but, on the other hand, the monopolars are stratified, which points to an ordered segregation of inputs from visual fibers among monopolars. The comparison of the monopolars of both nocturnal species shows that the lamina of *Phausis* is substantially more differentiated than that of *Sphinx*.

Another remarkable insect in this context is the male mayfly, *Cloeon dipterum* (Ephemeridae), which probably is the oldest phylogenetic group of pterygote insects. Male *Cloeon* possess two separate types of compound eyes: a dorsal superposition eye covering the whole dorsal head, and a lateral apposition eye (Wolburg-Buchholz, 1977). But females of this species have only lateral apposition eyes that are smaller than those of the males. In males, the lamina is therefore divided into a dorsal and lateral neuropil (Wolburg-Buchholz, 1977). In the dorsal superposition eye, all seven visual fibers of one ommatidium terminate in the lamina cartridge. But in the lateral apposition eye of both males and females, usually six visual fibers terminate in the lamina and the remaining two proceed on to the medulla. It is surprising that in both the lamina of the dorsal superposition eye and that of the lateral apposition eye of the male only small-field monopolars exist, as wide-field cells are typically for the superposition lamina. Three distinct types of monopolars can be identified by means of their branching pattern. All three cells, however, do not show clear stratification.

9.3.3. Insects With Neural Superposition Lamina

The neural superposition eye is basically different from both types of other eyes, in that one cartridge receives the axons of those receptor cells from within different ommatidia, which have parallel optical axes (Kirschfeld,

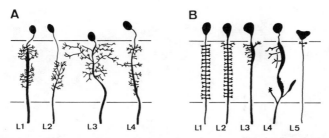

Figure 9.5. Summary diagram of the morphologies of the monopolar cells in two insects with neural superposition laminae. (A) Four monopolars (L1–4) in the waterbug *Notonecta glauca*. Their connectivity pattern is shown in Figure 9.6. (B) Five types of monopolars (L1–5) in the housefly, *Musca domestica*. (A from Wolburg-Buchholz, 1979; B from Strausfeld and Nässel, 1981.)

1967). Figure 9.5A,B shows the monopolar cells of two diurnal insects with a neural superposition lamina. In the tristratified lamina (three levels of receptor endings) of the waterbug *Notonecta glauca* (Wolburg-Buchholz, 1979), there are two types of small-field monopolars and two of wide-field monopolars (Fig. 9.5A). The receptor axons are connected with three of them in the following manner (Fig. 9.6). A typical divergence pattern of visual fibers is seen, in that each cartridge receives information originating

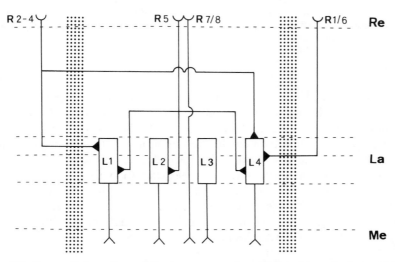

Figure 9.6. Connectivity pattern of the four monopolar cells L1–4 in a single cartridge of the lamina of the waterbug, *Notonecta glauca*. R5 and R7/8 = receptor cell axons of corresponding ommatidium; R2-4 and R1/6 = receptor cell axons of neighboring ommatidium. L1 receives synaptic inputs from R2-5; L2 is only postsynaptic to R5, whereas L4 receives inputs from all short receptor cells (R1–6). The intrinsic L1 and 4 contact each other. La = lamina; Me = medulla; Re = retina. (Redrawn from Wolburg-Buchholz, 1979.)

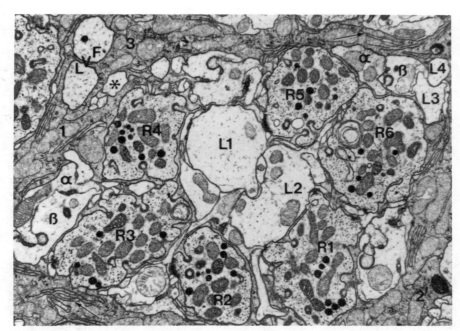

Figure 9.7. Cross section through a cartridge of the lamina of the housefly, *Musca domestica*. The two monopolar cells L1 and 2 are surrounded by the terminals of the six short visual fibers (R1–6). L3 and 4 lie at the periphery joined to α-processes of amacrine cells and β-elements of T1 centripetal cells. The pair of long visual fibers (LVF) runs between the cartridges. The small profile marked by an asterisk may belong to L5 (12,500×). (From St. Marie and Carlson, 1983.)

from receptors with the same visual axis belonging to different ommatidia. The receptor cells R1–6 of one ommatidium can be divided into groups according to their connectivity patterns in the lamina (R7 and 8 have no described synaptic involvement in the lamina). The three receptors of the first group terminate in a neighboring cartridge and there have presynaptic contacts with the small-field L1 and with the wide-field L4. The second group, consisting of two receptors, has contact only with the wide-field L4 in a neighboring cartridge. The remaining receptor cell projects to the underlying cartridge, where it makes synaptic contacts both with small-field L1 and 2 and with the wide-field L4. Furthermore, the intrinsic L1 and 4 contact each other. Thus, the connectivity pattern shows that the information of different ommatidia converges to a cartridge according to a determined pattern.

Five types of monopolars are found in the neural superposition lamina of the housefly, *Musca domestica* (Figs. 9.5B, 9.7). Strausfeld (1976a) presents an outline of the complex connectivity pattern:

1. Within the six different ommatidia six receptor cells with the same optical axis converge upon a single cartridge (convergence).
2. The six equivalent receptors are presynaptic to L1, 2, and 3 as well as other cells, such as α-processes of amacrine cells (divergence).
3. The six identical receptors are presynaptic to L2, and in the proximal lamina L2 is simultaneously presynaptic to the receptor cells.
4. L2 is presynaptic to L1 (lateral interaction in a single cartridge).
5. Receptor cells contact L4 via amacrine cells. L4 is again presynaptic to L1, 2, and 4 (lateral interaction across several cartridges).
6. L5 and the T-shaped centripetal neuron T1 have no direct synaptic input from receptor cells (Shaw, 1984).

The long visual fibers of the receptors R7 and 8 project, as do those of R7 and 8 in Hemiptera and of R8 in the crayfish to the medulla without synaptic contacts in the lamina. This is in contrast to Hymenoptera (Ribi, 1981; Meyer, 1979, 1984) and dragonflies (Armett-Kibel and Meinertzhagen, 1985), in which the long visual fibers are known to be synaptic in the lamina.

Male marchflies possess a specialized dorsal part of the eye, which is much larger than the ventral part. The facets there are significantly larger and receptor cells are longer than elsewhere (Zeil, 1983a,b). The dorsal eye lamina of the male is unilayered, whereas the ventral eye lamina in males and the lamina in females are multilayered (Fig. 9.8). The dorsal eye of males is a neural superposition eye where the axons of six receptors from six different ommatidia converge, projecting the same point in space onto one cartridge. In contrast to higher flies, however, the group of receptor cells with parallel optical axes do not lie in neighboring ommatidia but in two ommatidia separated by a third. Their terminals in the lamina appear to pass to cartridges separated by a third one. In the ventral eye of males and in the eye of females, the receptor cells with coincident optical axes, however, lie in neighboring ommatidia, but these are arranged in a horseshoe pattern. In contrast to the five monopolars in higher flies (Fig. 9.5B), only three types of monopolars (L1, 2, 5) have been found in the ventral lamina of males and in the lamina of female marchflies (Fig. 9.8A). These all branch in the region where short and long visual fibers terminate (in the ventral lamina of the male and in the lamina of the female long visual fibers have branches) so that their input field covers all three lamina layers. Four different types of monopolars (L1–3, 5) have been found, however, in the specialized dorsal lamina of males (Fig. 9.8B). Here, the long visual fibers pass the lamina without making contact with the monopolars.

Males of *Musca* and *Calliphora* also have a specialized dorsal part of the eye, but in contrast to marchflies it is usually only a small dorsofrontal region (Franceschini et al., 1981; Hardie et al., 1981; Hardie, 1983). In the frontal–dorsal region of the eye of the male *Musca*, the facets are much larger than elsewhere and therefore the interommatidial angles are smaller. The larger

(A)

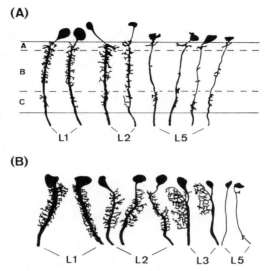

(B)

Figure 9.8. Summary diagram of the monopolar cells (L) in the lamina of the bibionid, *Dilophus febrilis*. (A) Ventral eye lamina in males and lamina in females. A, B, and C = horizontal layers of the neuropil. (B) Unilayered dorsal eye lamina of males. (Redrawn from Zeil, 1983b.)

facets are represented by a wide diameter and widely spaced cartridges in the corresponding part of the lamina. The larger cartridges cause a larger density of dendritic branches of L4 and a larger number of amacrine cells. As in the wiring diagram of the normal lamina, both elements are synaptically associated, L4 providing a network of presynaptic collaterals onto other monopolars such as L4 and the amacrines providing wide-field networks of pathways between receptors of many cartridges and the L4 elements (lateral interactions across several cartridges). Although L4 and 5 are similar in the dorsofrontal eye lamina of the male and in the normal lamina, the three cells L1–3 of the specialized part of the male lamina differ significantly from those of the normal lamina. A further striking feature can be observed, that is, the receptor r7, a sex-specific class of R7. This receptor r7 terminates in the underlying cartridge, not like R7 and 8 in the medulla (Hardie, 1984).

9.4. BASKET CELL

In the lamina of muscoid flies, there is, in addition to the main centripetal pathway composed of the monopolar cells, a secondary pathway, the so-called T-shaped neurons (T1), also known as β-fibers or basket fibers (Strausfeld, 1970, 1971; Campos-Ortega and Strausfeld, 1973; Strausfeld and Campos-Ortega, 1977). In Figure 9.9 one can see that, in the lamina, the main fiber of each bipolar T1 cell gives rise to a basket of six climbing fibers;

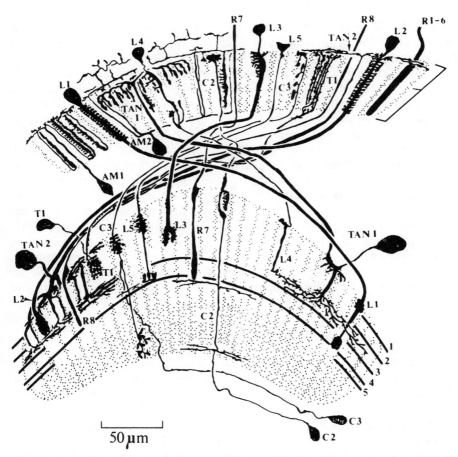

Figure 9.9. Summary diagram of short visual fibers (R1–6) and long visual fibers (R7, 8), monopolar cells (L1–5), centripetal cell (T1), amacrine cells (AM1; AM2 is now known to be TAN3, see Nässel et al., 1983), and centrifugal cells (TAN1, 2; C2, 3) in the lamina and between lamina and medulla of the housefly, *Musca domestica*. (From Strausfeld and Nässel, 1981.)

these baskets embrace the six short visual fibers in a cartridge. Although, it is assumed that T1 is in postsynaptic contact to L2 and to amacrine cells. It is in fact uncertain if these elements of T1 are really postsynaptic to the visual fibers, because there is no convincing evidence that T1 is a second-order cell (Shaw, 1984).

9.5. AMACRINE NEURONS

Amacrine neurons are understood to be intrinsic to the neuropil. These an-axonal cells have their cell body in the inner part of the lamina or proximal to it. In *Schistocerca* there are two types of horizontal cells (H1, 2), which may be amacrine cells (Nowel and Shelton, 1981). The main fiber of the H1 cell is oriented dorsoventrally, and the fine branches extend horizontally and anteriorly along the z axis between adjacent rows of cartridges. H2 differs from H1 in that it possesses no main fiber. It forms a tangle of fibers ranging along the entire anteroposterior extent of the lamina.

One amacrine cell (AM1) have been identified in the lamina of muscoid flies (Fig. 9.9) (Campos-Ortega and Strausfeld, 1973; Strausfeld and Campos-Ortega, 1973, 1977). The second, AM2, is now known to be the tangential cell, TAN3, described from Golgi and 5-HT immunocytochemistry (Nässel et al., 1983). The processes of AM1 are called α-fibers. Each AM1 supplies one to three α-fibers to any cartridge. Furthermore, AM1 has, besides the descending and ascending α-fibers, thin superficial tangential processes that represent a regular network of interconnections between cartridges. AM1 shows the following synaptology: AM1 is presynaptic to L4. Furthermore, AM1 is pre- as well as postsynaptic to AM1 and to TAN2, respectively. There are also postsynaptic contacts of AM1 to short visual fibers. In addition, α-fibers and β-fibers are reciprocally pre- and postsynaptic.

9.6. CENTRIFUGAL NEURONS

A comparison of the laminae of *Periplaneta* (Ribi, 1977), *Schistocerca* (Nowel and Shelton, 1981), *Cataglyphis* (Meyer, 1979), *Apis* (Ribi, 1975a,b, 1984) and *Musca* (Strausfeld, 1970; Campos-Ortega and Strausfeld, 1973; Strausfeld and Campos-Ortega, 1973; Strausfeld, 1976a,b) shows that basically two categories of centrifugal cells can be distinguished with respect to their terminal domains in the lamina neuropil. In the first group (type A), after entering the lamina, the centrifugal main fiber ramifies into an indeterminable number of branches, which diffusely innervate a wide field of the neuropil. The terminals can usually be assigned to distinct layers of the neuropil, but not to distinct cartridges. The second group (type B), however, consists of cells whose main fiber ramifies into a limited number of branches, whereby each branch can be assigned to a certain cartridge. These cells have a unique morphology. The perikarya of all of them lie beneath the lamina or proximal to the crossover of the fibers in the first optic chiasma. The main fiber connects the medulla with the lamina.

In the lamina of *Periplaneta*, Ribi (1977) documented several types of centrifugal cells, probably all type A (Fig. 9.10A). Type a innervates the distal layer of the neuropil. Type b possesses diffusely arranged wide-field end arborizations either in the medial or in all layers of the neuropil. Finally, type c is a T-shaped cell, the secondary branches of which are found only in the proximal neuropil.

In the lamina of *Schistocerca* (Nowel and Shelton, 1981), the centrifugal cell C_1 has a wide-field network within the distal stratum, which covers more

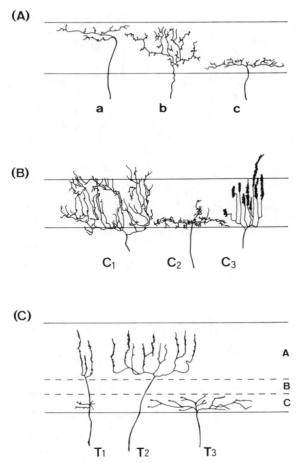

Figure 9.10. Summary diagram of the terminals of centrifugal neurons in the lamina of three insect species. (A) Lamina of the cockroach, *Periplaneta americana*. Cells with terminals (a) in the distal layer of the neuropil; (b) with wide-field end arborizations in the medial layer, and (c) with T-shaped horizontal elements in the proximal layer. (B) Three types of centrifugal cells in the locust, *Schistocerca gregaria*. C_1 = "wide-field" centrifugal neuron; C_2 = "inferior plexus" centrifugal neuron; C_3 = "candelabra" centrifugal neuron. (C) The three types of tangential fibers (T1, T2, T3) in the lamina of the desert ant, *Cataglyphis bicolor*. (A from Ribi, 1977; B from Nowel and Shelton, 1981; C from Meyer, 1979.)

than 10 cartridges (Fig. 9.10B). The C_2 cell is similar to C_1, but has its network only in the proximal stratum. Both cells may belong to type A. C_3 is quite different from both other types, in that its main fiber ramifies into only a relatively small number of branches after entering the lamina. Each branch is apparently associated with a single cartridge so that the cell may actually belong to type B.

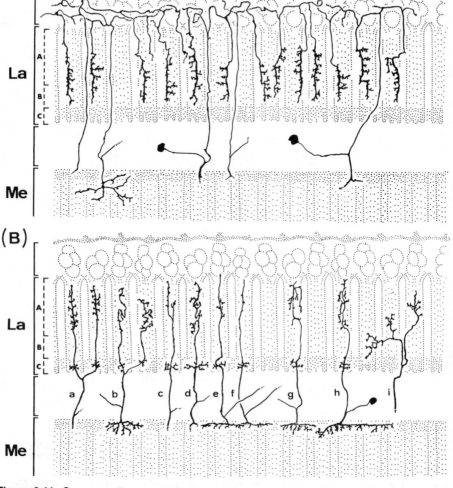

Figure 9.11. Summary diagram of the four types of centrifugal cells in the lamina of the honeybee, *Apis mellifera*. (A) Five different components of garland cells. La = lamina; Me = medulla; A, B, and C = the three horizontal strata within the neuropil. (B) Different components of Y-cells (a, b), single bottle-brush cells (c-h), and triptych cell (i). (From Ribi, 1984.)

In the ant, *Cataglyphis* (Meyer, 1979), the centrifugal fiber T1 may appear to be similar to C_3 in *Schistocerca* (Fig. 9.10C). The branches of this cell are restricted to only one cartridge (type B). T2 and 3 in the ant are, like C_1 and C_2 in *Schistocerca*, typically wide-field elements (type A). Their dendritic network spreads over about half of the whole length of the lamina. However, T2 covers the distal stratum and T3 only the proximal stratum.

In the lamina of *Apis*, Ribi (1984) described four types of centrifugally arranged interneurons (Fig. 9.11). Here, one wide-field cell is remarkable, which, like the remaining types, belongs to type B. Ribi calls this wide-field cell a "garland" cell (Fig. 9.11A). The centrifugal main fiber runs between the cartridges to the distal surface of the lamina. There, it ramifies and the second branches run tangentially between the monopolar cell body layer and the neuropil. From these branches, several tertiary side branches, the so-called "garlands" again project centripetally to the medial neuropil. Interestingly, each garland only goes into one cartridge. The second type of centrifugal cell is called the triptych cell (Fig. 9.11B). Upon entering the lamina, the main fiber divides into three branches. Each branch runs into a cartridge. Thus three neighboring cartridges are innervated. The Y cell (Fig. 9.11B) innervates two cartridges. The bottle-brush cell runs into one cartridge (Fig. 9.11B).

In the lamina of flies, five types of centrifugal cells can be detected: three wide-field cells (TAN1, 2, and 3) and two narrow field cells (C2 and 3; Fig. 9.9, TAN 3 is not illustrated) (Strausfeld, 1970; Campos-Ortega and Strausfeld, 1973; Strausfeld and Campos-Ortega, 1973; Nässel et al., 1983). The fibers of centrifugal cells TAN1 and TAN2 are derived from perikarya lying above the medulla. They then project to the distal surface of the lamina, where they form bi- and tristratified dendritic areas. TAN1 (type B) has an obvious similarity to the garland cell in the lamina of *Apis*, but the "garland" in *Musca* reaches only the distal lamina. In contrast, TAN2 (type A) has only thin tangential fibers. The terminals of both cells cover up to 30 cartridges. TAN3 may run above the lamina and may release its presumed transmitter serotonin nonsynaptically (Nässel and Klemm, 1983; Nässel et al., 1983). The two narrow-field centrifugals (C2 and 3; type B) are derived from cell bodies lying beneath the medulla. C2 is characterized by a thin fiber that has its terminal in the distal lamina (Fig. 9.9). C3 has also a thin fiber, which, however, gives rise to a set of unilateral tuberous outgrowths that enter into a cartridge. The terminals of these branches come to lie between L1 and 2.

9.7. SUMMARY

The gross morphology of the lamina monopolar cells is similar in all insects; however, there are significant structural and fine structural variations among various species.

In comparing the connectivity pattern of the monopolars among the few well-studied species (*Sympetrum, Apis, Musca*), it is clear that there are many similarities: two monopolars (L1, 2 or their counterparts) with general receptor input, feedback synapses between L2 and receptors, one monopolar with selective receptor input (except *Musca*, in which R1–6 are equivalent), and one monopolar cell with input only from other monopolar cells. Differences arise because of (1) the neural superposition in the fly, which ensures that R1-6 are equivalent and in which R7, 8 have no synaptic involvement; (2) the absence of a described fifth monopolar in *Apis*, whereas M5 in *Sympetrum* receives selective input from the long visual fibers R6, 7; and (3) the absence of described connections between cartridges in *Sympetrum*.

The second part (Sections 9.4–9.6) of this chapter shows that the lamina contains, besides the monopolars and visual fibers, three further classes of neurones:

1. Centripetal cells derived from cell bodies situated above the medulla. These cells project from the lamina to the outer medulla. However, the centripetal cells are presumably not second-order neurons, like the monopolars, but third-order neurons.

2. Amacrine cells, intrinsic to the lamina. Their morphology suggests that they may be involved, on the one hand, in local circuits within single cartridges and, on the other hand, in the connection of different cartridges. They may, therefore, play a role in lateral inhibition and neural adaptation.

3. Centrifugal cells derived from the medulla. The endings of these cells are either restricted to single cartridges or are diffusely extended over a large area. They may relay information from the outer medulla and/ or from more central areas to the lamina, possibly in a feedback manner.

The synaptic relationships of the three cell classes with monopolar cells and visual fibers can be summarized as follows: 1) inputs to monopolars via amacrine and/or centrifugal cells from visual fibers; 2) inputs to T1 centripetal cells from amacrine cells and monopolars; and 3) inputs to visual fibers from centrifugal cells.

ACKNOWLEDGMENTS

I am very grateful to Drs. Ian Meinertzhagen, Willi Ribi, Roger Hardie, and Jochen Zeil for their useful comments, and to Eugenia Lamont and Dr. Peter Holzer for translating this paper. Supported by Grant P4832 from the Austrian Science Foundation.

REFERENCES

Armett-Kibel, C. and I. A. Meinertzhagen, 1985. The long visual fibers of the dragonfly optic lobe: Their cells of origin and lamina connectivities. *J. Comp. Neurol.* **242:** 459–474.

Campos-Ortega, J. A. and N. J. Strausfeld. 1973. Synaptic connections of intrinsic cells and basket arborizations in the external plexiform layer of the fly's eye. *Brain Res.* **59:** 119–136.

Franceschini, N., R. C. Hardie, W. A. Ribi, and K. Kirschfeld. 1981. Sexual dimorphism in a photoreceptor. *Nature (London)* **291:** 241–244.

Hardie, R. C. 1983. Projection and connectivity of sex-specific photoreceptors in the compound eye of the male housefly (*Musca domestica*). *Cell Tissue Res.* **233:** 1–21.

Hardie, R. C. 1984. Properties of photoreceptors R7 and R8 in dorsal marginal ommatidia in the compound eyes of *Musca* and *Calliphora*. *J. Comp. Physiol. A* **154:** 157–165.

Hardie, R. C., N. Franceschini, W. A. Ribi, and K. Kirschfeld. 1981. Distribution and properties of sex-specific photoreceptors in the fly *Musca domestica*. *J. Comp. Physiol.* **145:** 139–152.

Kirschfeld, K. 1967. Die Projektion der optischen Umwelt auf das Raster der Rhabdomere im Komplexauge von *Musca. Exp. Brain Res.* **3:** 248–270.

Meinertzhagen, I. A. and C. Armett-Kibel. 1982. The lamina monopolar cells in the optic lobe of the dragonfly *Sympetrum. Philos. Trans. R. Soc. Lond. B* **297:** 27–49.

Meinertzhagen, I. A., R. Menzel, and G. Kahle. 1983. The identification of spectral receptor types in the retina and lamina of the dragonfly *Sympetrum rubicundulum. J. Comp. Physiol.* **151:** 295–310.

Menzel, R. 1977. Farbensehen bei Insekten—ein rezeptorphysiologischer und neurophysiologischer Problemkreis. *Verh. Dtsch. Zool. Ges.* **1977:** 26–40.

Menzel, R. 1979. Spectral sensitivity and color vision in invertebrates, pp. 503–580. In H. Autrum (ed.), *Handbook of Sensory Physiology* vol. VII/6A. Springer, Berlin, Heidelberg, New York.

Menzel, R. and M. Blakers. 1976. Colour receptors in the bee eye—Morphology and spectral sensitivity. *J. Comp. Physiol.* **108:** 11–33.

Meyer, E. P. 1979. Golgi-EM study of first and second order neurons in the visual system of *Cataglyphis bicolor* Fabricius (Hymenoptera, Formicidae). *Zoomorphologie* **92:** 115–139.

Meyer, E. P. 1984. Retrograde labelling of photoreceptors in different regions of the compound eyes of bees and ants. *J. Neurocytol.* **13:** 825–836.

Mote, M. I. and R. Wehner. 1980. Functional characteristics of photoreceptors in the compound eye and ocellus of the desert ant, *Cataglyphis bicolor. J. Comp. Physiol.* **137:** 63–71.

Nässel, D. R. and N. Klemm. 1983. Serotonin-like immunoreactivity in the optic lobes of three insect species. *Cell Tissue Res.* **232:** 129–140.

Nässel, D. R., M. Hagberg, and H. S. Seyan. 1983. A new, possibly serotonergic neuron in the lamina of the blowfly optic lobe: An immunocytochemical and Golgi-EM study. *Brain Res.* **280:** 361–367.

Nowel, M. S. and P. M. J. Shelton, 1981. A Golgi-electron microscopical study of the structure and development of the lamina ganglionaris of the locust optic lobe. *Cell Tissue Res.* **216:** 377–401.

Ohly, K. P. 1975. The neurons of the first synaptic region of the optic neuropil of the firefly, *Phausis splendidula* L. Coleoptera. *Cell Tissue Res.* **158:** 89–109.

Ribi, W. A. 1974. Neurons in the first synaptic region of the bee, *Apis mellifera*. *Cell Tissue Res.* **148:** 277–286.

Ribi, W. A. 1975a. The neurons of the first optic ganglion of the bee, *Apis mellifera*. *Adv. Anat.* **50(4):** 1–43.

Ribi, W. A. 1975b. The first optic ganglion of the bee. I. Correlation between visual cell types and their terminals in the lamina and medulla. *Cell Tissue Res.* **165:** 103–111.

Ribi, W. A. 1977. Fine structure of the first optic ganglion (lamina) of the cockroach *Periplaneta americana*. *Tissue Cell* **9:** 57–72.

Ribi, W. A. 1981. The first optic ganglion of the bee. IV. Synaptic fine structure and connectivity patterns of receptor cell axons and first order interneurones. *Cell Tissue Res.* **215:** 443–464.

Ribi, W. A. 1984. The first optic ganglion of the bee. V. Structural and functional characterization of centrifugally arranged interneurones. *Cell Tissue Res.* **236:** 577–584.

Shaw, R. S. 1984. Early visual processing in insects. *J. Exp. Biol.* **112:** 225–251.

St. Marie, R. L. and S. D. Carlson. 1983. Glial membrane specializations and the compartmentalization of the lamina ganglionaris of the housefly compound eye. *J. Neurocytol.* **12:** 253–275.

Strausfeld, N. J. 1970. Golgi studies on insects. Part II. The optic lobes of diptera. *Philos. Trans. Soc. Lond. B* **258:** 175–223.

Strausfeld, N. J. 1971. The organization of the insect visual system (light microscopy). I. Projections and arrangements of neurons in the lamina ganglionaris of Diptera. *Z. Zellforsch.* **121:** 377–441.

Strausfeld, N. J. 1976a. Mosaic organizations, layers, and visual pathways in the insect brain, pp. 245–279. In F. Zettler and R. Weiler (eds.), *Neural Principles in Vision*. Springer, Berlin, Heidelberg, New York.

Strausfeld, N. J. 1976b. *Atlas of an Insect Brain*. Springer, Berlin, Heidelberg, New York.

Strausfeld, N. J. and A. D. Blest. 1970. Golgi studies on insects. Part I. The optic lobes of Lepidoptera. *Philos. Trans. R. Soc. Lond. B* **258:** 81–134.

Strausfeld, N. J. and J. A. Campos-Ortega. 1973. The L4 monopolar neurone: A substrate for lateral interaction in the visual system of the fly *Musca domestica*. *Brain Res.* **59:** 97–117.

Strausfeld, N. J. and J. A. Campos-Ortega. 1977. Vision in insects: Pathways possibly underlying neural adaptation and lateral inhibition. *Science (Washington, D.C.)* **195:** 894–897.

Strausfeld, N. J. and D. R. Nässel. 1981. Neuroarchitectures serving compound eyes of crustacea and insects, pp. 1–132. In H. Autrum (ed.), *Handbook of Sensory Physiology*. vol. VII/6B. Springer, Berlin, Heidelberg, New York.

Wolburg-Buchholz, K. 1977. The superposition eye of *Cloeon dipterum*: The organization of the lamina ganglionaris. *Cell Tissue Res.* **177:** 9–28.

Wolburg-Buchholz, K. 1979. The organisation of the lamina ganglionaris of the hemipteran insect *Notonecta glauca, Corixa punctata, Gerris lacustris. Cell Tissue Res.* **197:** 39–59.

Zeil. J. 1983a. Sexual dimorphism in the visual system of flies: The compound eyes and neural superposition in Bibionidae (Diptera). *J. Comp. Physiol.* **150:** 379–393.

Zeil. J. 1983b. Sexual dimorphism in the visual system of flies: The divided brain of male Bibionidae (Diptera). *Cell Tissue Res.* **229:** 591–610.

CHAPTER 10

Comparative Aspects of the Retinal Mosaics of Jumping Spiders

A. David Blest
Developmental Neurobiology Group
Australian National University
Canberra City, Australia

10.1. INTRODUCTION

Jumping spiders stalk prey by sight and perform elaborate sexual and social displays that depend on vision (Forster, 1982; Jackson, 1982); they distinguish, therefore, among various classes of small moving objects in their vicinity. An adult *Portia fimbriata*, roughly 1 cm long, can discriminate between conspecifics, other species of jumping spider, and flies at distances up to 28 cm (Jackson and Blest, 1982a). It is not surprising that the principal eyes of *Portia* can sustain spatial acuities of approximately 2.4 arcmin in terms of geometrical optics (Williams and McIntyre, 1980).

Phylogenetically advanced jumping spiders have six functional eyes, and some primitive species eight (Fig. 10.1B). How they collaborate was elucidated by Land (1969a,b). Retinae of the secondary eyes are immobile, and their fields of view, which are contiguous with little overlap, cover much of the space around a spider (Fig. 10.1A,C,D). A spot subtending a small angle at one of the secondary retinae elicits turning movements of a whole spider that position it so that the target can be seen by the principal eyes, the latter alone mediating object recognition. Fields of view of the principal eyes correspond to the boomerang-like shapes of their retinae (Figs. 10.1C,D, 10.2B,C), a "boomerang" being oriented vertically so that its lateral field of view is small.

Like the principal retinae of all spiders (Land, 1985a), those of salticids are motile, but a potential to perform orderly movements has been much elaborated. After the image of a spot has been brought into the overall visual fields of the two principal eyes, it is located on a principal retina as a result of random movements; further directed saccades bring it to the retinal fovea, allow it to be tracked should it alter its position, and, apparently, scan it for diagnostic features (Land, 1969b).

The secondary retinae contain single sheets of receptors. The retinae of the principal eyes, however, are tiered; at the region of highest acuity at the center of the narrow retina, there are four layers of receptors, each layer lying at a different distance from the dioptrics (Fig. 10.2A). Land (1969a) showed that this arrangement might compensate for the chromatic aberration of the corneal lens, or, alternatively, allow focused images to be received from a wide range of distances from a spider in object space. Subsequent work has revealed a rather more complex situation that will be described. Meanwhile, it should be noted that only the receptor mosaic farthest from the corneal lens (layer 1) is organized to sustain fine visual discriminations and that the functional roles of the three more distal layers have not been resolved.

The principal retinae remain enigmatic: their organization is complex and fails to reflect precedents in the eyes of other spiders. No other family of spiders depends upon vision to the same degree, and the phylogenetic affinities of the Salticidae are obscure. A discussion of the evolution of the

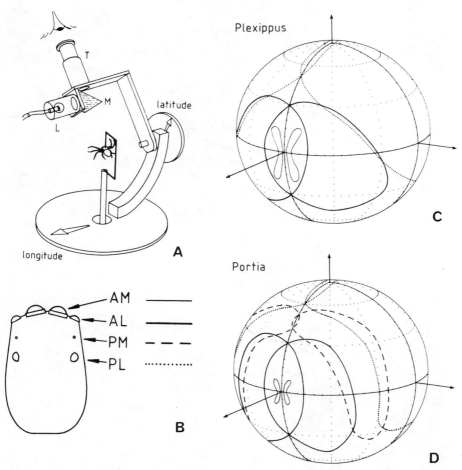

Figure 10.1. (A) An apparatus used to measured the fields of view of spider eyes. T is a small telescope, M a half-silvered mirror, and L a lightsource with a collimating lens. (B) The carapace of a jumping spider, seen from above. *Arrow* indicates the four pairs of eyes: AM = anterior median; AL = anterior lateral; PM = posterior median; PL = posterior lateral. Conventions for the fields of view of these eyes are given at right. (C and D) The visual fields of *Plexippus* (C) and *Portia* (D) orthographically projected onto a sphere around each animal, observed from 15° above the equator and 30° to the animal's left. (From Land, 1985a.)

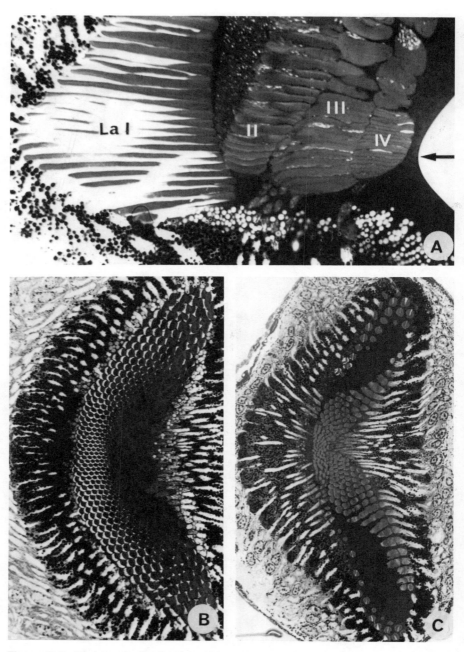

Figure 10.2. (A) A longitudinal, axial section through a principal retina of *Itata*, showing the tiers of receptors (La I–IV) at the fovea. The apex of the diverging component of the telephoto system (the "pit") and the optical axis are indicated by the *arrow* (700×). (B) Transverse section of layer I of *Itata*, taken near the tips of the rhabdomeres (280×). (C) Transverse section of layer II of *Plexippus validus* to illustrate the irregular mosaic at the fovea (280×).

salticid visual system suffers from the disadvantage that nothing is known of its antecedents, and that within the family there remain only traces of the later stages of its development. It is with these traces that the present chapter is concerned. Retinae of other families of spiders are discussed by Blest (1985a).

10.2. SYSTEMATICS AND PHYLOGENY OF THE SALTICIDAE

10.2.1. Taxonomy

A detailed account of the infrafamilial taxonomy of the Salticidae cannot be attempted here, not least because it is currently undergoing revision. Briefly: (1) the Dendryphantinae discussed by Land (1969a,b) consist of phylogenetically advanced forms, although categorizations at subfamily levels are now in some state of flux. (2) the Spartaeinae (Wanless, 1984) are regarded as primitive. Some genera are web-dependent, and most possess functional posterior median (PM) eyes, these being vestigial and probably nonfunctional in advanced forms. (3) the Lyssomaninae are also considered to be primitive, in part from ethological evidence (Crane, 1949), but also on morphological grounds (Platnick, 1971; Homann, 1971; Galiano, 1976; Wanless, 1978).

10.2.2. A Model for the Evolution of Vision-Dependent Behavior

Jackson and Blest (1982a) proposed that immediate ancestors of the Salticidae were web-building spiders that, like the contemporary *Portia fimbriata*, acquired the strategy of invading the neighboring webs of other species to capture and eat the occupants. We suggested that such temporary excursions may have led to the emancipation from webs altogether. The hypothesis was held to account for the remarkable sequence to be described: an elaborate, tiered principal retina is found in all contemporary Salticidae so far examined, yet the secondary retinae vary greatly between primitive and advanced species, achieving a fully efficient optical design only in the advanced forms (Blest, 1983, 1984). Notionally, an invader advancing across a web toward an occupant can be supposed to orient itself kinesthetically, but to need acute vision to evaluate the dangerous prey toward which it points. The hypothesis suggests that natural selection acted first on the principal eyes, substantial development of the secondary eyes following emancipation from webs and the mediation of orientation by visual stimuli.

10.3. FUNCTIONAL ASPECTS OF RETINAL DESIGN: ADVANCED SALTICIDAE

10.3.1. Anatomy of the Secondary Retinae

The secondary eyes are simple ocelli (Land, 1969a; Eakin and Brandenburger, 1971). A corneal lens provides focused images to a receptor mosaic whose properties are discussed in Section 10.4.1. Comparative accounts of the secondary ocelli of spiders are given by Blest (1985a) and Land (1985a).

10.3.2. Anatomy of the Principal Retinae

A principal retina is remote from its corneal lens. It lies at the end of a long, effectively fluid-filled, motile tube (Land, 1969a,b). In transverse section, a principal retina is shaped like a boomerang (Fig. 10.2B,C). At the central point of a retina (the "fovea"), there are four tiers of receptors, shown in longitudinal section in Figure 10.2A, and in transverse section in Figures 10.2B and 10.3A-C. The classification of receptors given follows that of Blest et al. (1981). For layers II–IV, receptors lying at the outer side of the retina are designated "a," and those on the inner side, "b."

Layer I. Farthest from the corneal lens, this tier consists of receptors whose rhabdomeres are some 45–90 μm in length. At the fovea, a receptor contains a single rhabdomere surrounded by a medium of low refractive index that enables it to function as a light guide (Williams and McIntyre, 1980; Blest et al., 1981). Peripherally, along the arms of the boomerang, receptors are larger, and each contains two rhabdomeres (Figs. 10.2B, 10.9B). Overall, the mosaic is extremely regular.

Layer II. These receptors have short rhabdomeres (usually < 20 μm). Mosaic patterns tend to be complex and do not correspond closely to those of the layer I receptors that they overlie over most of the retina (Fig. 10.3A), although recent results (Blest, unpublished data) suggest that there may be exact correspondence in a very small foveal region. Both layers I and II extend over the whole retina.

Layer III. This layer is confined to the central retina (Fig. 10.3B). Receptors contain short (< 20 μm) rhabdomeres with large, irregular, and contiguous transverse cross-sectional profiles (Fig. 10.4C).

Layer IV. Layer IV is complex (Figs. 10.3C and 10.4A,B). 4a receptors lie as a vertical strip or strips along the anatomical outer side of the retina; 4b lie centrally in an irregular mosaic on the inner side; 4c cells lie peripherally, but do not extend far along the arms of the boomerang.

10.3.3. Spectral Sensitivities of Identified Receptors

Hardie and Duelli (1978) made the only successful intracellular recordings from a salticid lateral eye, and found the receptors to have spectral peaks at $\lambda = 535$ nm in the green. All models that provide functional explanations for the tiered principal retinae demand that the four layers have different spectral sensitivities (Land, 1969a), and attempts have been made to test the issue directly. Unfortunately, the results have been contradictory.

De Voe (1975) obtained intracellular recordings from unmarked cells, and found only green, ultraviolet (UV), and mixed green-UV receptors, the latter

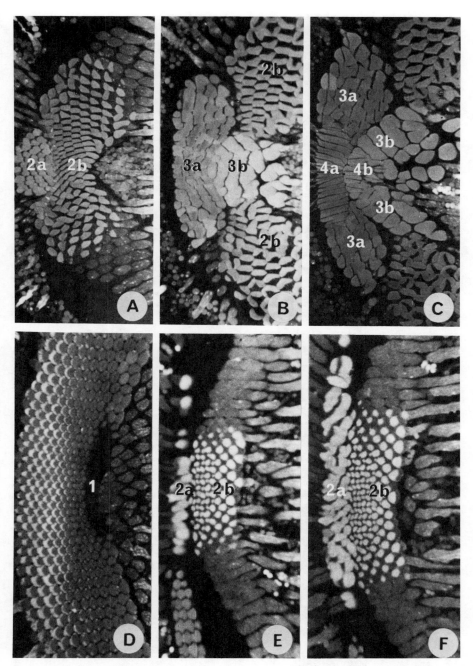

Figure 10.3. Transverse sections through the foveal receptor mosaics of an advanced salticid (*Lagnus*: A–C) and a primitive salticid in the Spartaeinae (*Portia*: D–F). (A) 2a and 2b mosaics, near to the midpoints of their receptors. (B) 3a and 3b mosaics. (C) 4a and 4b mosaics flanked by peripheral 3a receptors. (D) Layer I. (E) Layer II sampled proximally. (F) Layer II sampled more distally (A–C, 680×; D–F 720×).

209

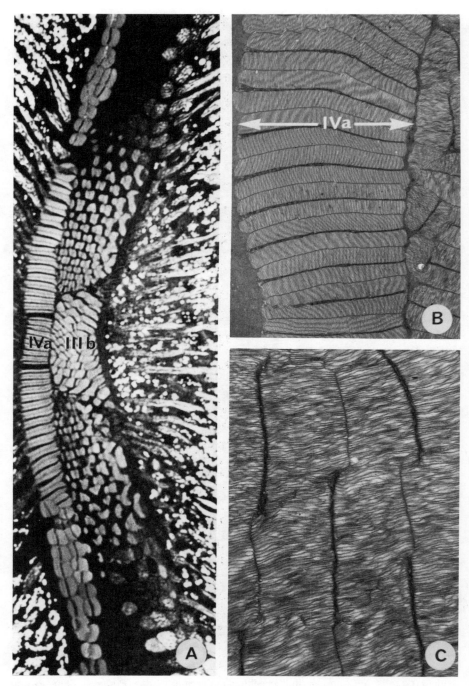

Figure 10.4. (A) Transverse section through the outer layer IV mosaic of *Portia* to show the linear, vertical array of receptors. Medially, the fovea is occupied by 3b receptors of layer III (850×). (B) Layer IV (4a) receptors of *Spartaeus*, near to the fovea (3700×). (C) Layer III receptors of *Spartaeus*, near to the fovea (10,300×).

class possibly being an artifact of recording (Blest et al., 1981). Yamashita and Tateda (1976) found cells with sensitivities at $\lambda = 360$ (UV), 480–500 (blue), 520–540 (green), and 580 nm (yellow), in a sample of only seven receptors. Blest et al. (1981) obtained intracellular records from more than 30 cells, of which roughly one third were marked by injection of Lucifer yellow; only green and UV cells were encountered, the former being divided between layer I and II, and the latter found only in layer IV. No marked cells were recorded from layer III, which for anatomical reasons must be difficult to sample; it may, quite plausibly in terms of optics, consist of blue receptors. The single yellow cell obtained by the Japanese workers remains an enigma. It should be remembered that the three groups of authors dealt with different and rather distantly related species.

10.3.4. Optical Designs

Land (1969a) described the optics of the secondary eyes of some advanced Salticidae. A simple, corneal lens focuses an image onto a single receptor mosaic. The nonpigmented glial sleeve that surrounds each pair of rhabdomeres would now be supposed to generate a refractive index difference such that light from focused images received by the tips of the paired rhabdomeres must be transmitted down them by total internal reflection (Snyder, 1979).

The minimum receptor spacing estimated by Land for two Californian species, *Phidippus* and *Metaphidippus*, was approximately 30 arcmin for the AL eyes and 60 arcmin for the PLs.

The principal eyes are constructed as miniature Galilean telescopes (Fig. 10.5A,B; Williams and McIntyre, 1980). In addition to the corneal lens, which has a long focal length, there is a diverging component placed just in front of the focal plane in image space (Fig. 10.5B). It is formed by the interface between the dense retinal matrix and the material of low refractive index that fills the retinal tube (Fig. 10.2A) and magnifies the image received by the receptor mosaics. The discovery that these eyes have telephoto optics may invalidate the estimates of optimal acuities made by Land (1969a). For foveal layer I mosaics of some advanced neotropical species, Blest (1985) obtained optimal spatial acuities ranging from approximately 2.3 arcmin for the large open-space species, *Phiale magnifica*, to 16.8 arcmin for a small forest floor species, *Fluda princeps*. Acuity is limited by three major considerations: (1) a rhabdom cannot be of lesser diameter than about 1 μm, for physical reasons (Snyder, 1979); (2) whatever its diameter, it must receive an adequate supply of photons, so that in dimly lit habitats it may need to be relatively large; and (3) the optical construction implies that a small species whose corneal lens has a short focal length will achieve a poorer spatial acuity than a large one, other parameters being equal.

Land (1969a) offered two alternative functional models to explain retinal

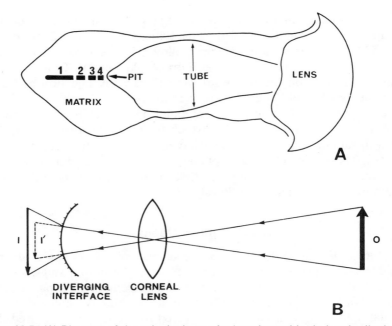

Figure 10.5. (A) Diagram of the principal eye of a jumping spider in longitudinal optical section. The four tiers of receptors at the fovea are numbered 1–4. The corneal LENS is separated from the retina by a long TUBE, and the PIT, formed by its interface with the MATRIX in which the receptors are embedded, acts as the diverging component of the telephoto system. (B) An optical diagram of the telephoto system. A corneal lens by itself would provide an inverted image, I′, of an object, O. The diverging component provided by the pit allows a magnified image, I, without substantial lengthening of the retinal tube. (B from Blest and Sigmund, 1985.)

tiering, and the later elucidation of the telephoto system does not prejudice them, although it modifies their precise application:

1. Tiering of the retina compensates for the inability of the dioptrics to accommodate, objects at different distances from a spider forming focused images on mosaics at different depths.
2. Tiering compensates for the measured chromatic aberration of the system, such that in Land's model (which ignores the diverging component) layer I would receive focused images from infinity in the red and layer IV in the ultraviolet.

All attempts to determine the "real" optics of the principal eye depend on modeling, because direct measurements of the parameters of the various optical components are insufficiently accurate to determine how they interact, and ophthalmoscopic measurements are prejudiced by uncertainty about exactly what retinal landmarks are being observed (Blest et al., 1981; Blest and Sigmund, 1985).

Assuming a "best model," the principal retina can be interpreted as follows (Blest et al., 1981):

1. Layer I receives focused images from green light. Because layer I rhabdomeres are designed as light guides, it can be inferred that images are focused on their tips. Tips of foveal layer I rhabdomeres are "staircased" horizontally across the retina; they lie at different distances from the dioptrics, such that receptors at the anatomical outer side of the retina can be supposed to receive focused images from distant objects, and those on the inner side of the retina from nearby objects. If we assume that fine-image analysis depends on layer I (because of the quality of its receptor mosaic), and accept the results of optical modelling, then it would appear that layer I is able to compensate for the inability of the principal eye to accommodate. Despite the "staircase," layer I cannot be used as a range-finder unless there is substantial higher-order processing, because the depth of field at any one point of the "staircase" is too large.

2. Layer IV receives focused images in the ultraviolet. Its mosaic quality is poor, other than the arrangement of 4a receptors in such a way that they could function as a simple line detector. An additional problem is presented by a degree of "staircasing" of layer IV receptors in some species (Blest and Sigmund, 1985), but it will not be explored here.

Thus, the two models proposed by Land (1969a) are not, strictly speaking, alternatives. The principal eye *does* need some mechanism to compensate for lack of accommodation, and this is provided by the "staircasing" of the tips of layer I receptors. Equally, there is at least bichromatic color vision that depends upon green and ultraviolet receptors; whereas the evidence is equivocal, it is possible that spectral sensitivities may include blue and perhaps even long wavelength receptors, although the latter would not readily fit our present optical models.

That the layer I mosaic is constructed in such a way as to allow it to mediate fine visual discriminations cannot seriously be questioned. The corollary—that layer III and IV are not—seems just as unambiguous. Layer II is more difficult to interpret, because the mosaic is in one way or another degraded in most species, and our intracellular recordings from layer II cells in *Plexippus* showed them to have the same spectral sensitivities as receptors of layer I.

For reasons that will become apparent, the subsequent discussion must concern the evolution of layer I, the other three layers being conservative and differing little between advanced species and the few primitive forms so far available to us.

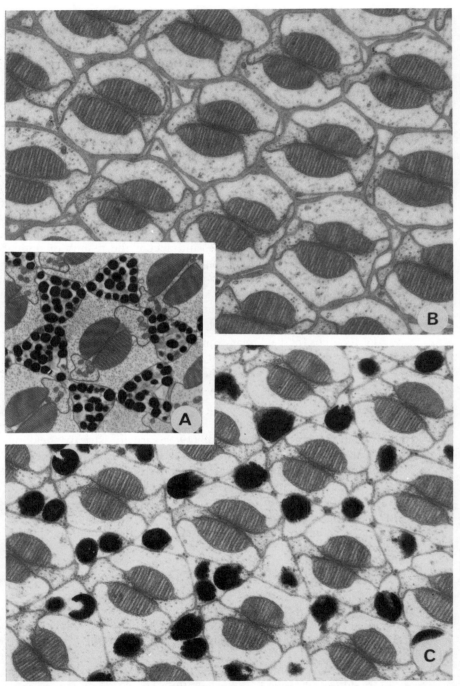

Figure 10.6. Mosaics of the anterior lateral retinae of some advanced salticids. (A) Adult *Plexippus*. A receptive segment contains two rhabdomeres (Rh) ensheathed by two non-

10.4. RETINAL ULTRASTRUCTURE

10.4.1. Retinal Ultrastructure of the Secondary Eyes

Eakin and Brandenburger (1971) first described the secondary eyes of advanced salticids, subsequent authors merely elaborating their account (e.g., Blest, 1983).

The mosaic organization of the AM retina of second-instar *Phiale magnifica* is illustrated in Figure 10.6B,C and compared with a single AM receptor of an adult *Plexippus* (Fig. 10.6A).

A receptor contains two closely apposed rhabdomeres. Together, they constitute a single unit—a rhabdom. Each receptor is ensheathed by two processes of a nonpigmented glial cell that similarly supplies a large local population of other receptors. The processes contain little more than microtubules and provide an effective refractive index difference between a rhabdom and its surround, such that it can perform as a light guide. In one species, *Plexippus*, the nonpigmented glial processes participate in the turnover of rhabdomeral membrane (Blest and Maples, 1979).

The nonpigmented glial processes are, in turn, ensheathed by six processes of pigmented glial cells whose somata lie proximal to the receptive segments (Fig. 10.6A). Each such process contains pigment granules, so that a receptor is shielded from light scattered from its neighbors.

10.4.2. Retinal Ultrastructure of the Principal Eyes

Layer I is composed of only one type of receptor. At the fovea, each receptor contains only one rhabdomere, and rhabdomeres are narrow. The cross-sectional area of a rhabdomere bears some relationship to retinal illuminance (Blest, 1985b). For example, a small species, whose activities are confined to bright sunlight can have rhabdomeres that approach the physical limit of about 1 μm (Fig. 10.7A). A large species occupying a dimly lit habitat will have large foveal rhabdoms (Fig. 10.7B). It should be noted that the greater the optical power of the diverging component of the telephoto system provided by the pit, the lower the retinal illuminance, all other factors being equal. Species living in dim habitats also have pits of relatively low optical power, presumably because retinal illuminance must be maintained at some critical level (Blest, 1985b).

---

pigmented glial processes containing microtubules, and enclosed, in turn, by six processes of the pigmented glia, creating a "Star of David" symmetry. (B) The far distal mosaic of an anterior lateral eye of a second-instar *Phiale magnifica*. At this level, partitions between receptors are derived solely from the nonpigmented glia. (C) The same mosaic, sampled at the midpoint of the receptive segments. The adult arrangement of processes is already clear, although a pigmented glial process can only accommodate a longitudinal array of single pigment granules. (A 6600×; B,C 17,3000×.)

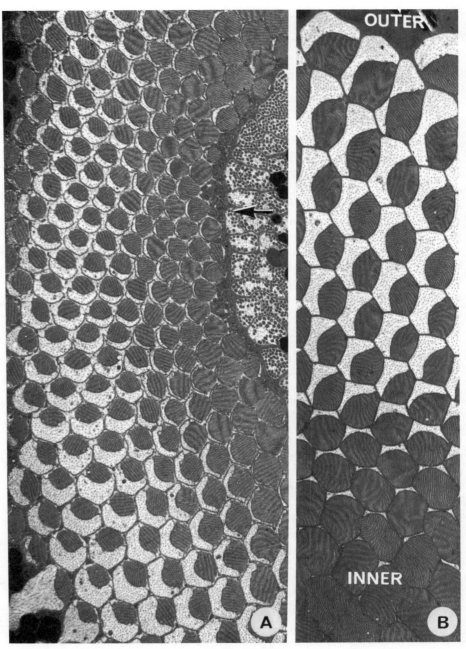

Figure 10.7. (A) The foveal layer I mosaic of a second-instar *Phiale magnifica*, a Panamanian species active in bright sunlight. A foveal rhabdomere is about 1 μm in diameter. The *arrow* indicates the inner side and the foveal midpoint of the retina. (B) An equivalent strip to that marked by the *arrow* in A, from foveal layer I in *Itata*, a Panamanian species that occupies a dimly illuminated forest niche. (both, 4500×.)

216

The cytoplasm of the receptive segments bearing the rhabdomeres contains almost no organelles other than microtubules. We suppose that this serves to maximize the refractive index difference between a rhabdomere and its surround, so that the former can act as a light guide. The long rhabdomeres (45–90 μm in most species) will favor the efficiency of photon capture, and also adequate signal-to-noise ratios in the electrical responses after transduction (Laughlin, 1980). In the peripheral retina, rhabdomeres become very much larger. Throughout layer I, the receptor mosaic is extremely regular.

Layer II receptors can be seen to be derived from cell bodies either on the outer (2a) or inner (2b) sides of the retina, the mosaics of receptive segments from the two sources usually appearing rather distinct (Fig. 10.3A). Compared with the mosaics presented by foveal layer I receptors, those of layer II at the fovea are to various degrees irregular, or exhibit patterns that are coarser than and do not correspond to those of the layer I mosaics that they overlie. Layer II receptive segments are short (< 20 μm).

Layer III receptors barely extend beyond the fovea (Land, 1969a). Receptive segments are extremely large, filled with rhabdomeres, and have irregular profiles (Figs. 10.3B,C; Fig. 10.4C). Characteristically, the microvilli present irregular, often sinuous, longitudinal profiles, and they are narrow (Fig. 10.4C).

Layer IV receptors belong to three categories. From the standpoint of image analysis, 4a receptors are the most interesting: they are disposed as a strip that lies along the outer lateral side of the retina (Figs. 10.3C; 10.4A), allowing the possibility that they cooperate to form a simple line detector. As Eakin and Brandenburger (1971) suggested, the disposition of their microvilli would also allow them to be polarization analyzers in the ultraviolet, although there is as yet no ethological evidence to indicate that polarization sensitivity is required by salticids. 4b receptors are ovoid, and form a group on the inner side of the foveal retina; 4c receptors lie in the peripheral arms of the "boomerang," close to the fovea. Neither forms a regular mosaic.

Receptors of layers II–IV are interposed between the regular, fine mosaic of layer I and the dioptrics. None of these receptors is designed as a light guide, but if there were substantial refractive index differences between their rhabdomeres and surrounding material, images arriving at layer I might be substantially degraded. The three distal layers are designed so as to avoid image degradation by refractive index differences: (1) the cytoplasm of the receptive segments is packed with mitochondria in the condensed configuration; (2) the matrix in which the receptive segments are embedded is dense and has a high refractive index ($n_M = \sim 1.40$); and (3) The refractive indexes of rhabdomeres, mitochondria-containing cytoplasm, and matrix are roughly equilibrated (Williams and McIntyre, 1980; Blest et al., 1981).

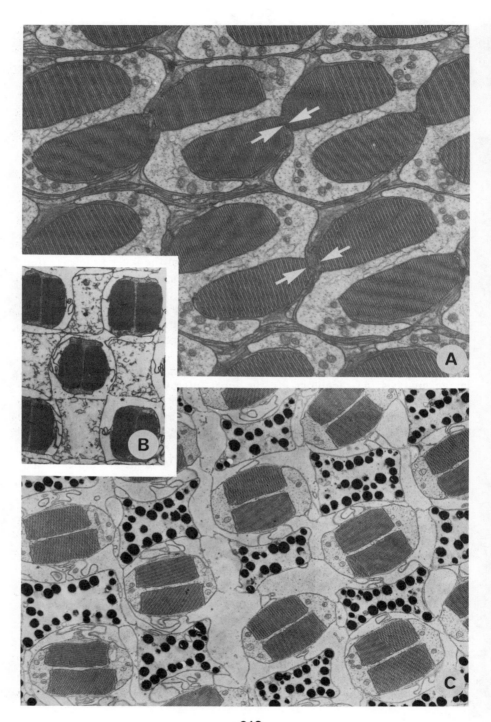

218

10.5. "PRIMITIVE" SALTICIDAE AND THE EVOLUTION OF RETINAL MOSAICS

The morphological pathway that led to the complex, tiered principal retinae of advanced Salticids has not been elucidated, and it is possible that contemporary species do not include forms that preserve the early stages of the sequence. Nevertheless, examination of a few species in the primitive Spartaeinae, and in the presumptively primitive Lyssomaninae, has led to a remarkable conclusion (Blest and Sigmund, 1984): the evolution of high-resolution principal eyes preceded the development of high-resolution secondary eyes.

10.5.1. Secondary Eyes of the Lyssomaninae and Spartaeinae

A transverse section through a secondary eye of *Lyssomanes* is shown in Figure 10.8A. A receptive segment contains two rhabdomeres, but is not surrounded by an ordered complex of glial processes. Instead, receptors are separated by thin partitions of nonpigmented glia which, in effect, allow contiguities between the rhabdomeres of adjacent cells, and thus considerable optical pooling between adjacent receptors along strips. Although the real phylogenetic position of the Lyssomaninae is uncertain, the organization of these secondary retinae can certainly be stigmatized as optically inefficient and unlikely to have been achieved by secondary reduction.

Secondary retinae of the Spartaeinae are organized in a manner that is much closer to that of advanced Salticids. They differ (from three species so far examined) in two major respects: (1) the nonpigmented glial processes that ensheath a receptive segment are much divided, especially distally, and do not contain microtubules; (2) there are four pigmented glial processes surrounding each receptor instead of six; and (3) in *Yaginumanis*, the latter processes do not contain pigment granules at the level of the receptive segments (Fig. 10.8B), although they are abundant proximally. In *Spartaeus*, there are a few, scattered pigment granules, but they are insufficient in density to screen the receptive segments from scattered light. Pigmented glial processes of *Portia* contain many large pigment granules (Fig. 10.8C). In all cases, the rhabdomeres appear to be short, those of *Lyssomanes* being ap-

---

Figure 10.8. (A) The receptor mosaic of an anterior lateral eye of *Lyssomanes viridis*. Receptive segments are separated by thin partitions of nonpigmented glia, and there is a pattern of contiguity between rhabdomeres of adjacent receptive segments (*arrows*); 8750×). (B) Receptors of an anterior lateral eye of *Spartaeus*: a receptive segment is ensheathed by divided processes of the nonpigmented glia. They are enclosed in turn by four processes of the "pigmented" glia that do not contain pigment granules (4000×). (C) Receptors of an anterior lateral eye of *Portia fimbriata*. The processes of the pigmented glia contain pigment granules (4800×).

proximately 16 μm in length for an AM eye, whereas foveal rhabdomere lengths in advanced salticids may approach 100 μm (unpublished data).

10.5.2. Principal Eyes of the Lyssomaninae and Spartaeinae

Layers III and IV are virtually identical in all salticids; differences of organization between primitive and advanced species are trivial and their significance is unknown. Layers III and IV of an advanced species are shown in Figures 10.3B,C, and of the primitive *Portia* in Figure 10.4A. The following account will concern layers I and II only. Tiering of the principal retina is fully developed in all contemporary forms known to us, a putative sequence in the evolution of layers I and II being represented by a series of dispositions within the Spartaeinae (Blest and Price, 1984; Blest and Sigmund, 1984, 1985). *Yaginumanis* is considered by Wanless (1984) to be the most primitive Spartaeine genus on morphological grounds: layer I is quite different from those of advanced Salticidae. Each foveal receptive segment bears two rhabdomeres on opposite faces of the cell, and the rhabdomeres of adjacent cells horizontally across the retina are closely contiguous (Fig. 10.9A). There is little cytoplasm in at least the distal half of a receptive segment; the optical pooling between receptors implied by the disposition of rhabdomeres would scarcely be consistent with the organization of rhabdomeres as light guides, but it is nevertheless satisfying that light-guide effects are precluded by the ultrastructure.

In *Lyssomanes*, whose primitive status is less certain, the foveal layer I mosaic predominantly conforms to that of advanced salticids in that receptors bear single rhabdomeres and there is provision for light guiding. Pooling, however, is implied near the fovea in the peripheral retina (Blest and Sigmund, 1984).

An intermediate condition is provided by *Spartaeus* (Blest and Sigmund, 1985): foveal receptors on the outer side of the retina each have a single rhabdomere with light-guide properties, whereas receptive segments on the inner side of the retina each bear two rhabdomeres so disposed that there must be optical pooling between receptors (Fig. 10.9C) for the same reasons exhibited by the entire foveal layer I mosaic of *Yaginumanis*. In terms of the optical consequences of "staircasing," the outer, distal tips of layer I receptors organized as light guides will receive focused images from distal objects, and the inner distal tips of layer I receptors, whose acuity is impaired by optical pooling, will receive focused images from objects nearby.

The mosaics of layer II exhibit analogous phylogenetic patterns. The interpretation is constrained by our inability even by strong inference to decide the possible nature of the tasks that layer II sustains. The foveal layer II mosaic of *Yaginumanis* appears to exactly replicate that of the layer I mosaic that it overlies. The same pattern is displayed by layer II in *Lyssomanes*. Again, there is optical pooling horizontally between receptors. In the primitive *Portia* (Fig. 10.3E,F), the advanced *Lagnus* (Fig. 10.3A), and other

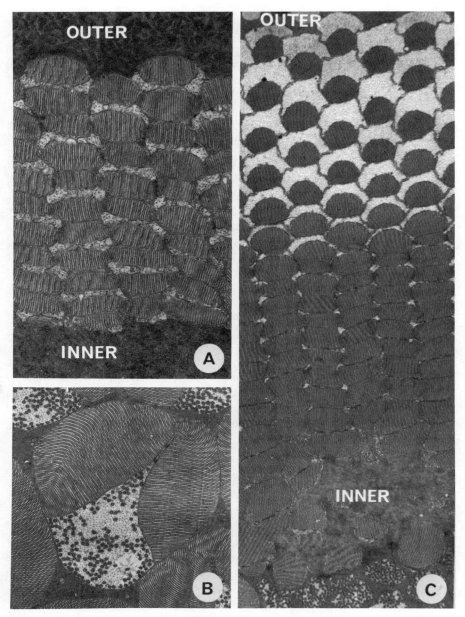

Figure 10.9. (A) The foveal layer I mosaic of *Yaginumanis*; each receptive segment contains two rhabdomeres with contiguity between those of adjacent receptive segments (8750 ×). (B) A far peripheral layer I receptive segment of *Portia* showing a rhabdomere which is horseshoe-shaped in transverse profile (6700 ×). (C) The foveal receptor mosaic of *Spartaeus*. At the outer side of the retina, receptive segments have single rhabdomeres organized as light guides; at the inner side, each receptive segment bears two rhabdomeres disposed as in *Yaginumanis* (4500 ×). (A modified from Blest and Sigmund, 1984.)

221

advanced Salticidae seen by us, foveal layer II receptors present either degraded mosaics unsuited to sustain precise image analysis or mosaics with eccentric patterns, that of *Lagnus* for example suggesting optical pooling (and possibly electrical coupling) along idiosyncratic linear arrays (Blest and Price, 1984).

10.5.3. Peripheral Principal Retinae

The foregoing account has noted that in all salticid retinae, layer I and II mosaics become coarser the more peripherally they lie along the arms of the retinal "boomerang." Interpretation of this apparent decline of mosaic resolution is difficult, because of the nature of the diverging interface of the telephoto system. The greater the magnification afforded by the diverging component, which at the fovea is provided by a hemispherical interface between the retinal matrix and the fluid-filled anterior chamber, the more acute will be the angle subtended by the sides of the pit remote from the fovea. Barlow (1980) discussed the optical implications of pits thus constructed, noting that the sides of a pit will generate an additional magnification factor. This alone may account for the coarser peripheral mosaics of layers I and II. Nevertheless, if the sides of a conical pit are straight, as they are in *Portia* and *Lyssomanes* (Blest, 1985b), there will be some peripheral smearing of images; this can be corrected if the sides of the pit are convexiclivate, as they are in advanced salticids (Blest, 1985b). Furthermore, the peripheral mosaics must be considered in terms of a discussion by Snyder et al. (1986): image quality should always be superior to potential anatomical resolution by receptor mosaics, but the degree of undersampling will be precisely determined.

10.6. THE EVOLUTION OF HIGH-ACUITY VISION

Optimistically, this survey of the retinal mosaics of jumping spiders might be supposed to lead to some general statement about the evolution of high-acuity vision as mediated by the principal eyes. The actual outcome is rather more confused but may follow from our evolutionary hypothesis (Jackson and Blest, 1982b):

1. The ancestors of contemporary Salticidae were web builders.
2. They acquired the tactic of invading the contiguous webs of unrelated spiders to prey on their occupants.
3. Such a strategy demanded that location of the occupant of a web by kinesthetic cues be supplemented by visual information about its size and nature. An element in the argument is that invasion of webs of

several different kinds precluded a common kinesthetic language that could characterize the occupants of all of them.

4. The supplementation implied, and its consequences in terms of natural selection, suggest that the relationships observed between the principal and secondary eyes are reasonable: the principal eyes evolved to their present, optically sophisticated state first, the secondary eyes following them as their owners became emancipated from their dependency on webs, and the kinesthetic orientation cues that webs provide.

Such an historical sequence is attractive, but can never be formally proved. At best, comparative ethological evidence can render it more (or less) probable. It is disappointing that there is no little difference between layers III and IV over the wide taxonomic range of species examined, and that we have no indication as to how the evolution of retinal tiering proceeded. Even so, it is satisfying that within the Spartaeinae, layer I presents a complete series that encompasses the undistinguished and functionally primitive foveal mosaic found in *Yaginumanis*, and the intermediate state represented by *Spartaeus* that both conforms to the constraints that optical considerations impose upon such a transition, and suggests how the high-resolution retina of *Portia* might have evolved.

The major conceptual hiatus concerns the origin of retinal tiering in the first place. Relict, primitive Salticidae persist in Madagascar, Sri Lanka, and Central Africa, and it is possible, but not likely, that they may illustrate earlier stages in the evolution of principal retinae. More probably, the evolution of the family from whatever ancestry was rapid, self-evidently successful, and left almost no traces to tell us in detail how it was achieved.

10.7. SPECULATIONS AND CONCLUSIONS

The tiered principal retinae of salticids are unique both in terms of their anatomy, and of how they function, although the latter issue is not yet satisfactorily resolved. It is apposite to look briefly at some possible interpretations of tiering that are speculative and go beyond the immediate matter of a trade-off between the chromatic aberration of the dioptrics and compensation for an inability to accommodate. Previous discussions have tended to emphasise the optical design and its limitations. Can their argument be inverted? Instead of presuming that *a principal retina is a compromise between the demands imposed by ethology and what an inefficient optical construction can sustain*, one can ask: *Why may the system be optimal, given that it can be seen to support precise and sophisticated patterns of vision-dominated behavior?*

If such a conceptual approach is plausible, it suggests that evolution of the principal retinae can be regarded in two ways:

1. The number of receptors in a principal retina is probably limited. Blest and Land (1977) found no evidence that receptors were added to a secondary eye of *Dinopis* (a web-casting spider) as it grows, and the same conclusion follows from more casual observations of the retinae of jumping spiders. It would follow that to evolve one satisfactory receptor mosaic for image analysis must prejudice the rest, given an optical requirement for retinal tiering: the more receptors deployed by a particular layer, the fewer will be available for others. This may explain the relatively feeble populations of receptors in layers III and IV, but does not account for the organization of layer II.

2. Given the complexity of the visual tasks that salticids perform, how most economically can a nervous system analyze the relevant input?

Our relative ignorance of the principles governing higher order visual processing even in insects does not encourage speculation, but one obvious possibility can be mentioned. Form recognition from the input provided by a single receptor mosaic makes formidable processing demands on the central nervous system, and the optic lobes supplied by the compound eyes of visually sophisticated insects are, indeed, massive. In contrast, the higher order visual neuropils of jumping spiders are economical in terms of volume, and appropriate for the limited space that an anterior cephalothorax can provide to house them. Perhaps jumping spiders have retinae whose tiers of receptors partition responsibility for the various visual tasks that they mediate, so that pattern recognition has become a function of the disposition of receptors at the fovea for each of the four mosaics. Thus, albeit intuitively, layer I is probably versatile, and can be presumed to be concerned with recognitions and discriminations at large distances. Layer II may mediate a different set of recognitions, given that in some species such as *Lagnus* (Blest and Price, 1984) foveal receptive segments are coupled in peculiar linear arrays. Layer III is so organized mosaically as to make its function incomprehensible unless one were to assume (on tenuous grounds) that salticids have an imperative need to recognize spatially unpatterned signals in the blue. Layer IV may plausibly be a polarization analyzer in the ultraviolet (Eakin and Brandenburger, 1971) and function in some simple context that demands line recognition.

This unsatisfactory position must be further qualified by two caveats:

1. Our present picture has been derived from several different species. The spectral characteristics of their receptors may not be uniform; some problems of interpretation may stem from interspecific diversity. Nevertheless, it would be difficult, now, to argue that in any species foveal layer I consists of long wavelength receptors, as proposed by one of Land's two alternative models. De Voe (1975) and Blest et al. (1981) found no evidence to indicate their presence and, if it is accepted that the mosaic quality of layer I in advanced salticids implies that it

mediates fine discriminations, spectral sensitivities in the green seem functionally likely in terms of the illumination of real habitats, where light is filtered through or reflected by leaves. Some salticids are substantially decorated by orange, vermilion, or crimson hairs that have been interpreted as epigamic signals (e.g., Crane, 1949). Whether such decorations are intraspecifically directed, or are in some manner deployed against predators (as pseudoposematic signals, or as flash coloration, for example), remains to be explored.

2. A simple assumption that because foveal layer I rhabdomeres present a highly ordered mosaic and foveal layer II rhabdomeres do not, layer I must alone be concerned with fine spatial discriminations, should not be accepted naively. Fortunately, the sampling properties of regular versus irregular mosaics have been examined by French et al. (1977) and Bossomaier et al. (1985): the output of irregular mosaics is prejudiced by noise, and the major disadvantages of regular mosaics— susceptibility to Moiré effects and to aliasing—can be discounted because in the real world eyes do not scrutinize optical gratings.

Some aspects of retinal tiering remain to be explored. The salticid visual system may have evolved discrete pathways for the recognition of different key stimuli by exploiting a limitation of its dioptrics. A contemporary designer of optical instruments would be unlikely to invent a complex sensor that depended upon chromatic aberration because, historically, instrument makers have devoted much ingenuity to eliminating it. Perhaps the salticids have turned a disadvantage into a strength.

10.8. SUMMARY

The retinal mosaics of jumping spiders have evolved along pathways leading to high spatial acuities. The secondary eyes are optically simple, a corneal lens supplying an unmodified image to a single layer of receptors. The principal eyes of all forms known to us (1) possess telephoto optics, a diverging component behind the corneal lens being provided by the glial matrix within which the receptive segments are embedded; and (2) have four tiers of receptors, layer I, farthest from the dioptrics, along being designed to sustain high spatial acuities.

In advanced forms, the individual rhabdoms of both the secondary eyes and of layer I of the principal eyes are long light guides, with an implication that for optimal performance images must be focused on their tips.

Jackson and Blest (1982b) proposed a model for the evolution of the Salticidae: it suggests that the principal eyes evolved first, the secondary eyes lagging behind them. The retinal mosaics of presumptively "primitive" forms support the model. All known extant forms have four tiers of receptors and telephoto optics in the principal eyes. However, layer I in *Yaginumanis* in the primitive Spartaeinae offers foveal receptive segments with twin rhabdomeres so disposed that acuity must be degraded by optical pooling. A transition between the condition of layer I in *Yaginumanis* and the high-acuity layer I of *Portia* can be traced within the Spartaeinae, to which they both belong. Secondary eyes in the Spartaeinae, and in *Lyssomanes*, also presumed to be primitive, are optically inferior to those of advanced forms. In *Lyssomanes*, receptive segments are disposed in linear arrays, such that there must be substantial optical pooling between adjacent receptors of an array. In the Spartaeines, there are various degrees of isolation between spatially separated, rather short rhabdoms: in *Yaginumanis*, there is no screening pigment at the level of the receptive segments, so that a rhabdom is unprotected from scattered light. In *Spartaeus*, the pigmented glia contains scattered pigment granules at the level of the receptive segments; in *Portia*, screening pigment is more substantial, but does not match the arrangement in advanced salticids, where screening pigment between receptive segments is sufficiently dense to provide protection against scattered light.

It is noted that the functions of layers II–IV of the principal eyes are obscure, and their phylogenetic origins unknown. Some possible properties of the regular receptor mosaics of layer I of the principal eyes are discussed in terms of recent theoretical treatments of the receptor mosaics of vertebrate retinea, and their implications for higher order visual processing discussed.

ACKNOWLEDGMENTS

I am indebted to M. F. Land P. McIntyre, and D. S. Williams for many helpful discussions, and to D. Silversten (California Institute of Technology) for a pertinent comment on distance perception. I particularly thank Claudia Sigmund for the preparation of thin sections, especially those that provide original figures for this account, and George Weston and the Staff of the Australian National University Transmission Electron Microscope Unit for support.

ADDENDUM

Since this account was completed, two informative studies have been undertaken and deserve mention.

First, A. D. Blest, P. McIntyre, and M. Carter (in preparation) have reexamined the principal retina of *Phidippus johnsoni*, one of the two species originally studied by Land (1969a,b). The discrepancies between Land's account of the retinal mosaics and our later findings from other species of notionally advanced Salticidae proved to be a consequence of histological artifacts generated by traditional procedures for light microscopy used in the earlier work. In particular, foveal layer I rhabdomeres are longer than Land described and presumably able to act as light guides. Foveal layers II–IV present short, confluent rhabdomeres; the mosaics of these distal layers are not well-organized to serve high spatial acuities, as we have found for other species (e.g., Blest and Price, 1984; Blest and Sigmund, 1984). Thus, our analysis of the receptor mosaics of principal eyes amounts to a "general case" that is probably applicable to all advanced Salticids.

Second, A. D. Blest and M. Carter (in press) have examined the post-embryonic morphogenesis of the principal eye of *Plexippus validus* by light and electron microscopy. The primordial retina consists of a simple hemisphere of nascent receptive segments, which initially lack rhabdomeral microvilli. Tiering and the boomerang-shaped mature retina are achieved by a dramatic narrowing of the hemisphere such that marginal receptive segments lying medially and laterally at the equator of the hemisphere come to overlie receptive segments at the retinal pole. Thus the mosaics of both layers III and IV are derived from receptive segments that originally lay around the margin of the early retina. The morphogenetic events are sharply distinct from an alternative model that would suppose tiering to be generated by a differential migration of receptive segments along a line parallel to the optical axis. Given the spectral sensitivities for three of the four tiers demonstrated by Blest et al. (1981), there is an intriguing implication that ancestrally, ultraviolet receptors were localized to the retinal margins. Any concept of "ontogenetic recapitulation" can only be cautiously invoked when attempting to infer a phylogenetic history from a contemporary morphogenetic sequence. In the case of the Salticidae, it is unlikely that any representative of their immediate ancestors is still available, so that a speculative approach to the evolutionary history of the extraordinary principal eye is, perhaps, justified.

REFERENCES

Barlow, H. B. 1980. Critical limiting factors in the design of the eye and visual cortex. *Proc. R. Soc. Lond. B* **212:** 1–34.

Blest, A. D. 1983. Ultrastructure of the secondary eyes of primitive and advanced jumping spiders (Salticidae, Araneae). *Zoomorphology* **102:** 125–141.

Blest, A. D. 1984. Ultrastructure of the secondary eyes of a primitive jumping spider, *Yaginumanis* (Araneae, Salticidae, Spartaeinae). *Zoomorphology* **104:** 223–225.

Blest, A. D. 1985a. Fine structure of spider photoreceptors in relation to function, pp. 79–102. In F. G. Barth (ed.), *The Neurobiology of Arachnids*. Springer-Verlag, Berlin, Heidelberg, New York.

Blest, A. D. 1985b. Retinal mosaics of the principal eyes of jumping spiders (Salticidae) in some neotropical habitats: Optical trade-offs between sizes and habitat illuminances. *J. Comp. Physiol. A* **157:** 391–404.

Blest, A. D. 1987. Post-embryonic development of the principal retina of a jumping spider. I. The establishment of receptor tiering by conformational changes. *Philos. Trans. R. Soc. Lond. B.*

Blest, A. D. and M. Carter. 1987. Post-embryonic development of the principal retina of a jumping spider. II. The acquisition and re-organisation of rhabdomeres and growth of the glial matrix. *Philos. Trans. R. Soc. Lond. B.*

Blest, A. D., R. C. Hardie, P. McIntyre, and D. S. Williams, 1981. The spectral sensitivities of identified receptors and the function of retinal tiering in the principal eyes of a jumping spider. *J. Comp. Physiol. A* **145**: 227–239.

Blest, A. D. and M. F. Land. 1977. The physiological optics of *Dinopis subrufus* L. Koch: A fish-lens in a spider. *Proc. R. Soc. Lond. B* **196**: 197–222.

Blest, A. D. and J. Maples. 1979. Exocytotic shedding and glial uptake of photoreceptor membrane by a salticid spider. *Proc. R. Soc. Lond. B* **204**: 105–112.

Blest, A. D. and G. D. Price. 1984. Retinal mosaics of the principal eyes of some jumping spiders (Salticidae: Araneae): Adaptations for high visual acuity. *Protoplasma* **120**: 172–184.

Blest, A. D. and C. Sigmund. 1984. Retinal mosaics of the principal eyes of two primitive jumping spiders, *Yaginumanis* and *Lyssomanes*: Clues to the evolution of Salticid vision. *Proc. R. Soc. Lond. B* **221**: 111–125.

Blest, A. D. and C. Sigmund. 1985. Retinal mosaics of a primitive jumping spider *Spartaeus* (Araneae: Salticidae: Spartaeinae): A phylogenetic transition between low and high visual acuities. *Protoplasma* **125**: 129–139.

Bossomaier, T. R. J., A. W. Snyder, and A. Hughes. 1985. Irregularity and aliasing: Solution? *Vision Res.* **25**: 145–147.

Crane, J. 1949. Comparative biology of salticid spiders at Rancho Grande, Venezuela. IV. An analysis of display. *Zoologica (NY)* **34**: 159–214.

De Voe, R. D. 1975. Ultraviolet and green receptors in principal eyes of jumping spiders. *J. Gen. Physiol.* **66**: 193–208.

Eakin, R. M. and J. L. Brandenburger, 1971. Fine structure of the eyes of jumping spiders. *J. Ultrastruct. Res.* **37**: 193–208.

Forster, L. 1982. Visual communication in jumping spiders (Salticidae), pp. 161–212. In P. N. Witt and J. S. Rovner (eds.), *Spider Communication: Mechanisms and Ecological Significance*. Princeton University Press, Princeton.

French, A. S., A. W. Snyder, and D. G. Stavenga. 1977. Image degradation by an irregular retinal mosaic. *Biol. Cybernet.* **27**: 229–233.

Galiano, M. E. 1976. Comentario sobre la categoria systematica del taxon Lyssomanidae (Araneae). *Revista Mus. Argent. Cienc. not. Bernadino Rivadavia Inst. Nat. Invest. Cienc. Nat.* **5**(3): 57–70.

Hardie, R. C. and P. Duelli, 1978. Properties of single cells in posterior lateral eyes of jumping spiders. *Z. Naturforsch.* **33c**: 156–158.

Homann, H. 1971. Die Augen der Araneae. *Z. Morphol. Tiere* **69**: 201–272.

Jackson, R. R. 1982. The behaviour of communicating in jumping spiders (Salticidae), pp. 213–247. In P. N. Witt and J. S. Rovner (eds.), *Spider Communication: Mechanisms and Ecological Significance*. Princeton University Press, Princeton.

Jackson, R. R. and A. D. Blest. 1982a. The distance at which a primitive jumping spider makes visual discriminations. *J. Exp. Biol.* **97**: 441–445.

Jackson, R. R. and A. D. Blest. 1982b. The biology of *Portia fimbriata*, a web-building jumping spider (Araneae: Salticidae) from Queensland: Ultilisation of webs and predatory versatility. *J. Zool Lond.* **196:** 255–293.

Land, M. F. 1969a. Structure of the principal eyes of jumping spiders (Salticidae: Dendryphantinae) in relation to visual optics. *J. Exp. Biol.* **51:** 443–470.

Land, M. F. 1969b. Movements of the retinae of jumping spiders (Salticidae: Dendryphantinae) in response to visual stimuli. *J. Exp. Biol.* **51:** 471–493.

Land, M. F. 1985a. The morphology and optics of spider eyes, pp. 53–78. In F. G. Barth (ed.), *The Neurobiology of Arachnids*. Springer-Verlag, Berlin, Heidelberg, New York, Tokyo.

Land, M. F. 1985b. Fields of view of the eyes of primitive jumping spiders. *J. Exp. Biol.* **119:** 381–384.

Laughlin, S. B. 1980. Neural principles in the visual system,. pp. 133–280. In H. Autrum (ed.), *Handbook of Sensory Physiology VII/6B*. Springer, Berlin, Heidelberg, New York.

Platnick, N. 1971. The evolution of courtship behaviour in spiders. *Bull. Br. Arachnol. Soc.* **2**(3): 40–47.

Snyder, A. W. 1979. The physics of vision in compound eyes, pp. 225–314. In H. Autrum (ed.), *Handbook of Sensory Physiology. VII/6A*. Springer-Verlag, Berlin, Heidelberg, New York.

Snyder, A. W., T. R. J. Bossomaier, and A. Hughes. 1986. Optical image quality and the cone mosaic. *Science (Washington, D.C.)* **231:** 499–501.

Wanless, F. R. 1978. A revision of the spider genus *Portia* (Araneae: Salticidae). *Bull. Br. Mus. Nat. Hist. Zool.* **34**(3): 83–124.

Wanless, F. R. 1984. A review of the spider subfamily Spartaeinae nom. nov. (Araneae: Salticidae) with descriptions of six new genera. *Bull. Br. Mus. Nat. Hist. Zool.* **46:** 135–205.

Williams, D. S. and P. McIntyre. 1980. The principal eyes of a jumping spider have a telephoto component. *Nature (London)* **288:** 578–580.

Yamashita, S. and H. Tateda. 1976. Spectral sensitivities of jumping spider eyes. *J. Comp. Physiol.* **105:** 29–41.

CHAPTER 11

The Architecture of the Mushroom Bodies and Related Neuropils in the Insect Brain*

Friedrich-Wilhelm Schürmann
Department of Cell Biology
Institute of Zoology
University of Göttingen
Göttingen, West Germany

* Dedicated to Prof. Dr. F. Huber on the occasion of his 60th birthday.

11.1. INTRODUCTION

Since their detection by Dujardin (1850), the mushroom bodies (corpora pendunculata) in the insect brain have been intensively investigated by morphologists, physiologists, and ethologists. Morphometric studies contributed to the exploration of evolutionary trends in mushroom body and brain development for different taxonomic groups of insects (Hanström, 1928; Rensch, 1959). Electrophysiological and stimulation experiments were performed to analyze the role of mushroom bodies and their neuronal elements within the brain (Huber, 1960; Otto, 1971; Wadepuhl, 1983; Homberg, 1984; Schildberger, 1984; Chapter 21 by Erber et al. in this volume). Recent studies on *Drosophila* wild types and mutants use genetic techniques to bridge the gap between the brain structure and behavior analysis (Heisenberg, 1980; Hall, 1984; Fischbach and Heisenberg, 1984; Heisenberg et al., 1985).

Because the mushroom body system imposes a number of limitations on the experimental investigation of its functions owing to the delicacy of its neuronal components, many insights and functional views are deduced or supported from descriptions of neural sets, their synaptic connectivity, and their links with the surrounding neuropil. Progress in the investigation of central brain parts other than mushroom bodies (Strausfeld and Bacon, 1983; Schürmann, 1985) and the description of identified large neurons (Strausfeld and Nässel, 1980) allow the formulation of some organizing principles and functional perspectives of central brain compartments and their interrelationships. This account focuses on structural details, which have largely been accumulated in recent years from studies based on classical histological methods or experimental anatomical techniques, such as the following: the cobalt procedures, Lucifer yellow fills of single neurons, electron microscopy of cell groups and identification of neuroactive substances by histo- and immunocytochemistry.

This chapter is not meant to be comprehensive (for older reviews on structure, see Bullock and Horridge, 1965; Klemm, 1976). Physiology of the mushroom bodies is reviewed in Chapter 21 of this volume, and developmental studies on the mushroom bodies and other brain parts can be found in the synoptic articles of Edwards (1969) and Pipa (1973).

11.2. GROSS MORPHOLOGY OF THE MUSHROOM BODIES

The mushroom bodies are found in most insect groups, as well as in other arthropods and annelids and must not be considered as homologous structures (Holmgren, 1916). They vary in shape and size in various pterygote insects, and are absent in most apterygotes (Hanström, 1940; Ratzendorfer, 1952; Korr, 1968). Some apterygotes, however, for example, *Japyx* (Diplura) (Hanström, 1940) exhibit corpora pedunculata. These prominent compartments of glomerular neuropil (the terms glomerular and nonglomerular neuropil are used for practical reasons. Glomerular-like neuropil parts can be detected in the nonglomerular neuropil surrounding the corpora pedunculata and central body by electron microscopy. All neuropilar parts of the corpora pedunculata are summarized here as glomerular) occur in adults, as well as in all immature stages of both hemi- and holometabolous insects (cf. Bullock and Horridge, 1965).

11.2.1. Form, Size, and Cell Numbers

The mushroom bodies comprise two symmetrically arranged parts of characteristic shape in the protocerebral brain hemispheres. Each corpus pedunculatum consists of a peripheral layer of the small globule cells (Kenyon cells), restricted to the corpora pedunculata, and, therefore, are named intrinsic cells. The neuropilar parts are typically divided into the calyx—often formed as a cup-shaped structure and organized as a double calyx—and a columnar system, consisting of the stalk (pedunculus) and two arbors, the α- and β-lobes (Figs. 11.1, 11.3). Size, form, and arrangement of these columns and their relations to body and brain axes vary in the taxonomic groups. Calyxes seem to be lacking in Odonata (Baldus, 1924) and are poorly developed in Diptera (Groth, 1971). Within a taxonomic group, the shape of the mushroom bodies may not vary significantly, as shown for Diptera (Groth, 1971) and for Coleoptera (Goosen, 1949). In a detailed study of 16 orthopteran species, Weiss (1981) documents form differences of the calyces and gives a hypothesis on mushroom body compartment evolution. The form of mushroom bodies is primarily due to the arrangement of their intrinsic fiber components, the Kenyon cells, and is only slightly modified by invading elements from other brain parts, the extrinsic fibers (Schürmann, 1973). A volume increase in closely related species is mainly due to augmentation of intrinsic Kenyon cells (Rensch, 1958). The expression of mushroom bodies in the insect groups in comparison to other brain parts and sense organs, has been thoroughly reviewed by Howse (1973, 1975).

Specific sex and caste volume differences have been repeatedly reported (Forel, 1874; Jonescu, 1909; Pandazis, 1930; Neder, 1959; Hinke, 1961; Rossbach, 1962; Witthöfft, 1967; Technau, 1984). Sex differences of other neuronal compartments and neuron types are documented in a number of studies, for example, male macroglomeruli in the antennal lobes (Boeckh et al., 1984) or male and female visual neurons (Hausen and Strausfeld, 1975; Ribi, 1985).

Sexual dimorphism resulting in considerable mushroom body volume differences in worker bees (0.05334 mm^3) and drones (0.04290 mm^3)—statis-

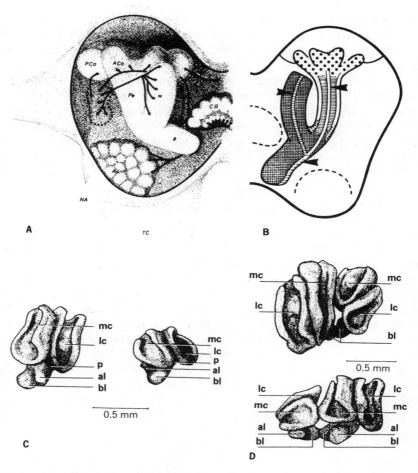

Figure 11.1. (A) *Acheta domesticus.* Mushroom body with main connections to the neuropilar brain compartments; posterior (PCa) and anterior calyx (ACa), stalk (Pe), α- and β-lobe; central body (CB), antennal lobe (AL), antennal nerve (NA), tritocerebrum (TC). (B) *Acheta domesticus.* Scheme of intrinsic fiber bundles in the stalk and lobes. Note transformation of a fiber bundle (*arrowhead*) centrally in the stalk and α-lobe column to the ventrolateral margin of the β-lobe. (C) *Apis mellifera.* Sexual dimorphismn of mushroom bodies. Worker bee (*left*) and male drone corpus pendunculatum (*right*). (D) Gynandromorph mushroom bodies. Worker bee-like left big corpus pedunculatum, drone-like right small corpus pedunculatum. Dorsocaudal view (*top*) and frontal aspect (*bottom*). (C, D) Reconstruction from serial sections; medial (mc), lateral calyx (lc), pedunculus (p), α-lobe (al), β-lobe (bl). (A from Schildberger, 1983; B from Schürmann and Klemm 1973; C, D from Buitkamp-Möbius 1975.)

tically significant—has been documented by Buitkamp-Möbius (1975). Depicted mushroom bodies (Fig. 11.1C) as well as cell body counts and measurements of the cell body rind/neuropil ratios (Witthöfft, 1967), reflect both fiber number and length differences in the mushroom bodies of male and female bees. Buitkamp-Möbius (1975) investigated the corpora pedunculata of bee gynandromorphs (Fig. 11.1D). While comparing the expression of mushroom bodies with different forms of worker bee behaviors, she did not observe elaborate behavioral sequences, such as complete foraging and dancing in gynandromorph individuals lacking the full mushroom body volume of workers.

Numbers of Kenyon cells are determined by cell body counts or calculated from fiber counts (Pipa, 1973; Klemm, 1976) for several species. The number of mushroom body cells of worker bees (340,000 total) (Witthöfft, 1967) is exceeded by that in *Periplaneta americana* males (403,600 total) (Neder, 1959), the species with the highest known number of Kenyon cells. Although belonging to different orders, both species show striking capabilities for learning (Menzel et al., 1973; Balderrama, 1980; Erber, 1983). Mushroom body intrinsic elements comprise about a 100,000 fibers in *Acheta* (Schürmann, 1973) and 42,000 in *Musca* (Strausfeld, 1976). We counted 4460 fibers in the upper stalks of *Drosophila* males (standard deviation 4.6%, 10 animals, wild type, Berlin strain). Differences between the left and right stalk are not statistically significant, reflecting a high degree of uniformity between individuals of the same stock, sex, and age. In a study on worker ants (*Formica rufa*), comparing sizes and measurement of the head and some brain compartments with foraging efficiency, including maze learning, Bernstein and Bernstein (1969) reported a positive correlation between navigation ability and the size of heads, external receptors (eyes and antennae), and the mushroom bodies (from measurements of 14 brain parameters). They tentatively related foraging efficacy to an increase of Kenyon cell number and mushroom body volume.

11.2.2. Plasticity of Mushroom Bodies

Developmental changes in mushroom body forms and volumes, have been described for both hemi- and holometabolous insects (Edwards, 1969; Pipa, 1973). Postlarval growth has been detected for the mushroom bodies of Coleoptera (Rossbach, 1962); volumetric determinations and cell counts of brain compartments of the rove beetle, *Aleochora* (Staphylinidae) have revealed a remarkable increase in the mushroom bodies, differing from other brain parts in young imagines during the first 20 days after eclosion (Bieber and Fuldner, 1979). The drastic augmentation of neuropil—more than 71%—comes from proliferation of persisting Kenyon cell neuroblasts. The number of Kenyon cells increases from 1632 (day 1 of imago) to 2797 (day 20). Thus, considerable imaginal growth of mushroom bodies could infer reactive plasticity to the environment in adults.

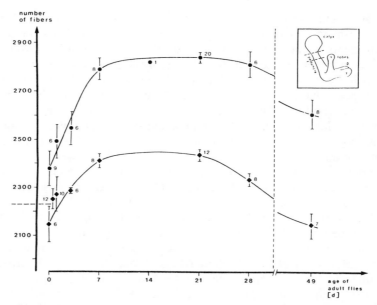

Figure 11.2. Number of fibers in the caudal mushroom stalk body of adult female *Drosophila melanogaster* as a function of age; wild-type Kapelle flies have significantly more fibers than those of wild type Berlin. Note rapid increase of fiber number during the first week after eclosion. (From Technau, 1984.)

Behavioral modifications due to individual experiences are known for worker bee adults (Menzel et al., 1974), but examples of structural changes in insect neurons induced by behavior or environmental conditions—as repeatedly shown for vertebrate nervous systems—are scarce. Recent studies by Technau (1984) and Technau and Heisenberg (1982), indicate considerable structural changes in the small corpora pedunculata of *Drosophila*. In the puparium stage, Kenyon cell fibers in the corpora pedunculata undergo a drastic decline and subsequent reorganization, whereas the overall morphology of the mushroom bodies is preserved. Technau (1984) demonstrated a considerable increase in fiber numbers within the mushroom body stalks for seven days after eclosion and a decline in 3–4 week-old flies (Fig. 11.2), probably due to aging accompanied by significant degeneration (Herman et al., 1971). 3-H-thymidine incorporation into mushroom body cells suggests neuronal cell division in adults. Moreover and most interestingly, olfactory and mechanosensory deprivation or isolation produced significant decreases of stalk fibers in wild types, whereas visual deprivation was ineffective (Technau and Heisenberg, 1982). However, it remains to be shown that stalk fiber number changes are solely due to Kenyon cell proliferation and fiber outgrowth, as invading extrinsic fiber branches running parallel to the intrinsic ones could contribute to alterations as well.

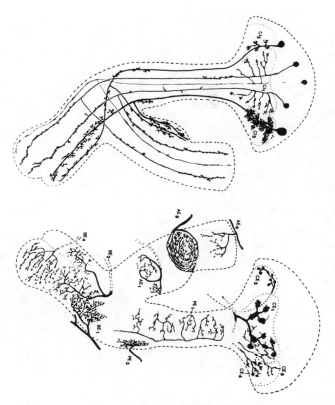

Figure 11.3. Intrinsic cells types (*left*) and extrinsic fiber typers (*right*) in the mushroom bodies of *Acheta domesticus.* (From Schürmann, 1973.)

11.3. CELL TYPES AND COMPARTMENTAL ARRANGEMENT IN THE MUSHROOM BODIES

The mushroom bodies comprise two main types of neurons: (1) Kenyon cells (globule/intrinsic cells) restricted to the corpora pedunculata; and (2) extrinsic cells, invading and linking the mushroom bodies with other brain areas and/or mushroom body parts (loop cells) (Figs. 11.3, 11.4, 11.5). According to Gestalt principles, these two basic types are classified into species-specific subtypes documented by a number of Golgi studies (Schürmann, 1973) and recent electrophysiological investigations of extrinsic neurons filled with cobalt or Lucifer yellow (Schildberger, 1983; Homberg, 1984; Gronenberg, 1984; Hörner, 1984; Schürmann, 1985).

11.3.1. Intrinsic Neurons

Kenyon cells are normally T-shaped neurons. Based on dendritic shape, fiber diameter, and position, three types of Kenyon cells were distinguished

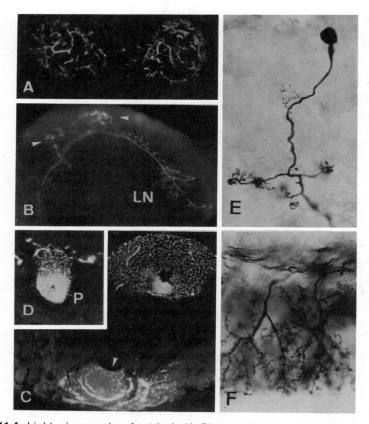

Figure 11.4. Light micrographs of extrinsic (A–D) and intrinsic cells (E, F) in mushroom bodies. (A) *Acheta domesticus.* Extrinsic fiber in the posterior calyx, Lucifer yellow stain. 150×. (B) *Acheta domesticus.* Single neuron of the antennoglomerular tract with boutons (*arrowheads*) in the anterior calyx and aborization in the lateral protocerebral neuropil (LN). Lucifer yellow stain. 120×. (C, D) *Acheta domesticus.* Loop cell with dense branching in the upper stalk (P), in the anterior calyx, and concentric marginal arborization in the α-lobe (*arrowhead*), Lucifer yellow stain. 120×. (E) *Acheta domesticus.* Kenyon cell with four dendritic branches in the anterior calyx. Golgi preparation. 560×. (F) *Apis mellifera.* Spiny overlapping dendrites in the calycal collar zone, Golgi preparation. 560×. (A from Schildberger, 1982; B–D from Hörner, 1984.)

in *Acheta* (Schürmann, 1973, 1974), four in *Formica* (Goll, 1967), and five in *Apis* (Kenyon, 1896; Mobbs, 1982). These types do not necessarily correspond to all calycal or lobe neuropil subcompartments, although a restriction to areas is clearly apparent. Thus, α-lobe subcompartmentation into at least six strata cannot be understood from the arrangement of intrinsic cell types (compare Gronenberg, 1984; Mobbs, 1982, 1984).

Micromorphology of calycal Kenyon cell arborization suggests their dendritic postsynaptic nature—shown by electron microscopy. Dendrites include small-field basket-like branchings with restricted input capacity, as

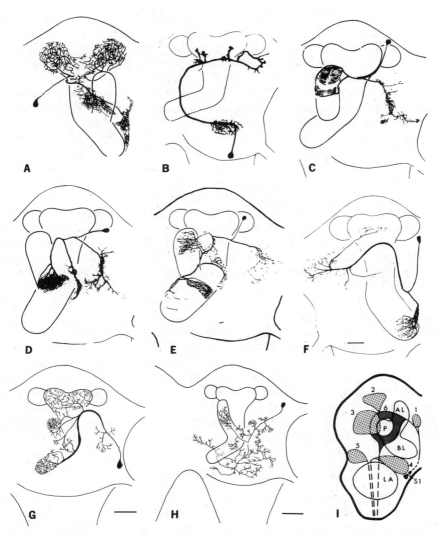

Figure 11.5. (A–H)Reconstructions of neurons stained intracellularly with Lucifer yellow, *Acheta domesticus*. (A) Cell with fine arborization in the posterior calyx and around the mushroom bodies. (B) Deutocerebral relay neuron with branching in one glomerulus and a few larger calycal boutons and collateral to the lateral protocerebrum. (C) α-lobe connecting cell. (D) α-lobe connecting cell with two fiber domains outside the mushroom bodies. (E) Cell with arborization in α- β-lobe and in protocerebral neuropil. (F, G) Two different types of loop cells connecting mushroom body parts. (H) Local interneuron around the mushroom body. (I) *Acheta domesticus*. Neuropilar compartments (1–6) in deuto- and protocerebral neuropil from cobackfillings via connectives demonstrating ascending neuronal fiber domains. Note partial overlap in mushroom body surrounding neuropil where extrinsic fibers project too. (A–F from Schildberger, 1982; F, G from Hörner, 1984; I from Rosentreter and Schürmann, 1982.)

239

well as wide-field two- or three-dimensional dense dendrites offering numerous spines for synaptic contact (Figs. 11.3, 11.4E,F). Dendrites of the same or different neighboring Kenyon cells overlap. Spine-stem length plasticity in intrinsic fiber dendrites correlated with flying experience is reported for worker bees (Coss et al., 1982; Coss and Brandon, 1984). Increase of spine–head width and stem shortening are discussed with respect to lowering dendritic input resistance and enhanced synaptic efficacy (Rall, 1978).

The parallel intrinsic fibers (diameters, 0.05–1 μm) form bundles mainly responsible for calycal and lobe subcompartments (Figs. 11.1B, 11.6). In the worker bee, a polar orientation of intrinsic cell types in the calyxes is transformed to a cartesian map in the lobes, convincingly demonstrated by Mobbs (1982, 1984) (but compare Strausfeld, 1970; Howse, 1974). In the cricket, *Acheta* (Schürmann, 1973; Schürmann and Klemm, 1973), transformation of intrinsic fiber bundles differs for the stalk and lobes. Concentric arrangement of cell types in the anterior calyx is maintained in the pedunculus and α-lobe, but not in the β-lobe (Fig. 11.1B). The length of fiber types from calyx to distal lobes may differ (Schürmann, 1973) and there is evidence that a portion of intrinsic cells does not branch into the α-lobe of *Drosophila* (Technau, 1984).

Intrinsic fibers develop blebs and branches, partially in the stalk and abundantly in the lobes, indicative of synaptic sites (Figs. 11.3, 11.8E). No evidence exists for direct synaptic interaction of fiber bundles of the compartments when spatial contact is produced by transition into the lobes. The functional significance of arborization of Kenyon cell fibers into the α- and β-lobe has been well demonstrated morphologically, but is by no means understood functionally. Increasing evidence for structural dissimilarities of the α- and β-lobe comes from the distribution of intrinsic fiber bundles, of extrinsic fiber types and from micromorphology (synaptic connectivity). This aspect is important to avoid oversimplification in models.

11.3.2. Extrinsic Neurons

According to their positions within the corpora pedunculata, extrinsic elements can be divided into three classes: (1) calyx neurons, (2) peduncle and lobe neurons, and (3) calyx-lobe connecting neurons (feedback or loop cells) (Figs. 11.3–11.6). All neurons of these classes share as a common feature arborizations in various brain parts, thus representing the functional link to intrinsic mushroom body cells. Extrinsic neurons invade the mushroom bodies via massive tracts (Weiss, 1981; Mobbs, 1984) or from the surrounding aglomerular neuropil as single elements not gathered to fiber bundles (Weiss, 1974; Schürmann, 1982; Schildberger, 1983). Golgi studies revealed at least 16 extrinsic fiber types in the cricket *Acheta* (Schürmann, 1973), but single-cell stains in crickets and bees (Schildberger, 1983; Homberg, 1984; Gronenberg, 1984) suggest a much greater number of extrinsic fiber types. While extrinsic calyx neurons, originating from tracts, show a high degree of uniformity, lobe cells—even when arising from the same cell cluster—tend to display their individuality by distribution and branching pattern. The

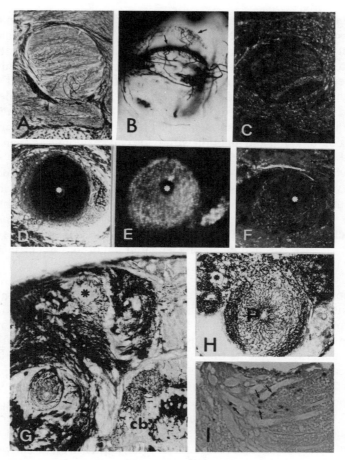

Figure 11.6. Subcompartments of mushroom body neuropil, light micrographs. (A) *Apis mellifera*. Strata in the cross-sectioned α-lobe. Bodian stain. 120×. (B) *Apis mellifera*. Selective stain of dendritic strata of extrinsic fibers (*arrows*) in the α-lobes. 144×. (C) *Apis mellifera*. Serotonin distribution in the strata of the α-lobe, immunocytochemistry. 90×. (D) Concentric fiber bundles in the stalk column, differentially stained with toluidine blue, central zone of fine intrinsic fibers (*asterisk*). 107×. (E) Dopamine indicating fluorescence in a concentric area around central zone (*asterisk*). 130×. (F) *Acheta domesticus*. Serotoninergic fiber invading stalk parts, immunocytochemistry. 140×. (G, H) *Acheta domesticus*. GABA-like immunoreactivity in fibers invading the same area as in E, marginal immunoreactivity in the α-lobe (*asterisk*). Strong reactivity in the central body (cb) and nonglomerular neuropil. 100×. (H) GABA-like immunoreactivity in the upper stalk (P) and calyx (*asterisk*). 130×. (I) *Apis mellifera*. Serotonin-like immunoreactivity (*arrows*) of extrinsic fibers with intimate spatial relation to others, α-lobe. 625 ×.

241

total number of extrinsic cells and cell clusters has not yet been described for any species. Detailed counts would be very valuable for a more exact judgment of input–output relation and intrinsic–extrinsic fiber ratio.

Extrinsic calyx elements mainly exhibit more or less prominent boutons of different size and shape (Figs. 11.4A,B,C; 11.5A,B) restricted to calycal compartments or with a wide-field distribution. The fine morphology of the extrinsic elements reflects different densities, remarkable size differences of boutons, as well as fine spiny and blebbed fibers, indicating a more complicated combination and junction within the calycal glomeruli than presently explored (Schürmann, 1985).

Knowledge of the extrinsic peduncle and lobe cells is far from being comprehensive as Golgi studies rarely exhibit complete neurons and iontophoretic stains of single neurons (Figs. 11.4, 11.5) are only available for the cricket, *Acheta* (Schildberger, 1983; Hörner, 1984; Schürmann, 1985), the cockroach (Ernst et al., 1983), and the worker bee (Homberg, 1984; Gronenberg, 1984). All pedunculi investigated incorporate extrinsic fibers. Mobbs (1982) describes axonless peduncule cells for the bee. Many fibers appear as parts of loop cells or of other extrinsic cells. The distribution of these fibers matches stalk column compartments (Fig. 11.6) as seen in single-cell stains or in histo- and immunocytological studies (Schürmann, 1973; Schürmann and Klemm, 1973, 1984).

Extrinsic fibers in the α- and β-lobes of various origins, mostly appear as spiny dendritic elements of different densities. Their arborizations characteristically form flattened bands, strata, or columns arranged perpendicularly to the intrinsic fiber strands, as pointed out by Schürmann (1974, 1985) and Mobbs (1982, 1984). They match one or several compartments or parts of them, lined up along the intrinsic fibers (Fig. 11.6). A further subcompartmentation is therefore indicated. Some extrinsic fibers display blebs or even bouton-like swellings, indicative of presynaptic sites, as shown by electron microscopy (Schürmann, 1972, 1973, 1974, 1985). The geometrical, precisely ordered array of the extrinsic and intrinsic fibers is the basis for an electrophysiological approach employing evoked potentials (Kaulen et al., 1984).

Extrinsic calyx-lobe neurons (feedback or loop neurons) promote information exchange between mushroom body parts, as first suggested by Vowles (1964) and first depicted by Goll (1967). Such neurons were later indicated or described for several species (Pearson, 1971; Schürmann, 1973; Homberg and Erber, 1979; Schildberger, 1981, 1983; Mobbs, 1982, 1984; Gronenberg, 1984; Hörner, 1984). These neurons (Figs. 11.5F,G; 11.12) exhibit, as a common feature, branching in the nonglomerular neuropil, but show different fine morphology and distribution, indicating functional diversity. In the cricket, *Acheta,* this class of neurons can either display boutons or very fine spine-like protrusions in the calyces. In addition to a feedback mode from lobe to calyx (Schildberger, 1981), a feed-forward flow in the opposite direction might occur as well. Though a distribution in corresponding calyx and lobe compartments is often observed, interaction with the same intrinsic fiber subgroups must not be assumed. Recent immuno-cytochemical studies show GABA-containing loop cells stemming from the protocerebral-calycal tract in the bee brain (Bicker et al., 1985). Loop cells show remarkable physiological properties, as shown for crickets (Schild-berger, 1981, 1984; Hörner, 1984) and bees (Homberg and Erber, 1979; Gro-nenberg, 1984). Their feedback or feed-forward functions remain to be ex-perimentally established.

11.4. CHEMONEUROANATOMY OF MUSHROOM BODIES

Biochemical, histochemical, and immunocytochemical methods are of in-creasing importance for the analysis of brain and mushroom body com-partmentation and connectivity (Fig. 11.6). Especially selective techniques to demonstrate putative transmitters (see also Chapter 23 by Klemm in this volume) and other neuroactive sustances complement classical neuromor-phological techniques.

11.4.1. Acetylcholine

Indirect evidence for cholinergic elements in the mushroom bodies comes from histochemical studies on acetylcholinesterase distribution. Reactivity is concentrated in the calyxes, but occurs to a minor extent in the lobes in cockroaches (Frontali et al., 1971; Hess, 1972) and in crickets (Lee et al., 1973). In *Musca* (Ramade, 1965) and *Pollistes* (Strambi, 1974) cholinesterase activity is reported to be restricted to the calyxes and lacking in the peduncle and the lobes. As Kenyon cell bodies appear devoid of cholinesterase ac-tivity, acetylcholine can be attributed to extrinsic cells, mainly to the deu-tocerebral relay neurons of the antennoglomerular tracts (see also Chapter 17 by Mercer in this volume).

11.4.2. Catecholamines and Serotonin

Diversity of distribution of biogenic amines in the mushroom bodies of dif-ferent insect orders—even in closely related groups—has been revealed by histofluorescence techniques (Klemm, 1974, 1976, 1983; Mercer et al., 1983) and more recently for 5-HT by immunocytology (Klemm and Sundler, 1983; Klemm et al., 1984; Schürmann and Klemm, 1984; Tyrer et al., 1984).

Catecholamines (dopamine) seem to be lacking mostly in the calyces, but have been detected in those of bees (Klemm, 1974; Mercer et al., 1983), cockroaches, and locusts (Klemm, 1983; Klemm and Sundler, 1983). Ser-otonin-containing boutons of extrinsic fibers within the calyces occur in locusts (Klemm and Sundler, 1983; Tyrer et al., 1984), cockroaches (Klemm et al., 1984), and crickets (Schürmann, 1985), as shown by immunocyto-chemistry, but are lacking in the bee (Schürmann and Klemm, 1984). These serotoninergic calycal fibers stem from extrinsic fibers in the nonglomerular neuropil around the mushroom bodies, and were never detected in the an-tennoglomerular tract.

Stalks can be devoid of catecholamines (e.g., in *Periplaneta, Lepisma, Trichoptera*) as pointed out by Klemm (1974, 1976). In the cricket, catecholamine-indicating fluorescence is correlated with the distribution of vesicles in the stalk and lobe subcompartments (Schürmann and Klemm, 1973). Patterns of catecholamines and 5-HT distribution, matching stalk compartments or their parts in the form of bands or stripes parallel or perpendicular to intrinsic fibers are most evident for the α- and β-lobes (Frontali, 1968; Klemm, 1976; Klemm and Sundler, 1983; Mercer et al., 1983; Schürmann and Klemm, 1973, 1984) (Fig. 11.6). 5-HT elements in the α-lobes in the bee appear presynaptic (Schürmann, unpublished data), as shown for other 5-HT-containing systems (Nässel et al., 1984). There is abundant evidence for the extrinsic nature of catecholamine- or serotonin-containing fibers. Catecholamines in intrinsic elements tentatively proposed for the cricket (Schürmann and Klemm, 1973) are indicated for a small group of Kenyon cells in the bee (Mercer et al., 1983; see also Chapter 17 by Mercer in this volume).

11.4.3. GABA

The occurrence of GABA (γ-aminobutyric acid) in the mushroom bodies could not be convincingly shown for the cockroach (Frontali and Pierantoni, 1973), but was recently detected immunocytochemically in bees and crickets. In the bee, massive fiber bundles with GABA-like immunoreactivity in the protocerebral-calycal tract interestingly represent feedback neurons with arborizations in the α-lobe compartments and restricted distribution of boutons in the calyces (Bicker et al., 1985). In the cricket, we found GABA elements of different extrinsic origin widely distributed in the calyces and exhibiting single fiber elements in the stalk and lobes (Fig. 11.6G,H).

Chemical diversity of extrinsic fibers has been established for all mushroom body parts. Up to four neuroactive substances (acetylcholine, GABA, dopamine, 5-HT) must at least be considered for extrinsic input and output fibers. Substantial overlap of extrinsic fibers with different putative transmitters is found in the lobes. Histo- and immunocytochemical studies have especially well demonstrated the intimate relationship of the mushroom bodies to the surrounding neuropil. This has been termed as accessory mushroom body neuropil (Schürmann, 1982, 1985). The synaptic equipment and polarity of the different extrinsic cell types remains to be investigated. The amounts of transmitters or other compounds in and around the mushroom bodies can only be roughly estimated at present and cannot be firmly attributed to extrinsic and intrinsic cell types (Mercer et al. 1984; also Chapter 17 by Mercer in this volume).

11.4.4. Octopamine

Kenyon cell somata contain considerable amounts of octopamine, as shown by biochemical methods for the cockroach (Dymond and Evans, 1979), the locust (Evans, 1980), and the bee (Mercer et al., 1984), suggesting octopaminergic nature of intrinsic elements. Further studies must clarify whether or not octopamine might be attributed to extrinsic fibers as well (Mercer et al., 1984).

11.4.5. Other Neuroactive Substances

No neuropeptides could be detected in the mushroom bodies of a number of insect species, except in the surrounding neuropil and some other glomerular neuropils (Bishop and O'Shea, 1982; El Salhy et al., 1983; Schooneveld et al., 1983; Veenstra, 1984; Veenstra and Schooneveld, 1984; Veenstra et al., 1984). This could mean that neuropeptide actions do not directly involve the mushroom bodies. Interestingly, the corpora pedunculata of *Limulus* contain substance P (Chamberlain and Engelbertson, 1982; see also Chapter 4 by Fahrenbach and Chamberlain in this volume). The lack of neuropeptides in insect mushroom bodies and the occurrence in those of some other arthropods might reflect nonhomologous structures.

11.5. ULTRASTRUCTURE OF THE MUSHROOM BODIES

Electron microscopical studies of the corpora pedunculata mainly deal with glia-neuron relations and synaptic organization of neuropil.

11.5.1. Pericaryal Layer

The peripheral rind of the mushroom bodies (Fig. 11.7) is composed of neuronal, glial, and tracheal elements (Landolt, 1965) and does not differ from other pericaryal regions. Few glial cells interdigitate between and envelop the small Kenyon cell perikarya (glia–neuron ratio 1 : 10–50). Glial invaginations (trophospongium) into neural cytoplasm is lacking. In ants, 42% of the nerve cell somata are not totally separated by glial elements, but have so-called windows for direct contact of neuronal somata (Landolt, 1965; Landolt and Ris, 1966). They have been interpreted as soma-somatic junctions for electrical contact, as they can appear with close membrane apposition sites, depending on fixation procedure. Such zones, with and without close membrane appositions, can also be observed in the pericaryal layer of mushroom bodies in the cricket (Fig. 11.7B,C). There is no experimental support for electric coupling of somata, which appear to be electrically passive. The cytoplasmic organization of globule cells does not differ from other small insect neurons. A thin glial sheath, interrupted by through running axons, separates the perikaryal layer from the deeper neuropil. A glial sheath surrounds neuropilar fiber compartments of the mushroom bodies (Figs. 11.8B, 11.9C) and areas within the lobes (Schürmann, 1970, 1971, 1972; Technau, 1984), but a complete isolation of fiber compartments was never observed.

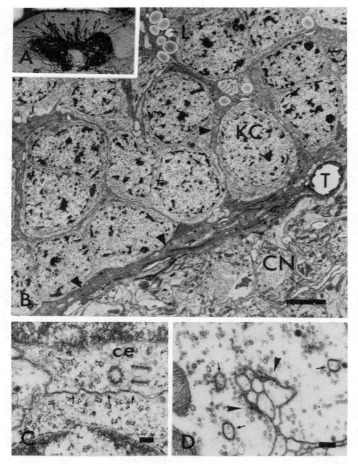

Figure 11.7. (A) *Acheta domesticus.* Somata of Kenyon cells filling the anterior calyx with their dendrites, Golgi impregnation. 63×. (B) *Acheta domesticus.* Pericaryal layer of Kenyon cells (KC) separated by glial layers (*triangles*), but partially with direct membrane contact (*arrows*), calyx neuropil (CN) with thick glial separation, lipid inclusion (L) in glia, trachea (T). Scale, 4 μm. (B, C) *Acheta domesticus.* Direct membrane contact of Kenyon cells in pericarya (*arrows*), centriol (ce). Scale, 100 nm. (D) *Apis mellifera.* Presynaptic bouton with synapses to intrinsic fibers (*arrowheads*), intrinsic fiber invaginations (*arrows*), calyx. Scale, 100 nm.

11.5.2. Synapses and Synaptic Connectivity

Synapses in the mushroom bodies (Figs. 11.8–11.10) have species-specific appearance and occur in different types (Steiger, 1967; Mancini and Frontali, 1967, 1970; Schürmann, 1970, 1971, 1972, 1974, 1985; for review Schürmann, 1980). A hexagonal presynaptic grid with variable extension, similar to vertebrate synaptic figures, was found in intrinsic fibers of the bee mushroom

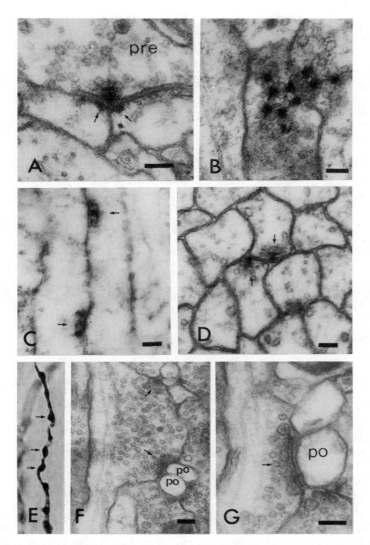

Figure 11.8. (A) *Apis mellifera.* Ultrastructure of a synapse in an extrinsic calyx presynaptic boutons (pre), subsynaptic electron dense membrane appositions (*arrows*) in postsynaptic intrinsic fibers. Scale, 100 nm. (B) *Apis mellifera.* Presynaptic grid of osmiophilic columns surrounded by clear vesicles, upper stalk. Scale 100 nm. (C) *Apis mellifera.* Reciprocal synapses with presynaptic figures (*arrows*) in presumably intrinsic fibers of the upper stalk, bismuth impregnation. Scale, 100 nm. (D) *Calliphora erythrocephala.* Synapses in stalk fibers with presynaptic figures (*arrows*) typical for flies. Scale, 100 nm. (E) *Acheta domesticus.* Blebbed intrinsic fiber in the α-lobe. 800 ×. (F) *Acheta domesticus.* Ultrastructure of a bleb, presynaptic (*arrow*) to perpendicular running fibers (po). Scale, 100 nm. (G) *Acheta domesticus.* Synapse (*arrow*) in an intrinsic fiber out blebs, α-lobe. Scale, 100 nm. (B, C from Schürmann, 1971; F, G from Schürmann, 1972.)

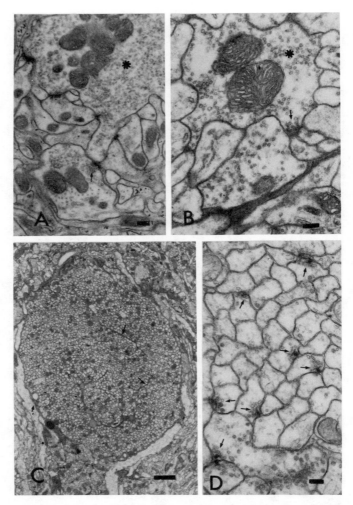

Figure 11.9. (A) *Acheta domesticus.* Extrinsic bouton (*asterisk*) presynaptic to vesicle-free fibers and postsynaptic to another vesicle-filled fiber (*arrow*), anterior calyx. Scale, 200 nm. (B) *Acheta domesticus.* Extrinsic bouton (*asterisk*) presynaptic (*arrow*) to a vesicle-containing fiber, anterior calyx. Scale, 100 μm. (C) *Drosophila melanogaster.* Glial sheath around and within the pedunculus (*arrows*), wild-type Berlin. Scale, 1 μm. (D) *Drosophila melanogaster.* Synapses in the stalk (*arrows* presynaptically), wild-type Berlin. Scale, 200 nm.

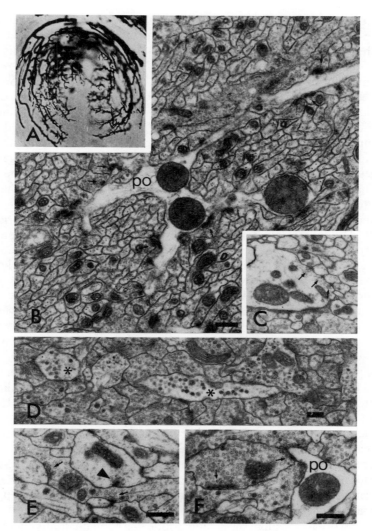

Figure 11.10. (A) *Acheta domesticus.* Dendritic part of an extrinsic fiber arranged perpendicular to the α-lobe, Golgi impregnation. 580×. (B) *Acheta domesticus.* Extrinsic fiber postsynaptic to intrinsic elements (*arrows*). Scale, 500 nm. (C) *Acheta domesticus.* Extrinsic fiber with presynaptic sites (*arrows*), α-lobe. Scale, 500 nm. (D) *Acheta domesticus.* Postsynaptic fiber (*asterisk*) with dense core vesicles, β-lobe. Scale, 500 nm. (E) *Acheta domesticus.* Fiber with clear synaptic vesicles (*triangle*) presynaptic to another filled with dense core vesicles and presumably intrinsic, synapses (*arrow*), β-lobe. Scale, 500 nm. (F) *Acheta domesticus.* Fiber with synapses (*arrow*) to vesicle-filled or vesicle-free postsynaptic element (po), β-lobe. Scale, 400 nm. (B, C from Schürmann, 1972; D from Schürmann and Klemm, 1973.)

249

bodies (Schürmann, 1971) (Fig. 11.8B). Differences in vesicle populations and pre- and subsynaptic membrane specializations might be used to distinguish intrinsic and extrinsic fiber elements from each other, but this awaits support from a study from marked cells.

Asymmetry of chemical synapses allows modeling of informational flow within the mushroom bodies and—in combination with light and electrophysiological studies—depicting a framework of interactions and functional capacities (Schürmann, 1974, 1982, 1985). The calyx is interpreted as the main input relay station of extrinsic to intrinsic elements. Calycal microglomeruli are composed of extrinsic presynaptic boutons with divergent connectivity on the intrinsic dendrites (Steiger, 1967) (Fig. 11.7D). The presynaptic nature of extrinsic fibers from the antennoglomerular tract has been confirmed by degeneration techniques (Ernst et al., 1977), showing presynaptic collaterals of the same neurons within the lateral protocerebral lobes as well. Our recent findings on a spiny nature of extrinsic fibers (Fig. 11.9A,B) and on elements presynaptic to extrinsic boutons (Schürmann, 1985) indicate a more complex wiring at least for calyx parts of crickets and bees than previously observed. Observations on close appositions of dendritic membranes (Steiger, 1967), as sites for electric coupling, have not yet been confirmed by use of special methods. Differences in bouton vesicle populations, indicating topographic and transmitter specialization, await further research on marked cells of known origin and morphometry of synaptic organelles.

Synapses in stalk compartments of cricket and bee mushroom bodies (Figs. 11.8B–D, 11.9C,D) have been interpreted as a basis for direct or indirect interaction of intrinsic cells (Schürmann, 1970, 1971, 1974, 1985). Electron microscopy of identified cells is still lacking. Electrical coupling or influence between parallel running intrinsic fibers cannot be excluded. Evidence for gap-like junctions between insect nerve cells is rarely given (Strausfeld and Bassemir, 1983; Killmann and Schürmann, 1985). Synapse distribution in the stalk appears nonrandom, and signals marked differences in informational flow of intrinsic fiber bundles (Schürmann, 1970, 1974; Schürmann and Klemm, 1973; Technau, 1984).

Synaptic organization in the α- and β-lobe has been qualitatively explored in *Periplaneta* (Mancini and Frontali, 1967, 1970; Frontali and Mancini, 1970) and in *Acheta* and *Apis* (Schürmann, 1972, 1974, 1985; Schürmann and Klemm, 1973). Intrinsic fibers appear densely packed with clear and dense-core vesicles along the lobe columns (size frequency at peaks of 30 and 47 nm, respectively) (Mancini and Frontali, 1967). Both vesicle types can be observed along the intrinsic fibers from the upper stalk to the top of the lobes, indicating vesicle migration. Extrinsic fibers can contain small translucent vesicles and dense-core vesicles (size distribution peak of 47 nm, diameters from 30–120 nm) as well. However, extrinsic fibers do not exhibit the massive accumulation and dense package of vesicles typical for intrinsic fibers (Schürmann, 1972). Presynaptic membrane specializations are found

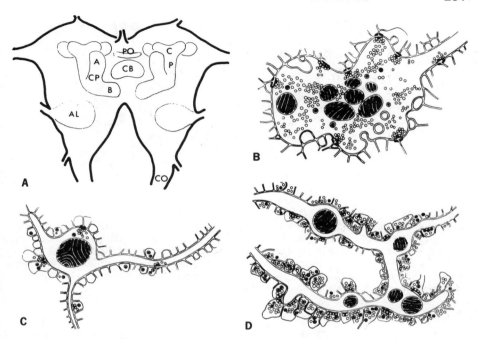

Figure 11.11. (A) Scheme of a cricket brain with mushroom body calyx (C), peduncle (P), α- β-lobe (A, B), and central body (CB), bridge (PO), antennal lobe (AL), connective (CO). (B–D) Main types of synaptic connectivity of intrinsic and extrinsic fibers in mushroom bodies in the cricket. (B) Presynaptic extrinsic bouton, calyx. (C) Reciprocal connection of an extrinsic fiber with a number of intrinsic elements in the stalk. (D) Convergence of intrinsic fibers to a dendritic extrinsic fiber in the α-lobe. (From Schürmann 1985.)

in both intrinsic and extrinsic fiber types (Figs. 11.8, 11.10). There is only conclusive evidence for convergence of presynaptic intrinsic onto extrinsic fibers which supports their proposed dendritic nature (Schürmann, 1972, 1974). Postsynaptic extrinsic elements with thick, dense-core elements do occur, indicating directional flow of vesicles into dendritic elements (Schürmann and Klemm, 1973). There is substantial evidence for coupling of intrinsic cells by chemical synapses (Frontali and Mancini, 1970; Schürmann, 1970, 1974) and circumstantial support for synaptic connectivity between extrinsic fibers which might be equipped with different neuroactive substances. This can be deduced from studies on serotonin distribution in the α-lobes (Fig. 11.6I). Presynaptic extrinsic fibers could also contact intrinsic fibers (Schürmann, 1972), but details on synaptic wiring require an electron microscopical investigation of marked identified cell types. Apparently α- and β-lobes do not represent mirror images of fiber types, fiber type distribution, and synaptic organization, although gross similarities (e.g., lobes as major output stations) cannot be neglected.

The dense package of vesicles within the intrinsic cells from the basal

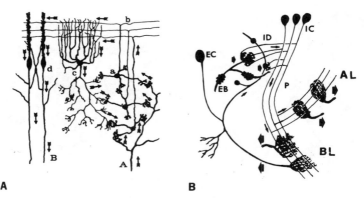

A   **B**

Figure 11.12. (A) Diagram of circuitry in the vertebrate cerebellum after Cajal with flow of information (*arrows*), mossy fibers (A), granule cell (a) with parallel fibers (b), Golgi cells (c), and Purkinje cells (d) with neurites (B). (B) Analogue circuitry in the insect mushroom body system. Intrinsic cells (IC) with parallel fibers in the stalk (P) and the α- β-lobes (A,B), input via extrinsic fiber boutons (EB) to intrinsic cell dendrites (ID), output in the lobes and extrinsic feedback cell (EC) corresponding to Golgi cell. Informational flow direction indicated by *arrows*. (From Schürmann, 1985.)

calyx up to their endings in the lobes, clearly sets them apart from all other known fiber elements of the central brain and even from similar-looking cells in optic lobe parts. The intrinsic cells restricted to their mushroom body compartments laden with synaptic organelles, have been tentatively considered as a special synaptic system of very fine fibers condensed in a limited space (Schürmann, 1970). Morphometry of synaptic distribution is needed for a comparison of brain and mushroom body compartments. Synaptic connectivity within the mushroom bodies displays features attributed to local circuits (Schmitt et al., 1976; Pearson, 1979).

The description of the geometry of the mushroom bodies and the partial analysis of the main synaptic arrangement of intrinsic and extrinsic elements (Fig. 11.11) permits pointing out prominent features relevant to functions. The intrinsic neurons with their fibers arranged in parallel represent a multichannel system having separated main input (calyx) and output (lobes) sites. Extrinsic neurons converge from different brain parts into the calyxes, where they are connected in a divergent mode on the abundant intrinsic elements. These converge to a smaller number of extrinsic elements within the lobes, which diverge around the corpora pedunculata into the nonglomerular neuropil. Crosstalk between the intrinsic elements and neuropil compartments is established via direct synaptic coupling or indirectly by extrinsic elements (Schürmann, 1974, 1985). The alignment of extrinsic fiber dendrites perpendicular to the intrinsic fibers, as well as other morphological and some physiological features, have led to the view of the mushroom bodies as an apparatus for sequential activation and signaling in the millisecond range along time delay lines (Schürmann, 1974, 1985). Thus, they show remarkable similarities with the vertebrate cerebellum (Fig. 11.12). The detection of the loop cells (Schildberger, 1981; Gronenberg, 1984) has even added a further analogue to that comparison. Although this sort of modeling cannot give more than a framework of the functional capacity, it allows some exclusion of functional properties and has provided information on which questions can be investigated at specific mushroom body sites.

11.6. RELATIONS OF MUSHROOM BODY NEUROPIL

The understanding of mushroom bodies is very much dependent on the knowledge of their connections to other parts of the brain and central nervous system. Though numerous investigations deal with mushroom body-connecting fiber tracts, only a few employing metal impregnation give information on the sites of mushroom body intrusion by extrinsic fibers, on their origin, and arborization outside the mushroom bodies (for critical review, see Weiss, 1981). Knowledge is mainly restricted to the brains of *Drosophila* (Power, 1943; Technau, 1982), butterflies (Pearson, 1973), bees (Mobbs, 1982, 1984; Homberg, 1984; Gronenberg, 1984), locusts, cockroaches (Ernst et al., 1977; Ernst and Boeckh, 1983) and crickets (Schürmann, 1973, 1974, 1985; Schildberger, 1983).

11.6.1. Calyx Connections

Connections are established by tracts or—probably more often—by single-fiber elements, forming sets of arborizations within and outside the mushroom bodies first detected by Weiss (1974). Massive tracts can be mainly traced as single- or multiple-fiber bundles (e.g., in the bee brain) from the antennal lobes to the calyces, which may give off branches to the lateral protocerebrum (Schürmann, 1973; Williams, 1975; Ernst et al., 1977; Mobbs, 1982). Antennoglomerular tracts (antennocerebral tracts, tractus olfactorioglobularis) comprise some hundred fiber elements (Boeckh et al., 1984). The dendritic arborizations of deutocerebral relay cells conveying information to the calyces are normally distributed in one glomerulus, seldom into two glomeruli of the antennal lobes (Mobbs, 1982; Burrows et al., 1982; Schildberger, 1984). Multiglomerular relay cells interpreted as centrifugal cells were demonstrated by Mobbs (1984). In addition, Mobbs mentions a structural basis for topical representation of distinct antennal glomeruli within the calyces of mushroom bodies in the bee. Gronenberg (1984) emphasizes that this might not be the unique principle for information flow from antennal glomeruli to the calyces. Immunocytochemical studies on GABA and 5-HT distribution give some hints for local restriction of fibers within the calyces of bee mushroom body (Bicker et al., 1985). Tritocerebral fibers joining the antennoglomerular tract have been documented for locusts, roaches (Aubele and Klemm, 1977; Ernst et al., 1977), and crickets (Schürmann, 1973; Weiss, 1981).

Good evidence for a direct optic lobe-calyx connection in crickets comes from cobalt backfilling experiments (Honnegger and Schürmann, 1975). More prominent, well-defined tracts from the optic glomerular neuropils with distinct projection fields in the calyces exist in the bee (Mobbs, 1982, 1984) and other social Hymenoptera (Howse, 1975). Cells of these optic tracts identified by single-cell iontophoresis exhibit mostly small-field arborizations within the optic lobe compartments (Gronenberg, 1984). Optic lobe projections are generally not as massive as antennal lobe connections to the mushroom bodies. It seems plausible to attribute antennal olfactory, mechanosensory, and optic information to the mentioned calyx-invading cells, and there is increasing evidence for a multimodal and multisensory nature of these centripetal relay cells. This indicates complex peripheral sensory information processing (Schildberger, 1984; Schürmann, 1985).

Many extrinsic fibers entering the calyces have wide- or small-field projections around the mushroom body lobes or in the lateral protocerebrum (Fig. 11.5) matching other brain interneurons, as well as ascending and descending neurons (Rosentreter and Schürmann, 1982; Schürmann, 1985).

11.6.2. Stalk and Lobe Connections

Gathering of extrinsic fibers into tracts entering the stalk and the lobes at defined areas occurs in the bee mushroom bodies (Mobbs, 1982, 1984), whereas in crickets extrinsic elements invade at different sites along the lobes (Schürmann, 1973; Schildberger, 1983). Most extrinsic output elements show dense branching around the lobes and in locations of the lateral protocerebrum and dorsal deutocerebrum (Fig. 11.5), in particular in the frontal protocerebrum around the α-lobes (Schürmann, 1982; Mobbs, 1982, 1984; Schildberger, 1983; Hörner, 1984). No direct connections of the mushroom bodies to ascending and descending brain neurons could be demonstrated up to now, although an intimate relationship was demonstrated for some frontal descending brain neurons in bees (Schürmann, 1982) and crickets (Rosentreter and Schürmann, 1982; Hörner and Gras, 1985). No direct contact between the mushroom bodies and the central body is noted in any description using modern anatomical techniques. A direct link between the calyces and the central bridge (pons) has been demonstrated for the cockroach (Klemm et al., 1984). Thus, massive direct informational flow between the central glomerular neuropils of the mushroom bodies, central body, and bridge should not be expected.

11.6.3. Connections Between Mushroom Body Compartments

Evidence for ipsilateral direct connectivity between the calyces, stalks, and the lobes via loop neurons (feedback cells) has been well documented in

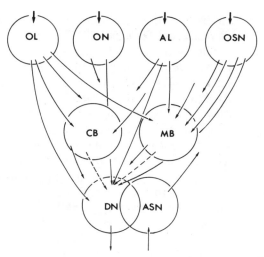

Figure 11.13. Simplified scheme of main pathways in the insect brain; informational flow indicated by *arrows*. Peripheral sensory neuropils (optic lobes, OL; ocellar neuropil, ON; antennal lobes, AL; and other sensory neuropils, OSN) represent primarily centers of convergence of their sense organs. The output of these neuropils diverges into surrounding nonglomerular and to glomerular neuropils such as the central body (CB) and mushroom bodies (MB), or is fed via direct pathways to descending brain neurons (DN). CB and MB represent areas of sensory convergence via direct or indirect input from the central nonglomerular neuropil. CB and MB are thought to feed information to descending neurons by nonglomerular interneurons (*broken lines*). Fiber domains of ascending and descending neurons show overlap. Pathways incorporating CB and MB are considered as "long loops" contrasting short circuits.

bees (Mobbs, 1982, 1984; Gronenberg, 1984) and crickets (Schildberger, 1981; Hörner, 1984). Interhemispherical direct connectivity has been shown for the β-lobes (Klemm et al., 1983, 1984; Mobbs, 1984), for α- and β-lobes (Mobbs, 1984), and for the calyces (Klemm et al., 1983). A number of indirect interhemispherical connections between the mushroom bodies might well be established by local interneurons with arborizations around the mushroom bodies in both hemispheres (Schildberger, 1982; Gronenberg, 1984; Hörner, 1984).

11.6.4. Mushroom Bodies and Brain Pathways

Information from head and body sense organs is processed by ascending plurisegmental and brain interneurons via a number of morphologically different channels—with and without central glomerular neuropils—(Fig. 11.13) and may converge upon the descending brain neurons (Schürmann, 1982, 1985; Strausfeld and Bacon, 1983). The main center of projection of ascending and descending neurons that partially overlap is the dorsal deutocerebrum and ventral parts of the caudal protocerebrum (Rosentreter and Schürmann, 1982; Strausfeld and Bassemir, 1985). Divergent spread of sensory excitation to all areas of the central brain is reflected by the predominance of multimodal reactivity in nonglomerular and mushroom body connected extrinsic neurons (Schildberger, 1984; Homberg, 1984) and descending nerve cells (Hörner and Gras, 1985).

In the scope of pathways to descending neurons, short circuits with few elements as well as long pathways can be found, which partially account for latency differences after natural stimulation (Hörner and Gras, 1985). Though the connectivity of mushroom body output neurons to descending neurons has not yet been established, we suggest the corpora pedunculata with their internal prerequisites for slow information flow (Schürmann, 1974, 1985) as an element in a chain of neurons modulating efferent brain cells. The mushroom bodies are regarded as part of a slow loop apparatus for elaborate information processing, in contrast to short pathways involved in stereotypic situations and reflex-like actions.

11.7. SUMMARY

The paired mushroom bodies (corpora pedunculata) in the central brain (protocerebrum) occur in most insects and vary in form and sizes. Sexual dimorphism and caste differences are pronounced in Hymenoptera and some other groups. Postlarval growth and structural plasticity are found in Coleoptera and Diptera. The intrinsic cells of the mushroom bodies (Kenyon cells) form neuropilar columns of parallel fiber bundles (stalk, α-, and β-lobe). Extrinsic neurons connect the mushroom body compartments with each other and with other areas of the brain. Compartmental and geometric arrangement is shown by anatomical studies and distribution of putative transmitters. Acetylcholine, dopamine, serotonin, and GABA have been attributed to extrinsic fibers, indicating complex synaptic connectivity. Electron microscopy reveals differential expression of synaptic organelles and synapses for the mushroom body neuropil compartments. The calyces represent the main input area with different extrinsic presynaptic boutons connected to intrinsic fiber dendrites in a divergent mode. There is evidence for complex glomeruli with presynaptic input to boutons. Chemical synapses in the stalk and probably in the lobes serve in direct or indirect coupling of intrinsic fibers. These elements converge upon postsynaptic extrinsic output fibers in the lobes. In addition, presynaptic contacts of extrinsic fibers also occur in the lobes. The corpora pedunculata are described as specialized neuropil for processing multisensory and multimodal input in parallel channels. In comparison to the variety of brain pathways leading to descending neurons, the mushroom bodies appear to be elements of a long loop for slow elaborate information processing.

ACKNOWLEDGMENTS

The investigation of the mushroom body system has been supported by the DFG (grants Schu 374/2,4,5). Ch. Buitkamp-Möbius, M. Hörner, K. Schildberger, and G. M. Technau provided published and unpublished data. I thank M. Klages and M. Knierim-Grenzebach for technical assistance in histology and photography.

REFERENCES

Aubele, E. and N. Klemm. 1977. Origin, destination and mapping of tritocerebral neurons of locust. *Cell Tissue Res.* **178**: 199–219.

Balderrama, N. 1980. One trial learning in the American cockroach, *Periplaneta americana. J. Insect Physiol.* **26**: 499–504.

Baldus, K. 1924. Untersuchungen über Bau und Funktion des Gehirns der Larve und Imago von Libellen. *Z. Wiss. Zool.* **121**: 557–620.

Bernstein, S. and R. A. Bernstein. 1969. Relationships between foraging efficiency and the size of the head and component brain and sensory structures in the red wood ant. *Brain Res.* **16**: 85–104.

Bicker, G., S. Schäfer, and T. G. Kingan. 1985. Mushroom body feedback interneurons in the honey bee show GABA-like immunoreactivity. *Brain Res.* **360**: 394–397.

Bieber, M. and D. Fuldner. 1979. Brain growth during the adult stage of a holometabolous insect. *Naturwissenschaften* **66**: 426.

Bishop, C. A. and M. O'Shea. 1982. Neuropeptide proctolin (H-Arg-Try-Leu-Pro-Thr-OH): Immunocytochemical mapping of neurons in the central nervous system of the cockroach. *J. Comp. Neurol.* **207**: 223–238.

Boeckh, J., K. D. Ernst, H. Sass, and U. Waldow. 1984. Anatomical and physiological characteristics of individual neurones in the central antennal pathway of insects. *J. Insect Physiol.* **30**: 15–26.

Buitkamp-Möbius, K. 1975. Strukturuntersuchungen an den Pilzkörpern im Oberschlundganglion von *Apis mellifica* -Gynandromorphen unter Berücksichtigung ihres Verhaltens. Inaugural-Dissertation, Universität Bonn.

Bullock, T. H. and G. A. Horridge. 1965. *Structure and Function in the Nervous System of Invertebrates,* vol. 2. Freeman. San Francisco.

Burrows, M., J. Boeckh, and J. Esslen. 1982. Physiological and morphological properties of interneurones in the deutocerebrum of male cockroaches which respond to female pheromone. *J. Comp. Physiol.* **145**: 447–457.

Chamberlain, S. C. and G. A. Engbretson. 1982. Neuropeptide immunoreactivity in *Limulus.* I. Substance P-like immunoreactivity in the lateral eye and protocerebrum. *J. Comp. Neurol.* **208**: 304–315.

Coss, R. G. and J. G. Brandon. 1982. Rapid changes in dendritic spine morphology during the honeybee's first orientation flight, pp. 338–342. In M. D. Breed, C. D. Michener, and H. E. Evans (eds.), *The Biology of Social Insects.* Westview Press, Boulder, Colorado.

Coss, R. G., J. G. Brandon, and A. Globus. 1980. Changes in morphology of dendritic spines on honeybee calycal interneurons associated with cumulative nursing and foraging experiences. *Brain Res.* **192**: 49–59.

Dujardin, F. 1850. Memoire sur le systeme nerveux des Insectes. *Ann. Sci. Nat. Zool.* **14**(3): 547–560.

Dymond, G. R. and P. D. Evans. 1979. Biogenic amines in the nervous system of the cockroach, *Periplaneta americana*: Association of octopamine with mushroom bodies and dorsal unpaired median (DUM) neurones. *Insect Biochem.* **9**: 535–545.

Edwards, J. S. 1969. Postembryonic development and regeneration of the insect nervous system. *Adv. Insect Physiol.* **6**: 97–137.

El-Salhy, M., S. Falkmer, K. J. Kramer, and R. D. Speirs. 1983. Immunohistochemical investigations of neuropeptides in the brain, corpora cardiaca, and corpora allata of an adult lepidopteran insect, *Manduca sexta* (L). *Cell Tissue Res.* **232**: 295–317.

Erber, J. 1983. The search for neural correlates of learning in the honeybee. In F. Huber and H. Markl (eds.), *Neuroethology and Behavioral Physiol.* Springer, Berlin, Heidelberg, New York.

Ernst, K.-D. and J. Boeckh. 1983. A neuroanatomical study on the organization of the central antennal pathways in insects. III. Neuroanatomical characterization of physiologically defined response types of deutocerebral neurons in *Periplaneta americana*. *Cell Tissue Res.* **229**: 1–22.

Ernst, K.-D., J. Boeckh, and V. Boeckh. 1977. A neuroanatomical study on the organization of the central antennal pathways in insects. II. Deutocerebral connections in *Locusta migratoria* and *Periplaneta americana*. *Cell Tissue Res.* **176**: 285–308.

Evans, P. D. 1980. Biogenic amines in the insect nervous system. *Adv. Insect Physiol.* **15**: 317–473.

Fischbach, K. F. and M. Heisenberg. 1984. Neurogenetics and behaviour in insects. *J. Exp. Biol.* **112**: 65–93.

Forel, A. 1874. Les fourmis de la Suisse. *Nouv. Mem. Soc. Helv. Sci. Natur.* **26**: 1–200.

Frontali, N. 1968. Histochemical localization of catecholamines in the brain of normal and drug-treated cockroaches. *J. Insect Physiol.* **14**: 881–886.

Frontali, N. and G. Mancini. 1970. Studies on the neuronal organization of cockroach corpora pedunculata. *J. Insect Physiol.* **16**: 2293–2301.

Frontali, N., R. Piazza, and R. Scopelliti. 1971. Localization of acetyl-cholinesterase in the brain of *Periplaneta americana*. *J. Insect Physiol.* **17**: 1833–1842.

Frontali, N. and R. Pierantoni. 1973. Autoradiographic localization of 3H-GABA in the cockroach brain. *Comp. Biochem. Physiol.* **44A**: 1369–1372.

Goll, W. 1967. Strukturuntersuchungen am Gehirn von *Formica*. *Z. Morphol. Okol. Tiere* **59**: 143–210.

Goosen, H. 1964. Untersuchungen an Gehirnen verschieden großer, jeweils verwandter Coleopteren- und Hymenopteren Arten. *Zool. Jahrb. Abt. Allg. Zool. Physiol.* **62**: 1–63.

Gronenberg, W. 1984. Das Protocerebrum der Honigbiene im Bereich des Pilzkörpers—Eine neurophysiologisch-anatomische Charakterisierung. Inaugural-Dissertation, Freie Universität Berlin.

Groth, U. 1971. Vergleichende Untersuchungen über die Topographie und Histologie des Gehirns der Dipteren. *Zool. Jahrb. Abt. Anat.* **88:** 203–319.

Hall, J. C. 1984. Complex brain and behavioral functions disrupted by mutations in Drosophila. *Dev. Genet.* **4:** 355–378.

Hanström, B. 1928. *Vergleichende Anatomie des Nervensystems der wirbellosen Tiere.* Springer, Berlin.

Hanström, B. 1940. Inkretorische Organe, Sinnesorgane und Nervensystem des Kopfes einiger niederer Insektenordnungen. *Kungl. Svenska Vetenskaps Akad. Handlingar* (Ser. 3B): 4–266.

Hausen, K. and N. J. Strausfeld. 1980. Sexually dimorphic interneuron arrangements in the fly visual system. *Proc. R. Soc. Lond.* B **208:** 57–71.

Heisenberg, M. 1980. Mutants of brain structure and function: what is the significance of the mushroom bodies for the behavior, pp. 373–390. In O. Siddiqui, P. Babu, L. M. Hall, and I. C. Hall (eds.), *Development and Neurobiology of Drosophila.* Plenum Press, New York.

Heisenberg, M., A. Borst, S. Wagner, and D. Byers. 1985. *Drosophila* mushroom body mutants are deficient in olfactory learning. *J. Neurogenet.* **2:** 1–30.

Herman, M. M., J. Miquel, and M. Johnson. 1971. Insect brain as a model for the study of aging. Age-related changes in *Drosophila melanogaster. Acta Neuropathol.* **19:** 167–183.

Hess, A. 1972. Histochemical localization of cholinesterase in the brain of the cockroach (*Periplaneta americana*). *Brain Res.* **46:** 287–295.

Hinke, W. 1961. Das relative postembryonale Wachstum der Hirnteile von *Culex pipiens, Drosophila melanogaster* und *Drosophila*-Mutanten. *Z. Morphol. Okol. Tiere* **50:** 81–118.

Holmgren, N. 1916. Zur vergleichenden Anatomie des Gehirns von Polycheten, Onychophoren, Xiphosuren, Arachniden, Crustaceen, Myriapoden und Insekten. *Kungl. Svenska Vetenskaps Akad. Handlingar.* **56:** 155–299.

Homberg, U. 1984. Processing of antennal information in extrinsic mushroom body neurons of the bee brain. *J. Comp. Physiol.* A **154:** 825–836.

Homberg, U. and J. Erber. 1979. Response characteristics and identification of extrinsic mushroom body neurons in the bee. *Z. Naturforsch.* **34c:** 612–615.

Honnegger, H. W. and F. W. Schürmann. 1975. Cobalt sulphide staining of optic fibres in the brain of the cricket, *Gryllus campestris. Cell Tissue Res.* **159:** 213–225.

Hörner, M. 1984. Physiologische und morphologische Charakterisierung bauchmarkverbindender und anderer Interneuronen im Gehirn des Heimchens *Acheta domesticus* L. Diplomarbeit, Universität Göttingen, Göttingen.

Hörner, M. and H. Gras. 1985. Physiological properties of some descending neurons in the cricket brain. *Naturwissenschaften* (in press).

Howse, P. E. 1973. Design and Function in the Insect Brain, pp. 180–194. In L. Barton Browne (ed.), *Experimental Analysis of Insect Behaviour.* Springer-Verlag, Berlin, Heidelberg, New York.

Howse, P. E. 1975. Brain structure and behavior in insects. *Annu. Rev. Entomol.* **20:** 359–379.

Huber, F. 1960. Untersuchungen über die Funktion des Zentralnervensystems und insbesondere des Gehirns bei der Fortbewegung und der Lautäußerung der Grillen. *Z. vgl. Physiol.* **44:** 60–132.

Jonescu, C. N. 1909. Vergleichende Untersuchungen über das Gehirn der Honigbiene. *Jena Z. Naturwiss.* **45:** 111–180.

Kaulen, P., J. Erber, and P. Mobbs. 1984. Current source-density analysis in the mushroom bodies of the honeybee (*Apis mellifera carnica*). *J. Comp. Physiol. A* **14:** 569–582.

Kenyon, F. C. 1896. The brain of the bee. *J. Comp. Neurol.* **6:** 133–210.

Killmann, F. and F. W. Schürmann. 1985. Electrical and chemical transmission between the lobula giant movement detector and the descending contralateral movement detector neurons of locust is supported by electron microscopy. *J. Neurocytol.* **14:** 637–652.

Klemm, N. 1974. Vergleichend-histochemische Untersuchungen über die Verteilung monoamin-haltiger Strukturen im Oberschlundganglion von Angehörigen verschiedener Insekten-Ordnungen. *Entomol. Ger.* **1:** 21–49.

Klemm, N. 1976. Histochemistry of putative transmitter substances in the insect brain. *Prog. Neurobiol.* **7:** 99–169.

Klemm, N. 1983. Monoamine-containing neurons and their projection in the brain (supraoesophageal ganglion) of cockroaches. A aldehyde fluorescence study. *Cell Tissue Res.* **229:** 379–402.

Klemm, N., H. W. M. Steinbusch, and F. Sundler. 1984. Distribution of serotonin-containing neurons and their pathways in the supraoesophageal ganglion of the cockroach *Periplaneta americana* (L.) as revealed by immunocytochemistry. *J. Comp. Neurol.* **225:** 387–395.

Klemm, N. and F. Sundler. 1983. Organization of catecholamine and serotonin-immunoreactive neurons in the corpora pedunculata of the desert locust, *Schistocerca gregaria* Forsk. *Neurosci. Lett.* **36:** 13–17.

Korr, H. 1968. Das postembryonale Wachstum verschiedener Hirnbereiche bei *Orchesella villosa* L. (Ins., Collembola). *Z. Morph. Tiere* **62:** 389–442.

Landolt, A. M. 1965. Elektronenmikroskopische Untersuchungen an der Perikaryenschicht der Corpora pedunculata der Waldameise (*Formica lugubris* Zett.) mit besonderer Berücksichtigung der Neuron-Glia-Beziehung. *Z. Zellforsch.* **66:** 701–736.

Landolt, A. M. and H. Ris. 1966. Electron microscope studies on soma-somatic interneuronal junctions in the corpora pedunculatum of the wood and (*Formica lugubris* Zett.). *J. Cell Biol.* **28:** 391–403.

Lee, A. H., R. L. Metcalf, and G. M. Booth. 1973. House cricket acetylcholinesterase: Histochemical localization and in situ inhibition by O,O-dimethyl S-aryl phosphorothioates. *Ann. Entomol. Soc. Amer.* **66:** 333–343.

Mancini, G. and N. Frontali. 1967. Fine structure of the mushroom body neuropil of the brain of the roach, *Periplaneta americana*. *Z. Zellforsch.* **83:** 334–343.

Mancini, G. and N. Frontali. 1970. On the ultrastructural localization of catecholamines in the beta lobes (corpora pedunculata) of *Periplaneta americana*. *Z. Zellforsch.* **103**: 341–350.

Menzel, R., J. Erber, and T. Masuhr. 1974. Learning and memory in the honeybee, pp. 195–217. In L. Barton Browne (ed.), *Experimental Analysis of Insect Behaviour*. Springer, Berlin.

Mercer, A. R., P. G. Mobbs, A. P. Davenport and P. D. Evans. 1983. Biogenic amines in the brain of the honeybee, *Apis mellifera*. *Cell Tissue Res.* **234**: 655–677.

Mobbs, P. G. 1982. The brain of the honeybee *Apis mellifera*. I. The connections and spatial organization of the mushroom bodies. *Philos. Trans. R. Soc. (B)* **298**: 309–354.

Mobbs, P. G. 1984. Neural networks in the mushroom bodies of the honeybee. *J. Insect Physiol.* **30**(1): 43–58.

Nässel, D. R., E. P. Meyer, and N. Klemm. 1985. Mapping and ultrastructure of serotonin-immunoreactive neurons in the optic lobes of three insect species. *J. Comp. Neurol.* **323**: 190–204.

Neder, R. 1959. Allometrisches Wachstum von Hirnteilen bei drei verschieden großen Schabenarten. *Zool. Jahrb. Anat.* **4**: 411–464.

Otto, D. 1971. Untersuchungen zur zentralnervösen Kontrolle der Lauterzeugung von Grillen. *Z. vgl. Physiol.* **74**: 227–271.

Pandazis, G. 1930. Über die relative Ausbildung der Gehirnzentren bei biologisch verschiedenen Ameisenarten. *Z. Morphol. Oekol. Tiere* **18**: 114–169.

Pearson, L. 1971. The corpora pedunculata of *Sphinx ligustri* L. and other Lepidoptera: An anatomical study. *Philos. Trans. R. Soc. (B)* **259**: 477–516.

Pearson, K. G. 1979. Local neurons and local interactions in the nervous systems of invertebrates, pp. 145–157. In F. O. Schmitt and F. G. Worden (eds.), *The Neurosciences*, 4th Study Program. The MIT Press, Cambridge, Massachusetts, London.

Pipa, R. L. 1973. Proliferation, movement and regression of neurons during the postembryonic development of insects, pp. 105–130. In D. Young (ed.), *Developmental Neurobiology of Arthropods*, Cambridge University Press, Cambridge.

Power, M. 1943. The brain of *Drosophila melanogaster*. *J. Morphol.* **72**: 517–559.

Rall, W. 1978. Dendritic spines and synaptic potency, pp. 203–209. In R. Porter (ed.), *Studies in Neurophysiology*. Cambridge University Press, Cambridge.

Ramade, F. 1965. L'action anticholinesterasique de quelques insecticides organophosphoes sur le systeme nerveux central de *Musca domestica*. *Ann. Soc. Entomol. Paris* **1**: 549–566.

Ratzendorfer, C. 1952. Volumetric indices for the parts of the insect brain. A comparative study in cerebralisation of insects. *J. N. Y. Entomol. Soc.* **60**: 129–152.

Rensch, B. 1959. Trends towards progress of brains and sense organs. *Cold Spring Harbor Symp. Quant. Biol.* **24**: 291–303.

Ribi, W. A. 1985. The first optic ganglion of the bee. VI. A sexually dimorphic receptor-cell axon. *Cell Tissue Res.* **240**: 27–33.

Rosentreter, M. and F. W. Schürmann. 1982. Topographie und Typologie von Verbindungsneuronen zwischen Gehirn und Bachmark bei der Grille *Acheta domesticus*. *Verh. Dtsch. Zool. Ges.* **75**: 255.

Rossbach, W. 1962. Histologische Untersuchungen über die Hirne naheverwandter Rüsselkäfer (Curculionidae) mit unterschiedlichem Brutfürsorgeverhalten. Z. Morphol. Okol. Tiere **50**: 616–650.

Rutschke, E., D. Richter, and H. Thomas. 1976. Autoradiographische Untersuchungen zum Eibau von 3H-Dopamin und 3H-5-Hydroxytryptamin in das Gehirn von *Periplaneta americana* L. *Zool. Jahrb. Anat.* **95**: 439–447.

Schildberger, K. 1981. Some physiological features of mushroom body linked fibers in the house cricket brain. *Naturwissenschaften* **67**: 623.

Schildberger, K.-M. 1982. Untersuchungen zur Struktur und Funktion von Interneuronen im Pilzkörperbereich des Gehirns der Hausgrille *Acheta domesticus*. Doctoral dissertation. Universität Göttingen, Göttingen.

Schildberger, K. 1983. Local interneurons associated with the mushroom bodies and the central body in the brain of *Acheta domesticus*. *Cell Tissue Res.* **230**: 573–586.

Schildberger, K. 1984. Multimodal interneurons in the cricket brain: properties of identified extrinsic mushroom body cells. *J. Comp. Physiol. A* **154**: 71–79.

Schmitt, F. O., P. Dev, and B. H. Smith. 1976. Electronic processing of information by brain cells. *Science (Washington, D. C.)* **193**: 114–120.

Schooneveld, H., G. I. Tesser, J. A. Veenstra, and H. M. Romberg-Privee. 1983. Adipokinetic hormone and AKH-like peptide demonstrated in the corpora cardiaca and nervous system of *Locusta migratoria* by immunohistochemistry. *Cell Tissue Res.* **230**: 67–76.

Schürmann, F. W. 1970. Über die Struktur der Pilzkörper des Insektengehirns. I. Synapsen im Pedunculus. *Z. Zellforsch.* **103**: 365–381.

Schürmann, F. W. 1971. Synaptic contacts of association fibers in the brain of the bee. *Brain Res.* **26**: 169–176.

Schürmann, F. W. 1972. Über die Struktur der Pilzkörper des Insektengehirns. II. Synaptische Schaltungen im Alpha-Lobus des Heimchens *Acheta domesticus* L. *Z. Zellforsch.* **127**: 240–257.

Schürmann, F. W. 1973. Über die Struktur der Pilzkörper des Insektengehirns. III. Die Anatomie der Nervenfasern in den Corpora pedunculata bei *Acheta domesticus* L. (Orthopt.). Eine Golgi-Studie. *Z. Zellforsch.* **145**: 247–285.

Schürmann, F. W. 1974. Bemerkungen zur Funktion der Corpora pedunculata im Gehirn der Insekten aus morphologischer Sicht. *Exp. Brain Res.* **19**: 406–432.

Schürmann, F. W. 1980. Methods for special staining of synaptic sites, pp. 241–262. In N. J. Strausfeld and T. A. Miller (eds.), *Neuroanatomical Methods—Experimental Entomology*. vol. 1A. Springer-Verlag, New York, Heidelberg, Berlin.

Schürmann, F. W. 1982. On synaptic connexions, tracts and compartments in the brain of the honey bee, pp. 390–394. In M. D. Breed, C. D. Michener, and H. E. Evans (eds.), *The Biology of Social Insects*. Westview Press, Boulder, Colorado.

Schürmann, F. W. 1985. Aspekte neuronaler Verknüpfung im zentralen Hirn der Insekten. pp. 51–93. In *Schriftenreihe der Westfälischen Wilhelms-Universität Münster*, Neue Folge, Heft 4.

Schürmann, F. W. and N. Klemm. 1973. Zur Monoaminverteilung in den Corpora pedunculata des Gehirns von *Acheta domesticus* L. (Orthoptera, Insecta)—Histochemische Untersuchungen mit Vergleichen zur Struktur and Ultrastruktur. *Z. Zellforsch.* **136:** 393–414.

Schürmann, F. W. and N. Klemm. 1984. Serotonin-immunoreactive neurons in the brain of the honeybee. *J. Comp. Neurol.* **225:** 570–580.

Steiger, U. 1967. Über den Feinbau des Neuropils im Corpus pedunculatum der Waldameise. *Z. Zellforsch.* **81:** 511–536.

Strambi, C. 1974. Histochemie des activités succinodesydrogenasiques et acetyl-cholinesterasiques dans les ganglions cephaliques de *Polistes gallicus* L. (Hymenoptère, Vespide). *Recherches Biologiques Contemporaines,* **4:** 139–153.

Strausfeld, N. J. 1970. Variations and invariants of cell arrangements in the nervous systems of insects. (A review of neuronal arrangements in the visual system and corpora pedunculata). *Verh. Zool. Ges.* **64:** 97–108.

Strausfeld, N. J. 1976. *Atlas of an Insect Brain.* Springer, Berlin.

Strausfeld, N. J. and D. R. Nässel. 1980. Neuroarchitecture of brain regions that subserve the compound eyes of crustacea and insects, pp. 2–133. In H. Autrum (ed.), *Handbook of Sensory Physiology,* vol. 7/6b. Springer, Berlin, New York.

Strausfeld, N. J. and J. P. Bacon. 1983. Multimodal convergence in the central nervous system of dipterous insects. *Fortschr. Zool.* **28:** 47–76.

Strausfeld, N. J. and U. K. Bassemir. 1983. Cobalt-coupled neurons of a giant fibre system in Diptera. *J. Neurocytol.* **12:** 971–991.

Strausfeld, N. J. and U. K. Bassemir. 1985. Lobula plate and ocellar interneurons converge onto a cluster of descending neurons leading to neck and leg motor neuropil in *Calliphora erythrocephala*. *Cell Tissue Res.* **240:** 617–640.

Technau, G. M. 1984. Fiber number in the mushroom bodies of adult *Drosophila melanogaster* depends on age, sex and experience. *J. Neurogenet.* **1:** 113–126.

Technau, G. and M. Heisenberg. 1982. Neural reorganization during metamorphosis of the corpora pedunculata in *Drosophila melanogaster. Nature (London)* **295**(5848): 405–407.

Tyrer, N. M., J. D. Turner, and J. S. Altman. 1984. Identifiable neurons in the locust central nervous system that react with antibodies to serotonin. *J. Comp. Neurol.* **227:** 313.

Veenstra, J. A. 1984. Immunocytochemical demonstration of a homology in peptidergic neurosecretory cells in the suboesophageal ganglion of a beetle and a locus with antisera to bovine pancreatic polypeptide, FMRFamide, vasopressin and a-MSH. *Neurosci. Letters* **48:** 185–190.

Veenstra, J. A. and H. Schooneveld. 1984. Immunocytochemical localization of neurons in the nervous system of the Colorado potato beetle with antisera against FMRFamide and bovine pancreatic polypeptide. *Cell Tissue Res.* **235:** 303–308.

Veenstra, J. A., H. M. Romberg-Privee, and H. Schooneveld. 1984. Immunocytochemical localization of peptidergic cells in the neuro-endocrine system of the Colorado potato beetle, *Leptinotarsa decemlineata,* with antisera against vasopressin, vasotocin and oxytocin. *Histochemistry* **81:** 29–34.

Vowles, D. M. 1964. Models in the insect brain. In E. Reiss (ed.), *Neural Theory and Modeling.* Stanford University Press, Stanford, California

Wadepuhl, M. 1983. Control of grasshopper singing behavior by the brain: Responses to electrical stimulation. *Z. Tierpsychol.* **63:** 173–200.

Weiss, M. J. 1974. Neuronal connections and the function of the corpora pedunculata in the brain of the American cockroach, *Periplaneta americana* (L.). *J. Morphol.* **142:** 21–70.

Weiss, M. J. 1981. Structural patterns in the corpora pedunculata of Orthoptera: A reduced silver analysis. *J. Comp. Neurol.* **203:** 515–553.

Williams, J. L. D. 1975. Anatomical studies of the insect central nervous system: A ground-plan of the midbrain and an introduction to the central complex in the locust, *Schistocerca gregaria* (Orthoptera). *J. Zool. London* **0176:** 67–86.

Witthöft. W. 1967. Absolute Anzahl und Verteilung der Zellen im Hirn der Honigbiene. *Z. Morphol. Tiere* **61:** 160–164.

CHAPTER 12

Functional Organization of the Subesophageal Ganglion in Arthropods

Jennifer S. Altman
Jenny Kien
Institute of Zoology
University of Regensburg
Regensburg, West Germany

12.1. INTRODUCTION

In arthropods, one or more ganglia innervate the mouthparts. These lie beneath the esophagus and form a complex known as the subesophageal ganglion or ganglia (SEG)*. The brain and SEG together are often termed the head ganglia, and in higher arthropods they may be fused. In this chapter, we review the current understanding of the structure, organization, and functions of the neurons that make up the SEG.

Apart from gross morphology, little is known about these ganglia in most arthropods, except in insects where some detailed studies have been made in recent years, mainly by cobalt sulphide staining, intracellular recording, and dye injection. After reviewing the evolution of the complex in the arthropods, we will concentrate on the insect SEG, particularly that of the Orthoptera, about which most is known.

The insect SEG appears to perform two quite different functions. As a complex of segmental ganglia, it carries out local motor functions, controlling the movements of the mouthparts in the same way that the thoracic ganglia drive the legs. But, unlike other segmental ganglia, the SEG has a higher motor function, collaborating with the brain in the initiation, maintenance, and coordination of behaviors, such as walking, flight, and stridulation. As the neuronal circuits that subserve these functions are to a large extent separate, we shall consider those neurons clearly involved in local functions before those involved in overall motor coordination.

12.1.1. SEG in Arthropods

The degree of fusion among the ganglia of the SEG complex varies enormously in the different groups of arthropods (Fig. 12.1). The primitive con-

---

* In the figures of this chapter the authors have used SOG to label the subesophageal ganglion.

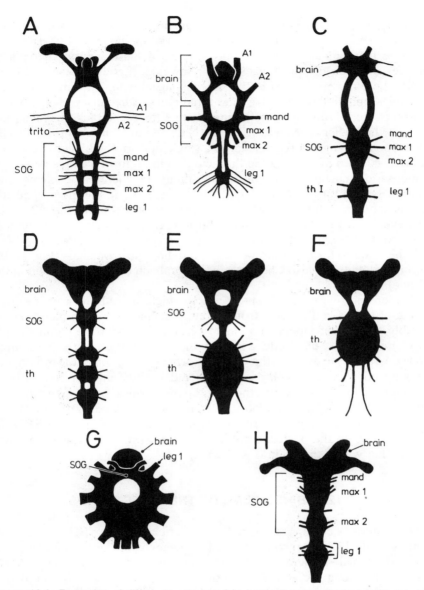

Figure 12.1. Examples of different patterns of fusion in the ganglia that make up subesophageal complex (SOG) in arthropods. (A–C) Crustacea. (A) *Triops* (Notostraca), (B) *Gigantocypris* (Ostracoda), (C) *Homarus* (Decapoda). (D–F) Insecta. (D) *Locusta* (Orthoptera), (E) *Musca* (Diptera), (F) *Hydrometra* (Heteroptera). (G) *Limulus* (aquatic Chelicerata), (H) *Thereupoda* (Myriapoda). Drawings are not to scale. A1 = antennule; A2 = antenna; trito = tritocerebrum; mand = mandible; max 1 = maxillule; max 2 = maxilla; th = thorax. (Adapted from Bullock and Horridge, 1965.)

dition of one ganglion in each postoral segment, seen in the lower orders of Crustacea (Fig. 12.1A) and the Chilopoda (Myriapoda) (Fig. 12.1H), seems to derive from annelids. In arthropods, each of these separate ganglia innervates one pair of appendages; in the lower decapod Crustacea, for example, there are six ganglia in the subesophageal complex, innervating the mandibles, maxillules, maxillae, and three pairs of maxillary legs, whereas in insects there are three ganglia, innervating the mandibles, maxillae, and labia. In the higher Crustacea (Fig. 12.1B,C) and in all insects (Fig. 12.1D), the segmental ganglia have fused into a single unit. In the higher Crustacea, the SEG is often fused with the thoracic ganglia to form a single mass, and condensation reaches its extreme in the Brachyura where the whole postoral nervous system forms a single structure. In insects, on the other hand, the SEG is sometimes fused with the tritocerebrum of the brain (Fig. 12.1E,F), but never with the thoracic ganglia. Fusion reaches its extreme in aquatic Chelicerata (*Limulus*) where the segmental ganglia migrate forward and fuse with the brain to form one structure surrounding the esophagus (Fig. 12.1G) (for more details see also Chapter 4 by Fahrenbach and Chaimberlain in this volume).

Condensing two or more ganglia into a single unit allows a more direct mixing of inputs and outputs from different segments, which increases the possible sources of inputs to a single neuron and, with the elimination of relay interneurons, may reduce integration times. Fusion of the SEG into a single unit seems to occur in those species where the action of the mouthparts is highly coordinated, such as the higher decapods and insects. In bees, bugs, and flies, for example, the mouthparts form a single structure, the proboscis. This includes the labrum, which is innervated by the most posterior part of the brain, the tritocerebrum. In these groups, the SEG has become fused to the posterior end of the brain (Fig. 12.1E) and may even be included within the brain (Fig. 12.1F).

12.2. BASIC PLAN OF THE INSECT SEG

In Orthoptera (locusts, grasshoppers, and crickets), the SEG is a separate ganglion (Fig. 12.1D), comprising the fused ganglia of the mandibular, maxillary, and labial segments, each known as a neuromere. These are aligned in sequence along the anteroposterior axis. The identity of the individual neuromeres is still recognizable from the repetition of the segmental commissures (Tyrer and Gregory, 1982), and homologies with ganglia innervating a single segment can be used as a key to the organization of neurons within the fused ganglion (Figs. 12.2, 12.3). In bees and flies, where the SEG neuromeres are fused to the brain, clear indications of segmentation are lost. An understanding of the orthopteran SEG may provide a stepping stone on the way to unravelling the complexities of SEG organization in higher insects.

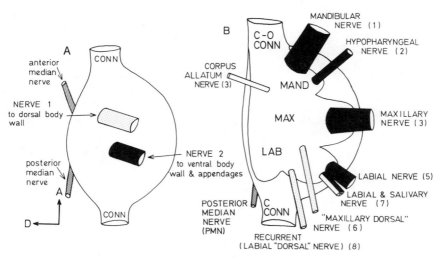

Figure 12.2. Proposed homologies in the locust between a hypothetical primitive segmental ganglion (A) and the SEG (B). Nerve roots shown black are probably homologous to the second (ventral) root in the primitive ganglion. In the SEG, each innervates a mouthpart. Nerve roots stippled are probably derived from the first (dorsal) root, innervating dorsal muscles and the dorsal body wall. The median nerves are hatched. The uncolored nerve roots (3 and 7, salivary) are involved in controlling secretion. c conn = cervical connective; c-o conn = circumesophageal connective; lab = labial neuromere; mand = mandibular; max = maxillary.

In the locust, the SEG is joined to the brain by paired circumesophageal connectives and to the prothoracic ganglion by paired cervical connectives (Fig. 12.2B). The connectives enter the ganglion dorsally at the anterior and posterior ends, with the tracts joining them occupying much of the dorsal part of the ganglion (Figs. 12.3, 12.7). There is a single posterior median nerve (pmN) in the labial neuromere, but no equivalent of the anterior median nerve. The nerve roots are numbered arbitrarily (1–8) from anterior to posterior (Fig. 12.2B; Altman and Kien, 1979; Tyrer and Gregory, 1982). They leave ventrally, with one pair of large nerves from each neuromere innervating the mandibles (nerve 1), maxillae (nerve 4), and labia (nerve 5), respectively. A second pair of finer nerves from the mandibular segment innervates the hypopharynx (nerve 2), and three pairs of fine nerves (6–8) from the labial neuromere innervate the back of the head, neck muscles, and salivary glands. A pair of fine nerves (3) from the dorsal surface of the mandibular neuromere run to the corpora allata (Fig. 12.8E).

Hypothetically, in a primitive ganglion innervating a basic insect segment there are two pairs of nerve roots, a more dorsal anterior one serving the dorsal body wall and the other more posterior and ventral that innervates the ventral body wall and the appendages (Fig. 12.2A). These are mixed nerves, containing both sensory and motor axons. The segmental nerves of

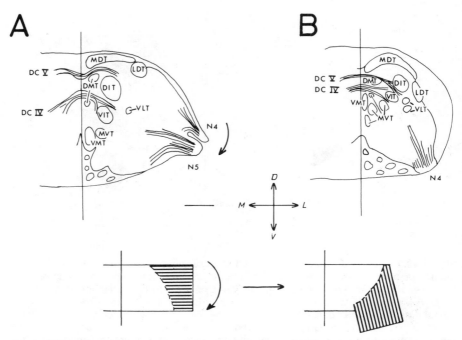

Figure 12.3. Transverse sections of the locust mesothoracic ganglion (A) and SEG in the maxillary neuromere (B), showing the major longitudinal tracts and dorsal commissures IV and V. In the SEG the flat structure of the typical thoracic ganglion has been distorted so that the nerve roots now leave ventrally instead of laterally. Inset shows the movement that could have taken place. Tract nomenclature after Tyrer and Gregory (1982). DC = dorsal commissure; DIT = dorsal intermediate tract; DMT = dorsal medial tract; LDT = lateral dorsal tract; MDT = medial dorsal tract; MVT = medial ventral tract; N = nerve; VIT = ventral intermediate tract; VLT = ventral lateral tract; VMT = ventral medial tract. Scale bars on this and all other figures, 100 μm.

the SEG are the equivalents of the ventral nerve; most of the dorsal roots have been lost, their only representatives being nerves 6 and 8 (Fig. 12.2B). These nerves innervate longitudinal muscles in the neck and appear to be the maxillary and labial homologues of the most posterior nerve (the recurrent nerve) in each thoracic ganglion. In fact, nerve 8 makes a similar recurrent loop, in this case joining nerve 1 of the prothoracic ganglion.

12.2.1. Tracts and Commissures

Like all segmental ganglia, the SEG has a scaffold of longitudinal tracts and transverse commissures (Fig. 12.3). A detailed identification of these is given in Tyrer and Gregory (1982). Figure 12.3B shows the main longitudinal tracts and gives the nomenclature and abbreviations used here. The longitudinal tracts are similar in appearance and arrangement to those in the thoracic ganglia (Fig. 12.3A), except in the ventral midline, where several additional tracts have been included in the MVT. Most of these do not run the whole length of the ganglion.

In a typical thoracic ganglion, there are six dorsal commissures (DC I–VI from anterior or posterior). These are all present in the labial and maxillary neuromeres, although maxillary DCI and II are very small. The dorsal part of the mandibular neuromere is reduced and compressed with much of the space taken up by the longitudinal tracts. Mandibular commissures I, II, and VI seem to be missing.

Ventrally, the neuromeres are rather less typical. Many of the commissures are not clearly homologous with thoracic ones. There is little trace of the neuropiles corresponding to the thoracic ventral association centers (VAC), except possibly at the anterior end of the labial neuromere (Fig. 12.6G). Boyan and Altman (1985) suggest that a small area of neuropile in the labial neuromere, dorsal to VMT and ventral to DCI and III, is the labial homologue of the anterior ring tract of the thoracic ganglia (Tyrer and Gregory, 1982).

Comparing the arrangement of the tracts and nerve roots in the mesothoracic ganglion and SEG emphasizes similarities that are not at first apparent. The lateral part of the flat sheet of the typical ganglion seems in the SEG to have become displaced through about 90° (Fig. 12.3, inset) in a ventral direction, with the result that the LDT is displaced ventrolaterally and the nerve roots with their attendant neuropiles lie under the ganglion core rather than to the side. The well-developed ventral neuropiles in each neuromere associated with the bases of the major paired peripheral nerves therefore correspond to the lateral neuropiles of the thoracic ganglia.

12.3. SEG AS A SEGMENTAL MOTOR CENTER

The SEG contains motor neurons innervating the muscles of the mouthparts and the neck, and related sensory inputs. In this section we detail what is known of the organization of these neurons and their relationships to intersegmental interneurons. The neurosecretory role of the SEG, through connections with the corpora allata and salivary glands, is also described.

12.3.1. Homologies Between SEG and Thoracic Ganglia in the Innervation of the Periphery

The mouthpart appendages show clear homologies with the legs. The cardo and stipes of maxilla and labium represent the coxopodite; the galea and lacinia are equivalent to the coxal basopodites and the palps to the teleopodites of the thoracic legs (Snodgrass, 1928). In the maxilla, nerve 4 divides into two major branches, labeled 4B and 4E (Ramirez, 1984), both containing motor and sensory axons, that enter the palp and anastomose. The innervation of hair plates, campaniform sensilla, and the tendon receptor in the first joint of the palp suggest that 4B is the equivalent of nerve 3B in the mesothoracic ganglion and 4E equivalent to mesothoracic nerve 5B (Bräunig, personal communication). The central projections from the sense organs, however, are rarely sufficiently similar to their thoracic equivalents to allow certain identification of their origins from their projections alone (Bräunig et al., 1983). Although it has been studied in less detail, the pattern of innervation of the labium appears to be similar to that of the maxilla. It is difficult to find any homologies in the innervation of the mandibles.

The neck is formed from components of the maxillary, labial, and prothoracic segments, although the segmental boundaries are no longer clear. Consequently, the neck muscles are innervated by both SEG and prothoracic ganglia (Fig. 12.4A; Honegger et al., 1984). There is a similar homology between the dorsal and dorsoventral muscles of the neck and the thoracic muscles that move the wings and coxae to that between the muscles of the mouthparts and legs.

Two pairs of muscles (50, 51 and 52, 65) seem from their attachments to be the dorsal longitudinal muscles of the maxillary and labial segments. The former are innervated by SEG nerve 6 ("maxillary dorsal" nerve; in cockroach, Davies, 1983), the latter by SEG nerve 8 (labial "dorsal" nerve), which is equivalent to the thoracic recurrent nerves. The dorsoventral muscles (54–56) are innervated by prothoracic nerve 3. They are probably homologous with the dorsoventral muscles of the thorax.

The innervation of muscle pair 62 by motor neurons with bilateral axons in the transverse nerve that branches off from the pmN (fig. 12.4A) suggests that these muscles are homologous with those of the spiracles in other segments (reviewed by Honegger et al., 1984).

12.3.2. Organization of Neurons Within the SEG

12.3.2.1. Neck Motor Neurons. For a detailed description of the innervation of neck muscles in cricket and locust see Honegger et al. (1984) and Shepheard (1973); Davis (1983) describes the innervation of the dorsal longitudinal muscles in the cockroach. We summarize the findings from the locust here (Fig. 12.4).

Muscles 50, 51, 52, 65 (dorsal longitudinal); 58, 59, 60 (dorsoventral); and 53, 62, 63 (ventral) have motor neurons in the SEG. Their axons run in nerve 6, 8, and pmN (Fig. 12.4A), and the cell bodies lie ventrally in the posterior half of the ganglion. Their arborizations extend dorsally and laterally in both the labial and maxillary neuromeres (Fig. 12.4B,C,D), and many have bilateral branches (Fig. 12.4C,D) so that they overlap with their contralateral counterparts. This suggests a strong functional interdependence between the muscles on both sides of the neck, because both left and right motor neurons can make direct contacts with a common set of neurons (Altman and Kien,

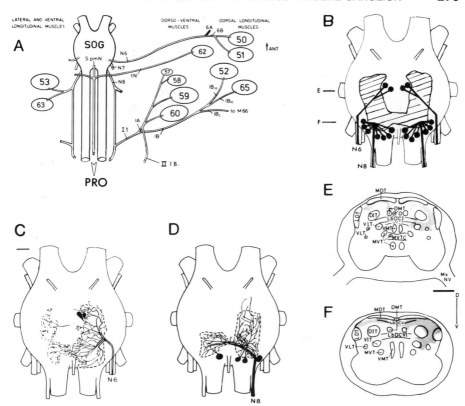

Figure 12.4. Innervation of the neck muscles from the SEG. (A) schematic representation of the main neck muscles that are innervated by motor neurons in the SEG. S.pmN = subesophageal posterior median nerve; tN = transverse nerve; I = prothoracic nerve; II = mesothoracic nerve. Nerve branch nomenclature follows the system of Campbell (1961). (B) Schematic composite summarizing the overall arborizations of neck motor neurons in the SEG. (C,D) Examples of motor neurons to individual dorsal longitudinal neck muscles; (C) to muscles 50, 51, and (D) to muscle 52. Drawn from whole mounts, dorsal view. (E,F) "Depth profiles"; that is, transverse sections of SEG at levels indicated in B to show the distribution in the depth of the neuropile of the arborizations of all the neck motor neurons. The same levels are used in subsequent figures. (After Honegger et al., 1984.)

1985). The need for bilateral interdependence clearly arises from the mechanics of the neck, where left and right members of a pair of muscles seldom function as simple antagonists (Shepheard, 1973).

The arborizations fill the area underneath the LDT and MDT, reaching medially to the VIT and as far ventrally as the VLT (Honegger et al., 1984; Fig. 12.4E). This area of the neuropile is equivalent to the part of the thoracic ganglia that is occupied by motor neurons to dorsal longitudinal and flight muscles, reinforcing the suggestion (Section 12.3.1) that the dorsal neck muscles are homologous to the thoracic muscles that move the wings.

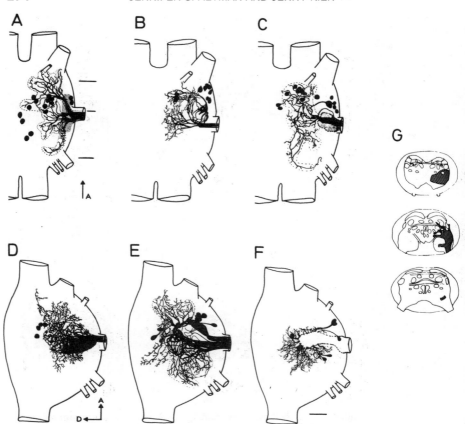

Figure 12.5. Motor neurons of the maxilla. *Top row,* wholemounts drawn from ventral; *bottom row,* lateral view. *Dashed lines* mark extent of arborization; *fine stipple* show areas too dense to resolve individual branches. (A,B,D) Motor neurons with axons in nerve 4A$_1$ (A), nerve 4A$_2$ (B), and both branches (D), innervating cardo promotor and adductor muscles. In (A), the coarsely *stippled area* indicates the position of motor neurons in nerve 4E. (C,E) Motor neurons with axons in nerve 4B, to lacinia, galea, and palp muscles. (F) Motor neurons with axons in nerve 4E, to palp. Position in plan view is indicated in (A). (G) Depth profiles of maxillary mouthpart motor neurons at levels indicated in (A).

12.3.2.2. Mouthpart Motor Neurons. Very little is known of the mouthpart motor neurons, except those with axons in the maxillary nerve of the locust (Ramirez, 1984; Kien and Altman, unpublished data). These innervate the muscles of the maxillary palp, the adductors of the stipes and cardo, and the flexors of the lacinia and galea (Albrecht, 1953). There are approximately 50 motor neurons with axons in nerve 4, with cell bodies in clusters, lying ventrally to laterally, and mostly within the maxillary neuromere, but some in the mandibular neuromere (Fig. 12.5). The main arborizations are in the ventral maxillary and mandibular neuromeres, but motor neurons with axons

TABLE 12.1. DISTRIBUTION OF MAXILLARY MOTOR NEURON ARBORIZATIONS IN THE SEG NEUROMERES

	Motor Neurons to Cardo, Nerve		Motor Neurons to Stipes and Palp, Nerve 4B	Motor Neurons to Palp, Nerve 4E
	$4A_1$	$4A_2$		
Mandibular	+ +	+ +	+ +	−
Maxillary	+ +	+ +	+ +	+
Labial	+	−	+ +	−

in nerve $4A_1$ (cardo) and B (stipes) also have branches extending into the labial neuromere (Fig. 12.5A–D; Table 12.1).

These motor neurons lie more ventrally than the neck motor neurons, with their branches forming a sheet on the lateral surface of the neuropile ventral to the LDT (Fig. 12.5G) in areas homologous to those in the thoracic ganglia containing leg motor neurons (Section 12.3.1; Altman and Kien, 1985).

Of the mandibular motor neurons, it is known that the mandibular adductor muscle is innervated by 12 neurons with large cell bodies lying anteriorly between the connectives. They branch bilaterally in the mandibular neuromere (P. Mobbs, personal communication). Six to eight of these react with antibodies for serotonin (Tyrer et al., 1984). The mandibular abductor is innervated by several motor neurons with large cell bodies lying laterally in the mandibular neuromere (P. Mobbs, personal communication). Their arborizations are unknown.

12.3.2.3. Mouthpart Sensory Neurons. In the locust, the projections from the sense organs in the labial submentum (Altman and Kien, 1979) and from individual mandibular and maxillary campaniform sensilla and hair plates (Bräunig et al., 1983) have been described. The central projections from the sensory structures innervated by the maxillary nerve (4) have also been reported (Ramirez, 1984). Projections from the campaniform sensilla in these segments show some similarity to those in thoracic segments, but the other projections are not clearly recognizable.

All five branches of the maxillary nerve contain sensory axons. The projection from the whole nerve forms six major plexi (Fig. 12.6A,B), most of which receive contributions from more than one branch (Table 12.2). So far, the only projections that have been identified are those of the hair plates and campaniform sensilla on the palp, with axons in branch 4B or E (Bräunig et al., 1983).

Apart from a large lateral plexus (1 in Fig. 12.6A), the projections tend to be oriented anteroposteriorly and most extend into the mandibular and labial neuromeres. The projections lie at three different levels in the ganglion

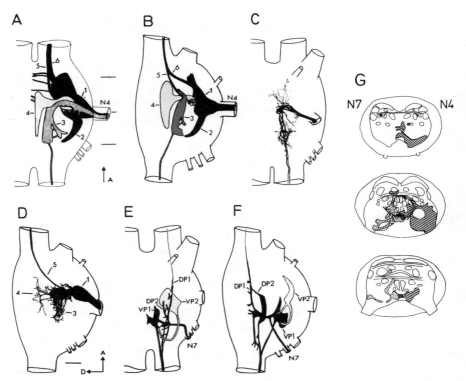

Figure 12.6. Sensory projections from the mouthparts. (A–D) Maxilla. (E,F) Labial sub-mentum. (A,B) Summaries of the major plexi of nerve 4: (A) plan view from dorsal; (B) lateral view; (C) dorsal view of the projection from nerve 4B into plexus 3; (D) lateral view of the projection from nerve 4E. (E,F) Summaries of the plexi of nerve 7 from ventral (E) and from lateral (F). VP2 is stippled as this projection consists of few and fine fibers. DP = dorsal plexus; VP = ventral plexus; (G) depth profiles at levels indicated in A. Nerve 7 is shown on the left (*stippled*), nerve 4 on the right (*hatched*). (Based on Altman and Kien, 1979; Ramirez, 1984; and unpublished observations.)

(Fig. 12.6B,D), from the ventral floor of the neuropil (plexi 1, 2 in Fig. 12.6A,B) to dorsal to the ventromedial tracts over the VIT and lateral MVT's (plexus 4; Fig. 12.6D).

Branches from plexi 1 and 3 cross the midline in the ventral commissures of the mandibular and maxillary neuromeres (Fig. 12.6A,C). Axons from plexi 5 and 6 ascend in the circumesphageal connectives—the bundle from plexus 6 includes fibers originating in the palp tip, some of which are chemosensory (Blaney and Chapman, 1969). These fibers ascend to the glomerular part of the antennal lobe in the brain (J. Boeckh, personal communication). In cockroaches some of these fibers cross the midline and ascend in the contralateral connective (J. Boeckh, personal communication). Only five to eight fibers originating from branch 4B (plexus 3 in Fig. 12.6A,C)

TABLE 12.2. CONTRIBUTION OF NERVE 4 BRANCHES TO THE VARIOUS SENSORY PLEXI (NUMBERED AS IN FIG. 12.6A)

	4A Cardo	4B Stipes, Palp	4C Lacinia	4D Galea	4E Palp
1		+	+	+	+
2	+	+	+	+	
3	+	+	+	+	+
4					+
5			+	+	+
6	+		+	+	

descend to the prothoracic ganglion. Mingled with the palp–tip projection are terminals that closely resemble those of campaniform sensilla in the mesothoracic ganglion (Hustert et al., 1981). Anterolaterally, on the ventral floor of the maxillary neuropil, are one to four large fibers that resemble stretch receptor projections (cf., Altman and Tyrer, 1977; Hustert, 1978).

Because the maxillary motor and sensory projections extend into both the mandibular and labial neuromeres, they probably overlap the sensory and motor projections from mandibles and labia (Fig. 12.7). Preliminary observations suggest that mandibular motor neurons do not extend much outside the mandibular neuromere, but Bräunig et al. (1983) describe a mandibular campaniform sensilla projection that reaches into the maxillary neuromere. Little is known about the projections of the labial nerve, but the

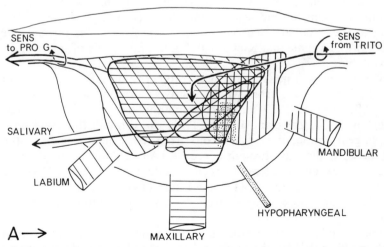

Figure 12.7. Summary of the main projections of mouthpart nerves showing their overlaps. The full extent of the hypopharyngeal projection is unknown; shown here is the neuropile around the nerve root seen in silver-stained sections.

projections from the labial submentum suggest that the sensory component extends at least to the anterior edge of the maxillary neuromere (Fig. 12.7, for summary).

Nerve 7. Nerve 7A contains sensory axons from the hairs, multipolar cells, and chordotonal organ in the labial submentum (Altman and Kien, 1979). As it enters the ganglion it merges with the root of nerve 5, suggesting that it was originally a branch of the labial nerve 5. The projections form longitudinal plexi in four layers, superficially similar to those from the maxilla (Fig. 12.6E,F). The two most dorsal plexi, DP1 and DP2, extend anteriorly into the maxillary and mandibular segments and posteriorly send fibers into the cervical connective, together with some from VP1. The ventral plexi lie in the labial and maxillary neuromeres, and VP1 extends into the contralateral labial neuropile. The area occupied by VP1 appears to be homologous to the VAC in the thoracic ganglia (Fig. 12.6G). The projections from the hair plate on the mentum innervated by nerve 5 (Bräunig et al., 1983) presumably overlaps VP2.

The SEG also receives an input from the labral nerve through the tritocerebrum and circumesophageal connective (Fig. 12.7). This terminates in the maxillary neuromere and some fibers cross to the contralateral side (Aubele and Klemm, 1977).

12.3.2.4. Other Primary Inputs. The SEG receives primary afferents from mechanoreceptors and proprioceptors in other segments, carrying information about head movements, head orientation, body movements, and position. There is an ipsilateral projection from antennal mechanoreceptors (Aubele and Klemm, 1977; Gewecke, 1979), including fibers from pedicellar campaniform sensilla and scapal hair plates (Bräunig et al., 1983). These project to the maxillary neuromere and some fibers pass through the SEG to the thorax (Aubele and Klemm, 1977), but no details of tracts and depths are available.

Several groups of hairs on the head project to the SEG and form a plexus in the MVT: afferents from wind-sensitive hairs on the frons and gena of locusts (Tyrer et al., 1979) and interommatidial hairs in crickets and mantids (Honegger, 1977; Zack and Bacon, 1981) run through the brain and circumesophageal connectives, and axons from hairs on the back of the head enter the SEG directly through nerve 6A (Altman and Kien, 1979). All the projections, except that from the gena (Anderson and Bacon, 1979), extend across the midline in the MVT commissure (Tyrer et al., 1979; Altman and Kien, 1979) and some axons from each group continue to the thoracic ganglia.

A variety of primary sensory inputs reaches the SEG from the thorax, including afferents from hairs on the neck membrane and pronotum (Kien, 1980), from the prothoracic myochordotonal and apodemal chordotonal organs and from the mesothoracic posterior chordotonal organ (Hustert, 1978;

Bräunig et al., 1981). Their terminations in the SEG are unknown. The projections from these chordotonal organs in the thoracic ganglia are in the medioventral area below the VIT and above the MVT and VMT tracts (Bräunig et al., 1981). If they occupy the same positions in the SEG, they would lie just dorsal to the hair tracts.

12.3.2.5. Interneurons. The dorsal longitudinal tracts consist mainly of axons of interneurons that connect the SEG and the brain with the thoracic ganglia. These comprise ascending interneurons (AINs), with posterior cell bodies and forward-running axons that terminate in the SEG or brain; descending interneurons (DINs), with cell bodies in the brain or SEG and posteriorly running axons; and T fibers, which have both ascending and descending axons. As well as axons of interneurons originating in other ganglia, some interneurons of all three types originate in the SEG (Figs. 12.10–12.12).

Most of the through-running fibers and all the known interneurons originating in the SEG have branches in one or more of the neuromeres, and many send processes to or across the midline. Unlike the through AINS and DINs, the SEG interneurons are mostly bilateral with the cell body on the opposite side of the ganglion to the axon, and arborizations on both sides. Their branches terminate in both dorsal and ventral neuropiles and so may contact any of the segmental neurons.

The ganglion also contains "local" interneurons (neurons with no axon leaving the ganglion). They generally have branches extending through more than one neuromere, so strictly they should be considered as interganglionic (see Burrows and Siegler, 1976). Local interneurons are very difficult to stain, except through intracellular electrodes, and so far very few have been found in the SEG. The morphology of the various interneurons is described in Section 12.4.3.

12.3.2.6. Overlaps Between Groups of Neurons in SEG. The intersegmental branching of both neck and mouthpart motor neurons is unusual in the locust CNS, where motor neuron arbors are usually confined to a single segment (Altman and Kien, 1985). Neurons that project into the same neuropils can share common inputs, thus facilitating the tight coordination between segmental structures required for both feeding and neck functions.

Examination of depth profiles (Figs. 12.5G, 12.6G; Table 12.3) shows that there are several places where the mouthpart projections intermingle with or are closely apposed to the mouthpart motor neuron arborizations, making direct synaptic contacts between them possible. The mouthpart sensory projections also overlap the more ventral branches of the neck motor neurons (cf., Figs. 12.4E, 12.6G), which probably promotes the coordination of mouthpart and head movements during feeding. The primary afferents from the mouthparts that descend to the prothoracic ganglion are presumably also

TABLE 12.3. BRANCHES OF MOUTHPART MOTOR AND SENSORY NEURONS IN THE SAME NEUROPILE

Sensory Plexi	Motor Nerves			
	A_1	A_2	B	E
1	+	+	+	+
2	+ ?		+ ?	+ ?
3			+	
4	+	+	+	
5		?	+ ?	
6		?	+ ?	

+, branches overlap; + ?, possible overlap; ?, unclear.

involved in this coordination. Both local and intersegmental interneurons have branches in the neuropiles that contain motor and sensory terminals. As in the thoracic ganglia, these areas can therefore be considered as sensorimotor integration neuropils (Altman, 1981), where the motor output patterns are generated.

12.3.3. Neurosecretion

The SEG also contains neurons that have been classified as neurosecretory, either because they terminate in secretory structures, such as the corpora allata and salivary glands, or because their cell bodies stain with paraldehyde fuchsin, the classic histological stain for neurosecretion.

12.3.3.1. Salivary Glands. Two pairs of large neurons in the SEG, SN1 and SN2 (Altman and Kien, 1979), innervate the salivary glands via nerves 7B (Fig. 12.8A,B). One pair has its cell bodies in the mandibular neuromere, the other in the maxillary neuromere, but their arborizations overlap almost completely on both sides of the ganglion. They branch predominantly in the areas of neuropil containing projections from nerves 1, 2, and 4 (Fig. 12.7), suggesting that there is direct coordination between the sensory and motor activities of the mouthparts and the production of saliva. The dense branching near the root of nerve 2, the hypopharyngeal nerve (Fig. 12.8A), is particularly interesting as the salivary duct opens through a pit on the labium, which is blocked by a knob projecting from the hypopharynx when the mouth is closed (Albrecht, 1953).

The salivary neurons give a strong reaction with antibodies to serotonin, and antiserotonin-reactive terminals are found on the salivary gland acini (Tyrer et al., 1984). Similar neurons innervate the salivary glands in cockroaches, but pharmacological experiments suggest that the transmitter they release is dopamine (House et al., 1973). In larval *Galleria mellonella,* there

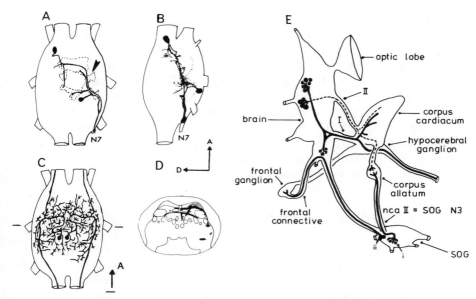

Figure 12.8. Neurons and pathways involved in the control of secretion. (A,B) Salivary neurons SN1 and 2 in the SEG drawn from ventral (A) and lateral (B). Branching around the root of the hypopharyngeal nerve (*arrow*). (C,D) The pair of neurons with ventral cell bodies reacting with antibodies to oxytocin and vasopressin; (C) dorsal view, (D) depth profile. (E) The neuronal pathways of the retrocerebral complex of the locust, which include (i) a group of neurons with cell bodies in the maxillary neuromere of the SEG and axons in the second corpus allatum nerve (nca II = SEG nerve 3), and (ii) a smaller group of SEG neurons innervating the frontal ganglion through the circumesophageal connective and frontal connective. (A,B adapted from Altman and Kien, 1979; C,D from N. M. Tyrer, unpublished; E modified from Mason, 1973, with additional data from Aubele and Klemm, 1977.)

are also cells in the SEG, pro-, and mesothoracic ganglia that innervate the prothor .cic glands (Granger, 1978).

12.3.3.2. Paraldehyde Fuchsin-Positive Cells. A pair of ventral cell bodies in the SEG, which stain with paraldehyde fuchsin and so are considered to be neurosecretory, have been reported in a variety of insects (reviewed by Weinbörmair et al., 1975). In the locust, a pair of cells with similar morphology reacts with antibodies to vasopressin (Rémy and Girardie, 1980) and oxytocin (Tyrer and Turner, personal communication). They have extensive arborizations in the SEG (Fig. 12.8C,D). Their axons extend to the brain and thoracic and abdominal ganglia where they also arborize extensively. In the thoracic ganglia, their branches terminate mainly in the dorsal neuropil in areas occupied by flight neurons (Tyrer, personal communication). Delphin (1965) has described other cells in the SEG that stain for neurosecretory products, but no details of their morphology are available.

12.3.3.3. Corpora Allata. Nerve 3 connects the SEG with the corpora allata; it contains 11 axons originating from cell bodies in the ventral maxillary neuromere (Mason, 1973; Fig. 12.8E) and terminating in the ipsilateral corpus allatum. The function of these neurons is unknown, but they are reported to stain for neurosecretion (see Mason, 1973). The corpora allata are neurosecretory structures that also receive innervation from the protocerebrum by way of the corpus cardiacum (Fig. 12.8E).

A small group of SEG neurons sends axons through the circumesophageal connective and tritocerebrum to the frontal ganglion (Fig. 12.8E), which is involved in control of the gut. These neurons may participate in coordinating gut activity with mouthpart movements and salivation.

12.4. SEG AS A CENTER FOR OVERALL COORDINATION

It has long been known that, in addition to its segmental motor functions, the SEG plays an important role in regulating the motor activity of the whole animal. In orthopterans, for example, when the brain is disconnected by cutting the circumesophageal connectives, walking activity may be enhanced, whereas cutting both cervical connectives, separating the SEG from the prothoracic ganglion, results in loss of spontaneous locomotion (Roeder, 1937; Roeder et al., 1960; Huber, 1960; Graham, 1979; Kien, 1983). This suggests that the output from the SEG has an excitatory influence on the thoracic and abdominal motor centers. Recent evidence has led us to the hypothesis that circuits in the SEG form part of loops between the brain and the body ganglia that are additional to, and parallel to, the direct connections between the brain and body ganglia (Fig. 12.13B). The output of the SEG combines the direct output from the brain with a modified version generated in the SEG, which appears to be both elaborated in detail and extended in time.

12.4.1. Clues to SEG Functions From Lesion and Stimulation Experiments

Connective-cutting experiments give a crude indication of the excitatory influences of the SEG on behavior. First, it plays a role in the control of posture: in the mantis its removal causes loss of leg extensor muscle tone (Roeder, 1937), and in locust extensor tone is lost in the prothoracic and to a lesser extent in the mesothoracic legs (Kien, unpublished data). Second, walking can still be initiated and maintained in an animal with the brain separated from the ventral nervous system by severing the circumesophageal connectives, but with the SEG still intact (Roeder, 1937; Huber, 1955; Kien, 1983). The ability to maintain walking is lost when even one cervical connective is then severed (Kien, 1983), so that the contribution of the SEG seems to be particularly important for maintaining walking. Neurons in the ganglion participate in the regulation of step size and walking speed (Kien, 1983) and, at least in stick insects, appear to coordinate stepping direction of the various legs (Bässler et al., 1985).

Stimulation studies confirm the importance of the SEG in maintaining walking, which can be evoked by focal electrical stimulation in the circumesophageal and cervical connectives and in the SEG (Kien, 1983). In the circumesophageal connectives only a short pulse train is required at many positions of the stimulating electrode to evoke a sequence of steps that outlasts the stimulus. In contrast, most positions in the cervical connectives and all in the SEG require continued stimulation, and walking lasts only for the duration of the stimulus.

Combined lesion, stimulation, and recording techniques have also revealed, in locusts, that the SEG plays a unique role in the bilateral distribution of information, at least in walking (Kien, 1983). In the thorax, the bilateral distribution of motor information takes place only within a ganglion or between neighboring ganglia (Kien, 1983; Altman and Kien, 1985); in contrast, the SEG contains circuits for redistributing descending information between right and left sides of the whole nerve cord (Kien, 1983; Kien and Altman, 1984).

12.4.2. Inputs to the SEG

To operate as a center for overall coordination, the SEG must receive information about the state and position of all parts of the body. The primary afferent inputs have been described in Sections 12.3.2.3 and 12.3.2.4. In addition, much ascending and descending sensory information is carried by interneurons, including the through-running AINs and DINs that connect the brain with the thoracic ganglia (Fig. 12.10).

Most of the input from the brain seems to be delivered to the SEG neuropils by collaterals from the axons of the brain DINs (Fig. 12.10B,C). The extent and destination of their branches in the SEG have so far been poorly documented, but preliminary observations indicate that most of the larger brain DINs send branches into the dorsal areas of the SEG, which contain arborizations both of SEG DINs and neck motor neurons (Kien and Altman, 1984; Altman, unpublished data). Similarly, in flies, small SEG DINs arborize only in areas containing branches from brain DINs (Strausfeld et al., 1984). Intracellular recording from SEG DINs (Section 12.4.4.3) confirms that many receive visual inputs that are most likely delivered by brain DINs (Ramirez, 1983; Kien and Altman, 1984; Boyan and Altman, 1985).

There are also interneurons from the brain that terminate in the SEG, but these have been very difficult to demonstrate. Evidence indicates that about 100 neurons on each side of the proto- and deutocerebrum terminate in the SEG, as well as a large number of small neurons in the tritocerebrum that are probably associated with control of mouthparts (Altman, unpublished data). Unfortunately, all attempts to fill the descending part of these neurons from the circumesophageal connectives to their terminations in the SEG have so far failed. They must have very fine axons and the only way to stain them will be by focal injections of cobalt chloride into the connective. Using this method, Kien and Williams (1983) obtained indirect evidence for a deutocerebral–SEG fiber involved in grooming.

Few AINs have been described. The best documented are the large auditory interneurons of Orthoptera (Rehbein et al., 1974; Hedwig, 1985). In locusts and grasshoppers, these run from the meta- or mesothoracic ganglia, where they synapse with primary afferents from the tympanum (Rehbein et al., 1974), to terminate in the lateral protocerebrum of the brain (Eichendorf and Kalmring, 1980; Boyan, 1983; Hedwig, 1985). In the SEG, collateral branches terminate in the dorsal neuropils, where SEG interneurons with acoustic inputs also arborize (Fig. 12.10A). At least one type of SEG AIN gets direct input from the acoustic AIN known as the G neuron (Boyan and Altman, 1985).

In the small grasshopper, *Omocestus viridulus,* other ascending AINs described include cercal giant fibers and several neurons with cell bodies in the prothoracic and mesothoracic ganglia that are phasically active during stridulation (Hedwig, 1985). These all branch ipsilaterally in the SEG, but no information on the distribution of the branches in the neuropil is available. Pflüger (1984) described a wind-sensitive AIN with its cell body in the fourth abdominal ganglion that receives its input from hairs on the prothoracic prosternum. Some premotor interneurons in the thoracic ganglia (e.g., Robertson and Pearson, 1983) extend to the SEG (Ramirez, unpublished data).

Intracellular recordings show that most SEG DINs receive various combinations of sensory information from sense organs on the body as well as on the head (Ramirez, 1983; Kien and Altman, 1984). Some receive precise information from single hairs on the legs, responding one-to-one with very short latency (Ramirez, 1983). As the primary afferents from leg sense organs do not extend beyond the ganglion of their own segment (Hustert et al., 1981), large, as yet undiscovered AINs must be involved.

12.4.3. Local Interneurons in the SEG

As so few local interneurons have yet been found in the SEG (Fig. 12.9), we have no idea of their roles. In the thoracic ganglia, local interneurons play a key role in segmental motor integration. They select both primary sensory and descending directional information and transfer it to the motor neurons (Siegler and Burrows, 1983; Rowell and Pearson, 1983; Reichert et al., 1985); and they modulate or set motor neuron firing levels (Burrows and Siegler, 1982). Thus, we would expect to find many more local interneurons in the SEG involved in neck and mouthpart motor integration. The local interneurons identified so far do not, however, appear to be involved in local integration, but instead integrate sensory information from other segments and produce widespread motor effects. For example, a pair of local inter-

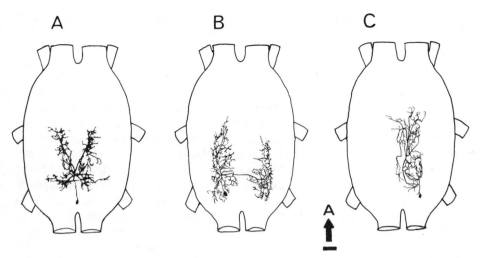

Figure 12.9. Local interneurons of the SEG. Wholemounts drawn from dorsal stained by intracellular injection of Lucifer yellow. (A) "Auditory" neuron SL1. The depth profile corresponds closely with that of the SA3 neuron in Figure 12.10G. (B,C) Two local interneurons involved in flight. (A from Boyan and Altman, 1985; B,C from J.-M. Ramirez, unpublished data.)

neurons with labial cell bodies and bilateral branches in both labial and maxillary neuromeres have been stained that respond to touch and movements of the legs (Ramirez, 1983).

Boyan and Altman (1985) describe an unpaired local interneuron (SL1) in the labial neuromere with a dorsal cell body and a bilaterally symmetrical arborization morphologically very similar to two SEG AINs (Section 12.4.4.2.; cf., Figs. 12.9A, 12.11B,D). It responds to auditory inputs with epsps and spikes and has the shortest latency of any of the SEG neurons with auditory inputs reported in their study. Its connections and role, however, are unknown.

Two local interneurons involved in flight have been found (J.-M. Ramirez, personal communication; Fig. 12.9B,C). Intracellular stimulation of one (Fig. 12.9C) excites a mesothoracic 404 flight neuron (Pearson et al., 1985) and can initiate flight activity, but it shows no reaction during wind-induced flight. Again, its direct connections are unknown. The simplest possibility for their wide-ranging output effects is that they are presynaptic to SEG DINs or to through-DINs, some of which, at least, receive synaptic inputs in the SEG (Kien and Altman, 1984).

12.4.4. Output Neurons of the SEG

The output of the SEG to both the brain and the body ganglia consists of two parallel pathways: the axons of through-running interneurons and of interneurons with cell bodies in the SEG. The SEG interneurons thus form an additional layer of connections between the brain and SEG, and SEG and body ganglia.

12.4.4.1. Through-Running Fibers. Most of the identified brain DINs have large diameter axons and respond to precise combinations of sensory stimuli, usually with a directional component. Examples in the locust are the tritocerebral commissure giant (TCG) (Bacon and Möhl, 1983); deviation detectors (Reichert et al., 1985); and the descending contralateral motion detectors (DCMD) (O'Shea et al., 1974). Although most are involved in the control of rapidly executed motor actions, such as escape jumping or flight steering, there is some evidence that they also participate in the control of walking (Kien and Williams, 1983; Kien and Altman, 1984). Hedwig (1985) has shown that the homologues of these, together with several other brain DINs, in *Omocestus viridulus* increase their firing rate during stridulation, although he found only one that was phase-coupled to the stridulatory movements.

Most brain DINs are paired; some have their axons ipsilateral to their cell bodies and main input arborization in the brain, whereas others have the axon crossing the midline in the brain to descend in the contralateral connective. Once the axons have left the brain, they remain on the same side throughout the nervous system. Although the DCMD (Fig. 12.10B; Altman and Kien, 1985) and some unidentified through-fibers (Kien and Williams, 1983) have a few branches that cross the midline in the SEG, most are restricted to the ipsilateral side (e.g., O_3, Fig. 12.10C). They usually branch in all three neuromeres, but some only in the maxillary and labial segments. There is little information about the depths of the branches in the neuropil, but most appear to lie in the dorsal half of the ganglion (see, for example, the TCG; Bacon and Tyrer, 1978).

So far only one brain DIN with a small-diameter axon has been described in detail—the tritocerebral commissure dwarf neuron (Tyrer and Bacon, unpublished data). Unlike the large axon brain DINs, it has branches on both sides of the SEG, penetrating to the ventral neuropils. Other small-axon brain DINs presumably exist but have not been revealed in cobalt chloride fills of connectives, either because their axons are too fine to transport cobalt far or because they are obscured by the heavy precipitates in their larger companions.

Through-AINs are described in Section 12.4.2. There have been no studies that could determine whether or not the signals in these neurons are in any way modulated as they pass through the SEG. It has generally been assumed this is not the case and that through-AIN branching in the SEG is entirely presynaptic.

12.4.4.2. Ascending and T-Interneurons Originating in the SEG. What little information we have about the SEG AINs suggests that they form an ad-

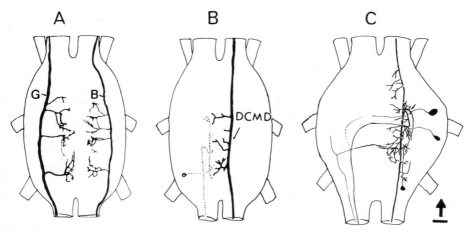

Figure 12.10. Through-fibers in the SEG. (A) Thoracic auditory AINs G and B; some branches occupy the same neuropils as those of the SA2 and SA3 neurons (Fig. 12.11G). (B) Course of the DCMD in the SEG showing its contralateral branches. The dotted neuron is a small SEG DIN stained by dye-coupling. (C) The O_3 neuron in the SEG, with three small dye-coupled SEG DINs. Depth profiles of G, B, and O_3 are in Figure 12.12D. (A after Boyan and Altman, 1985; B after Altman and Kien, 1985; C adapted from Kien and Altman, 1984.)

ditional step or loop in the ascending pathway. They integrate ascending information with information descending from the brain, which they relay back to the brain slightly delayed with respect to the signals in the through-AINs.

The best studied examples of SEG AINs are three morphological types that respond to auditory stimuli, SA1, 2, and 3 (Fig. 12.11A,B,D) (Boyan and Altman, 1985). SA1 (Fig. 12.11A) has an ipsilateral axon that terminates in the lateral protocerebrum, in the area containing terminals of the thoracic auditory AINs (Eichendorf and Kalmring, 1980), but its signals arrive some 20 msec after the spike in the largest thoracic AIN, the G neuron. The other two terminate in the mediodorsal protocerebrum, where most brain DINs arborize. The most interesting is the bilateral ascender SA3 (Fig. 12.11D), which has two axons, one in each connective, terminating in each side of the protocerebrum. It shows side-dependent habituation (Boyan, 1984): its response can be habituated by a repetitive stimulus presented to one ear without affecting the response to subsequent stimuli to the other ear. This means it can provide different inputs to the two sides of the brain. No records are available from the neuron in moving animals, so we do not know how ongoing activity alters its behavior.

Two AINs have been stained that respond to wind on the head and fire during flight muscle activity in the fixed preparation (J.-M. Ramirez, personal communication). One has a midline arborization (Fig. 12.11C), the other a branching area laterally on each side of the ganglion.

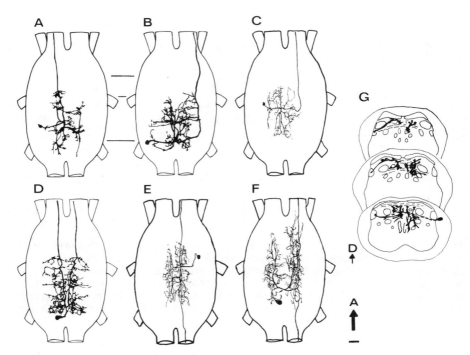

Figure 12.11. (A–D) SEG ascending interneurons (AINs). (E,F) T-fibers. (A,B) "Auditory" interneurons SA1 and SA2. (C) AIN involved in flight. (D) "Auditory" interneuron SA3. (E) T-fiber involved in flight. (F) T-fiber, function unknown. (G) Depth profiles: SA2 neuron (*left*) and SA3 (*right*). Sections at levels between (A) and (B). (A, B, D, G after Boyan and Altman, 1985; C, E, F from J.-M. Ramirez, unpublished.)

The T-fibers in the SEG so far described fall into two categories, those with medial arborizations (Fig. 12.11E) and those with two branching areas, one on each ganglion side (Fig. 12.11F). All T-fibers have the cell body contralateral to the axon, except that in Figure 12.11E. So far, no physiological correlation with structural type has been found.

12.4.4.3. Descending Interneurons Originating in the SEG: Form and Function.
Cobalt chloride infused through one cut cervical connective reveals about 120 cell bodies throughout the SEG. Most lie contralateral to the filled connective, but there are prominent groups in the midline and a few on the ipsilateral side (Altman, unpublished data). Probing the ganglion or cervical connections with dye-filled intracellular electrodes has enabled us to describe the morphology and physiology of several of these DINs (Kien and Altman, 1984; Hedwig, 1985; J.-M. Ramirez, personal communication).

Recording in a tethered locust, we found a number of SEG DINs active before or during walking or other spontaneous leg movements. Several of

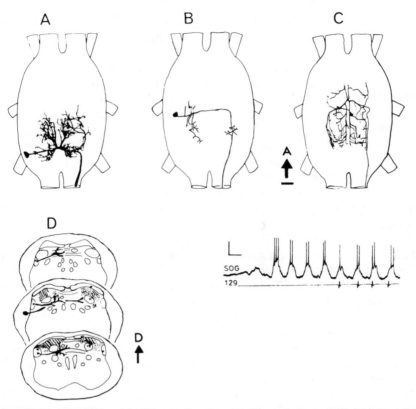

Figure 12.12. SEG DINs with different functions and morphologies. (A) "Auditory" DIN, SD1. Depth profile is similar to SA2 (Fig. 12.11G). (B) DIN active during walking. (C) DIN involved in flight, characterized by the medial pair of longitudinally oriented branches. *Inset,* recording from this neuron (*upper trace*) and myogram from the metathoracic wing depressor muscle, 129 (*lower trace*). Wind over the head, starting at trace onset, induces rhythmic inputs and activity in the DIN in phase with flight muscle activity. Calibration vertical, 10 mV; horizontal, 50 msec. (D) Depth profiles. Solid profiles on the left are of a walking-type SEG DIN, with cell body in the maxillary neuromere, together with the G and B neurons (Fig. 12.10A), which have branches in all neuromeres. Hatched areas contain the branches of such SEG DINs, stippled areas the branches of through DINs active during walking, including the O_3 neuron (Fig. 12.10C). (A after Boyan and Altman, 1985; B after Kien and Altman, 1984; C from J.-M. Ramirez, unpublished data; D adapted from Boyan and Altman, 1985, and Kien and Altman, 1984.)

these neurons fired before active movements, but also after passive movements of the same limb. Some neurons showed clear correlations with certain movements, others with walking in certain directions, but the majority fired irregularly during walking. All these neurons responded to visual inputs; tactile and proprioceptive stimulation of head, body, and legs; and sometimes weakly to white noise (Kien and Altman, 1984). Investigating tactile inputs in an immobilized preparation revealed that the SEG DINs receive and mix extremely precise information from many sense organs (Ramirez, 1983).

The cell bodies of these DINs lie in the maxillary and labial neuropils, usually contralateral to the axon. The branches are either bilateral (Fig. 12.12A,C) or mainly on the cell body side (Fig. 12.12B), but two neurons are entirely ipsilateral. The branches lie dorsally between the MDT and DIT (Fig. 12.12D).

We do not yet have details of the destinations of any of these neurons. One group with two large and one or two small cell bodies, which are active during walking, terminate in the prothoracic ganglion (Altman and Kien, 1985 and unpublished data), but, extrapolating from Hedwig's data (see following paragraph), most of them probably run at least to the posterior metathoracic ganglion.

Hedwig (1985) has described the morphology of 13 SEG maxillary and labial DINs that fire during stridulation in the grasshopper, *Omocestus viridulus*. These neurons fire phase-coupled with the stridulatory rhythm, the majority having their peak during the raising of the leg, but there is a wide range of coupling both in phase and peak duration. Some fire before or during other spontaneous leg movements, and four also show activity coupled to the ongoing respiratory rhythm. Neurons synergistic with the respiratory rhythm may show quite different phase-coupling in stridulation but are not completely uncoupled from the respiratory rhythm during stridulation; they have weak or no response to sensory stimulation (Hedwig, 1985).

About half of these neurons appear to be structurally homologous to those firing during walking and leg movements in locust (J.-M. Ramirez, 1983, and personal communication; Kien and Altman, 1984). Hedwig gives only plan views and no depth profiles, which is unfortunate, as in these small animals the thoracic arborizations can be filled with dye from an injection in the cervical connective, whereas they cannot in locust. Irrespective of their morphologies in the SEG, which vary from entirely ipsilateral, through symmetrical, to predominantly contralateral branching, they all have very similar arborizations in the thoracic ganglia, restricted to the ipsilateral side and with medial and lateral branching reminiscent of the thoracic projections of the TCG (Bacon and Tyrer, 1978) and optomotor fibers (Kien, 1980). It seems likely that the branches terminate in the neuropils containing wing and leg motor neuron branches (Tyrer and Altman, 1974), but sectioned material will be needed to confirm this. Most of the neurons have an axon leaving the metathoracic ganglion in the abdominal connective, but the projection in the free abdominal ganglia is not shown.

Most SEG DINs with firing related to the respiratory rhythm seem to have cell bodies in the ventral midline and lateral branching in the maxillary and labial neuromeres (Kien and Altman, unpublished data; J.-M. Ramirez, personal communication; Hedwig, 1985). Dye-coupling in one preparation suggested that seven to eight cells in the labial neuromere, not necessarily all DINs, may be coupled into a functional group.

Boyan and Altman (1985) report on the physiology of the largest DIN in the ganglion, which they name SD1 (Fig. 12.12A). It is excited by auditory and visual inputs, as well as receiving a barrage of unidentified ipsps. In the fixed preparation, pure tones and simple visual stimuli, presented separately, never elicited spikes unless the cell was depolarized by injecting a small amount of current. This suggests that it requires a particular combination or pattern of inputs to bring it to threshold. Its morphology (Fig. 12.12A) is unlike that of the "walking" DINs (e.g., Fig. 12.12B), which have their branches in the SEG mostly in the lateral neuropil, but it is very similar to the AINs and local interneurons receiving strong auditory inputs (Fig. 12.9A). There are morphological homologues to SD1 in the anterior labial and maxillary neuromeres (Ramirez, 1983; Boyan, personal communication; Altman and Kien, unpublished data). The maxillary homologue also receives auditory inputs (Boyan, personal communication).

Some SEG DINs involved in the generation of the flight motor output have also been found (J.-M. Ramirez, personal communication). They too have a distinctive morphology, characterized by the anteroposteriorly directed bilaterally symmetrical medial branches (e.g., compare Fig. 12.12C with Fig. 12.12A,B). In some, wind on the head evokes rhythmical activity that is correlated with and may precede the activity in the flight muscles. Intracellular electrical stimulation may even evoke flight activity (J.-M. Ramirez, personal communication), unlike SEG DINs that fire during walking and respiration (Kien and Altman, 1984).

The SEG DINs show great variability in their response properties. In walking animals, the firing pattern of some neurons changed dramatically, from quiescent to active, from random to rhythmic, without obvious changes in environment or behavior (Kien and Altman, 1984). Similar changes may be seen in the fixed animal, where a response of a neuron to repeated stimulation may vary considerably or change systematically with time (Ramirez, 1983).

Small SEG DINs. Another type of SEG DINs were fortuitously stained by dye-coupling in preparations where through-DINs had been injected with Lucifer yellow (Kien and Altman, 1984). Other examples have since been seen in cobalt chloride preparations (Kien and Altman, unpublished data). These DINs have small cell bodies and fine axons, with the axon usually contralateral to the cell body (Figs. 12.10B,C). Close contacts between their branches and those of the dye-filled through-DINs suggests physiological coupling between the two. Each small SEG DIN must also contact several other neurons, as each has many branches that are not juxtaposed to the filled through-fiber. The axon may run either ipsilateral or contralateral to the axon of the filled through-DIN. We suggest that these fine neurons transmit a modified version of the information carried by the through-DINs, and at a far slower rate than the through-DINs, which have large diameter axons. This would mean that they constitute a third class of descending output from the SEG, distinct from the through-DINs and large SEG DINs.

It is not clear whether the SEG DINs are unique or if they have serial homologues in each ganglion. Boyan (1984) reports a homologue to SD1 in the prothoracic ganglion, and the mesothoracic coactivation (C) neuron (Pearson and Robertson, 1981) is probably also homologous. The SEG neurons may have evolved from serially homologous interneurons, but they must now play a somewhat different role, for they have outputs in all the thoracic ganglia, unlike their thoracic homologues, which have outputs only in the ganglia more posterior to them. It seems likely that they have acquired their role in overall coordination because of their position, leaving those in the thoracic ganglia to subserve intersegmental coordination.

In summary, the SEG DINs seem to integrate sensory information from both sides of the body and head, relaying the resulting signals to the motor neuropils of the thoracic ganglia. Most collect information from both sides and send their outputs only to one side. The information the SEG DINs transmit will arrive later than that in the through DINs because of the synaptic delay in the SEG and, in the case of the small SEG DINs, because of the lower conduction velocity in their fine axons. A single stimulus complex could therefore result in three waves of signals from the SEG: the signals in the through-DINs, in the large SEG DINs, and then the small SEG DINs. We will need to know more about the distribution of these signals to the target neurons, before we can determine more about their functions in the control of motor activity.

12.5. CONCLUSIONS

12.5.1. Hypotheses for SEG Function

We have described briefly what is known of the organization and physiology of the SEG. On the one hand, this ganglion has the structure expected of a fused segmental ganglion; it innervates the segmental musculature of the head appendages and the neck, and is involved in the control and stabilization of head position, as well as coordinating behaviors, such as manipulating food and grooming. At the same time, the SEG functions as a higher motor center, involved in overall motor coordination and in the motor excitability of the whole animal. Its outputs can initiate, maintain, and coordinate walking and participate in the control of flight and respiration. Here, we develop a hypothesis to explain how the SEG could carry out its higher motor tasks, based on what is known of the circuitry in the ganglion and connections with other centers.

The conventional picture of locomotor control in arthropods (Pearson et al., 1985) and vertebrates (Grillner and Wallen, 1985) is of a linear hierarchy

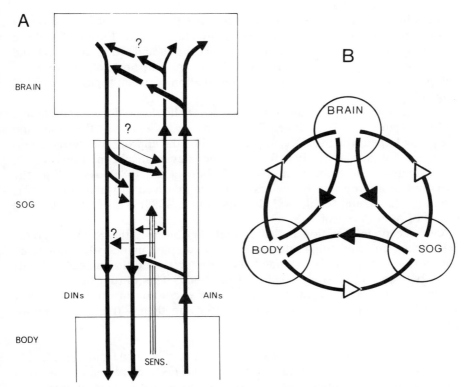

Figure 12.13. Pathways between brain, SEG, and body. (A) Summary of known and postulated neuronal pathways, drawn without considering bilaterality. Thin lines indicate very fine fibers. SENS, primary afferents. The various connections form three loops, schematized in (B). We suggest that (1) the loops are arranged nonhierarchically, (2) each center contributes its own specific function to behavior, and (3) each center is influenced by all other centers. Ascending arms of the loops are indicated by open arrows.

of centers, with the center or group of centers at each level regulating the activity of the ones below. But from recent work it has become apparent that the lower motor centers, that is the segmental ganglia in insects and spinal cord in vertebrates, are themselves capable of organizing fairly stereotyped motor programs, modulated by local segmental inputs. The higher centers seem to select the program most appropriate to the current external and internal conditions, and to direct and coordinate the activities of the segmental ganglia, so ensuring both smooth and coherent integration between the segments and an optimal fit to the momentary conditions.

We contend that in insects the brain and SEG operate in parallel, handling different parts of these functions rather than being different levels in a linear hierarchy. It is clear from what we already know of the anatomical connections of the brain and SEG, summarized in Fig. 12.13A, that there are two distinct pathways between the brain and the body ganglia, one direct and the other involving interneuronal circuits in the SEG. Essentially, these form loops between brain and body ganglia, brain and SEG, and SEG and body ganglia (Fig. 12.13B). The three loops are not, however, completely independent as there is exchange of information between them in the SEG (Fig. 12.13A).

We suggest that the brain determines *what* should be done, whereas the SEG regulates *how* it is done. The fast pathways between the brain and body, the large through-DINs, appear to prime the appropriate thoracic networks so that they are ready to fire when the relevant local conditions pertain (e.g., Pearson and Robertson, 1981; Reichert et al., 1985). As well as the initiation of appropriate behaviors, they are involved in rapid course corrections (Bacon and Möhl, 1983; Reichert et al., 1985). This is the *what* and *where* of motor control.

Our recent information indicates that, in contrast, the SEG integrates descending and ascending information in a complex manner to produce an output that is temporally extended and structured, and that makes comparisons between left and right sides of the body. This output helps both to initiate behaviors and maintain the level of excitation in the active motor networks, modulating the pattern of activity in them—the *how* of motor control.

12.5.1.1. Maintenance of Motor Excitability. As can be seen from Fig. 12.14A, inputs from both the brain and the rest of the body converge on the SEG DINs, so that a complete representation of the animal's external world and internal state can be built up in the SEG. Not only is sensory information from different sources mixed, but this complex information is combined with the already highly integrated output of the brain, through contacts between brain DINs and SEG DINs. The inputs may be combined in a variety of ways before being redistributed to the rest of the body.

Lesion experiments indicate that the SEG is especially important for maintaining motor activity, presumably through the SEG-body loop. Significantly, the SEG DINs project back to the systems from which they receive their inputs; for example, Kien and Altman (1984) report that DINs that fired preceding leg movements also responded to passive movements of the legs. There is some vidence that such loops may exist between individual neurons in the SEG and thoracic ganglia (J.-M. Ramirez, personal communication). Thus, during movements feedback of this type will reexcite the SEG DINs and so help to maintain their firing. In addition, the output of the DINs will be altered according to the performance of the motor network. As the SEG DINs also receive input from exteroceptors, they can monitor directly the effect of motor actions and assess the mismatch between command and performance. Such feedback loops could form the basis for a "fine tuning" control of activity in the thoracic motor networks.

The circuits of the SEG also build delays into both the ascending and

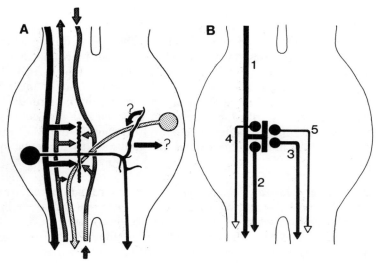

Figure 12.14. The functions of the SEG in overall coordination. (A) Schema of the anatomical substrates showing known and possible connections (*arrows*) between brain DINs,
thoracic AINs, the bilateral SEG DINs of each side, and descending and ascending primary
afferents (shown as one *hatched pathway; arrows* indicating information flow into the
SEG). Overlaps are shown on one side only; symmetrical overlaps are found on the other
side. (B) Schema, based on current anatomical data, suggesting how information passing
from the brain to the thorax may be delayed and extended in time in the SEG. The fastest
route from brain to thorax is the large brain DINs (pathway 1). Varying delays may be
achieved by synapses in the SEG (pathways 2–5), by neurons that cross the midline (pathways 3, 5) and by differences in axon diameter between neurons (shown here by line
thickness; pathways, 4, 5). Synaptic inputs to SEG DINs also provide further opportunities
for mixing information of different modalities. These SEG circuits promote the bilateral
distribution of the descending signals (pathways, 3, 5).

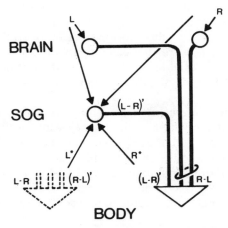

Figure 12.15. Hypothesis for the function of the bilateral distribution of descending information, based on the assumption that a right–left comparison differs from a left–right
comparison. The minus sign indicates comparison; arithmetical operations are not implied. L, R, information from left and right sense organs of the head; L*, R*, from left and
right sense organs of the body, respectively. See text for further explanation.

295

descending pathways, considerably extending the period during which particular information reaches its target. The sequence in which descending signals leave the ganglion is shown, much simplified, in Fig. 12.14B. As well as synaptic delay and axon diameter, the integration time in the SEG interneurons has to be considered. This depends partly on the structure of the neuron, but also varies with the other inputs the neuron is receiving. The circuits in the SEG thus produce a temporal layering of the SEG outputs, which may assist in maintaining excitability and help to sequence specific motor acts.

12.5.1.2. SEG as Bilateral Comparator. Motor coordination requires both intersegmental and left–right coordination. It is necessary that each motor center is informed on what all the effectors on both sides have just done and are about to do. Although much of this is done at an intersegmental level, the convergence of information in the SEG and its bilateral integration makes the ganglion an excellent candidate for an overall left–right comparator. This can be appreciated by considering the neural representation of the body sides in one cervical connective, as seen by a postsynaptic neuron in one of the thoracic motor networks (Fig. 12.15). We assume for this argument that there is a qualitative difference between a right–left and a left–right comparison.

Brain DINs have their axons in the connective, either contralateral or ipsilateral to their arborizations in the brain. One connective thus contains axons of neurons with identical receptive fields but on right (R) and left (L) sides of the brain, and a direct comparison can be made between their outputs (R-L). SEG DINs operate differently, for each integrates inputs from both sides of the head (L, R) and body (L*, R*), so a comparison (L-R) + (L*-R*) = (L-R)' is already present in its output. As most SEG DINs have their axons contralateral to their main inputs, their outputs will be biased in the opposite direction (L-R)' to those of pairs of brain DINs (R-L). The other connective will carry a mirror-image comparison. This information can be compared with the vectorial information carried by brain DINs to establish what each side of the body should do next.

12.5.1.3. Consensus in Motor Control. The picture that is emerging is of a number of fibers all carrying information relevant to the performance of a particular behavior, but with the information in each one being qualitatively, quantitatively, or temporally unique. This fits with the "consensus" hypothesis for the initiation and maintenance of behavior (Kien, 1983), which predicts that it is the "across fiber" firing pattern reaching the thoracic motor networks that determines the specific outputs of the local pattern generators.

12.5.2. Analogies With the Vertebrate Brain Stem

In both its local and its higher roles, the SEG seems to be closely analogous to the brain stem in vertebrates. Like the SEG, the brain stem innervates the exteroceptors, proprioceptors, and muscles of the head and neck; it is involved in the control of respiration, posture, and locomotion, and it regulates a variety of autonomic functions, such as heart rate, salivation, and gut activity. It also plays a large role in controlling overall excitability of the motor centers.

The brain stem is also necessary for coordinated walking. In cats, even the most reduced form of walking produced in experiments requires the integrity of the reticular formation to at least the rostral margin of the midbrain (Wetzel and Stuart, 1976). The brain stem exerts two types of control over walking, through several pathways: the more general function of activating the spinal stepping centers, and the specific augmentation and presumably adjustment of the degree of motor excitability in appropriate phases of the step (Wetzel and Stuart, 1976). The initiation and detailed control of walking through multiple pathways are precisely the functions we have suggested for the SEG. Furthermore, the locomotory areas within the brain stem receive inputs from both higher and lower regions of the CNS (Wetzel and Stuart, 1976), the descending pathways forming parts of multiple loops, similar to those we describe for the SEG.

Thus, in vertebrates, as in arthropods, there appears to be a separation of functions, with the higher centers deciding "what" and "where" and the brain stem determining "how" an action should be executed. In insects such as the locust, the SEG is separated from the brain by connectives, which allows one to examine its functions in isolation from those of the higher brain centers, simply by cutting the circumesophageal connectives. If our analogy between the SEG and the brain stem is acceptable, data obtained from studies in the locust may well be applicable to vertebrates, and vice versa. The locust could be a valuable model for determining the functional principles and detailed neuronal mechanisms of executive motor centers.

12.6. SUMMARY

The SEG has evolved from a series of segmental ganglia innervating head appendages. In higher arthropods, these ganglia are fused to form a single compound ganglion, which may become joined to the brain or thoracic ganglia. In insects, the SEG contains three neuromeres innervating mandibles, maxillae, and labium together with the neck muscles, as well as contributing to the innervation of the salivary glands, corpora allata, and frontal ganglion. The segmental homologies are established for the locust SEG and the local segmental functions in the control of the mouthparts and head movements are described. As well as these segmental functions, the SEG functions as a higher motor center, participating in the initiation, maintenance, and regulation of the motor output of the ventral nerve cord. Our current knowledge of the organization of the neurons involved in this function suggests that, rather than being one tier in a linear hierarchy of motor centers, they form separate loops between brain and SEG, and between SEG and ventral nerve cord ganglia, in parallel to the brain–ventral ganglia loop. We propose that the brain regulates *what* behavior is performed, whereas the SEG deals with *how* it is done. A functional comparison with the vertebrate brain stem is made.

ACKNOWLEDGMENTS

This work and JSA were supported by Sonderforschungsbereich 4, Projekt H2, JK by a Heisenberg-Stipendium. U. Roth provided technical assistance. We are particularly grateful to J.-M. Ramirez and N. M. Tyrer for allowing us to use unpublished data.

REFERENCES

Albrecht, F. O. 1953. *The Anatomy of the Migratory Locust*. Athlone Press, New York.

Altman, J. S. 1981. Functional organisation of insect ganglia. *Adv. Physiol. Sci.* **23:** 537–555.

Altman, J. S. and J. Kien. 1979. Suboesophageal neurons involved in head movements and feeding. *Proc. R. Soc. Lond. Ser. B.* **205:** 209–227.

Altman, J. S. and J. Kien. 1985. The anatomical basis for intersegmental and bilateral coordination in locusts, pp. 91–119. In B. Bush and F. Clarac (eds.), *S.E.B. Seminar Series,* **24**, Cambridge University Press, Cambridge.

Altman, J. S. and N. M. Tyrer. 1977. The locust wing hinge stretch receptors. I. Primary sensory neurons with enormous central arborisations. *J. Comp. Neurol.* **172:** 409–430.

Anderson, H. and J. Bacon. 1979. Developmental determination of neuronal projection patterns from wind-sensitive hairs in the locust *Schistocerca gregaria*. *Dev. Biol.* **72:** 364–373.

Aubele, E. and N. Klemm. 1977. Origin, destination and mapping of tritocerebral neurons of the locust. *Cell Tissue Res.* **178:** 99–122

Bacon, J. and B. Möhl. 1983. The tritocerebral commissure giant (TCG) wind-sensitive interneuron in the locust. I. Its activity in straight flight. *J. Comp. Physiol.* **150:** 439–452.

Bacon, J. and N. M. Tyrer. 1978. The tritocerebral commissure giant (TCG): A bimodal interneuron in the locust *Schistocerca gregaria*. *J. Comp. Physiol.* **126:** 317–325.

Bässler, U., E. Foth, and C. Breutel. 1985. The inherent walking directions differ for the prothoracic and mesothoracic legs of stick insects. *J. Exp. Biol.* **116:** 301–311.

Blaney, W. M. and R. F. Chapman. 1969. The anatomy and histology of the maxillary palp of *Schistocerca gregaria* (Orthoptera, Acrididae). *J. Zool. Lond.* **157**: 509–535.

Boyan, G. S. 1983. Postembryonic development in the auditory system of the locust. Anatomical and physiological characterisation of interneurones ascending to the brain. *J. Comp. Physiol. A,* **151**: 499–513.

Boyan, G. S. 1984. Neural mechanism of auditory information processing by identified interneurones in Orthoptera. *J. Insect Physiol.* **30**: 27–41.

Boyan, G. S. and J. S. Altman. 1985. The suboesophageal ganglion: A "missing link" in the auditory pathway of the locust. *J. Comp. Physiol.* **156**: 413–428.

Bräunig, P., R. Hustert, and H.-J. Pflüger. 1981. Distribution and specific central projections of mechanoreceptors in the thorax and proximal leg joints of locusts. I. *Cell Tissue Res.* **216**: 57–77.

Bräunig, P., H.-J. Pflüger, and R. Hustert. 1983. The specificity of central nervous projections of locust mechanoreceptors. *J. Comp. Neurol.* **218**: 197–207.

Bullock, T. H. and G. A. Horridge. 1965. *Structure and Function in the Nervous System of Invertebrates,* vol. 2. Freeman, San Francisco.

Burrows, M. and M. V. S. Siegler. 1976. Transmission without spikes between locust interneurones and motoneurones. *Nature (London)* **262**: 222–224.

Burrows, M. and M. V. S. Siegler. 1982. Spiking local interneurones mediate local reflexes. *Science (Washington, D.C.)* **217**: 650–652.

Campbell, J. I. 1961. The anatomy of the nervous system of *Locusta migratoria migratorioides. Proc. Zool. Soc. Lond.* **137**: 403–432.

Davis, N. T. 1983. Serial homologues of the motor neurons of the dorsal intersegmental muscles of the cockroach, *Periplaneta americana* (L.). *J. Morphol.* **176**: 197–210.

Delphin, F. 1965. The histology and possible functions of neurosecretory cells in the ventral ganglia of *Schistocerca gregaria* Forster (Orthoptera: Acrididae). *Trans. R. Entomol. Soc. Lond.* **117**: 167–214.

Eichendorf, A. and K. Kalmring. 1980. Projections of auditory ventral-cord neurons in the supraoesophageal ganglion of *Locusta migratoria. Zoomorphology* **94**: 133–149.

Gewecke, M. 1979. Central projection of antennal afferents for the flight motor in *Locusta migratoria* (Orthoptera: Acrididae). *Entomol. Gen.* **5**: 317–320.

Graham, D. 1979. Effect of circumoesophageal lesion on the behaviour of the stick insect *Carausius morosus.* II. *Biol. Cybern.* **32**: 147–152.

Granger, N. A. 1978. Innervation of the prothoracic glands in *Galleria mellonella* larvae (Lepidoptera, Pyralidae). *Int. J. Insect Morphol. Embryol.* **7**: 315–324.

Grillner, S. and P. Wallen. 1985. Central pattern generators for locomotion, with special reference to vertebrates. *Ann. Rev. Neurosci.* **8**: 233–261.

Hedwig, B. 1985. Untersuchungen zur Kontrolle des Feldheuschreckengesangs durch intersegmentale Neurone. Doctoral Thesis, Universität Göttingen, Göttingen.

Honegger, H.-W. 1977. Interommatidial hair receptor axons extending into the ventral nerve cord in the cricket *Gryllus campestris. Cell Tissue Res.* **182**: 281–285.

Honegger, H.-W., J. S. Altman, J. Kien, R. Müller-Tautz, and E. Pollerberg. 1984.

A comparative study of neck motor neurones in a cricket and a locust. *J. Comp. Neurol.* **230:** 517–535.

House, C. R., B. L. Ginsborg, and E. M. Silinsky. 1973. Dopamine receptors in cockroach salivary gland cells. *Nature (London)* **245:** 63.

Huber, F. 1955. Sitz and Bedeuntung nervöser Zentren für Instinkthandlung beim Männchen von *Gryllus campestris* L. *Z. Tierpsychol.* **12:** 12–48.

Huber, F. 1960. Untersuchung über die Funktion des Zentralnervensystems und insbesondere des Gehirns bei der Fortbewegung und der Lauterzeugung der Grillen. *Z. Vgl. Physiol.* **44:** 60–132.

Hustert, R. 1978. Segmental and interganglionic projections from primary fibres of insect mechanoreceptors. *Cell Tissue Res.* **194:** 337–351.

Hustert, R., H.-J. Pflüger, and P. Bräunig. 1981. Distribution and specific central projections of mechanoreceptors in the thorax and proximal leg joints of locusts. *Cell Tissue Res.* **216:** 97–111.

Kien, J. 1980. Morphology of locust neck muscle motorneurons and some of their inputs. *J. Comp. Physiol.* **140:** 321–336.

Kien, J. 1983. The initiation and maintenance of walking in the locust. An alternative to the command concept. *Proc. R. Soc. Lond.* B **219:** 137–174.

Kien, J. and J. S. Altman. 1984. Descending interneurones from the brain and suboesophageal ganglia and their role in the control of locust behaviour. *J. Insect Physiol.* **30:** 59–72.

Kien, J. and M. Williams. 1983. Morphology of neurons in locust brain and suboesophageal ganglion involved in initiation and maintenance of walking. *Proc. R. Soc. Lond.* B **219:** 175–192.

Mason, C. A. 1970. New features of the brain-retrocerebral neuroendocrine complex of the locust *Schistocerca vaga* (Scudder). *Z. Zellforsch. Mikrosk. Anat.* **141:** 19–32.

O'Shea, M., C. H. F. Rowell, and J. L. D. Williams. 1974. The anatomy of a locust visual interneurone: The descending contralateral movement detector. *J. Exp. Biol.* **60:** 1–12.

Pearson, K. G. and R. M. Robertson. 1981. Interneurons coactivating hind leg flexor and extensor motoneurons in the locust. *J. Comp. Physiol.* **144:** 391–400.

Pearson, K. G., D. N. Reye, D. W. Parsons, and G. Bicker. 1985. Flight-initiating interneurons in the locust. *J. Neurophysiol.* **53:** 910–925.

Pflüger, H.-J. 1984. The large fourth abdominal intersegmental interneuron: A new type of wind-sensitive ventral cord interneuron in locusts. *J. Comp. Neurol.* **222:** 343–357.

Ramirez, E. 1984. Die zentralen Projektionen des Maxillarnervs in dem Unterschlundganglion der Heuschrecke *Schistocerca gregaria*. Diplom thesis, Universität Regensburg, Regensburg.

Ramirez, J.-M. 1983. Untersuchung der sensorischen Eingänge von multimodalen Neuronen im Unterschlundganglion der Heuschrecke (*Schistocerca gregaria*). Diplom thesis, Universität Regensburg, Regensburg.

Rehbein, H.-G., K. Kalmring, and H. Römer. 1974. Structure and function of acoustic neurons in the thoracic ventral nerve cord of *Locusta migratoria* (Acrididae). *J. Comp. Physiol.* **95:** 263–280.

Reichert, H., C. H. F. Rowell, and C. Griss. 1985. Course correction circuitry translates feature detection into behavioural action in locusts. *Nature (London)* **315:** 142–147.

Rémy, C. and J. Girardie. 1980. Anatomical organization of two vasopressin-neurophysin-like neurosecretory cells throughout the central nervous system of the migratory locust. *Gen. Comp. Endocrinol.* **40:** 27–35.

Robertson, R. M. and K. G. Pearson. 1983. Interneurons in the flight system of the locust: Distribution, connections, and resetting properties. *J. Comp. Neurol.* **215:** 33–50.

Roeder, K. D. 1937. The control of tonus and locomotor activity in the praying mantis (*Mantis religiosa*). *J. Exp. Zool.* **76:** 353–374.

Roeder, K. D., L. Tozian, and E. A. Weilant. 1960. Endogenous nerve activity and behaviour in the mantis and cockroach. *J. Insect Physiol.* **4:** 45–62.

Rowell, C. H. F. and K. G. Pearson. 1983. Ocellar input to the flight motor system of the locust: Structure and function. *J. Exp. Biol.* **103:** 265–288.

Shepheard, P. 1973. Musculature and innervation of the neck of desert locust, *Schistocerca gregaria* (Forskåal). *J. Morphol.* **139:** 439–464.

Siegler, M. V. S. and M. Burrows. 1983. Spiking local interneurons as primary integrators of mechanosensory information in the locust. *J. Neurophysiol.* **50:** 1281–1295.

Snodgrass, R. E. 1928. Morphology and evolution of the insect head and its appendages. *Smithson. Misc. Collect.* **81:** 1–158.

Strausfeld, N. J., U. K. Bassemir, R. W. Singh, and J. P. Bacon. 1984. Organisational principles of outputs from dipteran brains. *J. Insect Physiol.* **30:** 73–93.

Tyrer, N. M. and J. S. Altman. 1974. Motor and sensory flight neurones in a locust demonstrated using cobalt chloride. *J. Comp. Neurol.* **157:** 117–138.

Tyrer, N. M. and G. E. Gregory. 1982. A guide to the neuroanatomy of locust suboesophageal and thoracic ganglia. *Philos. Trans. R. Soc. Lond. B* **297:** 91–123.

Tyrer, N. M., J. Bacon, and C. A. Davies. 1979. Sensory projections from the wind-sensitive head hairs of the locust *Schistocerca gregaria*. Distribution in the central nervous system. *Cell Tissue Res.* **202:** 79–92.

Tyrer, N. M., J. Turner, and J. S. Altman. 1984. Identified neurons in the locust central nervous system which react with antibodies to serotonin. *J. Comp. Neurol.* **227:** 313–330.

Weinbörmair, G., K. Pohlhammer, and H. Dürnberger. 1975. The axonal system of the ventromedian pair of neuro-secretory cells in the suboesophageal ganglion in *Teleogryllus commodus*. Staining of whole mount preparations with resorcin fuchsin. *Mikroskopia* **31:** 147–154.

Wetzel, M. C. and D. G. Stuart. 1976. Ensemble characteristics of cat locomotion and its neural control. *Progress in Neurobiol.* **7:** 1–98.

Zack, S. and J. Bacon. 1981. Interommatidial sensilla of the praying mantis: Their central neural projections and role in head-cleaning behaviour. *J. Neurobiol.* **12:** 55–65.

CHAPTER 13

Control of Mouthparts
by the Subesophageal Ganglion

W. M. Blaney
Behavioural Entomology Group
Department of Biology
Birkbeck College
University of London
London, England

Monique S. J. Simmonds
Behavioural Entomology Group
Jodrell Laboratory
Royal Botanic Gardens
Kew, Richmond
Surrey, England

13.1. INTRODUCTION

The subesophageal ganglion, consisting of the fused ganglia of the primitive mandibular, maxillary, and labial segments, controls activity occurring in these segments and modulates activity occurring elsewhere in the central nervous system (CNS). In modern arthropods, these three segments are variously reduced and fused to form part of the head structure (Chapter 12 by Altman and Kien in this volume). Their principal, or only, overt activity is to participate in feeding and, by the actions of their appendages, largely or completely to control the mechanics of that process. Other activities of mouthparts, such as grooming, have received relatively little attention, but feeding by arthropods, and especially by insects, impinges so greatly on the life and well-being of humans that for decades it has been the subject of numerous, extensive, and varied studies. Yet surprisingly, relatively little is known about the precise nature of the neuronal control exerted over that activity by the subesophageal ganglion (SEG).

13.2. ROLE OF THE SUBESOPHAGEAL GANGLION

This review concentrates on our knowledge of the integration and control of feeding activity in arthropods, under the control of the SEG and its associated neurons. Two aspects of this control, although in practice interlinked, may for convenience and clarity be considered separately. Thus, on the one hand, we may consider the movements of various mouthparts and the ways in which sensory information describing the movement affects the motor output causing it. Even here, where fairly direct input–output relationships are recorded, the integrating activity of the SEG is inferred, rather than directly investigated. The second aspect involves the initiation, continuation, and termination of feeding activity insofar as that is influenced by sensory information obtained by mouthpart sensilla and processed, at least initially, by the SEG. Here, a much greater body of knowledge on sensory activity exists, but the integrating role of the SEG is, if anything, even more obliquely inferred. This is so partly because feeding behavior and the motor output driving it is not a simple correlate of the sensory input from the mouthpart receptors. Other central neuronal activity, such as that associated with hunger or satiety, plays a part in, sometimes dominating, the control of feeding. Nor does the SEG receive all the sensory input related to food selection and feeding: odor receptors on the antennae, synapsing in the brain, or taste receptors on the feet, synapsing in the ganglia of the ventral nerve cord, also have a role to play. Nevertheless, a study of the available literature on the functioning of mouthpart receptors reveals implications of the activity of the SEG.

13.3. FUNCTIONING OF MOUTHPARTS

The most complete studies to date have been made on the European lobster, the desert locust, and a blowfly. In the arthropods there is an enormous range of form and function in the mouthparts, reflecting the great diversity of food materials used by these animals. The mouthparts of modern arthropods are believed to have derived from segmentally arranged pairs of articulated ambulatory limbs (Snodgrass, 1952). This belief has implications for the innervation and control of the muscles operating the limbs. The lobster and locust have biting/chewing mouthparts where paired, limb-like structures are retained. The most complete information is available on the control of the mandibles, which are functionally similar in the two animals. In both cases, the lateral opening and closing used in biting and chewing has a rhythmicity superficially similar to that used by the walking or swimming limbs and has prompted the search for evidence of pre-existing central motor scores within the nervous system. In the blowfly, many of the primitive mouthparts have been modified by fusion into a tube-like proboscis (Snodgrass, 1943, 1953). During feeding, the distal portion of the proboscis, the haustellum, is extended by the action of two pairs of muscles (Dethier, 1959), and this activity may be considered in relation to sensory input.

13.4. CONTROL OF MANDIBLE MOVEMENTS

13.4.1. Lobster

The most extensive studies have been undertaken on the common European lobster, *Homarus gammarus* (L). The morphology of the mandibles, their movements, and their control have been described by Wales et al. (1976a,b) and Macmillan et al. (1976). The paired mandibles are the largest and the most powerful mouthparts. They bear cutting and crushing cusps used to reduce food to a suitable size for ingestion. Each mandible is a cup-shaped cuticular structure hinged along its anterior surface to the head capsule, and it has only two points of articulation. This restricts its movement to a single plane, at right angles to the hinge line, in which it moves in an arc of approximately 15° (Wales et al., 1976a). Muscles inserted on the mandible fall into two groups: adductors, whose contraction tends to close the mandibles on each other, and abductors or openers. Despite the limited scope for movement imposed by the articulation system, there is a surprisingly large number of muscles. Wales et al. (1976a) have described nine discrete muscles serving each mandible. It has been suggested (Wales et al., 1976a) that synergistic muscles show differential activity across the range of movement. Their activity is influenced by sensory feedback which is integrated in the SEG.

   There are at least two elements of the sensory system capable of monitoring the activities of the mandibles. The most completely studied is the mandibular muscle receptor organ (Wales and Laverack, 1972a,b), but in addition to this there are other proprioceptors, the so-called mouthpart receptors, which monitor movements of the mouthparts more generally. Further, the biting cusps of the mandibles are reported to have massive innervation, which could serve sense organs (Wales, 1982) acting in a similar manner to the intradental endings in mammalian teeth (Anderson et al., 1970).

The movements and neural control of the mandibles in normal feeding and in experimentally constrained situations have been investigated both in free-moving and restrained lobsters (Wales et al., 1976b; Macmillan et al., 1976). Electromyograms were recorded from up to six muscles simultaneously, and the movements of the mandibles were accurately monitored at the same time in restrained animals. Normal feeding activity was observed when animals were offered a standardized substrate, a *Nephrops* leg. Biting was a cyclic activity and each cycle fell into three parts: opening, closing, and biting. Mandibles could be moved independently of each other, but during rhythmic biting they were always bilaterally coupled. When engaging a new substrate, the mandibles could start moving from different positions and at different times but, on engaging the substrate, always adjusted closing velocity to grip the substrate symmetrically about the midline by the time that the bite phase was executed. The independence suggests that the closing movement is initiated by a central motor score, which is not dependent on the position of the mandibles (Wales et al., 1976b). When the substrate is centered during the bite, this calls for greater power output from one mandible than from the other. Loading one mandible artificially demonstrates the mechanism for this: electromyogram activity is greater in the adductor muscles of the restrained mandible (Wales, 1982). This capacity for adjustment calls for good sensory/motor feedback and intermandibular coordination. The mandibular muscle receptor organ is likely to be involved because it gives information on both velocity and position of the mandibles (Wales and Laverack, 1972a,b). Additionally, as it has its own muscle, the level of excitability could be set by efferent commands to allow for substrates of different hardness.

Within any given sequence of bites on a given substrate, the degree of mandible opening and velocity of closing vary in a compensatory manner so that the bite cycle duration is very constant. This is further evidence for the occurrence of a central motor score. With radically different substrates, the muscular activity can be altered profoundly. In addition to the standard substrate of a *Nephrops* leg, Macmillan et al. (1976) presented tethered lobsters with an incompressible substrate, a metal tube filled with *Nephrops* flesh, and an elastic substrate (a rubber tube). Both of these abnormal substrates presented an increased load on the mandibles, although in different ways, and resulted in increased motor output, presumably resulting in increased recruitment and/or increased tension. The effect on the structure of the biting cycle differed between the two abnormal substrates. With the incompressible substrate there was a decrease in cycle duration, that is, the biting frequency increased. When the mandible engaged the substrate, there was a sudden increase in closer muscle activity, but this was soon terminated, presumably as tension rapidly increased in the muscle and was monitored by the muscle receptor organ. Cross-correlation of the various elements of the biting cycle (e.g., closer burst duration, opener burst duration, interburst intervals) showed that with the incompressible substrate intracycle relationships were very strong, with significantly less variation than with the normal substrate. With the elastic substrate the biting frequency was much reduced and a larger proportion of each lengthened bite cycle was occupied by sustained closer muscle activity. Further, cross-correlation analysis showed that intracycle relationships were much weaker, with more variation than with the normal substrate (Macmillan et al., 1976).

Thus, the experiments with abnormal substrates show that there is relatively little variation in the time taken for the mandibles to close, but the main difference lies in the amount of time spent in compressing the substrate. From this, it may be concluded (Wales, 1982) that, in feeding, mandibular activity is governed by a central motor score, which is influenced by sensory input measuring position and stress in the mandibles. In this scheme, positional input would be of predominant importance before the mandibles engage the substrate, and information on tension would be of greatest importance after that.

13.4.2. Locust

A similar series of experiments has been conducted on adult females of the desert locust *Schistocerca gregaria* by Seath (1977a,b). The form and functioning of the locust's mandibles are similar to those of the lobster, except that the musculature is much simpler: the movement is controlled by two muscles, an opener and a closer. A similar dicondylic articulation restricts the movement to one plane so that it is possible to drive the mandibles artificially and mimic accurately their natural movements. In Seath's experiments, the locust was fixed to plasticene with the ventral side up, in which position it would readily eat grass if sufficiently hungry. Electromyograms were recorded by very fine copper wires inserted into the large closer muscles that occupy much of the head capsule. For some experiments, intrafiber activity was recorded by a glass capillary microelectrode inserted into the muscle through a small hole cut in the head cuticle. The mandibles were driven artificially by means of levers attached by fine stainless steel hooks to the mandibular cusps. The levers were driven so as to move the mandibles sinusoidally in the plane of their normal movement with controlled speed and amplitude. The device could be arranged so that the mandibles had to work against an artificial load, or, by means of a strain gauge, the tension developed by the mandibular closer muscles could be measured.

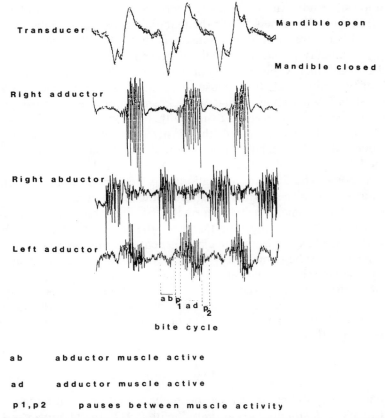

Figure 13.1. Three bite cycles (approximately 1 bite/sec) by a fifth instar nymph of *Locusta migratoria* eating grass. A light beam transducer attached to the right mandible shows opening and closing. As shown in the electromyograms, bursts of muscle activity are separated by pauses.

Further, the mandibles could be driven singly or together but at different speeds (Seath, 1977a).

During normal feeding, bursts of activity occurred synchronously in closer muscles of opposite sides. In a similar study with fifth instar nymphs of *Locusta migratoria* (Blaney, unpublished data), in which opener muscle activity was also recorded, there were bilaterally symmetrical alternating bursts of opener and closer muscle activity, with short interburst pauses after each burst (Fig. 13.1). As with lobsters (Wales et al., 1976b), there is good correlation between intracycle elements. When the locust is eating harder grass, the closer burst duration increases but the cycle length is fairly constant, suggesting the occurrence of a central motor score determining the overall rhythmicity.

In his study, Seath (1977a) drove the mandibles in synchrony in the absence of food and found synchronous bursts in the closers of both sides, initiated during the final opening or initial closing phase of the imposed movement. These bursts followed the artificial movement over a wide frequency range, and the burst duration was positively correlated with the wavelength of the imposed sinusoidal movement. As the driving frequency was lowered, the long closer bursts became broken into a series of shorter bursts, particularly while the mandibles were in a wide open position. When the mandibles were held in the fully open position, the closer muscle activity consisted of a long train of shorter bursts.

Situated in the cuticle adjacent to the mandibular closer apodeme is a group of campaniform sensilla (Thomas, 1966). Such sensilla monitor cuticle distortion (Pringle, 1938) and are well placed to monitor the build-up of tension due to the activity of the closer muscles. When these sensilla were cauterized and the mandibles held open, the closer muscles fired continuously for several seconds. Thus, it seems that when the mandibles are opened, closer muscle activity is initiated, but that if tension builds up unduly, as when an incompressible substrate is being bitten, the campaniform sensilla act as a safety device, perhaps preventing damage to the muscles, but at least allowing the cycle to be reset or the food otherwise manipulated.

When only one mandible was driven, the closer muscle serving it showed normal bursting activity, but electromyograms of the undriven paired mandible showed little or no activity (Seath, 1977a). The free mandible is reported to have opened in response to the opening of the driven mandible, but no recordings were made from opener muscles. Closing of the free mandible was apparently due to elasticity of the system: the normal relaxed position of the mandibles is closed. However, placing grass between the mandibles resulted in full, normal, synchronous closer muscle activity in the undriven side. This was not caused by chemical stimulation from the grass but was mechanical, as it could be obtained by placing slivers of polystyrene between the mandibles. The location of the mechanoreceptors involved is not certain. Mandibular receptors have been described in the thick cuticle of the biting cusps of *Locusta migratoria* (Le Berre and Louveaux, 1969). When Seath (1977a) cauterized the equivalent region of the undriven mandible, the presence of grass no longer produced synchronized firing in its closer muscle. Thus, this sensory input to the SEG would ensure that the mandibles move in synchrony.

Further insight into this feedback system is gained from experiments in which the two mandibles were driven independently and at different frequencies. The closer muscle firing bursts were of the same duration on the two sides and occurred synchronously. Interestingly, the repetition rate of the closer bursts was the same as the driving frequency of the faster driven mandible, but the length of the bursts was apparently determined by the position of the slower driven mandible. This is suggested (Seath, 1977a) as evidence that each mandible has its own position-detecting receptor and that

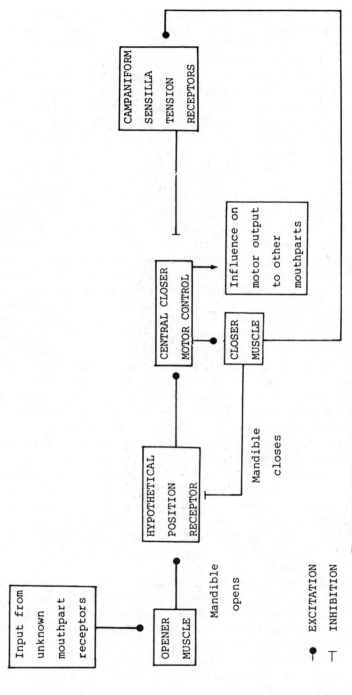

Figure 13.2. Diagram summarizing the control of mandibular movement in the locust. (Redrawn after Seath, 1977b.)

their outputs are summed to ensure that outputs to the closer muscles are synchronous.

Cine film analysis (Seath, 1977a; Blaney and Chapman, 1970) has shown that all the mouthparts move in synchrony with, but not necessarily in phase with, the mandibles. The mandibles are the largest of the mouthparts and thus synchrony is partly mechanical, but neural output from the SEG must also play a part because some of the movements, especially of the maxillae, are quite complex and are modulated by sensory feedback (Blaney, unpublished data). This coordination presumably allows the locust to eat foods of different textures efficiently. Because the mandibles are strongly and directly influenced by differing textures, chewing efficiency will be increased by dependence of the other mouthparts on movement of the mandibles.

In this system, there is strong evidence of a central motor score, presumably located in the SEG, continually influenced by sensory feedback. Thus, mandibular position is monitored, such that mandibular opening induces activity in the closer muscles. Because closer muscle activity is related to wavelength of imposed movements, it seems likely that the feedback system operates during the entire duration of mandibular opening. Presumed pressure receptors in the cusps ensure that the mandibles operate in synchrony. Tension receptors ensure that pressure from the two mandibles is equal and of appropriate strength for the substrate, while preventing excess tension with incompressible substrates. Within operating limits, the tension receptors would appear to exert an excitatory influence on the closer muscle motor center. This proposition would require that during a bite, as load on the mandibles increases, spiking frequency and power output would increase, that is, spiking frequency should increase during a closer burst. This is, in fact, shown to occur (Seath, 1977b). Factors initiating and terminating cycles have not been clearly elucidated but presumably involve input from other mouthparts (Blaney and Chapman, 1970; Blaney and Duckett, 1975). Within these limitations, Seath (1977a) has proposed a model to describe the overall control of mandibular movement in the locust (Fig. 13.2).

13.5. CONTROL OF MOUTHPART MOVEMENTS IN THE BLOWFLY

Investigation of the relationship between sensory input and motor output in blowfly mouthparts yields valuable information about the control of the mouthparts exercized by the SEG (Getting, 1971; Getting and Steinhardt, 1972; van der Starre, 1977; Pollack, 1977; van der Starre and Ruigrok, 1980). Food-deprived, water-satiated flies can be induced to extend the proboscis on stimulation of a single labellar sensillum with sucrose solution (Dethier, 1955). Each sensillum contains only four neurons and only one of these responds to carbohydrates, so that with appropriate choice of stimulating solution, it is possible to monitor the activity of single, identified sensory neurons. The motor response, extension of the proboscis, can be monitored by recording the motor output to two pairs of muscles, the extensors and adductors of the haustellum, the central part of the proboscis.

Adult blowflies, *Phormia regina* were starved for 62–72 hours, immobilized and secured with the proboscis held in a partly extended position (Getting, 1971). During a 1-hour postoperative period, flies were fed to satiation with distilled water. Sensory activity of individual labellar sensilla was recorded using the method of Hodgson et al. (1955). The stimulating capillary contained sucrose, in 50 mM LiCl to provide electric conductivity. This mixture stimulates only the neuron responding to sucrose and, at low sucrose concentrations, that responding to water. However, preliminary experiments showed that in water-satiated flies activity of the water neuron did not evoke proboscis extension, so its activity could safely be ignored. Muscle activity was recorded by a tungsten electrode inserted into the extensor muscle of the haustellum.

Stimulation of a single labellar sensillum with sucrose (100–400 mM) was adequate to induce motor output to the extensors of the haustellum. A single sugar spike was never adequate to initiate motor output nor did motor neurones fire when the interval between sensory spikes (ISI) exceeded 20 msec. The critical maximum ISI and number of spikes needed to trigger a motor response varied with the degree of starvation of the fly: the greater the starvation, the more readily the flies responded. Thus, flies starved 70–72 hours produced motor output with a constant latency after the second spike in the sensory train, provided the ISI between first and second spikes was less than approximately 20 msec. The threshold in flies starved 62–66 hours was higher: motor output was initiated by the first two sensory spikes only if the first ISI was less than 5 msec, but a third spike was needed if the interval was 5–10 msec. The motor response could outlast the sensory input causing it: a 50 msec stimulation with 200 mM sucrose initiated a motor response lasting 100 msec after the end of stimulation, well beyond the point at which sensory neurons would have ceased firing (Tateda and Morita, 1959). As frequency and duration of sensory input increased, the duration of motor output increased, but only up to a maximum, beyond which continuing sensory input was ineffective.

Repeated sensory stimulation reveals features of the neural organization in the SEG. Successive stimulations of a single sensillum at 1-min intervals produced a progressive decline in motor output, even though the sensory input was undiminished. At this point, stimulation of an adjacent sensillum produced a normal motor response, suggesting that habituation was occurring. Stimulation of the adjacent sensillum did not produce a dishabituating effect on the first sensillum. These facts suggest that habituation was occurring at a synapse before convergence of the input from different sensilla, that is, within the SEG because labellar sugar receptor axons do not synapse before the ganglion (Stürckow et al., 1967).

Stimulation of two sensilla simultaneously illustrates another aspect of sensory integration in this system, namely that the motor response is greater than that elicited by either sensillum alone. Thus, one sensillum firing at 30 spikes per second produced a motor output of 10 spikes, and another produced 12 motor spikes when firing at 43 spikes per second. When these two sensilla were stimulated together, such that their combined firing rate was 26 spikes per second, the motor output was 25 spikes. Not only is a double channel input more effective than the same input from one channel alone, but the summation is nonlinear: simultaneous stimulations of two sensilla produced motor outputs up to six times the sum of those produced by individual stimulations of the sensilla. The extent to which this enhanced motor output is due to a change in central excitatory state (CES) is uncertain. Getting (1971) had ruled this out as a possibility: prior stimulation with sucrose did not appear to cause a change in CES so far as proboscis extension is concerned. Fredman and Steinhardt (1972) came to a similar conclusion when salt was applied to one hair and sugar to another. They found no evidence for either stimulation or inhibition occurring centrally. However, Fredman (1975) in more extensive experiments demonstrated central inter-

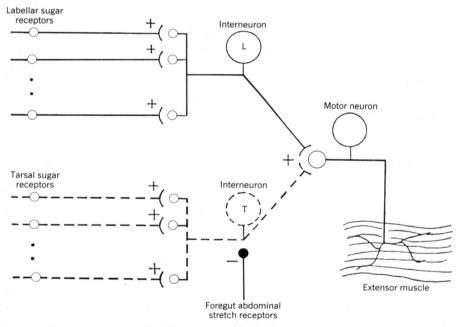

Figure 13.3. A neuronal model to account for motor activity in the extensor of the haustellum in response to sucrose stimulation of labellar and tarsal chemosensory hairs. The labellar sugar receptors converge on interneuron L permitting summation between receptor inputs. Activation of interneuron L causes proboscis extension via the motor neuron, thus interneuron L is the "decision-making" element of the network. A similar but independent pathway is shown for tarsal sugar receptor input. Tarsal threshold regulation is probably achieved by inhibitory input from the foregut and abdominal stretch receptors onto the "decision-making" interneuron T. Excitatory synapses are shown with + sign; inhibitory with − sign. (From Getting and Steinhardt, 1972.)

actions, both summation and inhibition, between input from sugar and water receptors and from salt receptors, and he postulated the involvement of a changing CES.

A feeding blowfly would normally contact the substratum with more than one labellar sensillum at a time and this has further implications for sensory integration. The summation effect allows lower concentrations of stimulant to be responded to. Further, habituation can occur as a result of a high rate of firing in a single sensillum, but the same overall input from two or more channels is less vulnerable as habituation occurs before convergence. Additionally, because two or more channels of input fire with independent spike frequency and pattern, excitatory postsynaptic potentials resulting from these spikes can occur with any intervals, due to convergence. Because some of these intervals will be very short, this increases the likelihood of summation adequate to trigger a spike in the postconvergence interneurone.

In a further, similar study Getting and Steinhardt (1972) investigated the role of this system in the termination of feeding. Distension of the gut due to feeding stimulates stretch receptors, the activity of which is thought to be inhibitory to sensory input from tarsal and labellar receptors (Gelperin, 1971). Getting and Steinhardt (1972) found no evidence, behavioral or electrophysiological, for alteration in labellar thresholds due to feeding, and it is assumed therefore that postfeeding inhibition acts only through the tarsal system. A minimal neuronal model for the labellar system would be as shown in Figure 13.3.

13.6. PROCESSING OF CHEMOSENSORY INFORMATION BY THE SEG

For decades the chemosensory systems of insects governing food selection have been studied, both behaviorally and electrophysiologically. So far as these studies shed light on the putative role of the SEG, the most complete information is available for lepidopterous larvae and locusts. Several sensory systems are involved, projecting into various parts of the central nervous system, but in all cases the most critical testing is carried out by sensilla whose input is processed in the SEG, and that ganglion provides motor output to those mouthparts immediately involved in the feeding process. A detailed analysis of input–output relationships continues to be hampered by a lack of direct recording from subesophageal interneurons. We have good information on the behavioral output and a wealth of evidence on the sensory input that allows informed speculation about the integrative mechanisms occurring in the SEG and sets guidelines for investigations of the interneurons. In all insects studied, the axons from sensory neurons project directly, without synapsing, into the SEG.

13.6.1. Locust

Gustation in locusts has been studied by Blaney (1974, 1975, 1980, 1981). Two species, with differing food selection mechanisms were chosen. *Schistocera gregaria*, the desert locust, is polyphagous, eating a wide range of grasses and many dicotyledons; *Locusta migratoria*, the migratory locust, is oligophagous, restricting its choice of food to the grasses. In locusts, which are feeding normally (Blaney et al., 1973), contact chemoreceptors on the maxillary palp tips play a key role in determining whether or not potential food material is acceptable (Blaney and Chapman, 1970).

There are approximately 350 of these sensilla on each palp tip, and each sensillum contains at least five chemosensory neurons: a large volume of information is transmitted to the SEG. The "message" contained in that information is not readily apparent. Neurons do not show clear specificities for stimulating compounds, such as sugars, salt, and water, but rather such specificity as does occur is more at the level of the sensilla than of the neurons (Blaney, 1974). Even so, most sensilla do not individually differentiate unambiguously between solutions that are distinguished behaviorally. Rather, the code for taste quality consists of the relative amounts of activity produced simultaneously in many different sensilla (Blaney, 1975). This type of coding occurs in many sensory systems and is known as across-fiber or ensemble coding. It allows a receptor population to respond differentially to a much wider range of compounds, and mixtures of compounds, than would be possible with a system in which individual receptors reported specifically on single compounds only, provided of course that second and higher order neuronal pathways exist which allow analysis of the data.

This mode of analysis also overcomes problems inherent in a "noisy" receptor system, that is one which is subject to a high degree of variability. In locusts the response of a given sensillum to a given stimulus can vary by as much as $\pm$ 10%. The effect of this was tested by stimulating about 20 sensilla on the palp of a locust with two solutions, which are discriminated behaviorally (e.g., 0.05 M NaCl, and 0.05 M NaCl with 0.025 M fructose) and giving each sensillum four separate tests with each solution (Blaney, 1975). The variance between responses to the two solutions was highly significant, indicating that, on the total information available, the insect could differentiate between solutions. The variance between the responses of individual sensilla to a given solution was highly significant, indicating that sensilla differ from each other in their responsiveness, one of the requirements for across-fiber analysis. Additionally, the variance due to interaction was highly significant, indicating that different sensilla respond in different ways to the two solutions, another requirement for across-fiber analysis. Thus, despite the variability, these sensilla generate a neural output that allows discrimination, provided a population of sensilla are in action at the same time.

Both species of locust rely on these unspecialized, or generalist neurons to assess nutritious, palatable substances, but their feeding range is determined partly by the occurrence of allelochemics in plants, which may or may not be assessed in the same way. Thus, *Schistocera* is inhibited from feeding by the presence of azadirachtin, which stimulates a specific "deterrent" neuron. Whenever this neuron is stimulated adequately, the message is unambiguously "do not feed," that is, this is a "labeled line" and its information is processed differently in the SEG.

The input of the palp receptors to the CNS has been modeled by stimulating a population of receptors with a salt and sugar solution as a control, then the same solution with an allelochemic added (Blaney, 1980). That addition can affect the firing rate elicited by the control in one of three ways: it produces no change, it produces increased firing, or it produces decreased firing. The magnitude of these changes and the numbers of sensilla in each of the categories was used to construct a model of the sensory input that correlates well with the observed behavioral response to the test compounds (Fig. 13.4).

It is also instructive to compare the output of sensilla in each of the three categories, either individually or in various combinations, with the behavioral response (Blaney and Winstanley, 1980). In *Locusta*, the best correlation with behavior is given by the summed output of "increase" firers and "no change" firers. Strangely, behavior correlates well with "no change" firers but not at all with "decrease" firers, which nevertheless account for nearly 50% of all sensilla. This would be explained if a minimum output level was required from each sensillum before it could contribute toward palatability assessment. This would eliminate the potentially disruptive influence of low-firing sensilla. Feeding is associated with high firing in *Locusta*; nutrients increase firing and deterrent allelochemics generally depress it. Feeding will not occur if the overall firing rate is too low. If a similar threshold level existed at a central neuronal level, "increase" firers alone (indicating palatability) might not get over this barrier, but if their input were summed with the generally quite high firing rates of the "no change" sensilla, this would lift the overall level above the barrier and allow it to be modulated by the "increase" firers. Future analysis of subesophageal neuron activity may of course require this proposal to be revised.

13.6.2. Lepidopterous Larvae

A similar approach has recently been taken with lepidopterous larvae (Blaney and Simmonds, 1983; Blaney et al., 1984; Simmonds and Blaney, 1984). *Spodoptera littoralis* is polyphagous, accepting a very wide range of host plants, whereas *Spodoptera exempta* is oligophagous and is restricted to a much narrower range of plants. An investigation of the role of antifeedant allelochemics has revealed that the same compound may elicit very different sensory responses in the two species, so that the SEG of each has to deal with very different types of information.

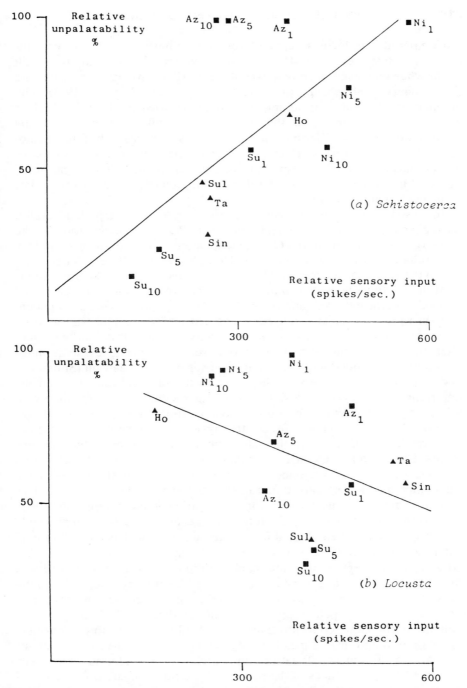

Figure 13.4. Correlation between relative unpalatability and relative sensory input. (From Blaney, 1980.)

Compared with locusts, lepidopterous larvae have a much smaller complement of taste receptors. It has been shown that discrimination can still occur when all chemoreceptors, except the maxillary styloconic sensilla and the epipharyngeal sensilla, are ablated (Hanson and Dethier, 1973). The epipharyngeal sensilla appear only to monitor swallowing (Ma, 1972; de Boer et al., 1977), and input from the two pairs of maxillary styloconic sensilla is adequate to allow discrimination. Each sensillum has only four chemosensory neurons, none of which is highly specific, but most have a class of compounds to which they respond best. Thus, there may be a "sugar best" neuron (or sugar neuron), also a salt neuron, an allelochemic neuron, and, possibly, a water neuron.

In these two *Spodoptera* species, sugars promote feeding, and, generally, allelochemics from nonhost plants inhibit it. Presumably therefore, in the convergence occurring in the CNS they must be associated with excitatory and inhibitory synapses, respectively. Here, however, differences occur between the species at a peripheral level, which must have central implications as well. In *Spodoptera exempta* the taste neuron specificity is fairly well defined, and the deterrent neuron in particular responds in a dose-dependent manner to a wide range of allelochemics that inhibit feeding. On the other hand, in *Spodoptera littoralis* the specificity is much less marked. Here allelochemics not only stimulate the deterrent neuron but, by some interaction not yet understood, they inhibit the sugar neuron as well (Simmonds and Blaney, 1984). Moreover, sugars exert a reciprocal inhibition on the deterrent neuron. The confusion (to the experimenter) is confounded with a high level of variability from one individual to the next so that good correlations between behavior and electrophysiology of receptor response are most readily obtained when the same individuals are used in both experiments. Clearly, although the same compounds produce the same behavioral responses in these two species with anatomically similar sensory systems, the compounds are perceived and interpreted in very different ways. This would seem to be an ideal system for investigation of integration in the SEG.

It would appear that in lepidopterous larvae, as in locusts, an appreciably better appraisal of the potential food material is obtained as a result of consideration by the CNS of several inputs simultaneously. Thus, Blom (1978), studying the perception of various concentrations of sugar by larvae of *Pieris brassicae*, correlated sensory input, recorded electrophysiologically (numbers of spikes in the first second) with behavioral response represented by dry weight of fecal pellets produced while feeding for 24 hours on diets containing the same concentrations of sugar. Each of the three sensillum types (lateral and medial maxillary styloconic and epipharyngeal) contains a sugar-sensitive neuron. Relative output (behavior) was plotted against relative input (as percentages of the maxima) for a range of sucrose concentrations, and it was only when the responses of all three neuron types were summed to represent a "total sensory input," that the relationship to relative output approximated to a straight line, with a correlation coefficient of 0.99.

Very similar results were obtained with larvae of *Mamestra brassicae* and a range of concentrations of inositol (Blom, 1978).

Dethier and Crnjar (1982) have investigated the sensory coding used by the lepidopteran larva, *Manduca sexta,* in discriminating between the plants tomato (*Lycopersicon esculentum*), tobacco (*Nicotiana tobaccum*), and Jerusalem cherry (*Solanum pseudocapsicum*). Freshly expressed leaf sap was used to stimulate the maxillary styloconic sensilla, and of the eight neurones available, only six fired with a high enough frequency to be analyzed. As is normally the case, a phasic response occupied the first 100 msec or so, and after some adaptation, a tonic response was produced for many seconds. The third second was taken as representative of the tonic phase and was used in analysis. The largest spike was produced by the "salt best" neuron, and only this neuron fired during the phasic period. The same neuron dominated the tonic phase but the other two neurons fired as well. The firing of the salt neuron in the phasic period could not be used as a means of discriminating between the plants and may simply be a means of alerting the CNS for information to follow. In the tonic phase the activity of no one neuron allowed discrimination between all three plants, again suggesting the operation of an across-fiber analysis.

Thus, the evidence strongly favors the operation of an across-fiber coding as the basis of the chemosensory information handled by the SEG, but this can certainly be overridden in some cases by the action of labeled lines signaling the presence of potent phagostimulants or deterrents. Much remains to be discovered on the precise nature of the coding and on the early stages of integration in the SEG. The topic is ripe for scientific exploration.

13.7. SUMMARY

The subesophageal ganglion (SEG) controls the principal activity of the mouthparts, namely feeding, in two distinct yet complementary ways. Movements made by the mouthparts in manipulating food rely on integration by the SEG of the sensory input describing the movements and the motor output causing them. Work on the activities of the mandibles of lobsters and locusts and on the proboscis of the blowfly is considered in detail. The speed and power of movements of the mandibles can be varied independently to achieve maximum efficiency in biting food material of varying size and texture, but the underlying rhythm of movement is fairly constant. The amount of sensory input from taste receptors needed to initiate proboscis extension depends on the state of hunger of the fly and on the number of receptors contributing to the input. We assess the evidence for feeding movements being driven by a central motor score and relate this to the modulating role of peripheral sensory feedback.

A different level of integration is involved in the analysis of sensory input reporting on the palatability of potential food material and in determining the appropriate feeding response. Rejection of food may sometimes result from the activity of receptors responsive to specific deterrents but generally the code for palatability consists of the relative activity of a range of receptors responding differentially to phagostimulants and deterrents concurrently. Less specificity, and therefore greater flexibility, is associated with species having a wider range of potential food material. Evidence from locusts and lepidopterous larvae is used to assess the integrative processes occurring in the SEG and to predict the properties and mode of operation of second-order neurons in the SEG.

REFERENCES

Anderson, D. J., A. G. Hannam, and B. Matthews. 1970. Sensory mechanisms in mammalian teeth and their supporting structures. *Physiol. Rev.* **50:** 171–195.

Blaney, W. M. 1974. Electrophysiological responses of the terminal sensilla on the maxillary palps of *Locusta migratoria* (L.) to some electrolytes and non-electrolytes. *J. Exp. Biol.* **60:** 275–293.

Blaney, W. M. 1975. Behavioural and electrophysiological studies of taste discrimination by the maxillary palps of larvae of *Locusta migratoria* (L.). *J. Exp. Biol.* **62:** 555–569.

Blaney, W. M. 1980. Chemoreception and food selection by locusts, pp. 127–130. In H. van der Starre (ed.), *Olfaction and Taste VII*. IRL, London.

Blaney, W. M. 1981. Chemoreception and food selection in locusts. *Trends Neurosci.* **Feb:** 35–38.

Blaney, W. M. and R. F. Chapman. 1970. The functions of the maxillary palps of Acrididae (Orthoptera). *Entomol. Exp. Appl.* **13:** 363–376.

Blaney, W. M. and A. M. Duckett. 1975. The significance of palpation by the maxillary palps of *Locusta migratoria* (L.): An electrophysiological and behavioural study. *J. Exp. Biol.* **63:** 701–712.

Blaney, W. M. and M. S. J. Simmonds. 1983. Electrophysiological activity in insects in response to antifeedants, pp. 1–219. Tropical Development and Research Institute, London.

Blaney, W. M. and C. Winstanley. 1980. Chemosensory mechanisms in locusts in relation to feeding: The role of some secondary plant compounds, pp. 383–389. In *Insect Neurobiology and Pesticide Action* (*Neurotox 79*). Chemical Industry, London.

Blaney, W. M., R. F. Chapman, and A. Wilson. 1973. The pattern of feeding of *Locusta migratoria* (L.) (Orthoptera, Acrididae). *Acrida* **2:** 119–137.

Blaney, W. M., M. S. J. Simmonds, S. V. Evans, and L. E. Fellows. 1984. The role of the secondary plant compound 2,5-dihydroxymethyl 3,4-dihydroxylpyrrolidine as a feeding inhibitor for insects. *Entomol. Exp. Appl.* **36:** 209–216.

Blom, F. 1978. Sensory input behavioural output relationships in the feeding activity of some lepidopterous larvae. *Entomol. Exp. Appl.* **24:** 58–63.

de Boer, G., V. G. Dethier, and L. M. Schoonhoven. 1977. Chemoreceptors in the preoral cavity of the tobacco hornworm, *Manduca sexta,* and their possible function in feeding behaviour. *Entomol. Exp. Appl.* **21:** 287–298.

Dethier, V. G. 1955. The physiology and histology of the contact chemoreceptors of the blowfly. *Q. Rev. Biol.* **30:** 348–371.

Dethier, V. G. 1959. The nerves and muscles of the proboscis of the blowfly *Phormia regina* Meigen in relation to feeding responses. *Smithson. Inst. Misc. Collect.* **137:** 157–174.

Dethier, V. G. and R. M. Crnjar. 1982. Candidate codes in the gustatory system of caterpillars. *J. Gen. Physiol.* **79:** 549–569.

Fredman, S. M. 1975. Peripheral and central interactions between sugar, water and salt receptors of the blowfly, *Phormia regina. J. Insect Physiol.* **21:** 265–280.

Fredman, S. M. and R. A. Steinhardt. 1973. Mechanism of inhibitory action by salts in the feeding behaviour of the blowfly, *Phormia regina. J. Insect Physiol.* **19:** 781–790.

Gelperin, A. 1971. Abdominal sensory neurones providing negative feedback to the feeding behaviour of the blowfly. *Z. Vgl. Physiol.* **72:** 17–31.

Getting, P. A. 1971. The sensory control of motor output in fly proboscis extension. *Z. Vgl. Physiol.* **74:** 103–120.

Getting, P. A. and R. A. Steinhardt. 1972. The interaction of external and internal receptors on the feeding behaviour of the blowfly *Phormia regina. J. Insect Physiol.* **18:** 1673–1681.

Hanson, F. E. and V. G. Dethier. 1973. Role of gustation and olfaction in food-plant discrimination in the tobacco hornworm *Manduca sexta. J. Insect Physiol.* **19:** 1019–1034.

Hodgson, E. S., J. Y. Lettvin, and K. D. Roeder. 1955. Physiology of a primary chemoreceptor unit. *Science (Washington, D. C.)* **122:** 417–418.

Le Berre, J. R. and A. Louveaux. 1969. Equipment sensoriel des mandibules de la larve du premier stade de *Locusta migratoria* (L.). *C. R. Acad. Sci. Paris* **268D:** 2907–2910.

Ma, W. C. 1972. Dynamics of feeding responses in *Pieris brassicae* Linn. as a function of chemosensory input: A behavioural, ultrastructural and electrophysiological study. *Med. Landbouhogeschool Wag.* **72**(11): 1–162.

Macmillan, D. L., W. Wales, and M. S. Laverack. 1976. Mandibular movements and their control in *Homarus gammarus.* III. Effects of load changes. *J. Comp. Physiol.* **106:** 207–221.

Pollack, G. S. 1977. Labellar lobe spreading in the blowfly: Regulation by taste and satiety. *J. Comp. Physiol.* **121:** 115–134.

Pringle, J. W. S. 1938. Proprioception in insects. II. The action of campaniform sensilla on the legs. *J. Exp Biol.* **15:** 114–131.

Seath, I. 1977a. Sensory feedback in the control of mouthpart movements in the desert locust *Schistocerca gregaria. Physiol. Entomol.* **2:** 147–156.

Seath, I. 1977b. The effects of increasing mandibular load on electrical activity in the mandibular closer muscles during feeding in the desert locust, *Schistocerca gregaria. Physiol. Entomol.* **2:** 237–240.

Simmonds, M. S. J. and W. M. Blaney. 1984. Some neurophysiological effects of azadirachtin on lepidopterous larvae and their feeding response. pp. 163–180. *In* H. Schmutterer and K. R. S. Ascher (eds.), *Proc. 2nd Int. Neem Conf. Ravischholzhausen 1983*. Gesellschaft für Technische Zusammenarbeit, Eschbom.

Snodgrass, R. E. 1943. The feeding apparatus of biting and disease-carrying flies. *Smithson. Inst. Misc. Collect.* **1**: 1–51.

Snodgrass, R. E. 1952. *A Textbook of Arthropod Anatomy*. Cornell University Press, Ithaca, New York.

Snodgrass, R. E. 1953. The metamorphosis of a fly's head. *Smithson. Inst. Misc. Collect.* **122**: 1–25.

van der Starre, H. 1977. On the mechanism of proboscis extension in the blowfly *Calliphora vicina. Neth. J. Zool.* **27**(3): 292–298.

van der Starre, H. and T. Ruigrok. 1980. Proboscis extension and retraction in the blowfly, *Calliphora vicina. Physiol. Entomol.* **5**: 87–92.

Stürckow, B., J. R. Adams, and T. A. Wilcox. 1967. The neurons in the labellar nerve of the blowfly. *Z. Vgl. Physiol.* **54**: 268–289.

Tateda, H. and H. Morita. 1959. Initiation of spike potentials in contact chemosensory hairs of insects. I. The generation site of the recorded spike potentials. *J. Cell Physiol.* **54**: 171–176.

Thomas, J. G. 1966. The sense organs on the mouthparts of the desert locust *Schistocerca gregaria* (Forsk.). *J. Zool.* **148**: 420–448.

Wales, W. 1982. Control of mouthparts and gut, pp. 166–191. In D. C. Sandeman and H. L. Atwood (eds.), *The Biology of Crustacea,* vol. 2. Academic Press, New York, London.

Wales, W. and M. S. Laverack. 1972a. The mandibular muscle receptor organ of *Homarus gammarus* (L.) (Crustacea, Decapoda). *Z. Morphol. Tiere* **73**: 145–162.

Wales, W. and M. S. Laverack. 1972b. Sensory activity and the mandibular muscle receptor organ of *Homarus gammarus* (L.). I. Response to receptor muscle stretch. *Mar. Behav. Physiol.* **1**: 239–255.

Wales, W., D. L. Macmillan, and M. S. Laverack. 1976a. Mandibular movements and their control in *Homarus gammarus*. I. mandible morphology. *J. Comp. Physiol.* **106**: 177–191.

Wales, W., D. L. Macmillan, and M. S. Laverack. 1976b. Mandibular movements and their control in *Homarus gammarus*. II. The normal cycle. *J. Comp. Physiol.* **106**: 193–206.

CHAPTER 14

Ultrastructure of the Arthropod Neuroglia and Neuropil

Stanley D. Carlson

Department of Entomology
Neurosciences Training Program
University of Wisconsin
Madison, Wisconsin

14.1. INTRODUCTION

Twenty years ago, Bullock and Horridge (1965) defined the word neuropil (also neuropile) as ". . . a tangle of fine fibers (of dendrites and axon arborizations) and their endings, quite or nearly devoid of nerve cell bodies . . . and forming much of the bulk of invertebrate ganglia." This definition

323

remains useful, except it fails to mention the glial and tracheal residents, and "tangle" is arguably inappropriate. Disorder or randomness of neurons in neuropil may be more apparent than real. When neuropil is finely (and sometimes serially) sectioned, only then does one derive a fundamental structural order between neurons and glia. At that resolution, the seeming structural chaos that "tangle" implies is not obvious, and one sees ultra-structurally neurons making particular structural alignments that ensure pre-cise functional associations. Developing neurons and glia (in neuropil) follow genetically preset and precise paths that ultimately produce a spatially con-servative and correctly connected network of neuronal processes. We can now conceive of neuropil in three dimensions as a series of matrices of hierarchically ordered neurons, wrapped, spatially isolated, and buffered by glia.*

After order, are there other characteristics of neuropil in arthropod ganglia that permit us to gauge the functional morphology and integrative effec-tiveness of this synaptic feltwork? The ganglia of most arthropods are vastly inferior in mass and volume to those of many higher invertebrates, and certainly to those of mammals. Were it possible, arthropods could take com-fort in the fact that their brains are proportional to their bodies, and such brains rank with the best insofar as neuronal complexity, intricacy of neu-ronal connections, and concentration of neurons per unit volume and mass. For example, the honeybee has a brain:body weight ratio of 1:98 as com-pared with that of the frog (Rana) with 1:92. In another analogy, neuronal density in the granule cell layer of the mammalian cerebellum is listed as $3-7 \times 10^6$ per mm^3, whereas the density of Kenyon interneurons in the corpora pedunculata of the dipteran fly is 1.6×10^7 (Strausfeld, 1976). The considerable density and complexity of neurons in arthropod neuropil un-derlie the sizeable sensory and behavioral capabilities of these animals with jointed legs. Based on many of the species studied, arthropod neuropil can scarcely be regarded as a simple system for neurobiology study. Neural networks of molluscs and annelids are current champions in this regard.

Within the past several decades, information has been rapidly accruing on the *functional* neuroanatomy of arthropod neuropil. We now have first impressions of neural circuitry in a number of arthropod ganglia, based on charting identified neurons and their synapses. Some of this advance began with the Stretton and Kravitz (1968) description of iontophoresing procion yellow into crustacean neurons whose electrophysiology was known. Those neurons could then be identified as to both structure and function. Working arthropod neurons can also be pinpointed by the radioactive deoxyglucose technique (Buchner and Buchner, 1983). Numerous other techniques for determining structure–function relationships in arthropod neurons are well

---

* For a general definition, the facile phrase "hierarchically ordered" is used, but a truer (but much more lengthy) explanation could involve concepts, such as local circuits and parallel processing (see also Section 14.7).

detailed by Strausfeld (1983). If humans and microtomes can finely divide arthropod neuropil, humans and computers can graphically reconstruct the same neuropil in three dimensions (e.g., Speck and Strausfeld, 1983). This recent flowering in functional neuroanatomy, of course, rests on the near century-old pioneering efforts of Golgi and Cajal, who first impregnated single neuronal and glial cells in neuropil, and went on to catalogue these cells on structural bases. Advances in ultrastructure of arthropod neuropil have come about from improvements in resolution in electron microscopy, cellular fixation, and new preservation techniques that go beyond (and often eliminate) chemicals using cryogenic methods (e.g., Heuser et al., 1979).

The glia of arthropod neuropil are also being better characterized and understood, but unfortunately, not at the pace nor with the intensity that has been devoted to neurons. However, concepts of neuroglia are evolving and these often complex cells are being found to be equally interesting in their intercellular associations and functions (Roots, 1978; Radojcic and Pentreath, 1979; Lane and Treherne, 1980; Treherne, 1981; Saint Marie et al., 1984).

This brief review on such a vast topic naturally has limitations. My interest and experience is in insect neuropil and this condition is a bias, strength, and weakness. Neuropil of adult cephalic ganglia will be a principal focus, with the trailing segmental ganglia, literally and figuratively, taking a subordinate position in this narrative. Structural commonalities (e.g., glial-neuronal contacts, synapses, membrane specializations, etc.) across the spectrum of arthropod neuropil will be stressed rather than emphasize or itemize differences in histology/ultrastructure among different taxa. Where appropriate, a morsel of physiology will be given with the structural data in order to flesh out a functional morphology concept. A final caveat is that the more recent findings are discussed and displayed. As a palliative for these necessary shortcomings, the reader is referred to seminal and comprehensive sources, such as: Snodgrass, 1935; Bullock and Horridge, 1965; Huber, 1974; Treherne, 1974, 1981; Strausfeld, 1976; Hoyle, 1977; Manton, 1977; Lane, 1981, 1985; Shaw, 1981; and Weevers, 1985.

14.2. NEUROPIL: BASIC PLANS

A general blueprint for arthropod neuropil exists. The *leitmotif* is that of a connective tissue sheath overlying a perineurium, which, in turn, surrounds the neuronal perikaryal rind, which is peripheral to the synaptic plexus (Figs. 14.1 and 14.2). Neuronal cell bodies appear to be little involved in direct neural integration (but read Saint Marie and Carlson, 1985); with few exceptions chemical synapses do not occur here. Nevertheless, glial cell processes and neuronal perikaya may be joined by gap junctions, which are probably low-resistance pathways for ions and molecules (Cuadras et al., 1985; Saint Marie and Carlson, 1985). The majority of synapses in arthropod

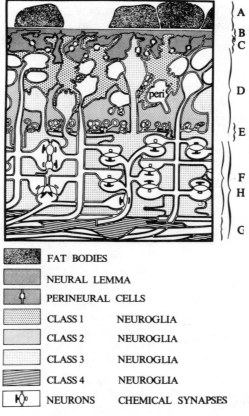

	FAT BODIES
	NEURAL LEMMA
	PERINEURAL CELLS
	CLASS 1 NEUROGLIA
	CLASS 2 NEUROGLIA
	CLASS 3 NEUROGLIA
	CLASS 4 NEUROGLIA
	NEURONS CHEMICAL SYNAPSES

Figure 14.1. Spatial distribution and stratification of neurons and glia in an arthropod (insect) neuropil (after Strausfeld, 1976). From distal to proximal are fat body cells (A) which lie over the neural lemma (B). Flattened perineurial cells (C) have lateral (and to some extent basal) borders and are sealed by tight junctions. Class 1 neuroglia (D) wrap neuronal perikarya; they form trophospongia and occasional gap junctions with neuron cell bodies. Class 2 glia (E) are between the nuclear layer and synaptic region. Class 2 (neuropilar) glia reside in the synaptic region and invaginate into neurons as gnarls or capitate projections. A great degree of plasma membrane folding may be seen and development of extensive networks of fine processes. In neuropil (H), the glial covering of neurons is interrupted at synaptic foci. Class 3 glia are interconnected, and these communicating junctions are also present between class 3 and the glia above and below. Class 4 (sheath) glia wrap neurons (G) exiting and entering neuropil. These glia may be tightly bound to each other via an extensive series of focal tight junctions.

326

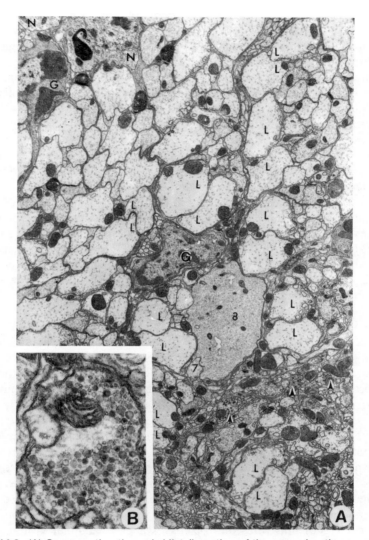

Figure 14.2. (A) Cross section through (distal) portion of the second optic neuropil (medulla externa) of *Musca domestica* (housefly). Most of the cellular elements of neuropil are present in this survey field. Rind glial cells (G) have a fusiform nucleus and an electron-dense cytoplasm. Long, thin glial processes insinuate through neuropil. The most recognizable of the axons are the nine pairs (in this field) of large diameter, lucent axons (L1/L2 [L] lamina monopolar cells). Each pair forms the core of a medullary column. At center right are moderately electron-dense photoreceptor axons (R[7], R[8]) which transverse the first optic neuropil without synapsing. These two color receptors (central cells of the ommatidium) soon separate and each terminates in a different stratum of medulla externa onto different follower interneurons, thus preserving certain spectral sensitivity and perhaps polarization information. R8 (a blue receptor) develops a transient swelling in its axon at this level. Interneuron cell bodies (N) are in the upper left of this field. Cell bodies of about a half dozen different interneurons serving the medulla externa are found in the perikaryal rind of this neuropil. Arrowheads point to three small diameter axons containing neurosecretory droplets. Similar appearing but fewer such neurosecretory fi-

neuropil is axoaxonal with vastly fewer axodendritic and dendrodendritic ones known (Lane, 1974).

Neurons in neuropil are a heterogeneous lot. These include the terminals of sensory afferents (from cuticular sense cells and/or sensory interneurons), local internuncials, other interneurons whose cell bodies reside in another ganglion, and motoneurons. Studies by Goodman and co-workers (1984) on embryonic neurons in *locusta* and *Drosophila* are beginning to elegantly demonstrate lineage, developmental perigrinations, ultimate destinations, cell interactions, recognition of landmarks, and the fate of identified neurons in particular neuropils (see also Chapter 3 by Keshishian in this volume). Current depictions of neuropil need no longer show only masses of anonymous neurons as some neurons now have identities along with pedigrees, destinies, and known functions. Lane and Treherne (1980) provide instructive graphics on neuropil organization of *Limulus*, crayfish, insects, and arachnids. Those sketches should be consulted. The following synopses, with Figures 14.1, 14.2, and 14.4, will acquaint the general reader with the basics of these plans.

In insects, the major ganglia are covered with a neural lamella (about 1 μm thick), consisting of elongate collagenous fibrils embedded in a mucopolysaccharide matrix. Underlying this ion-permeable, acellular layer is a monolayer of modified glia (perineurial cells). These can be rather electron-opaque, thin cells which interdigitate laterally. Tight junctions on those surfaces prevent paracellular passage of small molecules and water-soluble ions to and from the inner neuropilar glia. (Perineurial cells never directly appose neurons, but are neighbors to the neuropilar glia that surround neurons in the synaptic plexus.) Beneath the perineurium and within the first layer of neuropilar glia are the neuron cell bodies, which are often closely packed and form a rind around the core synaptic plexus. Neuropilar glia exhibit a splendid variety of ultrastructural types, and in the larger neuropils, glia are stratified (Fig. 14.1; Saint Marie and Carlson, 1983a). There are no channels within neuropil that directly communicate with hemolymph, although Wigglesworth (1960) illustrates and discusses a glial lacunnal system, within which metabolites are probably transported throughout neuropil.

As to aquatic Chelicerala (Arachnida), Fahrenbach (1977) describes the cytoarchitecture of the corpora pedunculata of *Limulus*. (see also Chapter 4 by Fahrenbach and Chamberlain in this volume). A relatively robust muscular sheath covers the central nervous system, and beneath this covering is a space through which hemolymph circulates bathing cell surfaces of the vascular neuroglia. Several other glial cell types are also described by Fah-

---

bers have been found in the next more distal neuropil, the lamina ganglionaris. It is not known whether these are axons of passage to the lamina or end in the medulla externa (8250×). (B) Magnified view of a cross-sectioned neurosecretory axon of this neuropil. Four or five types of granules/vesicules are noted (66, 390×).

renbach (1976). Satellite glia ensheath and form trophospongia in Kenyon cell bodies and other interneurons of this neuropil. It is interesting that velate astrocytes may not completely surround Kenyon cell bodies, ". . . hence many Kenyon cells are in direct contact with one another . . ." (Fahrenbach, 1976). Fahrenbach (1977) quantifies the corpora pedunculata of *Limulus* as 54% rind and 46% synaptic plexus. In crayfish neuropil, an extensive blood sinus ramifies throughout the perineurium (far more so than in *Limulus*). Extracellular channels pervade the neuropilar glia, whereas clusters of "transglial channels" and tubular lattice systems ". . . are continuous from one side of the glial cell membrane to the other . . . and are patent to tracers" (Lane and Treherne, 1980).

Fahrenbach (1976) and Lane and Treherne (1980) depict arachnids as having perineurial cells whose outer surfaces form peduncular processes that apparently anchor the neural lamella. Sizeable labyrinthine blood sinuses meander throughout the neuropil. Apart from those features spider neuropil adheres to the general schema.

14.3. NEURONAL STRUCTURE

When seen in toto (as in Golgi impregnations), arthropod neurons exhibit a wide variety of forms, although nearly all are unipolar; that is, a single neurite emanates from the perikaryon to enter and synapse in one or more neuropils. Cell bodies in the neuropil rind are often far distant from their respective dendritic arbors and axonal terminations. In addition, the perikaryal neck may be very long, narrow, and tortuous. Strausfeld (1976) asserts ". . . there is no simple structural paradigm of an interneuron or internuncial," but then draws some basic plans which are instructional. These schemata involve five "categories" within which 15 types are defined. Anyone interested in the diverse geometrics of arthropod neurons should also consult Bullock and Horridge (1965, vol. 1, Chapters 1 and 2).

The ultrastructure of neurons is another sweeping topic and only a few generalities can be given here. Each spherical-to-ovoid cell body has a sizable nucleus, so perinuclear cytoplasm is reduced. Large patches of heterochromatin are often present. Randomly oriented pores traverse the perinuclear envelope, whose outer membrane is often in continuous with the rough endoplasmic reticulum. Golgi stacks are usually few, small, appressed, and found near the proximal pole of the nucleus adjacent to the neurite. Mitochondria are generally abundant and evenly apportioned. Microtubules are first sighted in the perikaryonal neck, which may be extremely narrow (as slender as 0.2 μm for some monopolar cells in *Musca*; Saint Marie and Carlson, 1985) and can be scores of micrometers long. The restricted diameter of the neurite neck may have a bearing on reducing the likelihood of antidromically propagated signals from reaching the cell body. Neither cell body nor perikaryonal necks are considered as synaptic foci, but it is in-

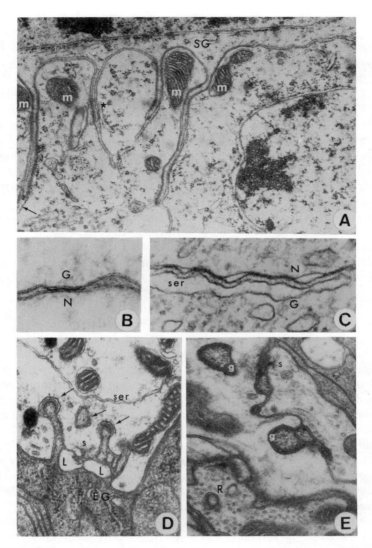

Figure 14.3. (A) Trophospongia. Four processes from a satellite glial cell (SG) invaginate into a type 1 (L1 or L2) monopolar cell body (N). Note mitochondria (m) at the border cleft area in neuron cell body; clefts are deep (*arrow*). Rough endoplasmic reticulum (*) is often situated close to the trophospongial process (19,500×). (B) Apparent gap-like junction between a neuronal perikaryon (N) and a glial (G) process in the crustacean decapod, in an abdominal ganglion of *Procambarus* (135,000×). (C) *Procambarus*; three gap-like junctions are seen between a neuronal perikaryon (N) and a glial cell (G). (D) The expanded, smooth-surfaced endoplasmic reticulum (ser) is in proximity to these junctions and a relationship between junction and this cistern is suggested (74,700×). (B and C reprinted from Dr. Jordi Cuadras.) (D) Several capitate projections (*arrows*) flank a T-bar chemical synapse(s) of a photoreceptor cell axon onto dyad processes from two monopolar (L) interneurons, first optic neuropil, housefly. Such processes from an epithelial glial cell (EG) randomly insert into the photoreceptor axon terminal, not always in proximity to

teresting that gap junctions (anatomical correlates of electrical synapses) have been described between the cell bodies of various monopolar relay interneurons and their adjacent perikaryonal necks in an optic neuropil of *Musca* (see Section 14.5.5; Saint Marie and Carlson, 1985). Dendrites and axons do not possess a wealth of organelles; mitochondria, microtubules, microfilaments, and some smooth-surfaced cisternae are the main features. Pre- and postsynaptic structures and synaptic vesicles are also present, and these are discussed in Section 14.7, as are various membrane specializations associated with contiguous neurons and glia (Section 14.5).

14.4. GLIAL STRUCTURE

Of all the cells in any arthropod, it is very likely that glial cells are absolute leaders in morphological diversity. One reason for this may be their insulating–isolating capacity in which glial processes fill in the highly irregular extraneuronal volumes and invaginate into and wrap neurons. Glia range in form from simple stellate cells to ones whose three-dimensional structure is only slightly less complex than brain coral. Fahrenbach (1976) speculates that glial cell diversity is proportional to the given species' evolutionary history. "The (*Limulus*) disposition toward anatomical stability is exemplified by the limited diversity of cell types." Indeed, the morphological variability in *Limulus* neuropil between stellate astrocytes, velate astrocytes, and vascular neuroglia is not nearly as remarkable as that found in the dipteran fly, which is more highly evolved and genetically more plastic. Glial cells of fly neuropil exhibit flamboyant morphological expressions ranging from the simple pillow-shaped marginal glial cells (fly optic neuropil) to such cytological extravaganzas as radiate glia (insect neuropil) whose myriad of elongate fine processes are each adorned with microlamellate extensions.

Ultrastructural studies of glial cells show that diversity of cell form continues internally. Commonly, adjacent heterologous glial cells show distinct fine structural differences. Part of the fine structural variability in arthropod glia undoubtedly relates to the putative task(s) each glial cell type subserves, for example, structural support, blood–brain barrier, metabolic assistance, phagocytosis, and neuron development. One striking example of this structure–function correlation is the cytological character of pseudocartridge glia versus fenestrated glia in fly optic neuropil. These two types oppose each

---

synaptic sites (64,620×). (E) Gnarl and "gliapse." Gnarls (g) are peculiar serial, bulbous invaginations into the β-fibers of the T-cell by an epithelial glial cell. Alpha-processes of an amacrine cell from lamina and β-fiber (cell body in medulla externa) pair, entwine, and project upward along the sides of the optic cartridge between descending retinular (R) axons. Some gnarls are also found interposed between α- and β-cells where T-bar synapses(s) are located. Thus, it appears that the postsynaptic cell is glial, and thus a "gliapse" is designated (31,100×).

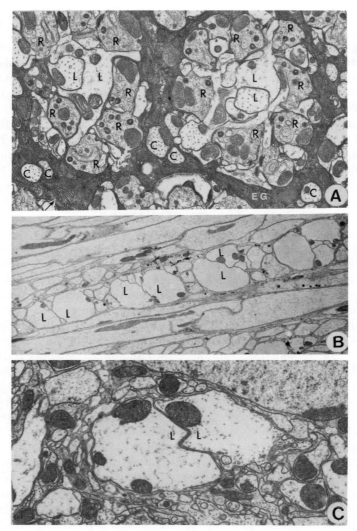

Figure 14.4. (A) Iteratively organized neuropil; lamina ganglionaris (first optic neuropil) of *Drosophila*. Two cross-sectioned optic cartridges. Each cartridge is a columnar (and synaptic) aggregation of some 20 neurons. Several electron-lucent monopolar (L) axons are surrounded by six moderately dense axons from the peripheral (R) retinular cells, which make extensive chemical synaptic contact with the monopolar relay neurons. Pairs of central (C) retinular axons pass outside each cartridge en route to the second optic neuropil without synapsing in the lamina. Each cartridge is surrounded by several epithelial glial cells (EG), which are very electron dense and whose plasma membrane folds to form parallel, stacked lamellae (*arrow*). At high magnification and resolution, lamellae appear as extensive gap junctions (reflexive gap junctions; 10,308 ×). (B) External chiasm. Axons of passage between first to second optic neuropil. Sections through this chiasm show alternately longitudinally and cross-sectioned fibers (some monopolar axons, L) because the axons are projecting at right angles to each other (10,500 ×). Glial processes are attenuated. Naked axons abound. Resident glia of chiasm are different from the neigh-

other: fenestrated glia is in the distal stratum. Photoreceptor axons from each ommatidium begin to twist in the layer of pseudocartridge glia and diverge into different optic cartridges of their neuropil. Pseudocartridge glia are filled with microtubules, all oriented parallel to the long axis of the axons they surround. Possibly such a cytoskeletal array facilitates and enforces axonal twisting. Where there is no twist, microtubules in those glial cells (e.g., fenestrated glia) are neither numerous nor oriented (Saint Marie, 1981; Saint Marie and Carlson, 1983b; Saint Marie, et al., 1984). Epithelial glia (Fig. 14.4) of the synaptic neuropil are of a different ultrastructural character than that of the glia immediately above and below—which types do not contact synaptic surfaces. Epithelial glia are replete with numerous stacked arrays of infolded membrane, which structures (reflexive gap junctions) may be a means for lateral electrotonic interaction between adjacent cartridges (Saint Marie, 1981). Membrane specializations of glia and the other structural associations they make with adjoining neurons are discussed in Sections 14.5 and 14.6.

As for fine structure of the cytoplasm, the glial nucleus is usually less spherical than that of associated neurons and glial chromatin is more uniformly distributed. In other contrasts to neurons, glial cells have few mitochondria but possess metabolites (lipid and glycogen) greatly exceeding that held by neurons. Such different inventories suggest a functional tradeoff between energy and nutrients shuttling between neurons and glia. Microtubules are conspicuous in some glia whose support and transport functions are easily guessed, and these organelles often extend into the finest glial processes. Ribosomes are ubiquitous in glia, either free, attached to the nuclear envelope, in clusters (polyribosomes), or as part of the widespread rough endoplasmic reticulum. Other organelles of neuropilar glia are various lysosomes, coated transfer vesicles, and smooth endoplasmic reticulum.

14.5. AXOGLIAL CONTACTS

There are at least five structural intimacies that glia make with neighboring neurons in arthropod neuropil: *viz.,* trophospongia, capitate projections, gnarls, gliapses, and gap junctions. Very likely these curious and well-formed structures do more than permit cellular adhesion. The curiosity of these

---

boring marginal glia covering proximal surface of first optic neuropil. (C) Cross section of the second optic neuropil showing prominent profiles of the L1/L2 fibers. Moderately dense fibers are probably those of the central retinular cells. The latter two cells separate and each terminates in a different stratum of neuropil. Although the two type 1 lamina (L) monopolar axons remain in tandem at this level, they eventually separate to synapse in different strata of neuropil. Interestingly, L1 and L2 are joined along their length with gap junctions, receive similar input, fire in synchrony, but go to different output areas and are thus presynaptic to different sets of relay interneurons (15,300×).

structures resides in their particular locations, their unusual geometries, their relatively high densities, and the fact that very little is presently known about their function.

14.5.1. Trophospongia

This system of channels into neuronal perikarya and axons was recognized in light microscopic sections as early as 1902 by Holmgren. Slender glial processes enter these clefts (Fig. 14.3A), and this propinquity is reputed to facilitate metabolite/ion transfer between neuron and glial cell. Bullock (1961) presents an early electron micrograph (prepared by Elizabeth Batham) of trophospongia from the mollusc, *Aplysia*. Bullock believes these channels achieve a near doubling in membrane surface area. Trophospongia of *Aplysia* were found in larger fibers surrounded by numerous glial lamellae. Smaller fibers with few (or one) glial winding(s) had little or no trophospongia, and such axons conducted small, fast spikes. Larger, slower spikes were characteristic of larger diameter, trophospongia-riddled axons. No axonal clefts are known in insects, but trophospongia have residence requirements in hexapods. For example, perikarya of type 1 lamina monopolar interneurons have trophospongia, whereas those of neighboring type 2 (L3, 4, 5) neurons do not (cf., Boschek, 1971; Saint Marie and Carlson, 1983a, 1985). In fly neuropil, trophospongia invade deeply and terminate near cisternae of rough endoplasmic reticulum. In Figure 14.3A, mitochondria are prominent within the sharply incised perikaryal clefts and often are found on both sides of such a channel (Saint Marie, 1981). One might presume mitochondria are supplying energy for active transport of metabolites between glia and neuron. Glial cells are often packed with lipid and glycogen, metabolites that are conspicuously absent from neurons.

14.5.2. Gnarls

Gnarls are generally spherical, glial processes (Fig. 14.3E) that often serially bud from each other. Boschek (1971) illustrates these structures in thin sections but fails to discuss or name them, a task completed by Campos-Ortega and Strausfeld (1973). Strausfeld (1976) graphically states gnarls ". . . are serially constructed like a fun fair balloon. . . ." These neuronal intrusions emanate from epithelial glia and enter β-fibers of the T1 cell in the first optic neuropil of *Musca*. At a number of these sites in *Musca* (not *Drosophila*), a T-bar synapse in an α-fiber lies opposite the "postsynaptic" membrane of the epithelial glial cell, and it is at this site that the bulbous gnarl forms. Thus, there is a "gliapse" (see Section 14.5.3) present and a continuing mystery as to: (1) the basic nature of the gnarl; (2) the functional implications of a gnarl opposite a presynaptic structure (T-bar); and (3) why the gnarl's appearance is curtailed to that of the β-fiber in only one neuropil. Carlson et al. (1983) showed that in freeze-cleaved gnarls intramembranal particles were randomly oriented and present on both P and E faces (cleaved glial and neuronal plasma membrane leaflets). That finding leaves the suggestion of delimited, focused metabolic activity.

14.5.3. Gliapse

Galambos (1961), in a short visionary paper, coined the term "gliapse," which refers to the functional apposition of a neuron and a glial cell (e.g., Fig. 14.3E). This definition might especially apply to a chemical presynaptic structure in the neuron lying opposite a glial cell. The gliapse gives the impression of an excitable cell synapsing onto a nonexcitable one. A host of questions spring to mind. Are quanta released here only to be "wasted" by the glial presence? In a neuron–glial association, might neurotransmitter be taken up by receptors in the adaxonal glial cell thereby altering the latter's membrane potential? (And to what specific integrative purpose?) In the neuromuscular preparation of the lobster the adjacent glia take up neurotransmitter (GABA; Orkand and Kravitz, 1971). Can neurotransmitter or a metabolite diffuse to postsynaptic receptors of nearby nerve cells? What kind of synaptic delay would the latter act entail?

A consistently occurring gliapse in insect (fly) optic neuropil is found between amacrine neurons (whose vertically disposed α-fibers contain presynaptic structures (T-bars)) and epithelial glial cells. Occasionally, β-fibers (the other half of the ascending fiber pair) make gliapses. Amacrine cells are intrinsic to the first optic neuropil and are wide-field cells by virtue of their horizontally ranging segments. As such, they are a route for inter(optic)cartridge communication. Original ultrastructural observations on these gliapses were made by Trujillo-Cenóz (1965) and Boschek (1971). More detailed accounts of the ultrastructure of the gliapse are in Campos-Ortega and Strausfeld (1973), Hauser-Holschuh (1975), and Burkhardt and Braitenberg (1976). Shaw (1981) suggests that ". . . an amacrine–glia synapse . . ." exists here, and that gnarls in the vicinity ". . . may be labile structures involved in recycling material from glia." Such a synapse might be involved in a "triggered release" of potassium by epithelial glial cells. The effluxed potassium would then diffuse laterally to depolarize other cartridge neurons whose mass response might be a basis of the slow field potentials in the lamina (Shaw, 1975, 1981). The supposition that a K^+ efflux from glial cells occurs and causes adjacent neurons to depolarize is in direct variance with the well-documented (cf., Lane and Treherne, 1980) K^+ uptake activity. One cannot assume that glial cells rapidly oscillate from K^+ absorber to K^+ secreter. The contemporary view is that glial cells are largely or exclusively engaged in the former pursuit.

Epithelial glia also make another approach to neurons that looks suspiciously like a synapse. In the dipteran fly first optic neuropil, usually one presynaptic site on an R axon subserves four to five small postsynaptic processes. Sometimes a process of an epithelial glial cell enters the area of

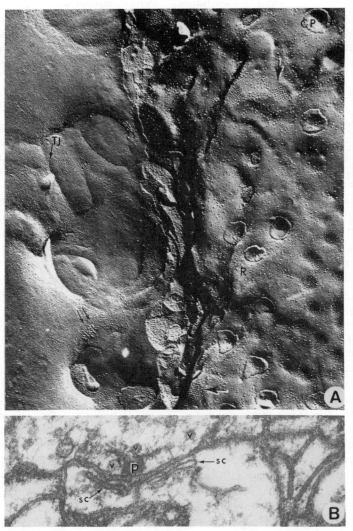

Figure 14.5. (A) Freeze-fracture replica. Chemical and electrical synapses in first optic neuropil, housefly. Relatively smooth surface at left is intramembranal surface (mostly E face leaflet) of one or both L1/L2 relay interneurons, core interneurons of an optic cartridge. Pockmarked surface at right is photoreceptor axon (R) terminal. "Pocks" are sheared-off capitate projections (CP). The photoreceptor (R) axon makes a chemical synapse with several (amacrine, L1, L2) interneurons whose individual processes fit into the triangle of depressions around the "bow tie" of P face particles (*arrow*). The latter particles are the base of a presynaptic structure. Another chemical synapse is at center bottom (*arrow*). An electrical synapse (*paired arrows*) lies between the two L axons. E face groves (TJ) above delineate focal tight junctions between L interneurons (54,000×). (B) Thin section of the T-bar synapse of the first optic neuropil of housefly. The presynaptic process (p) is viewed end on. Its base, if tipped 90° toward the viewer, would show a bow tie shape. (The T-table top is oval in en face view.) Synaptic vesicles (v) surround the presynaptic

336

a tetrad (Fröhlich and Meinertzhagen, 1982) and lies adjacent (postsynaptic?) to the photoreceptor terminal . . . "thus possibly constituting pentads."

14.5.4. Capitate Projections

More structured than the gnarl, capitate projections (Fig. 14.3D) are mushroom-shaped (usually with head and stalk) processes of epithelial glial cells that insert into photoreceptor axons of the first optic neuropil of the fly (Trujillo-Cenóz, 1965). Capitate projections profusely stud the axon terminals of photoreceptor cells and, as such, are excellent cell markers in freeze-fracture replicas (Chi and Carlson, 1980a,b), where the cleaved stalks look like pockmarks (Fig. 14.5A). Capitate projections often insert rather deeply within the neuron, and several may branch off from one invading process. There is, however, no regular pattern of axonal insertion and, depending on the plane of section, any part (or the whole) of a capitate projection may be seen. Our (Carlson et al., 1983) freeze-fracture views revealed numerous unoriented intramembranal particles on both E and P faces of the glial cell at the site of its modification into a capitate projection. These particle densities may relate to the considerable electron densities noted in thin sections. Thus, based on their depth of penetration, branching, hammer-headed morphology, and intramembranous particle-rich nature, one might presume capitate projections to be enduring structures, providing at least excellent cohesion between neuron and adjoining glia. On the other hand, Shaw (1981) presents the notion that capitate projections may sometimes be stalkless, have an ephemeral existence, are periodically lysed, and are possibly associated with ". . . recovery of liberated receptor neurotransmitter." If that is true, it would seem disadvantageous that there be a continuing integrity of structure. Our thin-sectioned material, taken at various times during the day, show no discernible circadian variation in form. Serial thick sections for HVEM reveal that capitate projections invariably have a glial stalk and persist throughout the life of the fly (Stark and Carlson, 1986).

14.5.5. Gap Junctions

Another kind of gliapse, the electrical synapse, is present between neurons and satellite glia in the optic neuropil of houseflies (Carlson et al., 1983, Saint Marie and Carlson, 1985) and in the abdominal ganglion of the crayfish, *Procambarus* (Cuadras et al., 1985). Although this kind of coupling (Fig. 14.3B,C) infers some sort of electrotonic union between neuron and glia, there are other substantial functions to consider first. Such low-resistance

---

ribbon base (P); one vesicle appears near the presynaptic membrane. The active zone apparently follows the perimeter of the "bow tie." In this plane of section, two postsynaptic processes are present, each with a prominent subsynaptic cistern (sc) (111,333 ×).

junctions presumably pass ions and metabolites in a trophic relationship between neurons and glia. Cuadras et al. (1985) believe that such gap junctions ". . . could constitute a short circuit route for ionic exchange, especially K^+ from neurons to glia; or from glia to neurons." Gap junctions are also found between satellite glia, and both types 1 and 2 relay interneurons in the nuclear layer of the fly first optic neuropil. These low resistance pathways may largely relate to the metabolic requirements of the nuclear layer of this avascular neuropil (Saint Marie and Carlson, 1985). Based on neuronal geometries and known electrophysiological responses of the fly L cells in this neuropil, it is possible to hypothesize that gap junctions between all the glia in the lamina and those between monopolar somata and glia could be a major factor in the summation of sustained potentials across the entire visual field. In that event such neuronal–glial communications would ". . . also influence the processing of visual signals by providing general range-setting and light adaptation effects, which are likely to involve all the coupled glia of the lamina as well" (Saint Marie and Carlson, 1985). More about "coupled glia" is in Section 14.6.

14.6. MEMBRANE SPECIALIZATIONS

Membrane specializations effect cellular cohesiveness (desmosomes, tight-septate and scalariform junctions); paracellular blockade of ions and small molecules (tight, and perhaps, septate junctions); and intercellular passage of ions and metabolites (gap junctions). Well-illustrated and documented accounts of these structures in arthropod neuropil greatly augment the brisk coverage here (e.g., Lane and Skaer, 1980; Noirot-Timothée and Noirot, 1980; Lane and Treherne, 1980; Lane, 1981, 1982; Saint Marie and Carlson, 1983b).

With high resolution, and the considerable magnifications afforded by transmission electron microscopy, plus the use of extracellular electron-dense tracers, we now understand that the surfaces of arthropod glia and neurons bear numerous types of junctions. Glial cells ionically communicate with each other over vast expanses of neuropil by means of a network of gap junctions. These low-resistance pathways connect similar and dissimilar types of glia in a syncytium-like array, which is of great importance in ionic (K^+, Na^+) homeostasis of extracellular fluid. Lane and Treherne (1980) advocate a model for such cation normalization, which proposes (among other things) that Na^+ and K^+ move through gap junctions of glial (and perineurial) cells, ensuring a "spatial buffering" especially of K^+ effluxed from depolarized neurons. Thus, K^+ does not accumulate in extracellular fluid, but is absorbed, translocated, and quickly equilibrated throughout neuropil. In crustacea, there is a similar but more lethargic movement of K^+ (Lane and Treherne, 1980) which is probably effected by the widespread glia-glia gap junctions recently reported by Cuadras and Garcia-Verdugo (1985).

Insects are the only arthropods possessing a blood–brain barrier in which water soluble ions and small molecules in hemolymph are excluded from neuropil. The anatomical correlates of this barrier are the tight junctions between perineurial cells and those between the neuropilar glia. A number of phytophagous species develop high K^+ titers in hemolymph; concentrations that would be ruinous to nerve conduction should K^+ reach axonal and synaptic surfaces (reviews by Treherne and Pichon, 1972; Treherne and Schofield, 1981). From Lane's laboratory come numerous reports (review by Lane, 1981) about tight junctions sealing perineurial cells to outer glial cells in embryos of arachnids (*Limulus*, ticks, spiders, scorpions). Interestingly, interglial septate junctions are absent, a situation quite unlike that found in the CNS of most insects.

The cohesiveness of the glial network and that encountered between glia and neurons is due in large part to spot and belt demosomes and septate and scalariform junctions. Axons may take twisting paths to a synaptic destination, as do fly photoreceptor axons en route to optic cartridges in the first optic neuropil. Apposing glial and neuronal cells in the area of greatest twisting (pseudocartridge glial cells) are bound by extensive septate junctions, which may enforce the twisting, as well as impede ion passage between cells in this area (Saint Marie and Carlson, 1983a,b). Glial cells make strong anchorages to basement membranes with hemidesmosomes (Saint Marie, 1981). Scalariform junctions are present between glial cells or folds of a glial cell near nerve cell bodies in locust ganglia (Lane, 1982). There these junctions lie near, and parallel to, smooth endoplasmic reticulum.

Tight junctions impart intercellular cohesiveness, but more interestingly, these junctions are points of membrane fusion that tend to ionically compartmentalize portions of neuropil. Areas of high resistance within certain sections of neuropil thus form. Sizable standing potentials recorded in the dark from locust (Shaw, 1975) and fly (Zimmerman, 1978) neuropil are structurally underlaid by these barriers to extracellular ion flow. Far greater details on the electrophysiological consequences of tight junctions in neuropil are given in Saint Marie and Carlson (1983b).

14.7. THE SYNAPSE

The primary and ultimate business of neuropil is synaptic transmission, which enables neural integration. Chemically mediated and modulated synapses remain in the majority, although each year more electrical synapses are physiologically identified or proffered, based on the presence of gap junctions.

Presynaptic structures taken as the sine qua non of the chemical synapse assume a variety of forms, such as an elaborate T-bar synapse (Fig. 14.5) in the fly (which in three dimensions resembles a pedestal table; Burkhardt and Braitenberg, 1976), bars in the antennal neuropil of grasshoppers (Schürmann and Wechsler, 1970), or more punctate densities "buttons" found in the α-lobes of the corpora pedunculata in the cricket (Schürmann, 1972).

There are also putative synapses in which no definable presynaptic structures are resolved, except for appreciable electron densities along both neuron's appositional membranes (e.g., between amacrine cells in optic neuropil; Campos-Ortega and Strausfeld, 1973). A variation in the T-bar theme is that in which three T-bars are joined at their (table top) edges (Strausfeld, 1976). "Dense (presynaptic) bodies" at the contact zone between two motoneurons in the stick insect give a T-table appearance (Osborne, 1970).

Indicator structures for postsynaptic processes include ill defined, electron-dense floculations that line the postsynaptic membrane, for example, in the ant prothoracic ganglion (Lamparter et al., 1969) and postsynaptic "buttons" in the antennal lobe of the locust (Schürmann and Wechsler, 1970). In the dyad-tetrad synapses in fly optic neuropil, each postsynaptic process usually contains a flattened saccule (subsynaptic cistern; Fig. 14.5B) that lies just beneath and parallel with the postsynaptic membrane. The function of these structures can only be guessed and one wonders about these structures being infallible signs of competent connectivity. It was some comfort to me when Saint Marie (1981) drove fly photoreceptor cells in light, during chemical fixation, freeze-cleaved the operant neuropil and found exocytotic holes or pits (presumably from neurotransmitter release) in an "active zone" around the perimeter of the T-bar base. Clearly, some of the structurally demarked loci were doing the expected work (Saint Marie and Carlson, 1982). There is also the possibility of neurotransmitter release from nonvesiculate cytoplasmic stores.

A variety of vesicle types abound in and around the synaptic sites in arthropod neuropil; sometimes several forms are present at a single synapse. Two or three dozen neurotransmitters and neuromodulators are probably present in arthropods. The lucent spherical vesicle is often found in Insecta, and such vesicles are rather homogeneously distributed throughout the terminal. Most of these pleomorphic or round vesicles measure about 30–40 nm in diameter (housefly optic neuropil data). In arthropod neurons, there is no clear-cut correlation between vesicle geometry and the electrical sign of the postsynaptic response, as was purported for CNS neurons in mammals (Uchizono, 1965). The latter dictum appears overthrown by Nakajima and Reese (1983), who found round vesicles in both excitatory and inhibitory synapses in the abdominal stretch receptor organ of crayfish (*Procambarus clarkii*) and concluded that vesicle morphology ". . . must depend on some aspect of aldehyde processing."

Knowing the contents of a particular vesicle in an axon in a given arthropod neuropil is quite rare. We have been freed of Dale's principle* and now know that several neurotransmitters and/or neuromodulators may co-

---

* Shepherd (1983) paraphrases Dale's principle: "the metabolic unity of the neuron would seem to require that it release the same transmitter substance of all its synapses." This law refers to *pre*synaptic unity only, as one transmitter may evoke several kinds of postsynaptic activity depending on the receptors of the follower cell.

exist in and be released from a presynaptic process. A glance at Figure 14.2B should provide the impression of a variety of neuroactive agents packaged in vesicles and being transported along the axon for possible release. In arthropods, there are accounts (e.g., O'Shea and Adams, 1981) of an identified neuron containing a biogenic amine and neuropeptide (proctolin).

More attention is being directed to the ultrastructure of dendritic arbors in arthropod neuropil. One result is that a whole constellation of synaptic arrangements is now known, which provides insights as to local processing of signals by single neurons and within small local circuits in delineated areas of neuropil. As examples, in particular pairs or trios of neurons that functionally associate, one might see dyad, triad, or tetrad (or more?) postsynaptic processes being serviced by a single presynaptic terminal as in the fly optic lobes (Fröhlich and Meinertzhagen, 1982). Other triadic arrangements are figured (Schmitt, 1979); reciprocal synapses are found between α- and β-fibers as well as between photoreceptor axons and type 1 lamina monopolar cells in the fly. By definition serial synapses connect a series of dendrites such that "a dendrite is presynaptic to one dendrite and postsynaptic to another" (Schmitt, 1979).

Electrotonic (electrical) synapses that are electrophysiologically and pharmacologically defined are uncommon, but O'Shea and Fraser Rowell (1975) provide a useful example from locust neuropil. Some sightings of gap junctions between axons in neuropil (cf., Chi and Carlson, 1976; Cuadras et al., 1985) have been reported recently, and electrical synapses may be more common than previously suspected. Their role in arthropod neuropil has yet to be assessed to any degree. Speaking generally, Schmitt (1979) supposes that electrical coupling might be involved in oscillatory behavior (among other things), and from the findings of O'Shea and Fraser Rowell (1975) an electrical synapse certainly hastens transmission of some visual signals, which command a prompt behavioral reflexive response. The intriguing finding of gap junctions and chemical synapses side by side was recently made by Strausfeld and Bassemir (1983). This "mixed synapse" connects the giant descending interneurons with column A visual interneurons in *Drosophila*. Those workers speculate that such a synaptic arrangement might effect a response with minimum delay (electrical synapse) and, at nearly the same time but slightly later, the chemical synapse would release neurotransmitter that would then inhibit the postsynaptic cell. These vignettes on synaptic structure should instill an appreciation for the exquisitely quick and vital reactions that take place at these special sites in neuropil.

14.8. SUMMARY

The ultrastructure of neurons and glia in arthropod neuropil is briefly discussed and illustrated. Neuropil is that synaptic sector of a ganglion without neuronal perikarya, which consists of a series of matrices of hierarchically ordered neurons wrapped and spatially isolated and ionically buffered by glia. Arthropod neuropil is renowned for neuronal complexity, intricacy of neuronal connections (and neuronal-glial associations), and concentrations of neurons per unit volume and mass. The basic structural plan of arthropod neuropil is that of a connective tissue sheath overlying a perineurium, which, in turn, surrounds the neuronal perikaryonal rind. The latter is adjacent to the synaptic plexus. Neuronal ultrastructure is seemingly simpler than that of the neuropilar glia. Neurons possess such main features as mitochondria, microtubules, microfilaments, and smooth surfaced cisterna. Glial cells contain most of the organelles found in neurons and in some cases appear to have a denser cytoskeleton and (often) copious footstuffs. Strangely enough, both chemical and electrical synaptic structures (gliapses?) are found in the apposition between nerve and glial cells. In some neuropils, the glia are extensively coupled to each other via numerous gap junctions, a feature that could effect spatial buffering of potassium. Other membrane specializations bond neuron to neuron, glial cell to glial cell, and neuron to glia. Tight junctions between neurons and glia (and glia-glia) partition neuropil into high-resistance sectors, and account for blood–brain barrier properties. Peculiar and intriguing structural intrusions by glia into neurons are made by capitate projections, trophospongia, and gnarls. The axoaxonal synapses of arthropod neuropil are, for the most part, chemically mediated and modulated by more than a score of neurotransmitters and small neuropeptides. Greater understanding of neural integration within the neuropil is being gained, based on descriptions of circuitry involving serial, reciprocal, and dyad-triad and tetrad synapses, along with advances in functional neuroanatomical techniques.

ACKNOWLEDGMENTS

Original micrographs for illustrations were from my laboratory and specifically provided by Dr. Che Chi, Mr. John Gustella, and Dr. Richard L. Saint Marie. These co-workers are heartily thanked. Research support was from Hatch grant 2100 and National Institutes Health (EYO 1686). Dr. Nicholas J. Strausfeld, European Molecular Biology Laboratory, Heidelberg, Germany, is appreciated for stimulating discussions in Hamburg and Madison and for permission to use previously published material. Earlier drafts were read and commented upon by Ms. Sonya Zetlan, Neuroscience Training Program, University of Wisconsin; Dr. David C. Post, Entomology Department, University of Wisconsin; and Dr. Richard L. Saint Marie, Department of Anatomy, University of Conneticut, Farmington.

REFERENCES

Boschek, C. B. 1971. On the fine structure of the peripheral retina and lamina ganglionaris of the fly, *Musca domestica. Z. Zellforsch. Mikrosk. Anat.* **118:** 369–409.

Buchner, E. and S. Buchner. 1983. Anatomical localization of functional activity in flies using ³H-2-deoxy-D-glucose, pp. 225–238. In N. J. Strausfeld (ed.), *Functional Neuroanatomy*. Springer-Verlag, Heidelberg, Berlin.

Bullock, T. H. 1961. On the anatomy of the giant neurons of the visceral ganglion of *Aplysia*, pp. 233–240. In E. Florey (ed.), *Nervous Inhibition*. Pergamon Press, Oxford.

Bullock, T. H. and G. A. Horridge. 1965. *Structure and Function in the Nervous System of Invertebrates*. Freeman, San Francisco.

Burkhardt, W. and V. Braitenberg. 1975. Some peculiar synaptic complexes in the first visual ganglion of the fly, *Musca domestica*. *Cell Tissue Res*. **173**: 287–308.

Campos-Ortega, J. A. and N. J. Strausfeld. 1973. Synaptic connections of intrinsic cells and basket arborizations in the external plexiform layer of the fly's eye. *Brain Res*. **59**: 119–136.

Carlson, S. D., R. L. Saint Marie, and C. Chi. 1983. Freeze fracture replication and the interpretation of glial and neuronal structures in insect nervous tissue, pp. 339–375. In N. J. Strausfeld (ed.), *Functional Neuroanatomy*. Springer-Verlag, Berlin.

Chi, C. and S. D. Carlson. 1976. Close apposition of photoreceptor cells axons in the housefly. *J. Insect Physiol*. **22**: 1153–1156.

Chi, C. and S. D. Carlson. 1980a. Membrane specializations of the first optic neuropile of the housefly, *Musca domestica*. I. Junction between neurons. *J. Neurocytol*. **9**: 429–449.

Chi, C. and S. D. Carlson. 1980b. Membrane specializations of the first optic neuropile of the housefly, *Musca domestica*. II. Junctions between glial cells. *Neurocytol*. **9**: 451–469.

Cuadras, J. and J. M. Garcia-Verdugo. 1985. Glial cells of ventral area of crayfish abdominal ganglia. *Morfol. Norm. Patol*. (in press).

Cuadras, J., G. Martin, G. Czternasty, and J. Bruner. 1985. Gap-like junctions between neuron cell bodies and glial cells of crayfish. *Brain Res*. **326**: 149–151.

Fahrenbach, W. H. 1976. The brain of the horseshoe crab (*Limulus polyphemus*). I. Neuroglia. *Tissue Cell* **8**: 395–410.

Fahrenbach, W. H. 1977. The brain of the horseshoe crab (*Limulus polyphemus*) II. Architecture of the corpora pedunculata. *Tissue Cell* **9**: 157–166.

Fröhlich, A. and I. A. Meinertzhagen. 1982. Synaptogenesis in the first optic neuropile of the fly's visual system. *J. Neurocytol*. **11**: 159–180.

Galambos, R. 1961. A glia-neuronal theory of brain function. *Proc. Nat. Acad. Sci. USA* **47**: 129–136.

Goodman, C. S., M. J. Bastiani, C. Q. Doc, S. duLac, S. L. Helfand, J. Y. Kuwada, and J. B. Thomas. 1984. Cell recognition during neuronal development. *Science (Washington, D.C.)* **225**: 1271–1279.

Heuser, J. E., T. S. Reese, M. J. Dennis, Y. Jan, L. Jan, and L. Evans. 1979. Synaptic vesicle exocytosis captured by quick freezing and correlated with quantal transmitter release. *J. Cell. Biol*. **81**: 275–300.

Hauser-Holschuh, H. 1975. Vergleichend quantitative Untersuchungen an den Sehganglien der Fliegen *Musca domestica* und *Drosophila melanogaster*. Doctoral dissertation, Eberhard-Karls Universität, Tübingen, Augsburg.

Holmgren, E. 1902. Beitrage zur Morphologie der Zelle. I. Nervenzellen. *Anat. Hefte* **18:** 267–325.

Hoyle, G. (ed.). 1977. *Identified Neurons and Behavior of Arthropods*. Plenum Press, New York.

Huber, F. 1974. Neural integration (central nervous system), pp. 3–100. In M. Rockstein (ed.), *The Physiology of Insecta,* vol. 4. Academic Press, New York.

Lamparter, H. E., H. Steiger, C. Sandri, and K. Akert. 1969. Zum Feinbau der synapsen im Zentralnervensystem der Insekten. *Z. Zellforsch.* **99:** 435–442.

Landolt, A. M. and H. Ris. 1966. Electron microscope studies on soma-somatic interneuronal junctions in the corpos pedunculatum of the wood ant. (*Formica lugubris* Zett). *J. Cell Biol.* **28:** 391–403.

Lane, N. J. 1974. The organization of insect nervous system, pp. 1–71. In J. E. Treherne (ed.), *Insect Neurobiology*. North-Holland, Amsterdam.

Lane, N. J. 1981. Invertebrate neuroglia-junctional structure and development. *J. Exp. Biol.* **95:** 7–33.

Lane, N. J. 1982. Insect intercellular junctions: Their structure and development, pp. 402–433. In R. C. King and H. Akai (eds.), *Insect Ultrastructure,* vol. 1. Plenum Press, New York.

Lane, N. J. and H. LeB. Skaer. 1980. Intercellular junctions in insect tissues. *Adv. Insect Physiol.* **15:** 35–213.

Lane, N. J. and J. E. Treherne. 1980. Functional organization of arthropod neuroglia, pp. 765–795. In M. Locke and D. S. Smith (eds.), *Insect Biology in the Future - VBW 80*. Academic Press, New York, London.

Lane, N. J. 1985. Structure of components of the nervous system, pp. 1–47. In G. A. Kerkut and L. I. Gilbert (eds.), *Comprehensive Insect Physiology Biochemistry and Pharmacology,* vol. 5. Pergamon Press, Oxford.

Manton, S. N. 1977. *The Arthropoda: Habits, Functional Morphology and Evolution*. Oxford University Press, Oxford.

Nakajima, Y. and T. S. Reese. 1983. Inhibitory and excitatory synapses in crayfish stretch receptor organs studied with direct rapid-freezing and freeze substitution. *J. Comp. Neurol.* **213:** 66–73.

Noirot-Timothée, C. and C. Noirot. 1980. Septate and scalariform junctions in arthropods. *Int. Rev. Cytol.* **63:** 97–140.

Orkand, P. M. and E. A. Kravitz. 1971. Localization of sites of gamma-amino butyric acid (GABA) uptake in lobster nerve muscle preparations. *J. Cell Biol.* **49:** 75–89.

Osborne, M. P. 1970. Structure and function of neuromuscular junctions and stretch receptors, pp. 77–100. In A. C. Neville (ed.), *Insect Ultrastructure*. Blackwell Scientific, Oxford, Edinburgh.

O'Shea, M. 1975. Spike transmitting electrical synapse between visual interneurons in the locust movement detection system. *J. Comp. Physiol.* **97:** 143–158.

O'Shea, M. and M. E. Adams. 1981. Pentapeptide (Proctolin) associated with an identified neuron. *Science (Washington, D.C.)* **213:** 567–569.

O'Shea, M. and C. H. Fraser Rowell. 1975. A spike-transmitting electrical synapse between visual interneurones in the locust movement detector system. *J. Comp. Physiol.* **97:** 143–158.

Radojcic, T. and V. W. Pentreath. 1979. Invertebrate glia. *Prog. Neurobiol.* **12:** 115–179.

Roots, B. I. 1978. A phylogenetic approach to the anatomy of glia, pp. 45–54. In E. Schoffeniels, B. Franck, L. Hertz, and D. B. Tower (eds.), *Dynamic Properties of Glial Cells.* Pergamon Press, Oxford.

Saint Marie, R. L. 1981. A thin section and freeze fracture study of intercellular junctions and synaptic vesicle activity in the first optic neuropil of the housefly compound eye. Doctoral dissertation, University of Wisconsin, Madison.

Saint Marie, R. L. and S. D. Carlson. 1982. Synaptic vesicle activity in stimulated and unstimulated photoreceptor axons in the housefly. A freeze fracture study. *J. Neurocytol.* **11:** 747–761.

Saint Marie, R. L. and S. D. Carlson. 1983a. Glial membrane specializations and the compartmentalization of the lamina ganglionaris of the housefly compound eye. *J. Neurocytol.* **12:** 243–275.

Saint Marie, R. L. and S. D. Carlson. 1983b. The fine structure of neuroglia in the lamina ganglionaris of the housefly, *Musca domestica* L. *J. Neurocytol.* **12:** 213–241.

Saint Marie, R. L. and S. D. Carlson. 1985. Interneuronal and glial-neuronal gap junctions in the lamina ganglionaris of the compound eye of the housefly, *Musca domestica. Cell Tissue Res.* **241:** 43–52.

Saint Marie, R. L., S. D. Carlson, and C. Chi. 1984. Glial cells, pp. 435–475. In R. C. King and H. Akai (eds.), *Insect Ultrastructure,* vol. 2. Plenum Press, New York.

Schmitt, F. O. 1979. The role of structural, electrical and chemical circuitry in brain function, pp. 5–20. In F. O. Schmitt and F. G. Worden (eds.), *The Neurosciences, Fourth Study Program.* MIT Press, Cambridge, Massachusetts.

Schürmann, F. W. 1972. Uber die Strucktur der Pilzkorper des Insek Tengehirns II. Synaptische Schaltungen in Alpha-lobus des Heimchens *Acheta domesticus* L. *Z. Zellforsch.* **127:** 240–257.

Schürmann, F. W. and W. Wechsler. 1970. Synapsen im Antennenhugel von *Locusta migratoria* (Orthoptera, Insecta). *Z. Zellforsch.* **108:** 563–581.

Shaw, S. R. 1975. Retinal resistance barriers and electrical lateral inhibition. *Nature (London)* **255:** 480–483.

Shaw, S. R. 1981. Anatomy and physiology of identified non-spiking cells in the photoreceptor-lamina complex of the compound eye of insects, especially Diptera, pp. 61–115. In A. Roberts and B. M. H. Bush (eds.), *Neurons Without Impulses: Their Significance for Vertebrate and Invertebrate Nervous Systems.* Cambridge University Press, Cambridge.

Shepherd, G. M. 1983. *Neurobiology.* Oxford University Press, Oxford.

Snodgrass, R. E. 1935. *Principles of Insect Morphology.* McGraw-Hill, New York.

Speck, P. T. and N. J. Strausfeld. 1983. Portraying the third dimension in neuroanatomy, pp. 156–182. In N. J. Strausfeld (ed.), *Functional Neuroanatomy.* Springer-Verlag, Heidelberg.

Stark, W. S. and S. D. Carlson, 1986. Ultrastructure and functional evaluation of capitate projections in optic neuropil of Diptera. *Cell Tissue Res.* **246:** 481–486.

Strausfeld, N. J. 1976. *Atlas of a Brain.* Springer-Verlag, Berlin.

Strausfeld, N. J. (ed.). 1983. *Functional Neuroanatomy*. Springer-Verlag, Berlin.

Strausfeld, N. J. and U. K. Bassemir. 1983. Cobalt-coupled neurons of a giant fiber system in Diptera. *J. Neurocytol.* **12:** 971–991.

Stretton, A. O. and E. A. Kravitz. 1968. Neuronal geometry: Determination with a technique of intracellular dye injection. *Science (Washington, D.C.)* **162:** 132–134.

Treherne, J. E. and Y. Pichon. 1972. The insect blood–brain barrier. *Adv. Insect Physiol.* **9:** 257–313.

Treherne, J. E. and P. K. Schofield. 1981. Ionic homeostasis of the brain microenvironment in insects. *J. Exp. Biol.* **95:** 61–73.

Treherne, J. E. (ed.). 1974. *Insect Neurobiology*. North-Holland, Amsterdam.

Treherne, J. E. (ed.). 1981. *Glial-Neuron Interactions*. Cambridge University Press, Cambridge.

Trujillo-Cenóz, O. 1965. Some aspects of structural organization of the intermediate retina of dipterans. *Ultrastruct. Res.* **13:** 1–33.

Uchizono, K. 1965. Characteristics of excitatory and inhibitory synapses in the central nervous system of the cat. *Nature (London)* **207:** 642.

Weevers, R. de G. 1985. The insect ganglia, pp. 213–297. In G. A. Kerkut and L. I. Gilbert (eds.), *Comprehensive Insect Physiology, Biochemistry and Pharmacology*. Pergamon Press, Oxford.

Wigglesworth, V. B. 1960. The nutrition of the central nervous system in the cockroach *Periplaneta americana* L. *J. Exp. Biol.* **37:** 500–512.

Zimmerman, R. P. 1978. Field potential analysis and the physiology of second-order neurons in the visual system of the fly. *J. Comp. Physiol.* **126:** 297–316.

CHAPTER 15

Structure and Functions of the Central Complex in Insects

Uwe Homberg
Division of Neurobiology
Arizona Research Laboratory
University of Arizona
Tucson, Arizona

15.1. INTRODUCTION

The mushroom bodies and the central complex are two structured neuropils in the protocerebrum of insects with no equivalents in the other ganglia. Therefore, both structures received early attention from neuroanatomists (Dietl, 1876; Flögel 1878; Viallanes 1887; Jonescu, 1909; Bretschneider, 1913), trying to understand their significance for the function of the insect brain. The relatively simple neuronal construction of the mushroom bodies, their prominent fiber connection with the antennal lobes, and their high differentiation in social insects led to hypotheses about their function (see Hanström, 1928), which could be tested in physiological investigations (see Chapter 21 by Erber et al. in this volume). The central complex, on the other hand, is probably the anatomically most complicated structure in the brain, and until today there exist only two investigations which in some detail tried to unravel its neuronal architecture (Williams, 1975; Strausfeld, 1976). The absence of significant connections to sensory neuropils or descending neurons and the invariability of its gross structure in insects of diverse habitats provide few clues for the function of the central complex that might serve as a basis for physiological experiments. Thus, despite its obvious importance in the function of the insect brain, the central complex is today still a largely unexplored area. This article reviews the anatomical, biochemical, and physiological studies on the central complex in the insect brain; in addition, it discusses some hypotheses for the function of the central complex as a possible guideline for future experiments.

15.2. NOMENCLATURE AND GROSS ANATOMY

The term central complex will be used here, according to Williams (1975), as a collective term for the protocerebral bridge and the central body. The central complex is situated in the center of the insect protocerebrum, bounded laterally and ventrally by the pedunculi and β-lobes of the mushroom bodies, anteriorly by the median bundle, and posteriorly by the ocellar and antennocerebral tracts (Fig. 15.1). The most posterior part, the protocerebral bridge, is a cylindrical neuropil with a transverse long axis and endings, which are curved posteroventrally. In some species, it is continuous (e.g., *Schistocerca;* Williams, 1975), whereas in others it consists of two symmetrically glomerular structures connected by a commissure (e.g., *Drosophila;* Power, 1943). The central body ("Centralkörper," Flögel, 1878; "corps central," Viallanes, 1887; "central complex," Power, 1943; "central body complex," Klemm, 1976) is a compact neuropilar structure in front of the protocerebral bridge, composed of several subunits (Fig. 15.1). Besides two small globular-shaped "noduli" ("tubercules du corps central," Viallanes 1887; "Knollen," Pflugfelder, 1937; "ventral bodies," Power 1943) situated posteroventrally, it consists of a dorsoposterior "upper division"

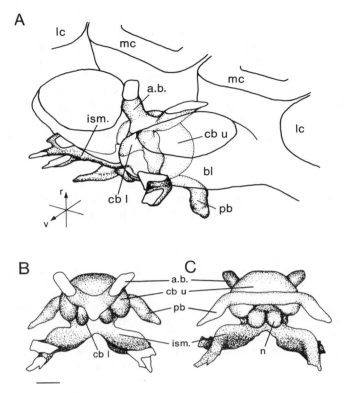

Figure 15.1. Diagram from computer reconstructions showing the compartments and fiber tracts of the central complex in the brain of the honey bee. (A) Stereodiagram showing the spatial relationship to the mushroom bodies. (B) Anterior view. (C) Posterior view of the central complex. r = rostral; v = ventral; a.b. = anterior bundles; bl = β-lobe; cb l = central body, lower division; cb u = central body, upper division; ism. = isthmus tract; lc = lateral calyx; mc = median calyx; n = nodulus; pb = protocerebral bridge. Horizontal bar: 100 μm.

("capsule supérieure," Viallanes, 1887; "central body," Power, 1943; "oberer Zentralkörperteil," Goll, 1967), which sits dome-like above a smaller "lower division" ("capsule inférieure," Viallanes, 1887; "ellipsoid body," Power 1943; "unterer Zentralkörperteil," Goll, 1967). The upper division of the central body can be further subdivided horizontally in a dorsal "superior arch" and a ventral "fan-shaped body" (Strausfeld, 1976; Homberg et al., 1987). In Diptera, Strausfeld (1976) includes under the term "central complex," in addition, the isthmi and the lateral accessory lobes, major arborization areas of central-complex neurons in the protocerebrum.

The size and gross anatomy of the central complex seem to be relatively constant in all insects, even in species with very different habits (Hanström, 1926; Howse, 1975). The central body is virtually invariant in different castes of social Hymenoptera, which contrasts with the large differences that can be found in the size of sensory neuropils and the mushroom bodies (Jonescu, 1909; Thompson, 1913; Lucht-Bertram, 1962). When comparing various orders, however, Bullock and Horridge (1965) noticed that the central body is large in primitive insects that have insignificant mushroom bodies (e.g., Apterygota) and smaller in highly evolved species, such as social Hymenoptera that have large mushroom bodies with well-differentiated calyces. This may support an earlier idea of Bretschneider (1913), who suggested that the central body is an evolutionary old and primitive brain center, whereas the mushroom bodies are more highly evolved secondary neuropils.

15.3. CELLULAR ARCHITECTURE AND CONNECTIONS TO OTHER BRAIN AREAS

In many, if not all species, the upper and lower divisions of the central body are subdivided like a fan into eight or 16 segments (Figs. 15.2, 15.3), as has been documented for *Gryllus* (Dietl, 1876), *Blatta* (Flögel, 1878), *Periplaneta* and *Tenebrio* (Bretschneider, 1914), *Trichoptera* (Klemm, 1968), *Schistocerca* (Williams, 1975), *Musca* (Strausfeld, 1976), and *Apis* (Mercer et al., 1983). A similar segmentation has been established for the protocerebral bridge in *Schistocerca* (Williams, 1975) and *Musca* (Strausfeld, 1976). In addition, Strausfeld demonstrated an anteroposterior and a dorsoventral stratification of the central body in flies, corresponding to innervation patterns in different depths and heights of the segments.

This three-dimensional neuronal network consists mainly of extrinsic neurons connecting the central complex with other brain areas. In *Schistocerca* (Williams, 1972; Tyrer et al., 1984) and *Apis* (Homberg, 1985; Mobbs, 1985), there are three main systems of tracts entering or leaving the central complex: (1) tracts entering at the end or along the protocerebral bridge; (2) the isthmus tracts, connections of the central body to the lateral accessory lobes and the posterior lateral protocerebrum (Fig. 15.1); and (3) the anterior bundles connecting the central body to the anterolateral protocerebrum (Fig. 15.1).

These three fiber systems spread out after leaving the central complex. In flies, Strausfeld (1976) describes a number of fiber tracts leaving the central body laterally, which are probably subsystems of the ones described earlier.

15.3.1. Neurons of the Protocerebral Bridge

Various connections have been reported between the protocerebral bridge and optic neuropils (Pflugfelder, 1937; Power, 1943; Williams, 1975; Strausfeld, 1976). A direct connection to the optic lobes has been demonstrated in *Gryllus* (Honegger and Schürmann, 1975), consisting of 11–15 neurons

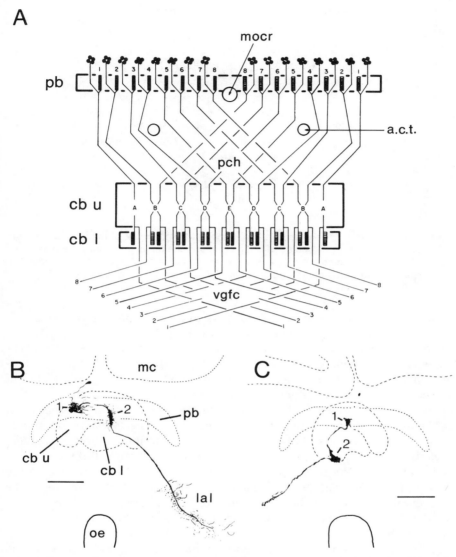

Figure 15.2. (A) Diagram showing the arrangement of 64 neurons in the central complex of *Schistocerca gregaria*. The fibers run in bundles of 4; their cell bodies lie in the pars intercerebralis. Hatched and shaded bars represent zones of arborization in the protocerebral bridge (pb) and the lower division of the central body (cb l). a.c.t. = antenno-cerebral tract; cb u = central body, upper division; lal = lateral accessory lobe; mocr = median ocellar nerve root; pch = posterior chiasma; vgfc = ventral groove fiber complex. (Adapted from Williams, 1975.) (B and C) Segmental interneurons of the central complex in the brain of the bee, reconstruction of cells injected with Lucifer yellow. mc = median calyx; oe = esophagus; 1 = arborizations in the protocerebral bridge; 2 = arborizations in the upper division (B), in the lower division (C) of the central body. Horizontal bars:100 μm. (After Homberg, 1985.)

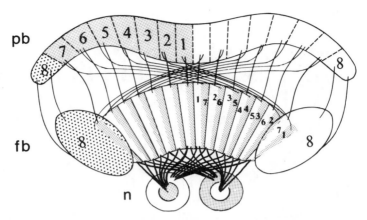

Figure 15.3. Diagram of fiber projections in the central complex of the fly, showing the chiasmata between the protocerebral bridge (pb) and the fan-shaped body (fb, part of the upper division of the central body), and between the fan-shaped body and the noduli (n). The lower division of the central body is omitted. (Adapted from Strausfeld, 1976.)

with cell bodies in the contralateral pars intercerebralis and arborizations throughout the protocerebral bridge. Many fibers from optic neuropils seem to innervate only the contralateral part of the bridge (Pflugfelder, 1937; Williams, 1975). Klemm et al. (1984) have described three fiber tracts in *Periplaneta* that connect the bridge to wide areas of the lateral protocerebrum, calyces, peduncle, and α-lobe of the mushroom bodies. A major innervation of the bridge by small ocellar interneurons has been shown in *Apis* (Pan, 1981), *Schistocerca* (Goodman and Williams, 1976), *Periplaneta* and *Acheta* (Koontz and Edwards, 1984). In *Schistocerca* the innervation is spatially organized, so that each part of the bridge gets input from all three ocelli.

Intrinsic neurons of the bridge have been described so far only in the honey bee (Homberg, 1985). These cells send a main fiber throughout the bridge with numerous collaterals, which are more dense in the lateral parts of the neuropil, and have their cell bodies in the lateral posterior protocerebrum.

The most prominent and numerous type of cells in the protocerebral bridge are interneurons connecting it to the central body. These cells innervate one segment in the bridge and the upper or lower division of the central body and have their cell bodies in the pars intercerebralis. In *Schistocerca,* a highly ordered arrangement of 64 interneurons has been described (CC I-system, Williams, 1975). Each of these cells has arborizations in one segment of the protocerebral bridge, one segment of the lower division of the central body, and the contralateral accessory lobe (Fig. 15.2). In addition, other types of segmental interneurons (CC II, CC III) have arborizations in the upper division of the central body, and unilateral (CC II) or bilateral projections (CC III) to the lateral accessory lobes (Williams, 1972). In

Musca, Strausfeld (1976) describes a system of segmental interneurons, which is slightly different from the CC I-system in *Schistocerca.* In his scheme, the 16 segments of the protocerebral bridge, except for the outer ones, are ipsi- and contralaterally connected to the central body (Fig. 15.3). In addition to these systems, segmental neurons connecting the bridge and the central body have been found in bees and flies, which lack a process to the lateral accessory lobe (Strausfeld, 1976; Homberg, 1985).

15.3.2. Neurons of the Central Body

Besides the segmental interneurons described, a great variety of small-field neurons have been found in flies (Strausfeld, 1976), which connect one or more segments especially of the upper division (fan-shaped body and superior arch) with other brain areas. In *Musca,* the noduli are connected to each of the 16 segments of the fan-shaped body. Ipsilateral connections innervate the core of the noduli, whereas contralateral projections from the fan-shaped body form a shell (Fig. 15.3). The small-field arborizations in the upper division of the central body often show various fiber specializations in discrete layers. In the fly (Strausfeld, 1976) and the bee (Milde, 1982), some neurons have been reported, that can best be characterized as bypassing elements, sending a small branch into the central body. These cells often have large arborization fields in different regions of the protocerebrum and run in fiber tracts passing near the central body.

Besides these small-field elements, the central body is innervated by many neurons with arborizations in all segments of the upper or lower division (Figs. 15.4 and 15.5). Some cells even innervate all segments of both divisions (Strausfeld, 1976; Schildberger, 1983). The branching pattern in the central body often looks like a fan (fan-shaped neurons), following the segmental compartments in horizontal or vertical strata (Williams, 1972; Strausfeld, 1976; Schildberger, 1983; Homberg, 1985; Mobbs, 1985). In the bee (Milde, 1982; Mobbs, 1985) and the fly (Strausfeld, 1976), some cells innervating the upper or lower division have additional projections in the noduli. In these species as well as in the moth *Manduca,* the posterodorsal part of the upper division (superior arch) is characterized by diffuse rather than fan-shaped branching patterns of large-field neurons (Strausfeld, 1976; Homberg, 1985; Mobbs, 1985; Homberg et al., 1987).

Fan-shaped neurons of the lower division project to the ipsilateral accessory lobe (Fig. 15.5) or to more lateral brain areas. Their somata lie in a cluster near the esophageal foramen between deuto- and protocerebrum (Goll, 1967; Williams, 1972; Strausfeld, 1976; Schildberger, 1983; Homberg, 1985). Fan-shaped neurons of the upper division have their cell bodies in the same cluster, the pars intercerebralis, or more lateral sites near the calyces (Fig. 15.4). They enter the central body either via the isthmus tracts or the anterior bundles (Williams, 1972; Homberg, 1985) and form large arborization fields around and between the lobes of the mushroom bodies,

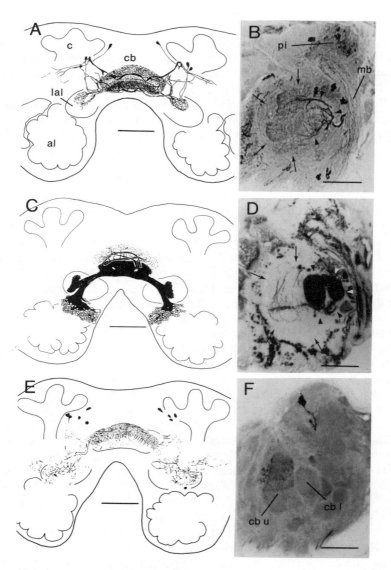

Figure 15.4. Immunoreactivity in the central body (cb) of *Manduca sexta* obtained with antisera against 5-HT (A and B, courtesy of S. G. Hoskins and J. Lauder), GABA (C and D, antisera prepared by T. G. Kingan), and adipokinetic hormone (E and F, antisera prepared by H. Schooneveld), visualized with the PAP technique. (A, C, and E) Horizontal reconstructions of immunoreactive cells. (B, D, and F) Light microscope photographs of midsaggital brain sections. *Arrows* point to upper division (cb u), *triangles* to lower division of the central body (cb l). (C) Open circles near antennal lobe indicate immunoreactive cell bodies. (E) Immunoreactive fibers and cell bodies in the lateral accessory lobes are suggested to be continuous with immunoreactive fibers in the central body. al = antennal lobe; c = calyces of the mushroom bodies; lal = lateral accessory lobe; mb = median bundle; pi = pars intercerebralis. Horizontal bars: 200 μm (A, C, E); 100 μm (B, D, F).

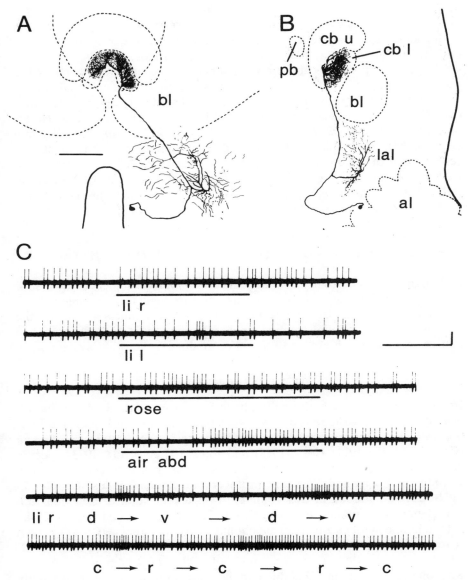

Figure 15.5. Frontal (A) and saggital (B) reconstruction of a cell injected with Lucifer yellow, which innervates the lower division of the central body (cb l) in the bee brain. al = antennal lobe; bl = β-lobe; cb u = central body, upper division; lal = lateral accessory lobe; pb = protocerebral bridge. Horizontal bar: 100 μm. (C) Electrical activity of the cell and responses to sensory stimuli: li r, li l = light to the right (r) or left (l) eye; rose = olfactory stimulus rose odor; air abd = stream of air over the abdomen; c = caudal; d = dorsal; r = rostral; v = ventral. The cell shows no response to stationary light flashes, but direction-specific excitation or inhibition to a moving light stimulus to the right eye. *Arrows* indicate direction of the moving light stimulus. Horizontal bar: 1 sec; vertical bar: 10 mV. (After Homberg, 1985.)

in the lateral accessory lobes, in the optic foci, or the tritocerebrum (Strausfeld, 1976; Schildberger, 1983; Homberg, 1985; Mobbs, 1985).

There are reports of central-body connections to nearly all parts of the protocerebrum, but some of these findings are poorly documented. In addition to the connections mentioned, neurons linking the central body with the α- and/or β-lobe have been reported in Geocorisae, terrestrial Heteroptera (Pflugfelder, 1937), *Formica* (Goll, 1967), and *Periplaneta* (Klemm et al., 1984). Strausfeld (1976) reported few connections to the mechanosensory part of the deutocerebrum in flies. Despite the large variety of projection areas, connections of the central complex seem to be largely restricted to the *non-glomerular protocerebrum*, with a major projection area in the lateral accessory lobes.

15.4. DISTRIBUTION OF PUTATIVE TRANSMITTER SUBSTANCES

Although there is currently no biochemical or pharmacological evidence for a role of any substance as a neurotransmitter in the central complex, there are indications from histochemical and immunocytochemical studies for the presence of acetylcholine, monoamines, γ-aminobutyric acid (GABA), and neuropeptides, which are transmitters or modulators in other systems.

15.4.1. Acetylcholine

Evidence for the distribution of cholinergic neurons in the insect brain has been obtained by variations of the histochemical method of Koelle and Friedenwald (1949) for the localization of cholinesterase activity. In *Periplaneta* (Frontali et al., 1971; Hess, 1972) and *Acheta* (Lee et al., 1973) specific acetylcholinesterase (AChE) activity has been found in all parts of the central complex including noduli and the protocerebral bridge (see also Chapter 19 by Booth and Larsen in this volume). Two recent studies used immunocytochemical techniques in *Drosophila* to investigate the distribution of the acteylcholine synthesizing enzyme choline acetyltransferase (ChAT), which is regarded as a more reliable indicator of cholinergic neurons than AChE (Buchner et al., 1986; Gorczyca and Hall, 1986). ChAT-immunoreactivity was found especially in the noduli and the upper division of the central body and seems to be distributed complementary to GABA-immunoreactive staining (see Section 15.4.3.).

15.4.2. Amines

Monoamines are apparently present in the central complex of all insects so far studied. Investigations based upon the aldehyde-histofluorescence technique, which demonstrates fluorogenic amines like dopamine (DA), norepinephrine (NE), and 5-hydroxy tryptamine (5-HT), revealed intense fluorescence in both divisions of the central body as well as in the noduli in a large variety of species (for review see Klemm, 1976; Evans, 1980). Microspectrofluorometric analyses showed a DA-like spectrum in the central body of the caddis fly (Klemm and Björklund, 1971), whereas evidence for both DA and NE was obtained in the locust (Klemm and Axelsson, 1973).

Immunocytochemical techniques, using antisera against amines or their synthesizing enzymes, led to a more specific and detailed description of the localization of these amines. In *Locusta,* Vieillemaringe et al. (1984) reported DA-like immunoreactivity in the posterior part of the central body. Dopamine-β-hydroxylase-like immunoreactivity, indicating the presence of NE, was detected in the fan-shaped body of *Calliphora* and *Periplaneta* (Klemm et al., 1985). In *Manduca* (Fig. 15-4A,B), *Apis* (Schürmann and Klemm, 1984), *Schistocerca* (Tyrer et al., 1984), and *Periplaneta* (Klemm et al., 1984) several large fan-shaped neurons of the central body are immunoreactive to 5-HT antisera. These cells have arborizations in certain layers in the upper or lower division (in *Schistocerca* and *Periplaneta* only in the upper division) and the lateral accessory lobes. In *Manduca,* at least eight pairs of 5-HT immunoreactive cells innervate the upper and one pair the lower division. In *Apis* and *Manduca,* immunoreactivity has also been found in the noduli. In addition to fan-shaped neurons, segmental neurons in *Schistocerca* (possibly type CC I) and *Periplaneta* connecting the upper division with the protocerebral bridge and the lateral accessory lobes are immunoreactive to 5-HT antisera (Klemm et al., 1984; Tyrer et al., 1984).

Earlier aldehyde-fluorescent studies in *Apis* (Mercer et al., 1983) and *Schistocerca* (Klemm and Axelsson, 1973) indicate that besides 5-HT, which shows a yellow fluorescence, green fluorescent catecholamines are also present in the central body of these species. In the bee and similarly in the cockroach (Frontali, 1968; Klemm, 1983), intense fluorescence especially has been found in the upper division, where it is distributed in particular layers inside the eight segments. An opposite distribution has been reported for Trichoptera (Klemm, 1968), where seven bars between the segments are highly fluorescent.

Fluorogenic amines could be detected in the protocerebral bridge of only certain species (Klemm, 1976). In *Periplaneta,* yellow histofluorescence in the bridge results from a dense innervation of approximately 30 5-HT-immunoreactive neurons per hemisphere. Three immunoreactive fiber tracts were found to connect the bridge to wide areas of the lateral protocerebrum and α- and β-lobes (Klemm et al., 1984).

15.4.3. GABA

Frontali and Pierantoni (1973) located sites of uptake of ^{3}H-GABA in the brain of *Periplaneta* by light-microscopic autoradiography. They found high uptake in the lower division of the central body and in a small portion of the protocerebral bridge. That these findings have significance for the putative role of GABA as a neurotransmitter is supported by dense GABA-immunoreactivity in the lower division of the central body in all species studied so far (Schäfer and Bicker, 1986; Meyer et al., 1986; Homberg et al., 1987). In the moth, *Manduca,* GABA-immunoreactivity is found in 200–260 neurons with aborizations in the lower division of the central body and in the lateral accessory lobes and somata near the esophageal foramen. In addition, in *Manduca* two pairs of fan-shaped neurons of the upper division and several small-size fibers, which especially innervate a narrow band in the upper division of the central body, are GABA-immunoreactive (Fig. 15.4C,D).

15.4.4. Peptides

Neuropeptides have been detected in neurosecretory cells in the insect brain (El-Salhi et al., 1983; Duve and Thorpe, 1984) and in other parts of the protocerebrum, including the central body (Duve and Thorpe, 1984; Schooneveld et al., 1985). Neurons immunoreactive to antisera against gastrin-releasing peptide/bombesin are located in the lower division of the central body of the stick insect and the Colorado potato beetle (Veenstra and Yanaihara, 1984). Gastrin/cholecystokinin- and pancreatic polypeptide-immunoreactivities have been detected in the central body of *Calliphora* (Duve and Thorpe, 1984), and substance P-immunoreactivity in the central body of locusts (Benedeczky et al., 1982). A substance immunologically related to locust adipokinetic hormone (AKH I) has been demonstrated in the central body and protocerebral bridge of various insect species (Schooneveld et al., 1985). In *Manduca,* AKH-immunoreactivity has been found in a few neurons with wide-field arborizations in several layers in the upper division of the central body (Fig. 15.4E,F).

15.5. PHYSIOLOGICAL INVESTIGATIONS

15.5.1. Electrical Stimulation and Lesion Experiments

Attempts to study the function of the central complex have been made, in which the behavior of an animal is monitored after brain lesions or during electrical stimulation. The interpretation of these experiments, however, is difficult, because it has been impossible to affect the central body without damage to other parts of the brain.

Steiniger (1933) reported a loss of catalepsy, a frequently observed state of tonic immobility, in the stick insect after removal of the central body. A similar result has been found in *Periplaneta* (Drescher, 1960). After a mid-saggital brain section, the animals showed increased excitability, hyperactivity, and dramatic hyperphagia. In accordance with Steiniger's findings, Drescher suggested from these results that the central body is a site of tonic inhibition of locomotor activity. Roeder (1937) demonstrated in the praying mantis that a midsaggital section results in behavioral effects that are opposite to those observed after removal of the brain. In contrast to results with *Periplaneta,* a separation of the two halves of the brain of the mantis resulted in a loss of spontaneous locomotion, an increase in the tonus of neck and prothoracic muscles, and an increase in visual excitability. Similar observations had been made earlier by Bethe (1897) on *Hydrophilus,* an aquatic bug. From these experiments, Roeder suggested an inhibitory influence of the brain on walking behavior and reciprocal inhibitory interactions of both halves of the brain. After removal of the brain, he proposed that prolonged periods of walking are owing to a loss of inhibition from the brain, whereas splitting of the brain increases the inhibitory action of the two independent brain halves and walking is greatly reduced. Roeder's scheme is supported by Huber's work with *Gryllus* and *Gomphocerus* (Huber, 1955, 1960a,b). He showed that electrocoagulation of the whole central body results in a similar inhibition of walking and acoustic behavior as after midsaggital section (Huber, 1960b, 1962). This suggests that a reciprocal inhibition of the two halves of the brain might act through the central body to modulate a descending inhibition of locomotor behavior.

Electrical stimulation of the central body gave further evidence for its role in the control of behavior in crickets and grasshoppers. Local stimulation often has excitatory effects on the animal's behavior. Elements of escape or feeding behavior could be elicited (Huber, 1960b), and ventilatory frequency could be enhanced or reduced (Huber, 1960a). Both normal (Otto, 1971) and atypical patterns of song (Huber, 1960b; Wadepuhl, 1983) were observed after electrical stimulation in the central complex region. Otto (1971) showed that excitatory and inhibitory loci for the control of acoustic behavior in the central body lie close together. The conclusion from these and other experiments that the central complex might generate the singing pattern under the control of the mushroom bodies, however, had to be revised after Otto (1971) demonstrated that patterns similar to normal song could be elicited with electrical stimulation of the neck connectives in a headless cricket. Moreover, Kutsch and Otto (1972) observed spontaneous singing in crickets with cervical connectives cut. Therefore, the role of the central complex in the control of acoustic behavior is still unclear. It seems, however, that together with the mushroom bodies, the central complex has an influence on the initiation and perhaps selection of the song type.

15.5.2. Single-Cell Analyses

Recordings from single cells of the central complex with subsequent injection of Lucifer yellow have been performed in crickets (Schildberger, 1982) and bees (Milde, 1982; Homberg, 1982, 1985). These experiments show that neurons of the central complex often exhibit only small modulation in the frequency of action potentials when a variety of sensory stimuli is presented to the animal. Compared with mushroom body cells, neurons innervating the central complex show a lesser degree of multimodality, and a relatively high percentage of neurons does not respond to any stimulus. The experiments do show, however, that the central complex overall gets weak indirect inputs from many sensory organs. In the bee, intrinsic cells of the protocerebral bridge respond predominantly to stationary and moving optical stimuli (Homberg, 1985), which is in good agreement with the anatomical evidence of a fiber connection with optic neuropils. Segmental and fan-shaped neurons also respond to stimulation of the ocelli with light (Milde, 1982). No systematic differences could be found between cell groups, such as segmental and large-field neurons of the upper or lower division. In a few recordings, only specific stimuli, such as a moving visual stimulus (Fig. 15.5) or a specific odor, elicited responses in these cells. This might indicate that the central complex is involved only to a small degree in the processing of sensory information and may have a more general modulatory function in the brain. Another possibility, however, is that neurons of the central complex respond only to combinations of stimuli that are relevant in a behavioral context, a situation, which could not be simulated under the conditions of the recording experiments.

15.6. HYPOTHESES ABOUT THE FUNCTIONAL ROLES OF THE CENTRAL COMPLEX IN THE INSECT BRAIN

The central complex is the only unpaired neuropil in the center of the insect brain. Therefore, its functions are probably related to an integration of information from the right and left halves of the brain. Its cytoarchitecture has many similarities to that of the optic neuropils. Both systems show fiber chiasmata and a neuronal organization in columns and strata, which are oriented in two or even three planes in the central complex compared with the simple saggital strata and centrifugal columns in optic neuropils (Strausfeld, 1976). In the optic system, this feature serves to maintain a retinotopic organization over several stages of visual integration. In the central complex, only some of the fibers cross over in each chiasma, so that information from and to the right and left side of the brain is in a different way combined at the level of the protocerebral bridge, the upper and lower divisions of the central body, and the lateral accessory lobes.

Inputs to the central complex seem to derive mainly from higher order neuropils in the protocerebrum. The low response rate of central-complex neurons to sensory stimuli suggests that there are only a few direct connections between sensory neuropils and the central complex. Several authors have reported connections to the mushroom bodies (e.g., Klemm et al., 1984) but the number of cells connecting these two neuropils seems to be small. Therefore, the functional scheme drawn by Huber (1960b) to model the action of the brain in locomotion of crickets, which assumes a hierarchical relationship between mushroom bodies and the central complex, is unlikely. Neuroanatomical evidence suggests instead that both neuropils are independently connected to the surrounding protocerebral lobes. At present, there is no evidence for direct connections to descending cells, many of which terminate in the lateral protocerebrum. A chain of several interneurons from the central complex to the thoracic ganglia is, therefore, required to explain its role in the control of cricket behavior found by Huber and coworkers.

Evidence suggests that different classes of transmitters are located in the central complex. The majority of fan-shaped neurons of the lower division appear to be GABAergic, whereas many neurons innervating the upper division and noduli might be cholinergic. In addition, amines and peptides are contained in some fan-shaped and possibly some types of segmental neurons. The physiological role of most transmitter candidates in the insect CNS has not yet been investigated. Peptides may have a role as neurotransmitters or neuromodulators, as has been shown, for example, in crustaceans and molluscs (O'Shea and Schaffer, 1985; Marder and Hooper, 1985). A modulatory function, demonstrated for amines in the stomatogastric nervous system of crustaceans (Marder and Hooper, 1985; Harris–Warrick and Flamm, 1986), has also been proposed for aminergic neurons in the insect brain (Evans, 1980; Mercer et al., 1983), but so far this has been established only at peripheral synapses (Evans, 1982). Fan-shaped neurons immunoreactive to 5-HT antisera connect the central complex with various parts of the protocerebral lobes and therefore might be suited to regulate the neuronal activity of larger brain areas.

Good evidence exists that GABA acts as an inhibitory transmitter in the insect nervous system (reviewed by Usherwood, 1978). GABA-immunoreactive cells of the lower division are probably fan-shaped neurons similar to the type shown in Fig. 15.5. Sagittal brain sections show that these are the most prominent GABA-immunoreactive neurons connecting the two brain halves. It is, therefore, tempting to speculate that GABA-immunoreactive neurons of the central complex are the neuronal substrate of the mutual inhibition of the two halves of the brain that has been found with lesion experiments in several species. A direct reciprocal inhibition of the calyces of the mushroom bodies as suggested by Huber (1960b) cannot be ruled out, but seems unlikely at least in *Manduca* and *Apis,* where no GABA-immunoreactive cells connecting the calyces have been detected. If fan-shaped neurons of the lower division are outputs of the central body, as might be speculated from their innervation pattern in the lower division (Fig. 15.5), the mutual inhibition of the two halves of the brain might be accompanied by self-inhibitory loops on both sides, because the bilateral dendritic field in the central body suggests inputs from both ipsi- and contralateral sources. If one assumes that segmental interneurons of the central complex and fan-shaped neurons of the upper division act in an excitatory way, then the central complex might function as a well-balanced system of inhibitory and excitatory interactions of the two halves of the brain, which is a pre-requisite for behavioral outputs that are influenced by many internal and external factors.

Some of the most promising attempts to study the functional significance of the central complex are investigations on *Drosophila* mutants with structural defects in the central complex. In a preliminary report, Heisenberg et al. (1985) described deficiencies in olfactory learning in central complex mutants. Further investigations will have to show if the central complex has a special role in learning, or if this defect is due to a changed olfactory perception or other impairments of brain interactions. Immunocytochemical investigations can help in studying the neuroanatomy and physiology of neurochemical subsystems. Antisera against putative neurotransmitters are valuable tools for ultrastructural investigations of input and output relationships of the different classes of central-complex cells. Finally, new techniques of microlesions and injections of specific neurotoxins or transmitters into the subunits of the central complex may lead to a better understanding of its organization and role in the control of insect behavior.

15.7. SUMMARY

The central complex is a system of interconnected neuropils in the insect brain. Its neuroanatomy is characterized by several systems of right-left fiber chiasmata and arborization patterns in the central body in vertical segments and in strata intersecting these segments. The central complex has connections to many parts of the protocerebrum, and its major projection areas are the lateral accessory lobes. Histochemical and immunocytochemical evidence suggest that acetylcholine, GABA, amines, and neuropeptides are possible transmitters in the neurons of the central complex. The central complex receives mostly indirect inputs from many sensory organs, and is involved in the control of motor activity. The precise right-left fiber geometry and the results of lesion experiments, together with the neurochemical data, suggest that the lower division of the central body plays a major part in the inhibitory interaction found between the two halves of the brain in several species. Together with excitatory pathways, the central complex may serve a role in balancing the information outflow of the right and left sides of the brain.

ACKNOWLEDGMENTS

I am grateful to J. Erber, J. G. Hildebrand, S. G. Hoskins, and T. G. Kingan for their valuable comments on the manuscript. I thank N. J. Strausfeld and J. J. Milde for their help with a computer reconstruction of the central complex; T. G. Kingan, H. Schooneveld, and J. Lauder for gifts of antisera; and

S. G. Hoskins for providing me with immunocytochemical preparations of 5-HT antisera. The work was supported by a DFG grant to J. Erber (ER 79/ 2-1), and to the author (Ho 950/2-1).

REFERENCES

Barker, J. L. 1976. Peptides: Roles in neuronal excitability. *Physiol. Rev.* **56:** 435–452.

Benedeczky, I., J. Z. Kiss, and P. Somogyi. 1982. Light and electron microscopic localization of substance P-like immunoreactivity in the cerebral ganglion of locust with a monoclonal antibody. *Histochemistry* **75:** 123–131.

Bethe, A. 1897. Vergleichende Untersuchungen über die Funktionen des Central-nervensystems der Arthropoden. *Pflügers Arch. Gesamte Physiol. Menschen Tiere* **68:** 449–545.

Bretschneider, F. 1913. Der Centralkörper und die pilzförmigen Körper im Gehirn der Insekten. *Zool. Anz.* **41:** 560–569.

Bretschneider, F. 1914. Über die Gehirne der Küchenschabe und des Mehlkäfers. *Jena. Z. Nat.* **52:** 269–362.

Buchner, E., S. Buchner, G. Crawford, W. T. Mason, P. M. Salvaterra, and D. B. Sattelle. 1986. Choline acetyltransferase-like immunoreactivity in the brain of *Drosophila melanogaster*. *Cell Tissue Res.* **246:** 57–62.

Bullock, T. H. and G. A. Horridge. 1965. *Structure and Function in the Nervous System of Invertebrates,* vol. 2. Freeman, San Francisco.

Dietl, M. J. 1876. Die Organisation des Arthropodengehirns. *Z. Wiss. Zool.* **27:** 488–517.

Drescher, W. 1960. Regenerationsversuche am Gehirn von *Periplaneta americana* unter Berücksichtigung von Verhaltensänderung und Neurosekretion. *Z. Morphol. Ökol. Tiere* **48:** 576–649.

Duve, H. and A. Thorpe. 1984. Comparative aspects of insect-vertebrate neuro-hormones, pp. 171–196. In A. B. Bořkovec and T. J. Kelly (eds.), *Insect Neurochemistry and Neurophysiology.* Plenum Press, New York.

El-Salhi, M., S. Falkmer, K.-J. Kramer, and R. D. Speirs. 1983. Immunohisto-chemical investigations of neuropeptides in the brain, corpora cardiaca, and corpora allata of an adult lepidopteran insect, *Manduca sexta*. *Cell Tissue Res.* **232:** 295–317.

Evans, P. D. 1980. Biogenic amines in the insect nervous system. *Adv. Insect Physiol.* **15:** 317–437.

Evans, P. D. 1982. Properties of modulatory octopamine receptors in the locust, pp. 48–69. In D. Everet, M. O'Connor, and J. Whelan (eds.), *Neuropharmacology of Insects*. Pitman, London.

Flögel, J. H. L. 1878. Über den einheitlichen Bau des Gehirns in den verschiedenen Insektenordnungen. *Z. Wiss. Zool.* **30**(suppl): 556–592.

Frontali, N. 1968. Histochemical localization of catecholamines in the brain of normal and drug-treated cockroaches. *J. Insect Physiol.* **14:** 881–886.

Frontali, N., R. Piazza, and R. Scopoletti. 1971. Localization of acetylcholinesterase in the brain of *Periplaneta americana*. *J. Insect Physiol.* **17**: 1833–1842.

Frontali, N. and R. Pierantoni. 1973. Autoradiographic localization of ³H-GABA in the cockroach brain. *Comp. Biochem. Physiol.* **44A**: 1369–1372.

Goodman, C. S. and J. L. D. Williams. 1976. Anatomy of the ocellar interneurons of acridid grasshoppers. II. The small interneurons. *Cell Tissue Res.* **175**: 203–226.

Goll, W. 1967. Strukturuntersuchungen am Gehirn von *Formica*. *Z. Morphol. Ökol. Tiere* **59**: 143–210.

Gorczyca, M. and J. Hall. 1986. Immunocytochemical localization of choline acetyltransferase during development and in mutant *Drosophila melanogaster*. *Soc. Neurosci. Abstr.* **12**: 245.

Hanström, B. 1926. Untersuchungen über die relative Größe der Gehirnzentren verschiedener Arthropoden unter Berücksichtigung der Lebensweise. *Z. Mikrosk. Anat. Forsch.* **7**:135–191.

Hanström, B. 1928. *Vergleichende Anatomie des Nervensystems der wirbellosen Tiere*. Springer, Berlin.

Harris-Warrick, R. M. and R. E. Flamm. 1986. Chemical modulation of a small central pattern generator circuit. *Trends Neurosci.* **9**: 432–437.

Heisenberg, M., A. Borst, S. Wagner, and D. Byers. 1985. *Drosophila* mushroom body mutants are deficient in olfactory learning. *J. Neurogent.* **2**: 1–30.

Hess, A. 1972. Histochemical localization of cholinesterase in the brain of the cockroach (*Periplaneta americana*). *Brain Res.* **46**: 287–295.

Homberg, U. 1982. Das mediane Protocerebrum der Honigbiene (*Apis mellifica*) im Bereich des Zentralkörpers: Physiologische and morphologische Charakterisierung. Dissertation, Berlin.

Homberg, U. 1985. Interneurons of the central complex in the bee brain (*Apis mellifica*, L.). *J. Insect Physiol.* **31**: 251–264.

Homberg, U., T. G. Kingan, and J. G. Hildebrand. 1987. Immunocytochemistry of GABA in the brain and suboesophageal ganglion of *Manduca sexta*. *Cell Tissue Res.* **248**: 1–24.

Honegger, H. W. and F. W. Schürmann. 1975. Cobalt sulphide staining of optic fibers in the brain of the cricket *Gryllus campestris*. *Cell Tissue Res.* **159**: 213–225.

Howse, P. E. 1975. Brain structure and behavior in insects. *Ann. Rev. Entomol.* **20**: 359–379.

Huber, F. 1955. Sitz und Bedeutung nervöser Zentren für Instinkthandlungen beim Männchen von *Gryllus campestris* L. *Z. Tierpsychol.* **12**: 12–48.

Huber, F. 1960a. Experimentelle Untersuchungen zur nervösen Atmungsregulation der Orthopteren (Saltatoria, Gryllidae). *Z. Vgl. Physiol.* **43**: 359–391.

Huber, F. 1960b. Untersuchungen über die Funktion des Zentralnervensystems und insbesondere des Gehirns bei der Fortbewegung und Lauterzeugung der Grillen. *Z. Vgl. Physiol.* **44**: 60–132.

Huber, F. 1962. Lokalisation und Plastizität im Zentralnervensystem der Tiere. *Zool. Anz.* (suppl) *Verh. Dtsch. Zool. Ges.* **26**: 200–267.

Jonescu, C. 1909. Vergleichende Untersuchungen über das Gehirn der Honigbiene. *Jena. Z. Naturw.* **45**: 111–180.

Klemm, N. 1968. Monoaminhaltige Strukturen im Zentralnervensystem der Trichoptera (Insecta). *Z. Zellforsch. Mikrosk. Anat.* **92**: 489–502.

Klemm, N. 1976. Histochemistry of putative transmitter substances in the insect brain. *Progr. Neurobiol.* **7**: 99–169.

Klemm, N. 1983. Monoamine-containing neurons and their projections in the brain (supraoesophageal ganglion) of cockroaches. *Cell Tissue Res.* **229**: 379–402.

Klemm, N. and A. Björklund. 1971. Identification of dopamine and noradrenaline in the nervous structures of the insect brain. *Brain Res.* **26**: 459–464.

Klemm, N. and S. Axelsson. 1973. Determination of dopamine, noradrenaline and 5-hydroxytryptamine in the cerebral ganglion of the desert locust, *Schistocerca gregaria* Forsk. (Insecta, Orthoptera). *Brain Res.* **57**: 289–298.

Klemm, N., W. M. Steinbusch, and F. Sundler. 1984. Distribution of serotonin-containing neurons and their pathways in the supraoesophageal ganglion of the cockroach *Periplaneta amerciana* (L.) as revealed by immunocytochemistry. *J. Comp. Neurol.* **225**: 387–395.

Klemm, N., D. R. Nässel, and N. N. Osborne. 1985. Dopamine-β-hydroxylase-like immunoreactive neurons in two insect species, *Calliphora erythrocephala* and *Periplaneta americana. Histochemistry* **83**: 159–164.

Koelle, G. B. and J. S. Friedenwald. 1949. A histochemical method for localizing cholinesterase activity. *Proc. Soc. Exp. Biol. Med.* **70**: 617–622.

Koontz, M. A. and J. S. Edwards. 1984. Central projections of first-order ocellar interneurons in two orthopteroid insects *Acheta domesticus* and *Periplaneta americana. Cell Tissue Res.* **236**: 133–146.

Kutsch, W. and D. Otto. 1972. Evidence for spontaneous song production independent of head ganglia in *Gryllus campestris* L. *J. Comp. Physiol.* **81**: 115–119.

Lee, A. N., R. L. Metcalf, and G. M. Booth. 1973. House cricket acetylcholine esterase: Histochemical localization and in situ inhibition by O,O-dimethyl S-aryl phosphothioates. *Ann. Entomol. Soc. Am.* **66**: 333–343.

Lucht-Bertram, E. 1962. Das postembryonale Wachstum von Hirnteilen bei *Apis mellifica* L. und *Myrmeleon europeus* L. *Z. Morphol Ökol. Tiere* **50**: 543–575.

Marder, E. and S. L. Hooper. 1985. Neurotransmitter modulation of the stomatogastric ganglion of decapod crustaceans, pp. 319–337. In A. I. Selverston (ed.), *Model Neural Networks and Behavior.* Plenum, New York.

Mercer, A. R., P. G. Mobbs, A. P. Davenport, and P. D. Evans. 1983. Biogenic amines in the brain of the honeybee, *Apis mellifera. Cell Tissue Res.* **234**: 655–677.

Meyer, E. P., C. Matute, P. Streit, and D. R. Nässel. 1986. Insect optic lobe neurons identifiable with monoclonal antibodies to GABA. *Histochemistry* **84**: 207–216.

Milde, J. 1982. Elektrophysiologische und anatomische Untersuchungen an Interneuronen erster und höherer Ordnung des Ocellensystems der Biene (*Apis mellifica carnica*). Dissertation, Berlin.

Mobbs, P. G. 1985. Brain structure, pp. 299–370. In G. A. Kerkut and L. I. Gilbert (eds.), *Comprehensive Insect Physiology, Biochemistry, and Pharmacology,* vol. 5: *Nervous System: Structure and Motor Function.* Pergamon, Oxford.

O'Shea, M. and M. Schaffer. 1985. Neuropeptide function: The invertebrate contribution. *Ann. Rev. Neurosci.* **8:** 171–198.

Otto, D. 1971. Untersuchungen zur zentralnervösen Kontrolle der Lauterzeugung von Grillen. *Z. Vgl. Physiol.* **74:** 227–271.

Pan, K. C. 1981. The neuronal organization of the ocellar system and associated pathways in the central nervous system of the worker bee, *Apis mellifera.* Doctoral dissertation, University of London.

Pflugfelder, O. 1937. Vergleichende, anatomische, experimentelle und embryologische Untersuchungen über das Nervensystem und die Sinnesorgane der Rhynchoten. *Zoologica* **34:** 1–102.

Power, K. D. 1937. The brain of *Drosophila melanogaster. J. Morphol.* **72:** 517–559.

Roeder, K. D. 1937. The control of tonus and locomotor activity in the praying mantis (*Mantis religiosa*). *J. Exp. Zool.* **76:** 335–374.

Schäfer, S. and G. Bicker. 1986. Distribution of GABA-like immunoreactivity in the brain of the honeybee. *J. Comp. Neurol.* **246:** 287–300.

Schildberger, K. 1982. Untersuchungen zur Struktur und Funktion von Interneuronen im Pilzkörperbereich des Gehirns der Hausgrille *Acheta domesticus.* Dissertation, Göttingen.

Schildberger, K. 1983. Local interneurons associated with the mushroom bodies and the central body in the brain of *Acheta domesticus. Cell Tissue Res.* **230:** 573–586.

Schooneveld, H., H. M. Romberg-Privee, and J. A. Veenstra. 1985. Adipokinetic hormone-like immunoreactive peptide in the endocrine and central nervous system of several insect species: A comparative immunocytochemical approach. *Gen. Comp. Endocrinol.* **42:** 526–533.

Schürmann, F. W. and N. Klemm. 1984. Serotonin-immunoreactive neurons in the brain of the honeybee. *J. Comp. Neurol.* **225:** 570–580.

Steiniger, F. 1933. Die Erscheinung der Katalepsie bei Stabheuschrecken und Wasserläufern. *Z. Morphol. Ökol. Tiere* **26:** 591–708.

Strausfeld, N. J. 1976. *Atlas of an Insect Brain.* Springer, Berlin.

Thompson, C. B. 1913. A comparative study of the brains of three genera of ants with special reference to the mushroom bodies. *J. Comp. Neurol.* **23:** 515–572.

Tyrer, N. M., J. D. Turner, and J. S. Altman. 1984. Identifiable neurons in the locust central nervous system that react with antibodies to serotonin. *J. Comp. Neurol.* **227:** 313–330.

Usherwood, P. N. R. 1978. Amino acids as neurotransmitters. *Adv. Comp. Physiol. Biochem.* **7:** 227–309.

Veenstra, J. A. and N. Yanaihara. 1984. Immunocytochemical localization of gastrin-releasing peptide/bombesin-like immunoreactive neurons in insects. *Histochemistry* **81:** 133–138.

Viallanes, H. 1887. Études histologiques et organologiques sur les centres nerveaux et les organes des sens des animeaux articulés. 4eme memoire. Le cerveaux de la guêpe. (*Vespa crabro* et *V. vulgaris*). *Ann. Sci. Nat. Zool.* **7**(2): 5–100.

Vieillemaringe, J., P. Duris, M. Geffard, M. Le Moal, M. Delaage, C. Bensch, and

J. Girardie. 1984. Immunocytochemical localization of dopamine in the brain of the insect *Locusta migratoria migratorioides* in comparison with the catecholamine distribution determined by the histofluorescence technique. *Cell Tissue Res.* **237**: 391–394.

Wadepuhl, M. 1983. Control of grasshopper singing behavior by the brain: responses to electrical stimulation. *Z. Tierpsychol.* **63**: 173–200.

Williams, J. L. D. 1972. Some observations on the neuronal organization of the supra-oesophageal ganglion in *Schistocerca gregaria* Forskål with particular reference to the central complex. Doctoral dissertation, University of Wales.

Williams, J. L. D. 1975. Anatomical studies of the insect central nervous system: A ground-plan of the midbrain and an introduction to the central complex in the locust *Schistocerca gregaria* (Orthoptera). *J. Zool. (London)* **167**: 67–86.

CHAPTER 16

Insect Brain Metabolism Under Normoxic and Hypoxic Conditions

Gerhard Wegener

Institute of Zoology
Johannes Gutenberg University
Mainz, West Germany

Gratefully dedicated to Dr. Bernhard Rensch, Emeritus Professor of Zoology, University of Münster.

16.1. INTRODUCTION

Nervous tissue functions are based on the same principles in the entire animal kingdom and activation of neurons is connected with an increase in energy production (Wegener, 1981, for review). Among the three animal groups, arthropods (especially insects), cephalopods, and vertebrates, which possess highly efficient brains, only a few laboratory mammals and humans have been carefully studied with respect to brain energy metabolism. Although insects form the largest animal group and have eminent significance as pollinators, pests, or in spreading disease, little is known about their brain metabolism. This is primarily because of the difficulty of obtaining sufficient amounts of material for conventional biochemical analysis. If methods are scaled down, however, insect ganglia are suitable models, and studying their metabolism will add to both the basic knowledge on nervous systems and to comparative physiology. For instance, local metabolic activity can be mapped in the insect brain and has added to the knowledge of its functional organization (Buchner et al., 1979, 1984).

I shall review brain metabolism supplemented with data on nerve cord ganglia. Although the brain certainly plays a leading role with respect to sensory input, information processing, and hormonal coordination, its metabolic organization is not likely to differ qualitatively from the rest of the ganglia. Nevertheless, the picture we can draw is far from being complete. Only a few laboratory insects have yet been studied. Even this limited information clearly indicates that insect brain metabolism is not uniform, but subject to great variability and manifold adaptation.

16.2. INSECT BRAIN METABOLISM UNDER NORMOXIC CONDITIONS

16.2. Oxygen Supply

Energy metabolism of adult insects is very efficient and highly dependent on oxygen. Because the tracheal system transfers oxygen close to the sites of consumption in gaseous form, it can take advantage of the fast diffusion and high content of oxygen in air. The tracheation of the brain in *Musca domestica* is briefly described in Strausfeld (1976), and a thorough account is given for the CNS of the locust (*Schistocerca gregaria*) by Burrows (1980). The tracheae providing the nervous system are independent from those serving the flight muscle. Inspired air is first carried to the head region, supplying the brain, before it passes on to the rest of the CNS. Thus, preference is given to the brain with respect to oxygen supply. Inside the ganglia, the densest tracheation is in the synaptic regions of the neuropil, whereas tracheoles are scarce in cortical areas (Burrows, 1980) where the cell bodies are located, which in insects do not receive synaptic input. The mean distance between tracheoles is only 6–10 μm in the synaptic regions of the optic lobes and 17 μm in the thoracic ganglia (Burrows, 1980), whereas the corresponding data for capillaries in the brain cortex of humans and mammals are approximately 50 μm (Thews, 1960) and about twice this figure in brain tissue of lower vertebrates (Horstmann, 1960).

16.2.2. Fuel Supply, Substrates, and Energy Stores

16.2.2.1. Fuel Supply and Substrates. The insect brain is a compact organ, but it lacks capillaries. Consequently, all substrates have to cross the neural sheaths. Although some information is available on ion transport in the insect CNS (Treherne, 1974; Treherne and Schofield, 1979; Lane and Treherne, 1980), little is known about transport of substrates. In most insects, the predominant blood sugar is trehalose. Nevertheless, isolated nerve cord ganglia from *Periplaneta americana* take up glucose at a higher rate than trehalose, a difference that cannot be explained by the slightly different diffusion constants (Treherne, 1960). In incubated thoracic ganglia of the locust, *Schistocerca gregaria,* 10 mM glucose sustained maximal oxygen consumption, but 50 mM trehalose would not (Strang and Clement, 1980). Trehalose concentration in the ganglia was relatively low (6.4 μmol g^{-1}) compared with 54 mM in the hemolymph, and the Michaelis constant of trehalase was comparatively high (10 mM). It seems wise, however, not to draw too straightforward conclusions, because some of the results might be due to the in vitro conditions (cf., Section 16.2.4.). Another problem is the possible involvement of fat body cells adhering to the CNS (neural fat body). Although the neural fat body is conspicuous in many insects, next to nothing is known about its function in the nutrition of the CNS.

Glucose certainly is an adequate fuel for the insect CNS. Glycolytic enzymes have been demonstrated in brain tissue of various species (see Section 16.2.3.; Table 16.1), and ^{14}C-glucose is oxidized by isolated insect brains (Sections 16.2.4. and 16.4.2.). In insect CNS, as in nervous tissue from other sources (vertebrates and invertebrate phyla), glucose-carbon is rapidly incorporated into certain amino acids, such as aspartate, glutamate, and alanine (Treherne, 1960; Bradford et al., 1969). This must not be interpreted,

TABLE 16.1. METABOLIC ORGANIZATION OF INSECT BRAINS AS REFLECTED IN THE CATALYTIC CAPACITIES OF ENZYMES REPRESENTING THE MAIN PATHWAYS OF ENERGY METABOLISM*

	Locusta	Apis	Calliphora	Manduca	Bombyx	Mus
Glycogen phosphorylase (GPase, EC 2.4.1.1)	2.2	2.4	0.8	4.2	1.3	0.7
Hexokinase (HK, EC 2.7.1.1)	5.0	12.5	4.6	6.4	6.0	12.4
Glucose 6-P dehydrogenase (G6PDH, EC 1.1.1.49)	1.7	0.8	0.7	2.5	2.8	1.2
Phosphofructokinase (PFK, EC 2.7.1.11)	5.3	9.8	2.4	5.3	2.8	14.3
Glycerol 3 P dehydrogenase (GDH, EC 1.1.1.8)	3.0	15.4	8.4	9.6	7.4	4.2
Lactate dehydrogenase (LDH, EC 1.1.1.27)	5.6	0.4	1.3	9.9	9.0	97.1
Alanine aminotransferase (AlaAT, EC 2.6.1.2)	28.7	27.3	54.9	13.5	31.0	0.8[b]
Aspartate aminotransferase (AspAT, EC 2.6.1.1)	25.2	9.0	37.8	44.7	35.1	74.8[b]
Glutamate dehydrogenase (GluDH, EC 1.4.1.3)	11.2	38.4	65.1	41.3	62.2	24.3[b]
Citrate synthase (CS, EC 4.1.3.7)	11.0	19.0	16.5	12.8	16.3	13.0
3-hydroxyacyl-CoA dehydrogenase (HOADH, EC 1.1.1.35)	17.0	0.4	4.5	28.0	95.5	2.7
3-oxoacid CoA-transferase (KCT, EC 2.8.3.5)	3.5	2.5	6.9	5.7	7.9	9.8[a]

* Maximum activities in μmol substrate transformed per min and g fresh tissue at 25°C.

[a,b] Data for mouse brain are from Wegener and Zebe, 1971; Sugden and Newsholme, 1973[a], 1975[b]; data for insect brains from Wegener and Zebe, 1973; Wegener and Pfeifer, 1975; Diehl et al., 1980; Wegener, 1983; and unpublished results.

The following metabolic capacities are thought to be reflected by the enzymes investigated (abbreviations in brackets): glycogenolysis (GPase), glucose phosphorylation (HK), glycolysis (PFK), pentose phosphate pathway (G6PDH), lactate formation (LDH), transamination and oxidative deamination of amino acids (AlaAT, AspAT, GluDH), citric acid cycle (CS), ketone body oxidation (KCT), and oxidation of fatty acids (HOADH).

without additional evidence, as net synthesis of these amino acids from glucose. Nervous tissue has high activities of aminotransferases (cf., Table 16.1), which can rapidly equilibrate NH_3 between amino acids and the oxoacid-intermediates of glucose catabolism (Balázs and Haslam, 1965).

Fat is the most advantageous energy store in animals. Owing to its high energy content and its hydrophobic character, 1 g of fat is equivalent to about 8 g of glycogen (if the tissue water is not neglected). On first sight it is surprising that, with the notable exception of some insect groups, brains from both vertebrates and invertebrates have been found incapable of oxidizing fatty acids for energy production (Wegener and Zebe, 1971; 1973; Agardh et al., 1981). It sheds some light on the great adaptive potential of insects that among all the animals investigated, none, except a few insects, have been found capable of using fatty acids for brain metabolism (see Sections 16.2.4, 16.4.1; and Wegener, 1983).

The importance of ketone bodies for brain energy metabolism in humans and other mammals is now generally acknowledged (Sokoloff, 1973). Ketone bodies are present in insects; in contrast to vertebrates, however, acetoacetate, not 3-hydroxybutyrate, predominates. Acetoacetate is elevated in locust hemolymph during flight and after prolonged starvation (Hill et al., 1972; Strang, 1981 for review), but to what extent ketone bodies contribute to insect metabolism is not known. Interestingly, the activities of ketone body-oxidizing enzymes are high in brains that can also oxidize fatty acids (Table 16.1). We suppose that in those insects, whose mode of living or physiological activities (long distance fliers, nonfeeding adults) require large storage of fuels, fatty acids are of such paramount importance that they are made available to the CNS both directly and after "predigestion" in the form of water-soluble ketone bodies. Ketone body oxidation also has been demonstrated in vitro (Section 16.4.2). Isolated brains of the sphinx-moth, *Manduca sexta*, produced $^{14}CO_2$ from ^{14}C-acetoacetate (Knollmann and Wegener, 1983), whereas ^{14}C-hydroxybutyrate was not oxidized.

Amino acids are prominent constituents of insect hemolymph (for review Wyatt, 1961; Jeuniaux, 1971; Treherne, 1974) and also of CNS ganglia (Jabbar and Strang, 1984). Little is known, however, about their relevance as substrates for nervous tissue. Proline, in some insects a substrate for flight (Bursell, 1981), can also be oxidized by ganglia. $^{14}CO_2$ was produced from ^{14}C-proline by isolated brains of blowflies and other insects (Kern and Wegener, 1981, and unpublished results; Kern, 1982). It was also oxidized by isolated locust ganglia and mitochondrial preparations (Strang, personal communication). Proline has been suggested as an additional substrate in nervous tissue at times when carbohydrate is limited (Strang, 1981). Alanine is elevated in hypoxic insect ganglia (see Section 16.3.4).

16.2.2.2. Energy Stores in Insect Brains. Glycogen is abundant in the CNS of insects (Table 16.2) and it has been demonstrated by various methods. Wigglesworth (1960), in a thorough histochemical study on the last abdom-

TABLE 16.2. GLYCOGEN STORES AND RESPIRATORY CAPACITIES OF ISOLATED INSECT BRAINS (MOUSE/RAT BRAIN FOR COMPARISON)

	Locusta	Apis	Calliphora	Bombyx	Mus/Rattus
Glycogen (as μmol glucose·g^{-1})	20–30[a] 11.1[b]	15.7[c]	10.6[c]		2.2[d]
Oxygen consumption (μmol O_2·g^{-1} min^{-1})	4.5[b]	4.2	4.8	4.0	3.4[e]
Respiratory quotient (RQ = CO_2/O_2)		1.0	0.96	0.74	1[e]
$^{14}CO_2$ produced from ^{14}C fatty acids (dpm·mg^{-1} ww·h^{-1})		63	1082	5017	
Glycolytic rate (μmol·g^{-1}·min^{-1})	(0.75)[b]	0.7 (0.7)	0.76 (0.8)	0.08 (0.67)	0.6[e]

[a–e] The glycolytic rates of insect brains were calculated from oxygen consumption and RQ, assuming glucose and oleic acid as the only fuels; the values in parentheses would result if glucose was the only fuel (from Wegener, 1983). Determinations were performed as described in the text at 25°C unless otherwise indicated. From Wegener (1983) and the following sources: [a] Michel and Wegener, 1982; [b] Clement and Strang, 1978 (*Schistocerca gregaria*, thoracic ganglia, 37°C); [c] Kieffer and Wegener, 1983; [d] Lowry et al., 1964; [e] Siesjö, 1978 (rat, 37°C, lightly anesthetized).

inal ganglion of the cockroach, *Periplaneta americana,* demonstrated glycogen deposits in the perineurium cells surrounding the ganglion, and in the region of the axon cones of neurons, where glial invaginations (Trophospongium) are conspicuous. Glycogen was reduced during starvation and rapidly replenished on feeding. Carbohydrate is probably transferred to the neurons via perineurium and glia cells, but it is not known how this is achieved, particularly if and how the metabolic needs of the neurons are signaled to the glia and whether or not glycogen has to be degraded for transport. Glycogen has been demonstrated within fine branches of glia cells in the brain of *Musca domestica* (Fig. 16.3). Aggregations of glycogen in neurons were elicited by stressful interventions that had increased blood sugar (Wood et al., 1980). Glycogen was also found in synaptosomes from locust ganglia (Breer and Jeserich, 1984). In synaptosomal fractions glycogen appears to be abundant and its significance is obvious. Synaptosomes respire without additional substrate, thus showing, also on the subcellular level, the high degree of metabolic autarky of insect ganglia that has been emphasized (cf., Wegener, 1981; and Section 16.2.4.).

Information on lipid stores of the CNS is scarce. In many insects, the CNS is surrounded by fat body cells stuffed with lipid droplets, but their significance for the nervous tissue is not known. In *Periplaneta,* lipids seem to be transferred to the neurons from the perineurium and glia cells (Wigglesworth, 1960). Lipid droplets were also found in the perineurium of *Carausius morosus* (Smith, 1968).

16.2.3. Metabolic Pathways and Enzymes in Insect Brains

The metabolic capacities (but not the actual rates) of an organ can be inferred from its enzyme activities. Some information on insect brains is gathered in Table 16.1. (see also Strang, 1981; Rivera and Langer, 1983).

High activities of glycolytic enzymes have been found in all insect brains. PFK, a key regulatory enzyme of glycolysis, has a maximum activity in excess of the calculated glycolytic flux (Table 16.2). Significant activities have also been found of GPase and rather high activities of HK. The pentose phosphate pathway is also represented. Insect brains have low LDH activities, a characteristic reflecting the efficient oxygen-supplying tracheal system. It is therefore no surprise to find the Krebs cycle well represented. Cytosolic NADH can be shuttled into the respiratory chain via the cytosolic (GDH) and mitochondrial glycerol 3-phosphate dehydrogenase.

Striking differences exist with respect to lipid oxidation. The honey bee brain, for example, has virtually no capacity to use fatty acids in contrast to *Bombyx* (Table 16.1). The ketone body acetoacetate can be oxidized in insect nervous tissue. KCT is present in all species that have been assayed (Table 16.1; Hill et al., 1972; Sugden and Newsholme, 1973; Diehl et al., 1980), whereas 3-hydroxybutyrate dehydrogenase is negligible.

Apparently, all insect brains have high activities for the oxidative deamination of glutamate (GluDH) and for the transamination of amino acids (Bradford et al., 1969; Sugden and Newsholme, 1975). The latter enzymes rapidly shift radioactive label from ^{14}C-glucose into amino acids, mainly aspartate and glutamate (cf., Section 16.2.2.1). In locust thoracic ganglia, activity of proline dehydrogenase has been found (Strang, personal communication), and proline is oxidized by isolated insect ganglia (Sections 16.2.2.1, 16.2.4).

16.2.3.1. Subcellular Compartmentation of Enzymes in Insect Brains.

The subcellular localization of enzymes in the insect brain was achieved by the "fractionated extraction" (fraktionierte Extraktion, Pette, 1968). It was found identical to that in vertebrate brains, with the notable exception that mitochondrial binding of hexokinase could not be observed in insect brain (Wegener and Pfeifer, 1975) and thoracic ganglia (Strang et al., 1979). Enzymes of glycolysis as well as G6PDH and GDH could be classified as cytosolic proteins. CS, GluDH, and the enzymes of fatty acid and ketone body oxidation are located in the mitochondria, whereas AspAT is present in both compartments. Proline dehydrogenase was found in mitochondrial preparations (Strang, personal communication). Those enzymes, which differ markedly between different insect brains are of mitochondrial origin. The ability of insect brains to use different substrates, therefore, rests in the enzymatic specialization of their mitochondria.

16.2.4. Metabolism of Isolated Insect Brains

Their specialized organization (lack of blood vessels) render insect ganglia particularly suited for in vitro studies. Two approaches have been used. In the first, isolated ganglia were submersed in fluid media to study ion and sugar transport (Treherne, 1960, 1966, 1974) as well as oxygen consumption (Strang, 1981, 1985 for review). This method offers a fairly constant environment to the ganglia, but cannot use the respiratory system and is liable to loss of endogenous substrates (Hart and Steele, 1973; Strang et al., 1979, Strang and Clement, 1980) and probably other compounds from the tissue. Maximum respiration was only reached with 10 mM glucose and saturating the medium with 100% oxygen (Clement and Strang, 1978).

The second approach has aimed at using the tracheal system and at minimizing the loss of endogenous compounds. To this end, brains were isolated, placed on tiny pieces of filter paper, moistened with 1 μL of medium, and then transferred to a reaction vessel (volume, 150–200 μL) of a Hamdorf microrespirometer (Fig. 16.1; and Hamdorf and Kaschef, 1964 for details). Isolated brains can respire at a linear rate for up to 1 hour. The addition of substrate did not increase the oxygen uptake. This proves tissue storage sufficient for maintaining brain energy metabolism and reveals a remarkable metabolic stability of insect ganglia. With larger insects, maximum respiration in air may only be reached if the brain trachea are not obstructed. Whether or not this criterion is met can easily be tested by measuring at elevated oxygen concentration (Loskill and Wegener, unpublished data).

The microrespirometry of insect brains was first established by the author and further developed in collaboration with his former students R. Backes and M. Kern (cf., Wegener, 1975, 1981, 1983; Backes, 1978; Kern and Wegener, 1981, 1984; Kern, 1982). It can readily be adapted to problems not easily accessible with other methods. It has been used to determine respiratory quotients (RQ; Figure 16.1), to follow respiratory activity during brain development and aging (Section 16.4.3.), and to test substances for their potency to affect brain metabolism. A further refinement includes the use of ^{14}C-labeled substrates and the parallel determination of oxygen consumption and $^{14}CO_2$ production (radiomicrorespirometry: Kern, 1982; Wegener, 1983; Knollmann and Wegener, 1983, 1985). Application of ^{14}C-fatty acids revealed conspicuous differences of insect brains to oxidize these substrates (Wegener, 1983; Table 16.2).

16.2.5. Metabolic Rates of Insect Brains

Metabolic rates of brains can only be measured directly in humans and a few other mammals. Metabolic rates of insect brains in vitro are probably in the range of the in vivo values because (1) insect brains are completely aerobic, and oxygen uptake, therefore, is equivalent to metabolic rate; (2)

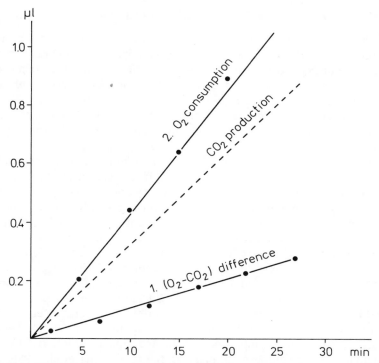

Figure 16.1. Respiration of an isolated silkmoth brain (*Bombyx mori*) in a Hamdorf microrespirometer at 25°C. Oxygen uptake and ($O_2 - CO_2$) difference were determined successively in the same ganglion in the indicated order yielding a respiratory quotient (RQ) of 0.75. The difference between O_2 consumption and CO_2 evolution (*broken line*) indicates the oxidation of fuels other than carbohydrates, most probably fatty acids (see text). (From Wegener, 1983.)

isolated ganglia show spontaneous electrical activity (Clement and Strang, 1978; Finlayson et al., 1976); and (3) cutting off the nerve connections does not affect respiration to a marked extent (Kern and Wegener, 1984, and unpublished results). Experiments on the isolated thoracic ganglia of locust would suggest similar conclusions (Clement and Strang, 1978; Strang, 1981 for review).

Isolated brains of small holometabolous insects at 25°C consume more oxygen than brains of small mammals at 37°C (Table 16.2). Because body temperature in active insects is usually above 30°C, holometabolous insects obviously have the most active brains of all animals. Most of the energy is apparently used for ion transport because ouabain, an inhibitor of the Na^+/K^+ ATPase, reduced oxygen uptake of isolated *Calliphora* brains by two-thirds (Loskill and Wegener, unpublished data).

16.2.6. Regulatory Features of Insect Brain Metabolism

Very little is known on the regulation of energy metabolism in insect brains. Glycogen was mobilized in the isolated nerve cord by octopamine (but not by catecholamines) and by a proteinase-labile factor from the corpora cardiaca. The effects are mediated through activation of glycogen phosphorylase, and cAMP seems to be involved (Steele, 1982 for review). In recent years several peptide hormones have been characterized in insects (Goldworthy and Wheeler, 1985 for review), but their effect on CNS metabolism is unknown.

Hexokinase from brain tissue of *Locusta migratoria* has been studied using affinity chromatography (Werner, 1980). Phosphofructokinase (PFK) is a key regulatory enzyme of glycolysis in many animal tissues. The enzyme from various mammalian organs has common regulatory features: cooperativity with respect to substrate binding; inhibition by physiological ATP levels and by citrate; and activation by AMP, inorganic phosphate, NH_4^+, and the reaction product, fructose 1,6-bisphosphate. Purification seems a prerequisite to fully assess the regulatory features of PFK. Up to now the enzyme has only been purified from the flight muscles of *Periplaneta* (Storey, 1985), *Apis* (Wegener et al., 1986a), and *Locusta* (Wegener et al., 1986b, 1987b). Phosphofructokinases from insects differ from their vertebrate counterparts in not being affected by citrate (Newsholme et al., 1977). The *Apis* PFK is, quite unusually, inhibited by the reaction product, fructose 1,6-bisphosphate, and by glucose 1,6-bisphosphate (Wegener et al., 1986a), and there is evidence that the PFK from honey bee brain shares this feature (Fig. 16.2). All phosphofructokinases seem to be highly activated by the recently detected effector fructose 2,6-bisphosphate, but further generalizations have to await more work.

The mammalian brain reduces glucose oxidation if ketone bodies are available. This saving of carbohydrate is essential to survive prolonged starvation and it is apparently based on inhibition of glycolysis (effect of citrate on PFK) and pyruvate oxidation. As isolated *Manduca* brains reduced glucose oxidation in the presence of acetoacetate, a glucose-saving mechanism might be present in insect brains, but the molecular basis is still unknown (Knollmann and Wegener, 1985). In flight muscle of locusts, which change from carbohydrate to lipid oxidation during flight, PFK and fructose 2,6-bisphosphate seem to be involved (Wegener et al., 1986b, 1987b).

16.2.7. Differences in Brain Metabolism Between Vertebrates and Insects

Among vertebrates, evolution has favored big brains (in homoiothermic animals) with a high capacity of learning and memory (Rensch, 1956). This was backed by metabolic adaptations as increased metabolic rate, reduction of energy stores within nervous tissue, and impressive regulatory features

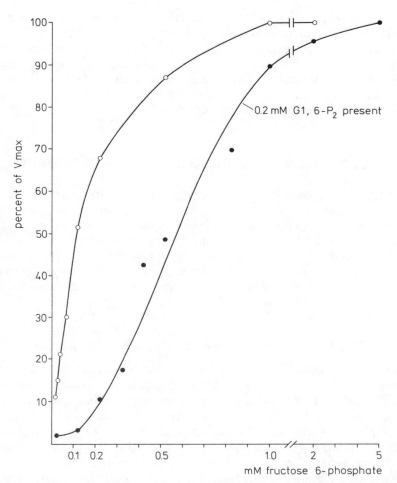

Figure 16.2. Regulatory features of phosphofructokinase (PFK) in tissue extract from honey bee brain (*Apis mellifera*). This PFK is unusual because at near physiological concentrations of ATP and in presence of AMP, the $S_{0.5}$ value (substrate concentration at 50% activity) is markedly increased by glucose 1,6-bisphosphate (G1,6-P_2). This inhibition, which in vivo might be effected mainly by the reaction product fructose 1,6-bisphosphate, can be counteracted by inorganic phosphate and fructose 2,6-bisphosphate (see text). Assay conditions: 50 mM triethanolamine, 50 mM KCl, 7 mM $MgCl_2$, 0.5 mM AMP, 5 mM ATP, pH 7.3, 25°C. (For details see Wegener et al., 1986a.)

of the body to ensure the uninterrupted supply of oxygen and substrate to the brain. This hierarchic organization makes brain function in birds and mammals very efficient but uniquely dependent on the caloric homoeostasis of the body and highly susceptible to disturbances (Wegener, 1981).

Due to the lack of capillaries, insects required a different organization of brain energy metabolism. Although the mode of oxygen supply favored an aerobic and active metabolism, stores within the ganglia seem inevitably to guarantee permanent function. Tracheation and fuel stores are supposed to confer a temporary independence to the ganglia, allowing in vitro studies. Brain energy metabolism in vertebrates is restricted to glucose, and to glucose plus ketone bodies in starvation, whereas a variety of substrates apparently can be used by ganglia of several insects.

16.3. INSECT BRAIN METABOLISM UNDER HYPOXIC CONDITIONS

The effects of hypoxia or anoxia have mostly been studied on larval and pupal stages of insects (von Brand, 1946; Gäde, 1985 for review), but specific information on the nervous system is not available. So we will restrict our consideration to adult insects. In humans and other mammals, the brain is the most vulnerable organ in hypoxia. This holds true also for lower vertebrates (Wegener et al., 1987a), but the question has not been answered for any insect. However, there are indications that in insects, too, the CNS is first affected by hypoxia (Section 16.3.2).

16.3.1. Reactions of Insects to Hypoxia and Anoxia

Insects can withstand degrees of hypoxia that could not be tolerated by vertebrates, let alone mammals. At 2% oxygen, adult locusts maintain their normal body posture (Michel and Wegener, unpublished data), and cockroaches can tolerate 2.2% oxygen (Walter and Nelson, 1975). In anoxia, however, insects will, after a short period of hyperactivity, rapidly become paralyzed. Locusts (*Locusta migratoria*), flushed with pure nitrogen at 24–25°C, immediately showed escape movements followed by heavy respiratory movements (abdominal pumping) and a loss of coordination. After about 30 sec, the animals fell on their side, and a little later, twitches, stretching, and tremor of the hind legs led to total immotility between 60–75 sec. Similar events have been described for the anoxic cockroach by Walter and Nelson (1975).

In *Calliphora, Musca,* and *Apis* (Ray and Heslop, 1963; Kieffer and Wegener, unpublished data) paralysis is reached even faster. Honey bees, for example, lost coordination within 10 sec, and after 30 sec all movements had ceased. A rapid effect on CNS function can also be inferred from the fact that flies will not establish a conditioned response if treated with nitrogen for 25 sec right after the training (Akahane and Amakawa, 1983; Menzel et al., 1974). Electrical activity was rapidly lost in the CNS of insects on hypoxic interference (thoracic ganglia of *Periplaneta;* Walter and Nelson, 1975). All observations prove that adult insects cannot maintain CNS functions anaerobically, not even for short periods.

16.3.2. Anoxic Survival (Revival) and CNS Functions in Insects

Recovery from anoxia appears, on an extended time scale, roughly as a reversal of the reactions to anoxia and has been described for *Musca domestica* (Ray and Heslop, 1963; Engels, 1968a). Some insects can completely recover from extended anoxic periods. In other species, a gradual increase of functional impairment with the length of the anoxic interval has been noted. Young *Musca domestica* regain full motility within 10 min after 1–2 hours of anoxia, and after 4 and 8 hours in nitrogen, 1 and 4 hours, respectively, are needed for recovery (Engels, 1968a). On the other hand, 8–15-day-old male *Calliphora erythrocephala* did not survive 4 hours of anoxia. Only approximately 50% were able to fly after 30 min of anoxia, and these had only a restricted command of their wings (Kieffer and Wegener, unpublished data). In *Locusta,* optomotoric reactions were changed after 2 hours of anoxia, and the animals failed to feed and died within 5–8 days if anoxia exceeded 3–4 hours (Michel and Wegener, unpublished data). Honey bees have been reported to survive 6 hours in 100% CO_2 without changes in behavior or loss of memory (Medugorac and Lindauer, 1967), but anoxia might not have been complete and an effect of CO_2 could have been involved. We noted that honey bees did not fully recover from anoxic periods exceeding 1 hour. After 60 min of anoxia, none of the 38 trained bees returned to the artificial food source, but some 25% returned after only 30 min of anoxia, whereas the corresponding figures were 44% and 63% for controls kept at 4°C (Kieffer and Wegener, 1983). This would suggest an impairment of complex CNS functions if anoxia exceeds a critical limit.

The conspicuous differences between the species with respect to revival from anoxia cannot be explained yet (Table 16.3). It is rather difficult to define complete recovery, but the data clearly show an enormous variation. Anoxic revival apparently does not depend on the type (predominant substrate) of energy metabolism, as can be seen from comparing Lepidoptera and Diptera. Larger animals tend to survive longer periods of anoxia (cf., the castes of the leaf-cutting ant, *Atta sexdens*) and females surpass males. This might reflect the different amounts of energy reserves as has been suggested by Engels (1968a,b). Considering their high metabolic rate and their aerobic energy metabolism, adult insects can survive surprisingly long anoxic periods.

16.3.3. Effects of Anoxia on Brain Ultrastructure

It is not known which mechanisms initiate irreversible anoxic changes in neurons, and which events mark the point of no return. The situation is further complicated in capillarized tissues, because changes in the vascular system certainly contribute to the damage. The insect CNS offers the unique opportunity to study anoxic cell changes in a highly active CNS that lacks capillaries, but work has only just been started (Matt, Koller, and Wegener,

TABLE 16.3. REVIVAL FROM ANOXIA IN ADULT INSECTS[a]

	Recovery after an Anoxic Period of Hours		
	Complete	Incomplete	No
Orthopteroidea[b]			
Locusta migratoria♀	3	4	5
Locusta migratoria♂	2	3	4
Heteroptera			
Dysdercus intermedius	0.5	1	2
Rhodnius prolixus	8	?	10
Lepidoptera			
Manduca sexta♂	6	?	8
Bombyx mori♀	8	12	14
Bombyx mori♂	8	12	14
Hymenoptera			
Apis mellifera, worker	1	8	10
Coccygominus turionellae♀	12	28	?
Coccygominus turionellae♂	10	12	14
Atta sexdens, soldier	17	26	28
Atta sexdens, large worker	8	20	22
Atta sexdens, small worker	4	6	8
Diptera			
Calliphora erythrocephala♂	0.25	2	4
Musca domestica♂	12	?	15
Drosophila melanogaster♀	4	?	6
Drosophila melanogaster♂	2	4	6

Source: Kieffer, Michel, and Wegener, unpublished results.
[a] Insects were subjected to an atmosphere of pure nitrogen for various periods of time at room temperature (22–23°C) and then returned to air for recovery. Depending on the length of the anoxic interval the effect could be classified as follows: Complete recovery: the majority of the animals recovered without noticeable functional impairment; incomplete recovery: the majority of the animals regained some function but suffered from damages that might eventually lead to death; no recovery: all animals died from the anoxic insult, and the majority of them showed no sign of life when returned to air.
[b] *Locusta* was tested at 24–25°C.

unpublished studies). Changes were first seen in the perikarya as distensions of the endoplasmic reticulum, including the nuclear envelope. Mitochondria were seriously affected and all degrees of disintegration could be found (Fig. 16.4). It is not known whether in insects (as in mammals) certain brain areas are particularly sensitive to anoxia. Changes could be found after relatively short anoxic intervals (5 min in *Calliphora,* 30 min in *Musca*), but the degree of disintegration and the percentage of afflicted organelles grew with the length of the anoxic interval. There were, however, even after extended anoxic periods, such as 6 hours in *Musca,* mitochondria that appeared quite normal (Fig. 16.4). The anoxic changes can be reversed, but the mechanisms involved have not been studied. *Musca domestica,* surviving for 2 days after 6 hours of anoxia, showed no signs of lasting damage in cell organelles, but a shrinking of axons was still visible.

Anoxic cell changes in insect brain are very similar to anoxic/ischemic cell changes in the CNS of mammals (cf., McGee-Russell et al., 1970), suggesting similar mechanisms. This renders insects promising models for the study of anoxic cell changes beyond the boundary of insect physiology.

16.3.4. Anaerobic Metabolism of Insect Brains

Work on anaerobic metabolism should not only describe the pathways involved, but reveal the mechanisms causing functional failure of and damage to the brain. Despite a plethora of studies, mainly on mammals, neither question can be answered for any system (cf., Wegener et al., 1987a).

In all animals, brain function is strictly connected with ATP production. A direct correlation is, however, difficult to establish. In the anoxic CNS of the cockroach, *Periplaneta,* electrical activity ceased after 2 min (at 25°C) when about 50% of the ATP was still present and the state of immobility had already been reached (Walter and Nelson, 1975). In mammals, no significant decrease in ATP was found at the time when brain function was lost (Duffy et al., 1972; Siesjö, 1978). The bulk of ATP in the whole organ, however, could well preclude changes in small localized pools, say in synaptic areas. The phosphagen level would appear a better indicator of a deranged energy status than ATP (unpublished). In *Musca,* recovering from anoxia, physical activity reappeared with the return of arginine phosphate to control levels (Ray and Heslop, 1963).

Maintaining CNS function in anoxia would require (1) availability of sufficient substrate, (2) sufficient catalytic activity to compensate for the less efficient anaerobic ATP production, and (3) removal of products arising from anaerobic metabolism. As we will see, these conditions are not met in insect brain tissue. Like mammals, adult insects cannot maintain CNS function without oxygen. While mammals, however, will be irreversibly damaged by anoxia exceeding 5–8 min, some insects recover from extended anoxia, thus demonstrating that revival is not necessarily connected with maintaining neuronal functioning and high ATP levels. Therefore, comparing anaerobic brain metabolism of insects and mammals is of special interest, particularly with respect to structural and metabolic features, that might be important for revival. This will, however, be a future task as information on insect brain metabolism is yet too patchy. We will first describe metabolic events in anaerobic locust and honey bee brain and then discuss these results in a wider context.

16.3.4.1. Anaerobic Metabolism in Locusta and Apis Brain. Changes in ATP, some intermediates, and products in the anoxic locust brain are given in Figure 16.5. There was a rapid decrease in ATP and a concomitant increase in AMP and phosphate (P_i). The carbohydrate stores were not depleted within 4 hours of anoxia. This suggests a markedly decreased metabolic rate, as well as shifting of material from the fat body to the brain. Among the anaerobic products, pyruvate rose initially, and lactate rose steadily,

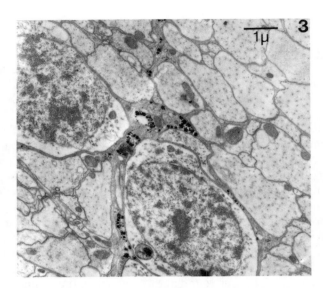

Figure 16.3. Perikarya and axonal neuropil of the optic lobe from a 9-day-old male housefly, *Musca domestica*. Glycogen can be seen in processes of glia cells surrounding the perikarya and being dispersed between axons (12,000×). (Koller and Wegener, unpublished data.)

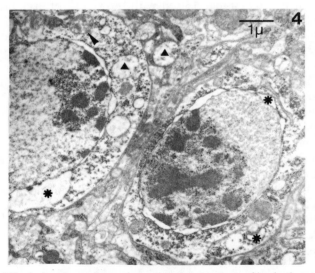

Figure 16.4. Effect of prolonged anoxia on the ultrastructure of brain tissue from a 9-day-old male housefly that had been subjected for 6 hours to an atmosphere of pure nitrogen at room temperature. Anoxic cell changes can be noted as: widening of the nuclear envelope and the endoplasmic reticulum (*stars*); shifting of chromatin material; various degrees of mitochondrial deterioration, ranging from apparently normal mitochondria over commencing disintegration of cristae (arrowheads) to vacuoles, devoid of ordered internal structures (*triangles*), or ruptured organelles. The brain was dissected under N_2 and fixed for 2 hours at 4–6°C in phosphate-buffered (0.1 M, pH 7.4) 2.5% glutaraldehyde that was adjusted to 430 mosmol by sucrose, and then postfixed in 2% OsO_4 for 1 hour at 0°C. Sections were stained with uranyl acetate and lead citrate (12,000×). (Koller and Wegener, unpublished data.)

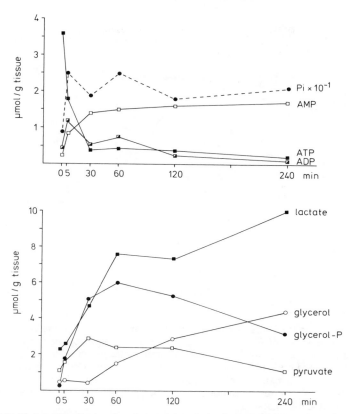

Figure 16.5. Metabolic effects of prolonged anoxia in the cerebral ganglia of female locust, *Locusta migratoria,* at 25°C (see Figure 16.6 and text for details). (From Michel and Wegener, 1982; and unpublished material.)

whereas glycerol 3-phosphate declined after 1 hour corresponding to an increase in glycerol, hence indicating a dephosphorylation of the former. When kept in air, after 60 min in nitrogen, the energy charge was back to normal within 10 min, lactate and glycerol phosphate declined rapidly, whereas glycerol continued to rise (Michel and Wegener, 1982).

In *Apis* brain ATP fell more rapidly (Fig. 16.6). Glycogen decreased from 15.7 to 10.5 and 4.8 $\mu mol \cdot g^{-1}$ brain tissue after anoxia of 30 and 60 min respectively (Kieffer and Wegener, 1983). As in the locust, the glucose content was increased in the anoxic bee brain. In accordance with the low LDH activity, little lactate, but alanine and some glycerol, was formed. The anaerobic products cannot account for the amount of glycogen degraded, thus indicating formation of additional compounds (e.g., Krebs cycle intermediates like malate, succinate, or polyols via the pentose phosphate pathway; Gäde, 1985 for review). Although depletion of ATP brings the animals close

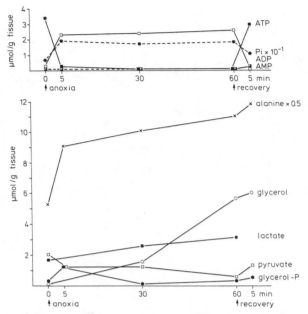

Figure 16.6. Metabolic effects of anoxia and postanoxic recovery in the cerebral ganglia of the honeybee, *Apis mellifera*. Collecting worker bees were subjected to pure nitrogen at 22°C and the tissue content (in μmoles per gram of tissue) of adenosine phosphates, inorganic phosphate and substrates, and possible anaerobic products were determined by specific enzymatic methods after rapidly freezing the animals in melting nitrogen (−210°C). Means of four determinations are given. (Data taken from Kieffer and Wegener, 1983; and unpublished material.)

to a state referred to as "rigor mortis," recovery is rapid if air is again available (Fig. 16.6).

16.3.4.2. Metabolic Pathways and Metabolic Rates in Anoxic Insect Brains. Among invertebrates (particularly endoparasites, annelids, molluscs), various metabolic adaptations to hypoxic/anoxic periods have recently been revealed (for review see von Brand, 1973; Hochachka, 1980). These include the coupled catabolism of glucose and amino acids, deviations from the usual citric acid cycle, combined with the formation of various anaerobic products to avoid congestion of metabolic pathways and to get rid of the surplus of reduction equivalents that would under aerobic conditions be transferred to the final electron acceptor oxygen. Whether or not mitochondrial reactions facilitating anaerobic energy production are operating in adult insects during anoxia is an open question. Carbohydrate is degraded anaerobically, but other pathways than glycolysis (e.g., pentose phosphate shunt) might be involved. As alanine is elevated in anaerobic insect brains, a stimulation of amino acid metabolism is likely to occur. For biochemical reasons, energy (ATP) cannot be gained anaerobically from fatty acids.

The enzymes of glycolysis in brain have maximal activities in excess of the aerobic glycolytic rate. If glycolysis was activated in anoxia (Pasteur effect), this could only be initial and short. Fructose 2,6-bisphosphate, a potent activator of glycolysis, was only initially increased in anoxic brain and flight muscle of *Locusta*. Later on it was considerably decreased (Michel and Wegener, unpublished data) as was glucose catabolism (as estimated from the decrease in substrate; Section 16.3.4.1). Studies combining biochemical, physiological, and histochemical methods are indicated to identify the responsible mechanisms.

16.3.4.3. Products of Anaerobic Metabolism. As substrates cannot be fully oxidized to CO_2 and H_2O in absence of oxygen, other products must be formed. The pattern of anaerobic products in different insects (whole animals or parts) is not uniform. I will restrict myself to a few examples of adult insects. Pyruvate, glycerol 3-phosphate, and glycerol have been demonstrated, often in addition to lactate, alanine, and succinate, in various species (Price, 1963; Wigglesworth, 1972; Conradi-Larsen and Sømme, 1973; Gäde, 1985 for review). ATP is broken down during anoxia; AMP, its deaminated product IMP (inosine monophosphate), and inorganic phosphate are accumulated (Ray and Heslop, 1963; Heslop et al., 1963).

Anaerobic products in insect nervous tissue comprise lactate and alanine (Figs. 16.5, 16.6; Ray, 1964; Strang et al., 1979), glycerol 3-phosphate, and subsequently glycerol (Figs. 16.5, 16.6), as well as succinate and malate (Strang, personal communication). A balance sheet has not yet been drawn for any anaerobic insect ganglion and additional products are likely to be formed (cf., Section 16.3.4.2). The total amount of adenosine phosphates is decreased during prolonged anoxia, but the products (probably IMP and adenosine) and their further metabolism have not been identified.

The end products cannot be removed from the brain, because the hemolymph flow comes to a complete standstill in anoxic insects (Michel and Wegener, unpublished findings). Nevertheless, the identified anaerobic products do not rise to particularly high levels. In the ischemic mouse brain, for example, lactate is, after only 2 min, as high as after 4 hours in the anoxic locust brain (Lowry et al., 1964). This difference cannot be explained solely on the basis of different LDH activities (Section 16.2.3), but does reflect the different, in many respects yet unknown, metabolic response to anoxia.

16.4. DEVELOPMENTAL CHANGES IN INSECT BRAIN METABOLISM

As the CNS, particularly the brain, organizes the sophisticated behavior of adult insects (compare, e.g., the blind and sluggish maggot with the agile fly), it is of special interest to follow the metabolic capacities of the nervous tissue during development. This can be achieved by determining enzyme activities and by measuring the respiration of isolated ganglia or the oxidation of added substrates (cf., Section 16.2.4).

16.4.1. Developmental Changes in Brain Enzymes

In cerebral ganglia of the locust, *Locusta migratoria,* enzyme activities have been measured, commencing at the third larval instar (Zimmermann and Wegener, unpublished data). Enzyme activities are rather constant throughout the life cycle, with one notable exception: LDH is decreased by two thirds around the imaginal moult, indicating that the highly aerobic organization of insect brain metabolism is restricted to the winged adults.

In the holometabolous species, the development of brain energy metabolism was different. Conspicuous was a depression in the activities of nearly all tested enzymes during the initial pupal stage in *Manduca* and *Bombyx* brain (Knollmann and Wegener, 1978, 1983, 1985; Knollmann, 1984). During the later phase of metamorphosis, the typical adult enzyme pattern was formed, which differed markedly from that of the larva, thus indicating the different modes of life of larvae and adults. Most conspicuous are the loss in LDH activity and the increase in the capacity to oxidize fatty acids (HOADH; Fig. 16.7). The latter feature is especially pronounced in *Bombyx mori,* clearly an adaptation to the life style, because these animals do not feed as adults, but live on their energy stores gathered during the larval period. The time courses of enzyme activities in blowfly brain differed from that in *Manduca* and *Bombyx.* In *Calliphora,* development is very rapid, with a pupal stage of approximately 10 days under laboratory conditions. The metabolic depression during the early pupa is less apparent, whereas the transition to the highly aerobic adult organization is drastic. LDH declines to less than 2% of the larval level, whereas the capacity of the respiratory chain, represented by cytochrome c oxidase, rises about tenfold (Bernd and Wegener, 1978).

16.4.2. Metabolism of Larval, Pupal, and Adult Brains In Vitro

Metabolism of insect brains has also been studied in vitro during metamorphosis. Oxygen uptake by isolated brains (kept in air) of *Manduca* (Loskill and Wegener, unpublished data), as well as $^{14}CO_2$ production from both ^{14}C-acetoacetate and ^{14}C-glucose followed a U-shaped time course (Fig. 16.8; Knollmann and Wegener, 1983, 1985; Knollmann, 1984). Similar results were obtained in the silkmoth, *Bombyx mori.*

16.4.3. Metabolism of Aging Insect Brains

The advantage of short-lived and easily bred insects for aging studies is obvious and has been emphasized by several authors (for review Rockstein

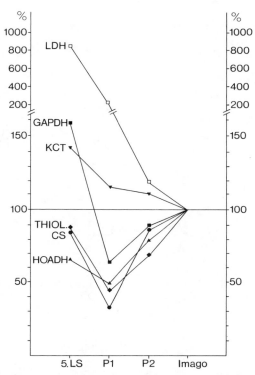

Figure 16.7. Developmental changes in the metabolic organization of brain tissue from *Manduca* as derived from specific activities of enzymes representing the main pathways of energy metabolism. P1 = early pupal stage, P2 = late pupal stage; GAPDH = glyceraldehyde-phosphate dehydrogenase, EC 1.2.1.12; see Figure 16.1 and text for further explanation. (From Knollmann and Wegener, 1985.)

and Miquel, 1973; Stoffolano, 1976; Kern, 1982; Sohal, 1985). Several aspects of age-related changes have been studied, mostly alterations in behavior, performance, or biochemical changes, using whole animals, parts of them (head, thorax), or big organs such as the flight muscle. The brain, however, which undoubtedly plays a leading role in developmental processes in insects, either directly or via hormones, has rather been neglected. Some authors have demonstrated structural changes (Herman et al., 1971; Sohal and Sharma, 1972; Collatz and Collatz, 1981), but information on the biochemistry or metabolism of the aging brain is wanting.

Recently, attempts have been made to follow brain metabolic rate through the life span of adult insects, mainly in the blowfly, *Calliphora erythrocephala* (Backes, 1978; Kern and Wegener, 1981, 1984; Kern, 1982). The respiratory rate of isolated brains (cf., Section 16.2.4) was not constant, but showed an age-dependent pattern characterized by low initial values, a peak at around half the maximal life span, followed by a rapid decrease to the

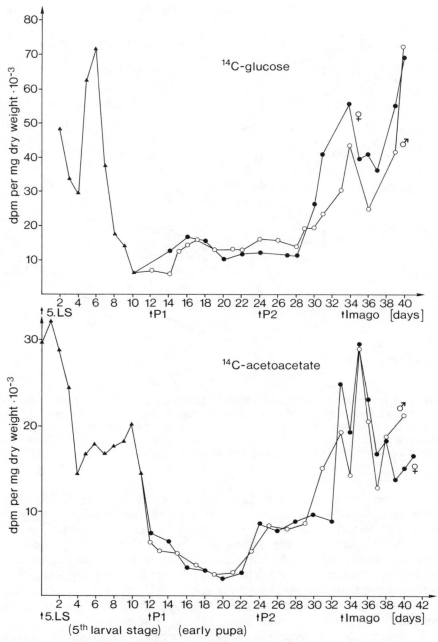

Figure 16.8. Oxidation of ^{14}C-glucose and ^{14}C-acetoacetate to $^{14}CO_2$ by isolated brains from *Manduca* during ontogenesis. The labeled substrate was applied directly to the isolated ganglia, and respiratory CO_2 was trapped during an incubation period of 30 min at 25°C. Each symbol represents the mean of four determinations. (From Knollmann and Wegener, 1985.)

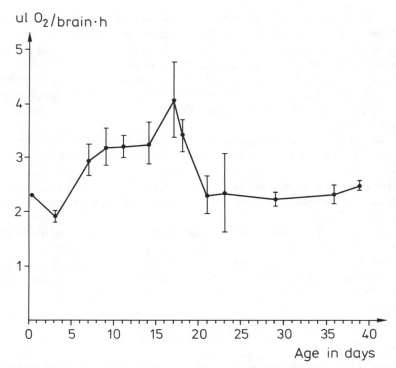

Figure 16.9. Age-related changes in the respiratory rate (microliter per brain and hour at 25°C) of isolated brains of female blowflies as measured in a microrespirometer. Means and standard deviations are given, each value representing three to five animals. (From Kern and Wegener, 1984.)

initial levels (Fig. 16.9). Interpretation of these observations is difficult. There is evidence that respiration of isolated brains reflects the in vivo metabolic activity (Section 16.2.5), indicating that respiratory efficiency of brain tissue might change during development (Kern and Wegener, 1984).

Due to their metabolic stability allowing in vitro studies, insect brains are doubtless interesting models for aging studies. But as different patterns of aging might exist in different insect orders, it would be premature to draw general conclusions at the present time.

16.5. SUMMARY

Insect brains are compact, avascular organs. Although oxygen is easily accessible to the brain cells via the tracheae, a rapid transport system for nutrients is lacking. These anatomical features, together with the all important and, with respect to metabolism and information processing, uniquely demanding ability of flying, are the key to appreciate in the pterygotic insect the basic features of brain energy metabolism, such as completely aerobic type of metabolism, very high metabolic rates, ample energy stores in the tissue, and, in some groups, use of various substrates. The latter ability reflects the mode of life, and is based on enzymatic specialization, mainly of the brain mitochondria. Insect brains are, for limited periods, metabolically independent from the body, thus favoring studies with isolated ganglia.

The insect CNS is not adapted to function anaerobically. The animals get rapidly paralyzed in anoxia, and the energy-rich phosphates decline to very low levels. Nevertheless, several species show revival and complete recovery after extended anoxic periods. Anaerobic metabolism is poorly understood; several observations suggest a marked reduction in metabolic rate during anaerobiosis.

In holometabolous forms, the drastic change in living conditions is reflected in the organization of brain metabolism. The transformation is accomplished during the pupal phase, when energy metabolism is reduced. Insects can conveniently be used to study aging of nervous tissue; in *Calliphora*, respiration of isolated brains was age dependent.

ACKNOWLEDGMENTS

I thank Professor W. Engels (Tübingen), Professor H. Langer, Dr. D. Zinkler (Bochum), and Professor E. Thomas (Mainz) for providing insects, and Dr. R. Strang (Glasgow) for making available to me unpublished material. The help from H. Bender, B. Matt, and R. Michel is gratefully acknowledged. The Deutsche Forschungsgemeinschaft, D-5300 Bonn, supported the work from the author's laboratory (grants We 494/2 to 494/6) and enabled the employment of M. Kern (now at Hoechst AG, Frankfurt) and U. Knollmann (now at the University of Bochum).

REFERENCES

Agardh, C. D., A. G. Chapman, B. Nilsson, and B. K. Siesjö. 1981. Endogenous substrates utilized by rat brain in severe insulin-induced hypoglycemia. *J. Neurochem.* **36:** 490–500.

Akahane, R. and R. Amakawa. 1983. Stable and unstable phase of memory in classically conditioned fly, *Phormia regina:* Effects of nitrogen gas anaesthesia and cycloheximide injection. *J. Insect Physiol.* **29:** 331–337.

Backes, R. 1978. Der Sauerstoffverbrauch des isolierten Cerebralganglions von *Calliphora erytrocephala.* Mikrorespirometrische Messungen während Metamorphose und Imaginalleben. Diplom dissertation, Johannes Gutenberg University, Mainz, West Germany.

Balázs, R. and R. J. Haslam. 1965. Exchange transamination and the metabolism of glutamate in brain. *Biochem. J.* **94:** 131–141.

Bernd, A. and G. Wegener. 1978. Änderungen der Stoffwechselorganisation von Insektengehirnen im Verlaufe der Ontogenese: Aktivitätsmuster von Enzymen des Grundstoffwechsels bei der Fliege (*Calliphora erythrocephala*). *Verh. Dtsch. Zool. Ges.* **1978:** 292.

Bradford, H. F., E. B. Chain, H. T. Cory, and S. P. R. Rose. 1969. Glucose and amino acid metabolism in some invertebrate nervous systems. *J. Neurochem.* **16:** 969–978.

Brand, T. von. 1946. *Anaerobiosis in Invertebrates.* Biodynamica, Normandy 21, Missouri.

Brand, T. von. 1973. *Biochemistry of Parasites.* Academic Press, New York, London.

Breer, H. and G. Jeserich. 1984. Invertebrate synaptosomes—implications for comparative neurochemistry. *Curr. Top. Res. Synap.* **1:** 165–210.

Buchner, E., S. Buchner, and R. Hengstenberg. 1979. 2-deoxy-D-glucose maps movement-specific nervous activity in the second visual ganglion of *Drosophila*. *Science (Washington, D.C.)* **205:** 687–688.

Buchner, E., S. Buchner, and I. Bülthoff. 1984. Deoxyglucose mapping of nervous activity induced in *Drosophila* brain by visual movement. *J. Comp. Physiol. A* **155:** 471–483.

Burrows, M. 1980. The tracheal supply to the central nervous system of the locust. *Proc. R. Soc. London B* **207:** 63–78.

Bursell, E. 1981. The role of proline in energy metabolism, pp. 135–154. In R. G. H. Downer (ed.), *Energy Metabolism in Insects*. Plenum Press, New York.

Clement, E. M. and R. H. C. Strang. 1978. A comparison of some aspects of the physiology and metabolism of the nervous system of the locust *Schistocerca gregaria* in vitro with those in vivo. *J. Neurochem.* **31:** 135–145.

Collatz, K. G. and S. Collatz. 1981. Age dependent ultrastructural changes in different organs of the mecopteran fly, *Panorpa vulgaris. Exp. Gerontol.* **26:** 183–193.

Conradi-Larsen, E.-M. and L. Sømme. 1973. Anaerobiosis in the overwintering beetle *Pelophila borealis. Nature (London)* **245:** 388–390.

Diehl, B., U. Knollmann, and G. Wegener. 1980. Die Bedeutung von Ketonkörpern für den Energiestoffwechsel in Insektengehirnen. *Verh. Dtsch. Zool. Ges.* **1980:** 331.

Duffy, J. E., S. R. Nelson, and O. H. Lowry. 1972. Cerebral carbohydrate metabolism during acute hypoxia and recovery. *J. Neurochem.* **19:** 959–977.

Engels, W. 1968a. Anaerobioseversuche mit *Musca domestica*. Alters- und Geschlechts-unterschiede von Überlebensrate und Erholfähigkeit. *J. Insect Physiol.* **14:** 253–260.

Engels, W. 1968b. Anaerobioseversuche mit *Musca domestica*. Verwertung normaler und experimentell erzeugter Kohlenhydratreserven. *J. Insect Physiol.* **14:** 869–879.

Finlayson, L. H., M. P. Osborne, and R. Anwyl. 1976. Effects of acetylcholine, physostigmine, and hemicholinum-3 on spontaneous electrical activity of neurosecretory nerves in *Carausius* and *Rhodnius*. *J. Insect Physiol.* **22:** 1321–1326.

Gäde, G. 1985. Anaerobic energy metabolism, pp. 119–136. In K. H. Hoffmann (ed.), *Environmental Physiology and Biochemistry of Insects*. Springer, Berlin.

Goldworthy, G. J. and C. H. Wheeler. 1985. Neurosecretory hormones in insects. *Endeavour* **9:** 139–143.

Hamdorf, K. and A. H. Kaschef. 1964. Der Sauerstoffverbrauch des Facettenauges von *Calliphora erythrocephala* in Abhängigkeit von der Temperatur und dem Ionenmilieu. *Z. Vgl. Physiol.* **48:** 251–265.

Hart, D. E. and J. E. Steele. 1973. The glycogenolytic effect of the corpus cardiacum on the cockroach nerve cord. *J. Insect Physiol.* **19:** 927–939.

Herman, M. M., J. Miquel, and M. Johnson. 1971. Insect brain as a model for the study of aging. Age related changes in *Drosophila melanogaster*. *Acta Neuropathol.* **19:** 167–183.

Heslop, J. P., G. M. Price, and J. W. Ray. 1963. Anaerobic metabolism in the housefly, *Musca domestica* L. *Biochem. J.* **87:** 35–38.

Hill, L., M. E. G. Izatt, J. A. Horne, and E. Bailey. 1972. Factors affecting concentrations of acetoacetate and D-3-hydroxybutyrate in the haemolymph and tissues of the adult desert locust. *J. Insect Physiol.* **18:** 1265–1285.

Hochachka, P. W. 1980. *Living Without Oxygen*. Harvard University Press, Cambridge, MA.

Horstmann, E. 1960. Abstand und Durchmesser der Kapillaren im Zentralnervensystem verschiedener Wirbeltierklassen, pp. 59–63. In D. B. Tower, J. P. Schadé (eds.), *Structure and Function of the Cerebral Cortex*. Elsevier, Amsterdam.

Jabbar, A., and R. H. C. Strang. 1984. The amino acids of the locust nervous system: Their concentrations and releases in vitro. *Comp. Biochem. Physiol.* **78B:** 453–460.

Jeuniaux, C. 1971. Haemolymph-Arthropoda, pp. 64–118. In M. Florkin, B. T. Scheer (eds.), *Chemical Zoology,* vol. 6. Academic Press, New York, London.

Kern, M. 1982. Das Insekt als Modell für Altersstudien. Doctoral dissertation, Johannes Gutenberg University, Mainz, West Germany.

Kern, M. and G. Wegener. 1981. Changes in cell structure and metabolic rate in an aging insect brain (*Calliphora erythrocephala*). 12th Int. Congr. Gerontol. Hamburg 1981: 274 (Abstract).

Kern, M. and G. Wegener. 1984. Age affects the metabolic rate of insect brain. *Mech. Ageing Dev.* **28:** 237–242.

Kieffer, S. and G. Wegener. 1983. Wirkungen von Anoxie auf Stoffwechsel und Funktion des Zentralnervensystems der Honigbiene (*Apis mellifera*) und der Schmeissfliege (*Calliphora erythrocephala*). *Verh. Dtsch. Zool. Ges.* **1983:** 295.

Knollmann, U. 1984. Energieliefernde Stoffwechselwege und Substrate in der Ontogenese von Insektengehirnen. Doctoral dissertation, Johannes Gutenberg University, Mainz, West Germany.

Knollmann, U. and G. Wegener. 1978. Zum Energiestoffwechsel in den Gehirnen von Insekten: Enzymaktivitätsmuster der Cerebralganglien von *Bombyx mori* in 3 Stadien der Ontogenese. *Verh. Dtsch. Zool. Ges.* **1978:** 292.

Knollmann, U. and G. Wegener. 1983. Entwicklung von Stoffwechselpotenzen in den Cerebralganglien von Insekten während der Ontogenese. *Hoppe Seyler's Z. Physiol. Chem.* **364**: 1161.

Knollmann, U. and G. Wegener. 1985. Energy metabolism in the developing brain of *Manduca sexta*. *Manduca Newslett.* **1**: 39–40.

Lane, N. J. and J. Treherne. 1980. Functional organization of arthropod neuroglia, pp. 765–795. In M. Locke, D. S. Smith (eds.), *Insect Biology in the Future*. Academic Press, New York, London.

Lowry, O. H., J. V. Passonneau, F. X. Hasselberger, and D. W. Schulz. 1964. Effect of ischaemia on known substrates and cofactors of the glycolytic pathway in brain. *J. Biol. Chem.* **239**: 18–30.

McGee-Russell, S. M., A. W. Brown, and J. B. Brierley. 1970. A combined light and electron microscopic study of early anoxic-ischaemic cell change in rat brain. *Brain Res.* **20**: 193–200.

Medugorac, J. and M. Lindauer. 1967. Das Zeitgedächtnis der Bienen unter dem Einfluss von Narkose und von sozialen Zeitgebern. *Z. Vgl. Physiol.* **55**: 450–474.

Menzel, T., J. Erber, and T. Masuhr. 1974. Learning and memory in the honeybee, pp. 195–217. In L. Barton Browne (ed.), *Experimental Analysis in Insect Behaviour*. Springer, Berlin.

Michel, R. and G. Wegener. 1982. Metabolic effects of anoxia in the central nervous system of the locust (*Locusta migratoria*). *Hoppe Seyler's Z. Physiol. Chem.* **363**: 1310.

Newsholme, E. A., P. H. Sugden, and T. Williams. 1977. Effect of citrate on the activities of 6-phosphofructokinase from nervous and muscle tissues from different animals and its relationship to the regulation of glycolysis. *Biochem. J.* **166**: 123–129.

Pette, D. 1968. Aktivitätsmuster und Ortsmuster von Enzymen des energieliefernden Stoffwechsels, pp. 15–52. In F. W. Schmidt (ed.), *Praktische Enzymologie*. Huber, Bern.

Price, G. M. 1963. The effects of anoxia on metabolism in the adult housefly, *Musca domestica*. *Biochem. J.* **86**: 372–378.

Ray, J. W. 1964. The free amino acid pool of the cockroach (*Periplaneta americana*) central nervous system and the effect of insecticides. *J. Insect Physiol.* **10**: 587–597.

Ray, J. W. and J. P. Heslop. 1963. Phosphorus metabolism of the housefly (*Musca domestica* L.) during recovery from anoxia. *Biochem. J.* **87**: 39–42.

Rensch, B. 1956. Increase of learning capability with increase of brain-size. *Am. Nat.* **90**: 81–95.

Rivera, M. E. and H. Langer. 1983. Enzyme pattern of energy releasing metabolism in eyes and ganglia of the blowfly *Calliphora erythrocephala* and the crab *Ocypode ryderi*. *Mol. Physiol.* **4**: 265–277.

Rockstein, M. and J. Miquel. 1973. Aging in insects, pp. 371–478. In M. Rockstein (ed.), *The Physiology of Insecta*, vol. 1. Academic Press, New York, London.

Siesjö, B. K. 1978. *Brain Energy Metabolism*. Wiley, New York.

Smith, D. S. 1968. *Insect Cells: Their Structure and Function*. Oliver and Boyd, Edinburgh.

Sohal, R. S. 1985. Aging in insects, pp. 595–631. In G. A. Kerkut, L. I. Gilbert (eds.), *Comprehensive Insect Physiology, Biochemistry and Pharmacology*, vol. 10. Pergamon Press, Oxford.

Sohal, R. S. and S. P. Sharma. 1972. Age-related changes in the fine structure and number of neurons in the brain of the housefly *Musca domestica*. *Exp. Gerontol.* **7**: 243–249.

Sokoloff, L. 1973. Metabolism of the ketone bodies by the brain. *Ann. Rev. Med.* **24**: 271–288.

Steele, J. E. 1982. Glycogen phosphorylase in insects. *Insect Biochem.* **12**: 131–147.

Stoffolano, J. G. Jr. 1976. Insects as model systems for aging studies, pp. 407–427. In M. F. Elias (ed.), *Special Review of Experimental Aging Research*. EAR, Inc, Bar Harbor, ME.

Storey, K. B. 1985. Phosphofructokinase from flight muscle of the cockroach, *Periplaneta americana*. Control of enzyme activation during flight. *Insect Biochem.* **15**: 663–666.

Strang, R. H. C. 1981. Energy metabolism in the insect nervous system, pp. 169–206. In R. H. Downer (ed.), *Energy Metabolism in Insects*. Plenum Press, New York.

Strang, R. H. C. 1985. Energy metabolism in the insect nervous system, pp. 182–206. In H. Breer, T. A. Miller (eds.), *Neurochemical Techniques in Insect Research*. Springer, Berlin, New York.

Strang, R. H. C. and E. M. Clement. 1980. The relative importance of glucose and trehalose in the nutrition of the nervous system of the locust *Schistocerca americana gregaria*. *Insect Biochem.* **10**: 155–161.

Strang, R. H. C., E. M. Clement, and R. C. Rae. 1979. Some aspects of the carbohydrate metabolism of the thoracic ganglia of the locust *Schistocerca gregaria*. *Comp. Biochem. Physiol.* **62B**: 217–224.

Strausfeld, N. 1976. *Atlas of an Insect Brain*. Springer, Berlin.

Sugden, P. H. and E. A. Newsholme. 1973. Activities of hexokinase, phosphofructokinase, 3-oxo acid coenzyme A-transferase and acetoacetyl-coenzyme A-thiolase in nervous tissue from vertebrates and invertebrates. *Biochem. J.* **134**: 97–101.

Sugden, P. H. and E. A. Newsholme. 1975. Activities of citrate synthase, NAD^+-linked and $NADP^+$-linked isocitrate dehydrogenase, aspartate aminotransferase, and alanine aminotransferase in nervous tissue from vertebrates and invertebrates. *Biochem. J.* **150**: 105–111.

Thews, G. 1960. Die Sauerstoffdiffusion im Gehirn. *Pflügers Arch. Ges. Physiol.* **271**: 197–226.

Treherne, J. E. 1960. The nutrition of the central nervous system in the cockroach *Periplaneta americana* L: The exchange and metabolism of sugars. *J. Exp. Biol.* **37**: 513–533.

Treherne, J. E. 1966. *The Neurochemistry of Arthropods*. Cambridge Monographs in Experimental Biology No. 14. Cambridge University Press, Cambridge.

Treherne, J. E. 1974. The environment and function of insect nerve cells, pp. 187–244. In J. E. Treherne (ed.), *Insect Neurobiology*. North-Holland, Amsterdam.

Treherne, J. E. and P. K. Schofield. 1979. Ionic homeostasis of the brain microenvironment in insects. *Trends Neurosci.* **2:** 227–230.

Walter, D. C. and S. R. Nelson. 1975. Energy metabolism and nerve function in cockroaches (*Periplaneta americana*). *Brain Res.* **94:** 485–490.

Wegener, G. 1975. Comparative biochemistry of energy metabolism in brain tissue from insects. Comm. 10th Meet. Fed. Europ. Biochem. Soc. 1975: 1554.

Wegener, G. 1981. Comparative aspects of energy metabolism in nonmammalian brains under normoxic and hypoxic conditions, pp. 87–109. In V. Stefanovich (ed.), *Animal Models and Hypoxia.* Pergamon Press, Oxford.

Wegener, G. 1983. Brains burning fat. Different forms of energy metabolism in the CNS of insects. *Naturwissenschaften* **70:** 43–45.

Wegener, G. and J. Pfeifer. 1975. Besonderheiten der Stoffwechselorganisation in den Hirnen von Vertebraten und Evertebraten. *Verh. Dtsch. Zool. Ges.* **1974:** 266–271.

Wegener, G. and E. Zebe. 1971. Zum Energiestoffwechsel des Gehirns. Eine vergleichende Untersuchung an Vertretern der verschiedenen Wirbeltierklassen. *Z. Vgl. Physiol.* **73:** 195–208.

Wegener, G. and E. Zebe. 1973. Zum Energiestoffwechsel in den Hirnen wirbelloser Tiere. *Naturwissenschaften* **60:** 551–552.

Wegener, G., H. Schmidt, A. R. Leech, and E. A. Newsholme. 1986a. Antagonistic effects of hexose 1,6-bisphosphates and fructose 2,6-bisphosphate on the activity of 6-phosphofructokinase from honey-bee flight muscle. *Biochem. J.* **236:** 925–928.

Wegener, G., R. Michel, and E. A. Newsholme. 1986b. Fructose 2,6-bisphosphate as a signal for changing from sugar to lipid oxidation during flight in locusts. *FEBS Lett.* **201:** 129–132.

Wegener, G., R. Michel, and M. Thuy. 1987a. Anoxia in lower vertebrates and insects: Effects on brain and other organs. *Zool. Beitr.* in press.

Wegener, G., I. Beinhauer, A. Klee, and E. A. Newsholme. 1987b. Properties of locust muscle 6-phosphofructokinase and their importance in the regulation of glycolytic Flux during prolonged flight. *J. Comp. Physiol. B* **157.** in press.

Werner, M. 1980. Vergleichende Untersuchungen an Hexokinasen des Nervengewebes von Vertebraten, Mollusken und Insekten. Doctoral dissertation, University Mainz, West Germany.

Wigglesworth, V. B. 1960. The nutrition of the central nervous system in the cockroach *Periplaneta americana* L: The role of perineurium and glia cells in the mobilization of reserves. *J. Exp. Biol.* **37:** 500–512.

Wigglesworth, V. B. 1972. *The Principles of Insect Physiology.* Chapman and Hall, London.

Wood, M. R., V. Argiro, P. Pelikan, and M. J. Cohen. 1980. Glycogen in the central neurons of insects: Massive aggregations induced by anoxia or axotomy. *J. Insect Physiol.* **26:** 791–799.

Wyatt, G. R. 1961. The biochemistry of insect haemolymph. *Ann. Rev. Entomol.* **6:** 75–102.

CHAPTER 17

Biogenic Amines
in the Insect Brain

Alison R. Mercer
Department of Zoology
University of Otago
Dunedin, New Zealand

17.1. INTRODUCTION

There is considerable evidence to suggest that biogenic amines function as neurotransmitters, neuromodulators, and neurohormones in the insect nervous system (for reviews see Klemm, 1976; Evans, 1980; Brown and Nestler, 1985); however, most of the information available is derived from studies of the peripheral nervous system. Only recently has attention focused on the central nervous system (CNS) and the roles of biogenic amines in the insect brain. Dopamine (DA), 5-hydroxytryptamine (5-HT), octopamine (OA), and noradrenaline (NA) have been detected in the brain of many species of insects (Klemm, 1976, and Chapter 23 in this volume; Evans, 1980). Aspects of the biochemistry, histochemistry, pharmacology, and physiology of aminergic systems in the insect brain have been examined, but there has been a tendency to consider each aspect in isolation. The need for an integrated multidisciplinary approach to the study of biogenic amines in insects has been emphasized by Evans (1980). Multidisciplinary studies are being used currrently to investigate the roles of biogenic amines in major sensory pathways and integrative centers in the brain of the honeybee, *Apis mellifera*. In the following chapter, work on the honeybee is compared with similar studies in other insects, and used as a reference point to examine our present knowledge of biogenic amines and their possible functional roles in the insect brain.

17.2. DISTRIBUTION OF BIOGENIC AMINES IN THE BRAIN

17.2.1. Concentration of Bioamines

Quantitative measurements of biogenic amine levels in the brain have been published for a wide variety of insects (Klemm, 1976; Evans, 1980). A selection of published values has been compared by Clarke and Donnellan (1982) who converted original data to nmole per milligram protein (assuming that CNS tissue contains 8 mg protein per 100 mg net weight of tissue) and found good correlation between amine levels reported by different workers for the cockroach and locust brain. The levels of biogenic amines in the brain of the honeybee estimated by radioenzymatic assay (Mercer et al., 1983) are similar to those reported in the locust and cockroach brain using similar techniques (Evans, 1978; Dymond and Evans, 1979; Clarke and Donnellan, 1982). As in the cockroach, the brain of the honeybee contains more DA than OA, approximately the same amounts of DA and 5-HT, and very low levels of NA. Adrenaline was not detected in the brain of the honeybee, locust, or cockroach in these studies. Indeed, adrenaline has not been conclusively identified in any region of the insect CNS (Evans, 1980). Recently, high levels of endogenous histamine, as well as the enzymes necessary for the biosynthesis and metabolism of histamine, have been detected in the nervous system of the locust (Elias and Evans, 1983, 1984). For any biogenic amine to be accepted as a neurotransmitter or neuromodulator it must be shown that the nervous system can synthesize the amine in question. Unfortunately, the metabolic pathways involved in synthesis and degradation of biogenic amines in the insect nervous system are only poorly understood. Critical assessment of work in this field can be found in excellent recent reviews by Evans (1980) and Brown and Nestler (1985).

In the antennal lobes of the honeybee, DA (0.75 pmole), 5-HT (3.03 pmole), and OA (0.48 pmole) are present, but no NA has been detected in this region of the bee brain (Mercer et al., 1983). DA (0.94 pmole), NA (0.44 pmole), 5-HT (1.93 pmole), and OA (0.51 pmole) are found in the double calyxes of the mushroom bodies of the bee (Fig. 17.1), and similar levels of these amines occur also within the α-lobes of the mushroom bodies in this

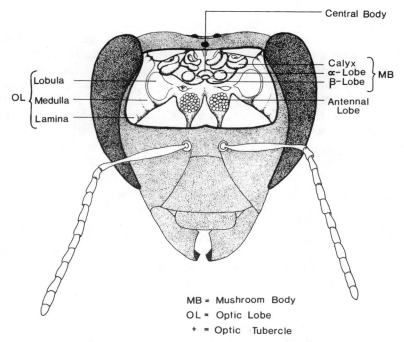

Figure 17.1. Diagram to show the major divisions of brain neuropil in the honeybee, *Apis mellifera.*

insect (Mercer et al., 1983). In the cockroach, *Periplaneta americana*, OA has been detected in the cell bodies of intrinsic mushroom body neurons (Dymond and Evans, 1979). There are many similarities in the distribution of biogenic amines in the brain of the locust, cockroach, and honeybee, but important differences between species also occur. In the optic lobes of the bee, for example, there is approximately three times as much DA as OA, whereas in the cockroach the opposite is true (Evans, 1978; Dymond and Evans, 1979; Mercer et al., 1983). The optic lobes of the locust contain six times as much OA as DA (Robertson, 1976). In all three insects, however, relatively high levels of OA are found in the medulla (refer Fig. 17.1) compared with other neuropil regions of the optic lobes (Evans, 1978, 1980; Mercer et al., 1983). In the honeybee, the level of 5-HT in the medulla is similar to that of OA, but neither DA nor NA have been detected in this region of the brain. Similarly, no NA and only small amounts of DA are found in the lamina and lobula neuropils of the optic lobes in the honeybee (Mercer et al., 1983). Some of the species differences in quantitative distribution of amines may be accounted for by differences in neuronal density between species (Klemm, 1976; Evans, 1980). Nevertheless, great care must be taken in extrapolation of results from one insect species to another.

Circadian changes in levels of biogenic amines in the insect brain have been reported (Evans, 1980; see Section 17.3) and changes in concentration of brain amines during development have been described for several insect species, including the locust, *Locusta migratoria* (Fuzeau-Breasch and David, 1979; Lafon-Cazal et al., 1982); the moth, *Mamestra configurata* (Bodnaryk, 1980); and the blowfly, *Calliphora erythrocephala* (Nässel and Laxmyr, 1983). In the blowfly, mean levels of 5-HT, OA, and DA are reported to be lower in larvae than in adult flies (Nässel and Laxmyr, 1983). During metamorphosis in the moth, levels of OA (per mg dry brain weight) increase 10-fold from 9.8 to 110 pmole OA (Bodnaryk, 1980). Levels of biogenic amines in the insect CNS can be altered pharmacologically also. Reserpine, for example, depletes levels of biogenic amines in the insect brain (Evans, 1980). In the vertebrate brain, reserpine is believed to inhibit the mechanism responsible for concentration of amines into storage granules (Guldberg and Broch, 1971). Depletion of catecholamines and partial reduction in the levels of OA in insects can be induced by treatment with the false transmitter 6-hydroxydopamine (Lafon-Cazal et al., 1982). In mammals, p-chlorophenylalanine (pCPA), or its methyl ester, causes severe depletion of 5-HT levels in the brain (Koe and Weissman, 1966). In insects, however, results to date with pCPA are contradictory. In the brain of *P. americana*, Omar et al. (1982) failed to detect any significant change in the concentration of 5-HT as a result of treatment with pCPA methyl ester (100 μg/g or 500 μg/g for 2 and 12 hours, respectively), whereas Pandey and Habibulla (1980) report a significant decrease in 5-HT levels in the cockroach brain after treatment with 15 μg/g pCPA. Pandey and co-workers (1983) present evidence to suggest that, in the brain of the cockroach, pCPA inhibits the activity of tryptophan hydroxylase, the enzyme responsible for conversion of tryptophan into 5-hydroxytryptophan, which in turn undergoes decarboxylation to form 5-HT. Drugs used to manipulate amine levels in the insect nervous system are usually chosen because of their effects on aromatic aminergic systems in the mammalian brain. However, the assumption that particular pharmacological agents will have the same effects in insects as they do in vertebrates must be made with caution.

17.2.2. Cellular Localization of Brain Amines

The presence of catecholamines in the brain of the cockroach, *P. americana*, was first demonstrated by Frontali and Norberg (1966) and Frontali (1968) using the aldehyde fluorescence techniques of Falck and Hillarp (Falck et al., 1962)). Since then fluorescence histochemistry has been used extensively to locate fluorogenic amines (catehol- and indoleamines) in the insect nervous system. Comparative studies reveal interesting differences between species in the distribution of fluorogenic amines in the insect brain (Klemm, 1976). These differences are discussed by Klemm (1976) in terms of the phylogeny of monoaminergic systems in the insect CNS. Biochemical and histochemical studies on the distribution of biogenic amines in the brain of the honeybee (Mercer et al., 1983) have been carried out in close conjunction with detailed neuroanatomical studies of the bee brain (Mobbs, 1982). The major divisions of brain neuropil in the honeybee can be seen in Figure 17.1. A series of frontal sections of the bee brain treated for fluorescence histochemistry with glyoxylic acid are shown in Figure 17.2A, C, and E. Silver-stained sections of the brain are included for orientation purposes (Fig. 17.2B, D, and F). As in most insects examined to date (Klemm, 1976; Evans, 1980), slow fading green fluorescence typical of catecholamines is found throughout the cerebral ganglion of the honeybee after treatment with glyoxylic acid (Fig. 17.2A, C, and E; Mercer et al., 1983) or formaldehyde gas (Klemm, 1976). Fluorescence is particularly intense in the highly structured neuropils of the central body and the corpora pedunculata or mushroom bodies (Fig. 17.2C). In the honeybee, unlike most other insects, catecholamine-containing neurons can be identified histochemically in the calyxes of the mushroom bodies (Fig. 17.2C). Fluorescence is strongest in a well-defined region between the collar and basal ring neuropils of each calyx (Mercer et al., 1983). Recent studies have shown the calyxal neuropils of the mushroom bodies in the honeybee to be devoid of neurons that stain with antibodies raised against the indolealkylamine, 5-HT (Schürmann and Klemm, 1984). The intense bands of fluorescence in the initial segment of the pedunculus (Fig. 17.2C), the α-lobes (Fig. 17.2A), and the β-lobes of the mushroom bodies in the honeybee are largely extrinsic in origin (Mercer et al., 1983). The cell bodies of intrinsic mushroom body neurons (Kenyon cells) that lie in and around the cup-shaped calyxes do not fluoresce after treatment with either formaldehyde gas (Klemm, 1976) or glyoxylic acid (Mercer et al., 1983), nor do they react to antibodies against 5-HT (Schürmann and Klemm, 1984). It is possible that some of the fluorescent fibers observed in the pedunculus and calyxal neuropils are derived from feedback neurons that connect the calyxes and pedunculus directly with the α- and β-lobes of the mushroom bodies in the honeybee (Mercer et al., 1983).

Monoamine-containing neurons have been identified histochemically in the central body complex of all insects examined to date (Klemm, 1976; Evans, 1980). In the honeybee, fluorescence is particularly intense in the anterior upper division of the central body (Fig. 17.2C) and in the ventral noduli (Fig. 17.2E; Mercer et al., 1983). 5-HT-immunoreactive neurons have been identified also in the central body neuropil (Schürmann and Klemm, 1984). Fibers that extend from the central body to the anterior protocerebrum where they encircle the α-lobes of the mushroom bodies contain fluorogenic monoamines (Mercer et al., 1983). However, no fluorescent fibers have been located in the protocerebral bridge of the honeybee (Klemm, 1976; Mercer et al., 1983), a region of brain neuropil closely associated with the central body complex (Mobbs, 1985). Yellow (indole) fluorescence has been demonstrated in the protocerebral bridge of locusts and dragonflies, and catecholamine fluorescence was reported in this region of the brain of the fly,

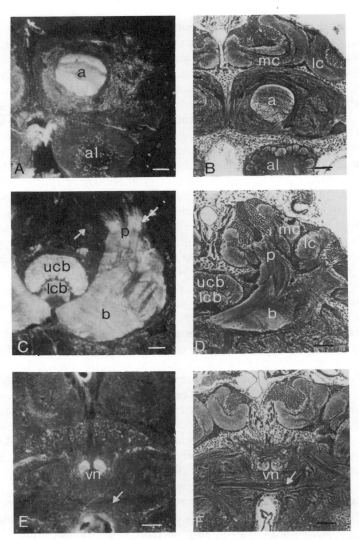

Figure 17.2. Frontal sections of the brain of the honeybee treated for fluorescence histochemistry with glyoxylic acid (A,C,E) or stained for normal histology with protargol (B,D,F). (A and B) Sections through anterior surface of the brain. (C and D) Sections approximately 330 μm from anterior surface of the brain. *Single-headed arrow* points to fluorescent fibers in the median calyx of the mushroom body. *Double-headed arrow* indicates intense fluorescence in the peduncular region of the mushroom body neuropil. (E and F) Posterior sections of the brain. Arrow points to the ventral posterior optic commissure. Scale bar = 100 μm. a = α-lobe of mushroom body; al = antennal lobe; b = β-lobe of mushroom body; lc = lateral calyx of mushroom body; lcb = lower division of central body neuropil; mc = medial calyx of mushroom body; p = pedunculus of mushroom body; vn = ventral noduli of mushroom body. (A, C, and E from A. R. Mercer and P. G. Mobbs, unpublished data.)

404

Calliphora (Klemm, 1976). The distribution of biogenic amines in the optic ganglia also varies considerably from one insect species to the next (Klemm, 1976; Evans, 1980). In the optic lobes of the locust, for example, amine fluorescence is intense, particularly in the medulla, after treatment of tissues for fluorescence histochemistry (Klemm, 1976), whereas in the optic lobes of the honeybee only a small number of amine-containing neurons have been identified using fluorescence histochemistry techniques (Mercer et al., 1983). Fluorescence in the honeybee optic ganglia is weak and fades rapidly under the excitation light, typical of 5-HT. Indeed, the distribution of these neurons correlates well with 5-HT-containing neurons identified recently in this region of the bee brain using immunocytochemistry techniques (Schürmann and Klemm, 1984). In all insects examined to date, 5-HT-immunoreactive fibers have been located in all three divisions of the optic lobe neuropil (Bishop and O'Shea, 1983; Nässel and Klemm, 1983; Tyrer et al., 1984; Nässel et al., 1985). The extensive distribution of 5-HT-immunoreactive fibers appears to be derived from a relatively small number of neurons with very wide-field arborizations. In the honeybee, aminergic fibers are found in the optic tubercles and in the anterior commissure that connects the two lateral protocerebra (Fig. 17.2C; Mercer et al., 1983). No amine-containing neurons have been identified in the ventral posterior optic commissure (Fig. 17.2E), the commissure that lies directly above the upper division of the central body (Fig. 17.2C), or in the large ocellar neurons of the honeybee (Mercer et al., 1983; Schürmann and Klemm, 1984). In the antennal lobes of the bee brain, amine-containing neurons are located at the periphery of the olfactory glomeruli (Fig. 17.2A; Klemm, 1976; Mercer et al., 1983; Schürmann and Klemm, 1984), where incoming sensory fibers converge on local interneurons and on extrinsic fibers that project to the calyxes of the mushroom bodies in the antennoglomerular tract (AGT; Mobbs, 1982). Amine-containing neurons have not been identified in either the AGT or in the antennal nerve of the honey bee (Klemm, 1976; Mercer et al., 1983; Schürmann and Klemm, 1984).

Cell bodies of amine-containing neurons have been located beneath the calyxes of the mushroom bodies, in the lateral protocerebrum between the proto- and deutocerebrum and in the posterior medial protocerebral rind in the honeybee, using fluorescence histochemistry techniques (Klemm, 1976; Mercer et al., 1983). It has not been possible, however, to trace these neurons using fluorescence histochemistry. An exciting development in recent years has been the application of immunocytochemical techniques for mapping the distribution of putative neurotransmitter compounds in the insect CNS. Antibodies raised to the putative neurotransmitter 5-HT have been used to locate 5-HT-immunoreactive neurons in the CNS of several insects, including cockroaches (Bishop and O'Shea, 1983, Nässel and Klemm, 1983), locusts (Klemm and Sundler, 1983; Nässel and Klemm, 1983; Tyrer et al., 1984, Lutz et al., 1985), flies (Nässel and Klemm, 1983; Nässel et al., 1983, 1985; Nässel and Cantera, 1985), and the honeybee (Schürmann and Klemm,

1984). Antiserum to DA has been used to investigate catecholamine-containing neurosecretory cells in the pars intercerebralis of the locust brain (Vieillemaringe et al., 1984). Processes of DA-immunoreactive fibers are found in the first pair of nerves (nervi corporis cardiaci I, NCCI) to the neural region of the corpora cardiaca, neurohemal organs involved in release of neurosecretory material from the brain (see Section 17.3). As yet, very little is known about the specific location of neurons that contain OA. Complementary studies using antibodies raised against OA are required, therefore, to provide a more complete picture of the distribution of biogenic amines in the insect brain.

17.3. ROLES OF BIOGENIC AMINES IN THE BRAIN

Biogenic amines have been implicated in the regulation of locomotor activity rhythms in insects (Evans, 1980). In the brain of the cockroach, *P. americana*, for example, cyclic changes in DA levels have been correlated recently with activity rhythms in this animal (Prée and Rutschke, 1983). High DA content in the cockroach brain is synchronized with maximum locomotor activity. Cyclic changes in 5-HT levels have been observed in the brain of the cricket, *Acheta domesticus* (Muszynska-Pytel and Cymborowski 1978a). Although changes in 5-HT levels show no direct correlation with circadian rhythms of locomotor activity in this insect, Muszynska-Pytel and Cymborowski (1978b) suggest, on the basis of histochemical studies, that release of 5-HT from the central body of the cricket brain at night reduces the activity of inhibitory cells in the pars intercerebralis, thereby increasing the overall locomotor activity in this insect during the night phase. Both 5-HT and DA have been localized in the central body neuropil of many insects (see Section 17.2.2). The pattern of distribution of fluorogenic monoamines in highly structured neuropils, such as the central body and mushroom bodies, suggests that biogenic amines may be involved in simultaneous control of the activities of whole assemblies of neurons in the insect brain (Evans, 1980). DA applied iontophoretically to the brain of *P. americana* is reported to elicit two types of responses: either an increase in frequency of neural activity in the central body or inhibition of background activity (Lapitskii et al., 1981). Spontaneous firing of units in the mushroom bodies of the ant are reported also to be inhibited by DA (Steiner and Pieri, 1969). In the honeybee, *A. mellifera*, both DA and OA reduce the size of potentials evoked by stimulation of the antennae with air or scent recorded in the α-lobe of the mushroom bodies (Mercer and Erber, 1983). Unlike DA, OA also causes a dramatic enhancement in the size of potentials evoked by light stimuli in this insect. Although mushroom body and central body neuropils have received a great deal of attention over the years, the specific functions of these distinctive regions of the brain neuropil remain unclear. The morphology of the mushroom bodies, as well as the distribution of biogenic amines therein, varies from species to species (Klemm, 1976). It is possible, therefore, that the function of the mushroom bodies and associated amines varies also from one insect species to another. The effects of DA and OA on neural activity in the mushroom bodies of the bee brain are long lasting (Mercer and Erber, 1983), and it is tempting to speculate that biogenic amines play a modulatory role in the brain of this insect (Mercer et al., 1983). However, further information concerning the specificity of receptors mediating the effects of biogenic amines in the insect brain is needed before the results of studies in which local microiontophoretic applications of biogenic amines to specific regions of the brain neuropil can be interpreted in any meaningful way.

Analysis of more accessible and better understood regions, such as primary olfactory and visual centers, would seem a promising strategy for any detailed investigation of the cellular mechanisms underlying responses to biogenic amines in the insect brain. This is particularly true in view of the fact that large amine-containing neurons accessible to detailed biochemical, physiological, and pharmacological analysis have been identified in the optic lobes (Bishop and O'Shea, 1983; Nässel et al., 1983) and antennal lobes (Tyrer et al., 1984) of the insect brain (Section 17.2.2). The basic pattern of 5-HT-immunoreactive neurons in the midbrain of adult flies is laid down during embryonic development (Nässel and Cantera, 1985). No dramatic changes in the organization of serotonergic neurons occurs during metamorphosis, except in the optic lobes, where a set of approximately 30 5-HT-immunoreactive neurons differentiate during pupal development (Nässel and Cantera, 1985; Nässel et al., 1985). It is interesting to note that in the moth, *Manduca sexta*, deafferentation early in adult development causes a reduction in the levels of 5-HT in the optic lobes of this insect (Maxwell and Hildebrand, 1981). In *M. configurata* levels of the phenolamine OA are reported to be much higher in the adult brain than in the brain of the pupal moth (Bodnaryk, 1980). Bodnaryk suggests that the establishment of an enhanced octopaminergic system during brain development may be required to process and/or modulate the greatly increased sensory input in the adult moth compared with the pupa, especially from the compound eyes. In the lamina of the optic lobes of the blowfly, *Calliphora erythrocephala*, large, possibly serotonergic neurons (Tan 3 neurons) are believed to exert centrifugal control on photoreceptors (or lamina neurons) in large areas of the visual field (Nässel et al., 1983). Similarly, wide-field lamina tangential neurons in the locust that react with antibody against 5-HT are thought to control information in sensory processing channels, or to act as level controllers by setting the responsiveness of the sensory channels (Tyrer et al., 1984). The extensive distribution of 5-HT-immunoreactive fibers in the optic lobes of the insect brain and their derivation from a small number of neurons (Section 17.2.2) is suggestive of a modulatory role for 5-HT in this region of the brain (Nässel et al., 1985). In the trilobite arthropod, *Limulus*, 5-HT influences circadian changes in the sensitivity of the eye (Barlow et al., 1977). Diurnal changes in the activity of visual interneurons have been demonstrated in the honeybee (Kaiser, 1979; Kaiser and Steiner-Kaiser, 1983). It is possible that 5-HT in the optic lobes of the honeybee is involved in diurnal regulation of neuronal activity. As in the optic lobes, the distribution of aminergic neurons in relation to glomeruli containing afferent terminals and first-order interneurons of the olfactory pathway, suggests that biogenic amines are involved in control of information in sensory processing channels and/or regulating the threshold for responses in such channels in the antennal lobes of the brain. Behavioral studies have shown that DA applied to the antennal lobes of the bee brain reduces the percentage of honeybees that respond to an olfactory stimulus that has been paired with a food reward in a single associative learning trial (Macmillan and Mercer, in preparation). The effects of biogenic amines on response characteristics of deutocerebral neurons in the brain of the honeybee are currently under investigation (Flanagan, personal communication).

There is good evidence that biogenic amines regulate hormonal control. Biogenic amines are closely associated with neurohemal organs in insects (Klemm, 1976; Evans, 1980; Raabe, 1982; Cassier, 1983). There are several populations of neurosecretory cells in the insect brain, the largest and most intensively studied of which are the median cell bodies in the anterior pars intercerebralis (Raabe, 1982). Neurosecretory cells in the pars intercerebralis send processes into the corpora cardiaca via NCCI (Section 17.2.2), whereas groups of neurosecretory cells in the lateral protocerebrum and tritocerebrum of the brain give rise to NCCII and NCCIII, respectively (Raabe, 1982). Parallels have been drawn between the pars intercerebralis–corpus cardiacum complex in insects and the hypothalmus–pituitary complex of vertebrates, both in terms of function and in the involvement of biogenic amines (Klemm, 1976; Evans, 1980). There appears to be a close association between biogenic amines and peptidergic neurosecretory systems in arthropods (Klemm, 1976; Evans, 1980; Mordue, 1982; Raabe, 1982). The following examples outline recent results implicating the phenolamine OA in the control of hormonal release in the locust and the cockroach. The presence of an aminergic link in the release of adipokinetic hormone (AKH) from intrinsic cells of the glandular lobes of the corpora cardiaca in the locust, *Schistocerca gregaria*, was established by Samaranayaka in 1976. In *Locusta migratoria* hormone release is regulated by a group of cells in the lateral protocerebrum of the brain, which send axons into the glandular lobe of the corpus cardiacum via NCCII (Rademakers, 1977). Evidence suggests that these cells release OA (Orchard et al., 1983a,b). Stimulation of NCCII results in an increase in cAMP levels within postsynaptic neurons. This effect is thought to be mediated by an OA-sensitive adenylate cyclase located on the postsynaptic membrane of the hormone-containing cells (Orchard et al., 1983a,b). The corpus cardiacum of the cockroach, *P. americana*, is innervated by cholinergic and adrenergic neurons that regulate the release of neurohormones from the organ (Gersch, 1972). Release of hypertrehalosemic factor appears to be regulated by octopaminergic neurons located within NCCII (Downer et al., 1984). As in the locust, evidence suggests that hormonal release is mediated by an increase in cAMP levels, which result from activation of adenylate cyclase by OA.

Cyclic nucleotides appear to play an important role in aminergic transmission in the insect brain. Harmer and Horn (1977) identified an OA-sensitive adenylate cyclase in the brain of the cockroach, *P. americana*. More recently, responsiveness of adenylate cyclase to putative neurotransmitters has been used to provide evidence for the presence of octopamine-, dopamine-, and serotonin-receptors in head homogenates of the fruitfly, *Drosophila melanogaster* (Uzzan and Dudai, 1982). It has been suggested that noradrenaline and adrenaline interact with the same receptor as octopamine in *Drosophila* (Uzzan and Dudai, 1982) and in the cockroach brain (Harmer and Horn, 1977). Direct binding studies using radioactive probes have been used to characterize putative aminergic receptors identified in *Drosophila* heads (Dudai, 1982; Dudai and Zvi, 1984a,b). As yet, however, the specific location of aminergic receptor sites in the brain of the fruitfly is unclear, and physiological studies are required to determine whether or not the sites identified in binding studies represent functional receptors for the neurotransmitters in question. The possibility of using information from binding studies to provide groundwork for genetic dissection of aminergic systems in *Drosophila* (Uzzan and Dudai, 1982; Dudai and Zvi, 1984a) is very exciting. Work with *Drosophila* learning mutants suggests that the monoamine-cAMP signaling system in insects is involved in a diverse range of learning behaviors, including associative, as well as nonassociative, forms of learning (Quinn, 1984). In the honeybee, DA or 5-HT injected directly into the brain affects the bee's ability to recall information stored in short-term memory (Mercer and Menzel, 1982). Both DA and 5-HT reduce the level of responses to an olfactory stimulus that has been paired with a sugar-water reward. OA, on the other hand, does not alter conditioned responses per se, but it enhances general responsiveness of the honeybee to olfactory stimuli. Pharmacological investigations indicate that biogenic amines influence both the storage and retrieval of information in vertebrates (Quartermain and Judge, 1982). Drugs that affect serotonergic systems in the mammalian brain such as pCPA, which inhibits synthesis of 5-HT, and the 5-HT-receptor antagonist LSD-25, have been tested on simple learning mechanisms in *P. americana* (Tarchalska-Krynska and Kostowski, 1977). Both pCPA and LSD-25 facilitated learning in the cockroach. Studies to date, however, tell us little about the precise role played by central aminergic systems in learning and memory in insects. This aspect of the function of biogenic amines in the insect brain also awaits further experimentation. The question of biogenic amines and their functions in the insect brain is a complex one. DA, OA, and 5-HT appear to be involved in many aspects of brain function, and it seems likely that the precise functions played by these amines may vary from one species of insect to another. It is important, therefore, that coordinated multidisciplinary studies be carried out on the same species, and that extrapolation of results from one species to the next be made with caution.

17.4. SUMMARY

Attention has been focused recently on the functional roles of central aminergic systems in insects. The catecholamines dopamine and noradrenaline, the indolealkylamine 5-hydroxytryptamine, and the phenolamine octopamine have been detected in the brain of many insects. The distribution of these amines throughout the brain is extensive. They are associated with primary sensory areas, higher brain centers, and with pathways linking major divisions of the brain neuropil. Brain amines have been implicated in a diverse range of functions, including modulation of information in sensory processing channels, regulation of hormone release, and the control of behavior. At present, our knowledge of the precise roles played by biogenic amines in the insect brain, and the mechanisms through which they operate, is limited. It is hoped, however, that with biochemical, physiological, and pharmacological analysis of specific amine-containing neurons identified in the insect brain, considerable progress will be made in this exciting and rapidly expanding field in the very near future.

REFERENCES

Barlow, R. B., S. C. Chamberlain, and E. Kaplan. 1977. Efferent inputs and serotonin enhance the sensitivity of the *Limulus* lateral eye. *Biol. Bull.* (*Woods Hole*) **153**: 414.

Bishop, C. A. and M. O'Shea. 1983. Serotonin immunoreactive neurons in the central nervous system of an insect (*Periplaneta americana*). *J. Neurobiol.* **14** (4): 251–269.

Bodnaryk, R. P. 1980. Changes in brain octopamine levels during metamorphosis of the moth *Mamestra configurata*. *Insect Biochem.* **10**: 169–173.

Brown, C. S. and C. Nestler. 1985. Catecholamines and indolalkylamines. pp. 435–497. In G. A. Kerkut and L. I. Gilbert (eds.), *Comprehensive Insect Physiology, Biochemistry and Pharmacology*, vol. 11. *Pharmacology*. Pergamon Press, Oxford.

Cassier, P. 1983. Cephalic neurohemal organs in Orthoptera, pp. 346–392. In A. P. Gupta (ed.), *Neurohemal Organs of Arthropods*. Charles C. Thomas, Springfield, IL.

Clarke, B. S. and J. F. Donnellan. 1982. Concentrations of some putative neurotransmitters in the central nervous system of quick frozen insects. *Insect Biochem.* **12**(6): 623–638.

Downer, R. G. H., G. L. Orr, J. W. D. Gole, and I. Orchard. 1984. The role of octopamine and cyclic AMP in regulating hormone release from corpora cardiaca of the American cockroach. *J. Insect Physiol.* **30:** 457–462.

Dudai, Y. 1982. High-affinity octopamine receptors revealed in *Drosophila* by binding of [^{3}H]octopamine. *Neurosci. Lett.* **28:** 163–167.

Dudai, Y. and S. Zvi. 1984a. High-affinity [^{3}H]octopamine-binding sites in *Drosophila melanogaster:* Interaction with ligands and relationship to octopamine receptors. *Comp. Biochem. Physiol.* **77C:** 145–151.

Dudai, Y. and S. Zvi. 1984b. [^{3}H]serotonin binds to two classes of sites in *Drosophila* head homogenate. *Comp. Biochem. Physiol.* **77C:** 305–309.

Dymond, G. R. and P. D. Evans. 1979. Biogenic amines in the nervous system of the cockroach, *Periplaneta americana:* Association of octopamine with mushroom bodies and dorsal unpaired median (DUM) neurones. *Insect Biochem.* **9:** 535–545.

Elias, M. S. and P. D. Evans. 1983. Histamine in the insect nervous system: Distribution, synthesis and metabolism. *J. Neurochem.* **41**(2): 562–568.

Elias, M. S. and P. D. Evans. 1984. Autoradiographic localization of ^{3}H-histamine accumulation by the visual system and the locust. *Cell Tissue Res.* **238:** 105–112.

Evans, P. D. 1978. Octopamine distribution in the insect nervous system. *J. Neurochem.* **30:** 1009–1013.

Evans, P. D. 1980. Biogenic amines in the insect nervous system. *Adv. Insect Physiol.* **15:** 317–473.

Falck, B., N. A. Hillarp, G. Thieme, and A. Torp. 1962. Fluorescence of catecholamines and related compounds condensed with formaldehyde. *J. Histochem. Cytochem.* **10:** 348–354.

Frontali, N. 1968. Histochemical localization of catecholamines in the brain of normal and drug-treated cockroaches. *J. Insect Physiol.* **14:** 881–886.

Frontali, N. and K. A. Norberg. 1966. Catecholamine-containing neurons in the cockroach brain. *Acta Physiol. Scand.* **66:** 243–244.

Fuzeau-Breasch S. and J. C. David. 1979. Octopamine levels during the moult cycle and adult development in the migratory locust, *Locusta migratoria. Experientia* **35:** 1349–1350.

Gersch, M. 1972. Experimentelle Untersuchungen zum Freisetzungs-mechanismus von Neurohormonen nach elektrischer Reizung der Corpora Cardiaca von *Periplaneta americana* in vitro. *J. Insect Physiol.* **18:** 2425–2439.

Guldberg, H. C. and O. J. Broch. 1971. On the mode of action of reserpine on dopamine metabolism in the rat striatum. *Eur. J. Pharmacol.* **13:** 155–167.

Harmer, A. J. and A. S. Horn. 1977. Octopamine-sensitive adenylate cyclase in cockroach brain: Effects of agonists, antagonists and guanylyl nucleotides. *Mol. Pharmacol.* **13:** 512–520.

Kaiser, W. 1979. Circadian variations in the sensitivity of single visual interneurones of the bee *Apis mellifera carnica. Verh. Dtsch. Zool. Ges.* **1979:** 211.

Kaiser, W. and J. Steiner-Kaiser. 1983. Neuronal correlates of sleep wakefulness and arousal in a diurnal insect. *Nature (London)* **301:** 707–709.

Klemm, N. 1976. Histochemistry of putative transmitter substances in the insect brain. *Prog. Neurobiol.* **7:** 99–169.

Klemm, N. and F. Sundler. 1983. Organization of catecholamine and serotonin immunoreactive neurons in the corpora pedunculata of the desert locust, *Schistocerca gregaria* Forsk. *Neurosci. Lett.* **36:** 13–17.

Koe, B. K. and A. Weissman. 1966. P-chlorophenylalanine: A specific depletor of brain serotonin. *J. Pharmacol. Exp. Ther.* **154:** 499–516.

Lafon-Cazal, M., J. F. Coulon, and J. C. Davis. 1982. Octopamine metabolism in the cephalic ganglion of *Locusta migratoria* L. *Comp. Biochem. Physiol.* **73C:** 293–296.

Lapitskii, V. P., A. A. Rusinov, and V. V. Kovaler. 1981. Microiontophoretic study of dopamine effects on electrical activity of the supraesphageal ganglion neurons in the cockroach *Periplaneta americana*. *Vestn. Leningr. Univ. Biol.* **0**(3): 81–85.

Lutz, E. M., N. M. Tyrer, J. S. Altman, and J. Turner. 1985. Some insect sensory neurones contain 5-hydroxytryptamine. *Brain Res.* **325:** 353–356.

Maxwell, G. D. and J. G. Hildebrand. 1981. Anatomical and neurochemical consequences of deafferentation in the development of the visual system of the moth *Manduca sexta*. *J. Comp. Neurol.* **195**(4): 667–680.

Mercer, A. R. and J. Erber. 1983. The effects of amines on evoked potentials recorded in the mushroom bodies of the bee brain. *J. Comp. Physiol.* **151**(4): 469–476.

Mercer, A. R. and R. Menzel. 1982. The effect of biogenic amines on conditioned and unconditioned responses to olfactory stimuli in the honeybee, *Apis mellifera*. *J. Comp. Physiol.* **145**(3): 363–368.

Mercer, A. R., P. G. Mobbs, A. P. Davenport, and P. D. Evans. 1983. Biogenic amines in the brain of the honeybee, *Apis mellifera*. *Cell Tissue Res.* **234:** 655–677.

Mobbs, P. G. 1982. The brain of the honeybee *Apis mellifera*. I. The connections and spatial organization of the mushroom bodies. *Philos. Trans. R. Soc. (B)* **298:** 309–354.

Mobbs, P. G. 1985. Brain structure, pp. 299–370. In G. A. Kerkut and L. I. Gilbert (eds.), *Comprehensive Insect Physiology, Biochemistry and Pharmacology*, vol. 5. *Nervous System, Structure and Motor Function*. Pergamon Press, Oxford.

Mordue, W. 1982. Neurosecretory peptides and biogenic amines, pp. 88–101. In D. Evered, M. O'Connor, and J. Whelan (eds.), *Neuropharmacology of Insects*. Ciba Foundation, Symposium 88, Pitman, London.

Muszynska-Pytel, M. and B. Cymborowski. 1978a. The role of serotonin in regulation of the circadian rhythms of locomotor activity in the cricket (*Acheta domesticus* L.) II. Distribution of serotonin and variations in different brain structure. *Comp. Biochem. Physiol.* **59C:** 17–20.

Muszynska-Pytel, M. and B. Cymborowski. 1978b. The role of serotonin in regulation of the circadian rhythms of locomotor activity in the cricket (*Acheta domesticus* L.). I. Circadian variations in serotonin concentrations in the brain and hemolymph. *Comp. Biochem. Physiol.* **59C:** 13–15.

Nässel, D. R. and N. Klemm. 1983. Serotonin-like immunoreactivity in the optic lobes of three insect species. *Cell Tissue Res.* **232:** 129–140.

Nässel, D. R. and L. Laxmyr. 1983. Quantitative determination of biogenic amines and dopa in the CNS of adult and larval blowflies, *Calliphora erythrocephala*. *Comp. Biochem. Physiol.* **75C:** 259–265.

Nässel, D. R. and R. Cantera. 1985. Mapping of serotonin-immunoreactive neurons in the larval nervous system of the flies *Calliphora erythrocephala* and *Sarcophaga bullata*. A comparison with ventral ganglia in adult animals. *Cell Tissue Res.* **239:** 423–434.

Nässel, D. R., M. Hagberg, and H. S. Seyan. 1983. A new, possibly serotonergic, neuron in the lamina of the blowfly optic lobe: An immunocytochemical and Golgi-EM study. *Brain Res.* **280:** 361–367.

Nässel, D. R., E. P. Meyer, and N. Klemm. 1985. Mapping and ultrastructure of serotonin-immunoreactive neurons in the optic lobes of three insect species. *J. Comp. Neurol.* **232:**190–204.

Omar, D., L. L. Murdock, and R. M. Hollingworth. 1982. Actions of pharmacological agents on 5-hydroxytryptamine and dopamine in the cockroach nervous system (*Periplaneta americana* L.). *Comp. Biochem. Physiol.* **73C:**423–429.

Orchard, I., B. G. Loughton, J. W. D. Gole, and R. G. H. Downer. 1983a. Synaptic transmission elevates adenosine 3'-5'-monophosphate (cyclic AMP) in locust neurosecretory cells. *Brain Res.* **258:**152–155.

Orchard, I., J. W. D. Gole, and R. G. H. Downer. 1983b. Pharmacology of aminergic receptors mediating an elevation in cyclic AMP and release of hormone from locust neurosecretory cells. *Brain Res.* **288:** 349–353.

Pandey, A. and M. Habibulla. 1980. Serotonin in the central nervous system of the cockroach, *Periplaneta americana*. *J. Insect Physiol.* **26:** 1–6.

Pandey, A., M. Habibulla, and R. Singh. 1983. Tryptophan hydroxylase and 5-HTP-decarboxylase activity in cockroach brain and the effects of p-chlorophenylalanine and 3-hydroxybenzylhydrazine (NSD-1015). *Brain Res.* **273:** 67–70.

Prée, J. and E. Rutschke. 1983. Zur circadianen Rhythmik des Dopamingehaltes in Gehirn von *Periplaneta americana* L. *Zool. Jahrb. Physiol.* **87:** 455–460.

Quartermain, D. and M. E. Judge. 1983. Recovery of memory following forgetting induced by depletion of biogenic amines. *Pharmacol. Biochem. Behav.* **18:** 179–184.

Quinn, W. G. 1984. Work in invertebrates on the mechanisms underlying learning, pp. 197–246. In P. Marler and H. S. Terrace (eds.), *The Biology of Learning*. Life Sciences Research Report 29. Springer, Berlin.

Raabe, M. 1982. *Insect Neurohormones*. Plenum Press, New York.

Rademakers, L. H. P. M. 1977. Identification of a secretomotor centre in the brain of *Locusta migratoria*, controlling the secretory activity of the adipokinetic hormone producing cells of the corpus cardiacum. *Cell Tissue Res.* **184:** 381–395.

Robertson, H. A. 1976. Octopamine, dopamine and noradrenaline content of the brain of the locust, *Schistocerca gregaria*. *Experientia* **32:**552–553.

Samaranayaka, M. 1976. Possible involvement of monoamines in the release of adipokinetic hormone in the locust. *Schistocerca gregaria*. *J. Exp. Biol.* **65:** 415–425.

Schürmann, F. W. and N. Klemm. 1984. Serotonin-immunoreactive neurons in the brain of the honeybee. *J. Comp. Neurol.* **225:**570–580.

Steiner, F. A. and L. Pieri. 1969. Comparative microelectrophoretic studies of invertebrate and vertebrate neurones. *Prog. Brain Res.* **31:** 191–199.

Tarchalska-Krynska, B. and W. Kostowski. 1977. Effects of drugs influencing serotonergic mechanisms on behavior of insects. *Pol. J. Pharmacol.* **28**(6): 635–639.

Tyrer, N. M., J. D. Turner, and J. S. Altman. 1984. Identifiable neurons in the locust central nervous system that react with antibodies to serotonin. *J. Comp. Neurol.* **227:** 313–330.

Uzzan, A. and Y. Dudai. 1982. Aminergic receptors in *Drosophila melanogaster:* Responsiveness of adenylate cyclase to putative neurotransmitters. *J. Neurochem.* **38:** 1542–1550.

Vieillemaringe, J., P. Duris, M. Geffard, M. Le Moal, M. Delaage, C. Beusch, and J. Girardie. 1984. Immunohistochemical localization of dopamine in the brain of the insect *Locusta migratoria migratorioides* in comparison with the catecholamine distribution determined by the histofluorescence technique. *Cell Tissue Res.* **237:** 391–394.

CHAPTER 18

Neurochemical Aspects of Cholinergic Synapses in the Insect Brain

H. Breer
Department of Zoophysiology
University of Osnabrück
Osnabrück, West Germany

18.1. INTRODUCTION

The function of the nervous system is based on a special interrelationship between neurons and synapses. In the insect central nervous system (CNS) synaptic contacts are restricted to the central neuropil region, where axons and dendrites, derived from the intrinsic motor- and interneurons, as well as the arborizing axons of the peripheral sensory neurons, form a mass of interwoven processes with numerous synaptic contacts. The cell bodies of intrinsic neurons, located outside the neuropil, appear to be free of synapses.

In contrast to the process in unraveling the neuroanatomy of the insect CNS and the neurophysiology of sensory afferences and central interneurons, our knowledge of the biochemical processes and molecular elements of insect synapses has lagged behind that of other animal groups. Detailed knowledge on the neurochemistry of synaptic function in the insect brain is a prerequisite for an understanding of the neurochemical development of the insect nervous system and for advanced studies of the molecular basis of neuroethological mechanisms in insects. Finally, if approaches to control insects by neuroactive chemicals are to be advanced toward higher selectivity and effectivity, a more detailed understanding of the molecular aspects of transmission at synapses in the CNS of insects will be required.

Synaptic communication between nerve cells is mediated by neurotransmitters, and most of the transmitter candidates in vertebrates may also serve as neurotransmitter in insects. Comparing the concentration of some classical neurotransmitters in the nervous system of, for example, *Locusta migratoria*, revealed that acetylcholine (ACh)) is by far the predominant transmitter; furthermore, a comparison of ACh-concentration in the central nervous tissue of some insect and vertebrate tissue, emphasizes the importance of ACh and cholinergic transmission in the ganglia of insects (Table 18.1; see also Chapter 19 in this volume).

TABLE 18.1. NEUROTRANSMITTER AND ACETYLCHOLINE CONTENT IN THE CENTRAL NERVOUS SYSTEM

Neurotransmitter content of *Locusta migratoria* central nervous system

	Acetylcholine	5-Hydroxytryptamine	Dopamine	Noradrenaline
μg/g	111	2.3	1.3	0.2
nmol/g	762	10.6	8.6	1.4

Acetylcholine content in the central nervous system of various species

Species	nmol/mg protein	μg/g weight	Reference
Fly (Musca)	9.0	150	Lewis, 1956
Cockroach	8.5	136	Smallman, 1961
Locust	7.7	111	Breer, 1981
Guinea pig	0.3	4.8	Barker, 1972
Rat	0.2	3.4	Saelens, 1973

NEUROCHEMICAL INNERVATION PATTERN

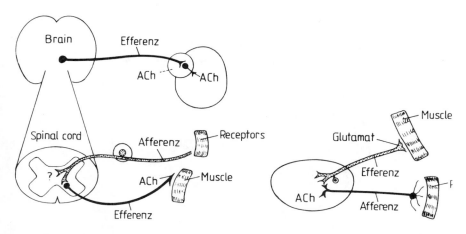

VERTEBRATES

efferent neurons : cholinerg
afferent neurons : ?

ARTHROPODS

efferent neurons : glutaminerg
afferent neurons : cholinerg

Figure 18.1. Diagram representing the neurochemical innervation pattern of vertebrates and arthropods. (Modified after Florey and Michelson, 1973.)

The preponderance of cholinergic synapses in the insect CNS may be due to the fact that in different phyletic lines, which have developed independently, the cholinergic synapses turned out to be localized in different parts of the nervous system (Fig. 18.1). In the peripheral nervous system of *vertebrates*, only the efferent, centrifugal nerves and corresponding neurons are cholinergic (motorneurons innervating the skeletal muscles, the pre- and postganglionic parasympathetic nerves, and some of the postganglionic sympathetic nerves). The sensory neurons are noncholinergic; their transmitter is not yet identified. In *arthropods*, in contrast to vertebrates, the sensory and associative neurons are cholinergic and the efferent neurons are *not* (Florey and Michelson, 1973).

18.2. CHOLINERGIC SYNAPSES

A detailed knowledge of the metabolism, the functional role of acetylcholine, and of biochemical processes and functional elements of cholinergic synaptic transmission may be one of the keys to the understanding of the nervous system of insects; it is of particular interest that an involvement of cholinergic synapses in learning phenomena in insects has been discovered (Ker-

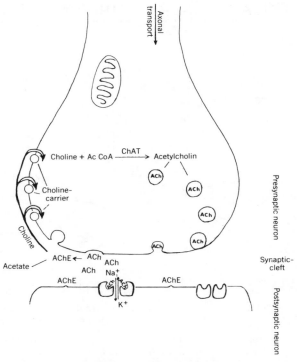

Figure 18.2. Schematic diagram of a cholinergic synapse.

kut et al., 1970). Furthermore, some interest in cholinergic synapses of insects is due to the fact that many insecticides act on the cholinergic system (Shankland et al., 1978).

In cholinergic neurons, acetylcholine is synthesized from choline and acetyl-CoA by means of choline acetyltransferase (ChAT). The neurotransmitter is translocated into synaptic vesicles, probably via H^+-antiporter mechanisms. When an action potential reaches a cholinergic nerve terminal, a series of cellular and molecular events occur that result in the release of ACh in a Ca^{2+}-dependent process. The released chemical messenger diffuses to the postsynaptic membrane and very transiently binds to a highly specific receptor, causing a change in membrane permeability. ACh is subsequently inactivated by the hydrolytic enzyme acetylcholinesterase (AChE). The resulting choline molecules are preferentially taken into the presynaptic terminal via specific high-affinity carrier mechanisms, and most of it is immediately converted to ACh. A schematic representation of the location of the various elements of a cholinergic synapse is shown in Figure 18.2. A typical cholinergic neuron is characterized chemically by an ACh concentration of 0.3–3 mM, by the presence of ChAT in its soma and axon, and by a specific choline carrier in synaptic membranes.

The first report of ACh in the insect nervous system was made by Gautrelet (1938), demonstrating the presence of the choline ester in the brain of several insect species. Although there are considerable variations in the data obtained by different authors, the ACh content in the insect nervous tissues investigated is very high and may be as much as 100 times higher than found in vertebrate central nervous tissue. Shortly after the discovery of ACh, the enzymatic activities for synthesis (Tobias et al., 1946), as well as for hydrolyzing (Mikalonis and Brown, 1941) the choline ester, were established. Thus, it is well documented that three essentials of the cholinergic system (ACh, ChAT, AChE) occur in several insect nervous systems. Advanced assay procedures for acetylcholine and the enzyme are described in recent publications (Kingan and Hildebrand, 1985; Breer and Knipper, 1985a).

18.3. ENZYMES OF THE CHOLINERGIC SYSTEM

18.3.1. Choline Acetyltransferase

In cholinergic nerve cells, acetyl-CoA provides acetyl groups and energy for the enzymatic acetylation of choline. Choline acetyltransferase is a key element and marker of cholinergic neurons. Smallman and Pal (1957) have demonstrated that ChAT activity is very high in the CNS of *Periplaneta americana*, for example, 53 mg/g per hour. Comparative studies have shown that the activity of choline acetyltransferase is more than 100 times higher in locust ganglia than in mouse brain (Breer, 1981a). As in other animals, in insects most of the enzyme activity is found in the soluble fraction, suggesting that ChAT is a cytoplasmic enzyme, although there are experimental evidences that a considerable portion of ChAT is bound to membranes.

The mechanisms of ACh synthesis and the kinetics of the reaction were studied in partially purified ChAT preparations from *Drosophila* and locusts (Driskel et al., 1978). More recently, the enzyme from *Drosophila* has been purified more than 10,000-fold, to a final specific activity of 500 μmol $\cdot$ min^{-1} $\cdot$ protein^{-1}, using a series of affinity chromatography procedures (Slemmon et al., 1982). The final preparation showed three major protein bands (67 Kd, 54 Kd, and 13 Kd) on SDS-polyacrylamide gel electrophoresis; however, structural and immunological studies showed that these proteins were closely related. Subsequent studies indicate that the ChAT enzyme is a monomeric globular protein and has a native molecular weight of approximately 67 Kd. In fact, there are evidences suggesting that *Drosophila* ChAT may exist in two forms. One is a continous 67 Kd polypeptide; the other form consists of associated polypeptide chains of 54 Kd and 13 Kd, respectively.

In genetic approaches a single gene, designed Cha (originally Cat), has been identified for ChAT (Hall and Greenspan, 1979); the structural gene was localized by aneuploidy techniques on the right arm of chromosome 3.

18.3.2. Acetylcholinesterase

At cholinergic synapses it is important that released neurotransmitter is rapidly removed from the synaptic cleft and from the site of action on the postsynaptic membrane, because prolonged exposure of cholinergic receptors to acetylcholine results in desensitization of those receptors and blockage of synaptic transmission. This essential prerequisite is realized by acetylcholinesterase, which hydrolyzes ACh very efficiently to physiologically inactive products. AChE is the target of widely used insecticides, such as organophosphates and carbamates, and has thus attracted much interest.

Numerous biochemical and histochemical studies have shown that the activity of acetylcholinesterase is very high in insect nervous tissue (Pitman, 1971), showing about the same capacity to catalyze ACh-hydrolysis as the electric tissue of *Electrophorus* (Metcalf and March, 1950). The cholinesterase in insect nervous tissue was shown to hydrolyze acetylcholine much more rapidly than butyrylcholine, and the enzyme activity is inhibited by high concentrations of the substrate acetylcholine.

The acetylcholinesterase in the nervous tissue of various insect species was shown to be a membrane protein, which was found in multiple forms (Eldefrawi et al., 1970; Tripathi et al., 1978; Silver and Prescott, 1982). The acetylcholinesterase polymorphism is obviously due to aggregation and it has been suggested that the enzyme exists as a high-molecular-weight aggregate in the membrane. The data for the minimum molecular weight of an active form of insect AChE varies between 80 kD and 110 kD (Silver and Prescott, 1982). The gene coding for AChE (Ace) was identified in *Drosophila*, (Kankel and Hall, 1976), and localized in a small segment (3-5 polytene chromosome "band") of the right arm of chromosome 3. Thus, the "cholinergic" gene Cha, coding for the ACh synthesizing enzyme, and Ace, coding for the enzyme that degrades the ACh, are both located in the same region of the 3 chromosome; however, they are at least 230 bands away from each other (Hall et al., 1979). It will be extremly interesting to evaluate whether or not genes coding for other cholinergic elements are located in the same chromosome region, possibly presenting a gene family.

18.4. ISOLATION OF CHOLINERGIC NERVE TERMINALS

Until recently, most of the approaches to characterize neurochemical aspects of cholinergic synapses and transmission in the insect nervous system have been concerned primarily with estimation of ACh concentration or with macromolecules involved in the synthesis and inactivation of the neurotransmitter. Only the development of an adequate subcellular fractionation procedure for insect nervous tissue and the opportunity to prepare intact synaptosomes in fairly high yield allowed us to study the dynamic processes at the cholinergic synapses, using appropriate biochemical approaches.

First attempts to gain subcellular fractions from insect neuronal tissue

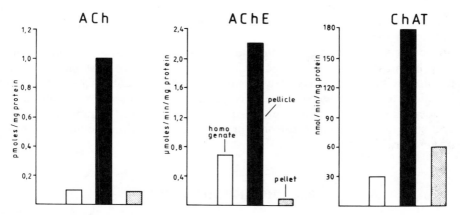

Figure 18.3. Distribution of cholinergic activities (ACh, ChAT, AChE) in the pellet and pellicle fraction after subcellular fractionation of locust nervous tissue.

were carried out by Telford and Matsumura (1970) by applying a classical step gradient, this was improved by introducing the osmotically inactive Ficoll to avoid damage of nerve terminals during centrifugation (Donnellan et al., 1976). A microprocedure especially designed for use with insect nervous tissue was developed, applying a mild homogenization procedure and using an isoosmotic Ficoll gradient in combination with a flotation technique to cope with the fragility of insect nerve terminals and to minimize artifacts caused by bulk sedimentation; the organelle suspension was homogeneously distributed in the Ficoll solution before centrifugation (Breer and Jeserich, 1980; Breer, 1981b). As documented by ultrastructural and biochemical analysis of the resulting fractions, the floating pellicle fraction contained numerous synaptosomal profiles, containing synaptic vesicles and intraterminal mitochondria. The vesicles were mostly spherical and electrolucent, approximately 50 nm in diameter. Evaluating the distribution of marker enzyme activity, as well as protein and lipid analysis of the resulting fractions, demonstrated that the mitochondrial markers were mainly located in the pellet, whereas synaptosomal markers were enriched in the pellicle fraction. Because the viability of the isolated nerve endings has proved of great importance for in vitro studies, several metabolic properties of insect synaptosomal preparation have been analyzed. The data obtained resemble very closely the values reported for synaptosomes from rat brain (Breer and Jeserich, 1984), and, furthermore, even a membrane potential of about 60 mV has been determined (Breer, 1982a; Breer, 1985b). Thus, it can be concluded that morphologically intact, viable, nerve terminals can be isolated from insect nervous tissue, and such insect synaptosomes may be considered as miniature cells, useful for neurochemical analysis of synaptic transmission.

When comparing the distribution of cholinergic elements in the subcellular fractions, it was found that all parameters of the cholinergic system (ACh,

ChAT, AChE) were highly enriched in the synaptosomal fraction (Fig. 18.3), indicating that synaptosomal fractions from insect ganglia contained high proportions of cholinergic nerve endings; thus, synaptosomes from insects may be most suitable for a neurochemical exploration of synaptic transmission in insects and may allow to unravel some molecular aspects of cholinergic synapses between nerve cells, in general, because most of our basic concepts regarding cholinergic transmission derives from studies on peripheral structures, such as electromotor synapses or neuromuscular junctions of vertebrates.

18.5. CHOLINE TRANSPORT

18.5.1. Transport Kinetics

Synaptosomes isolated from the nervous system of insects, accumulated exogenous choline in a temperature-sensitive process (Fig. 18.4). Analysis of the transport kinetics in concentration-dependent experiments revealed that there are two transport systems involved: a low-affinity choline transport, as found in all cells, and a high-affinity uptake system for choline K_T values of about 1 μM (Breer, 1982b). The high-affinity transport of choline is supported to be unique for cholinergic nerve endings and to be the rate-limiting, regulatory step for the synthesis of acetylcholine (Kuhar and Murrin, 1978); we have found that most of the choline taken up via the high-affinity pathway was converted to acetylcholine (Breer, 1982b). Accumulation of choline by insect synaptosomes appeared to be essential, Na^+-dependent, and competitively inhibited by hemicholinium-3.

Due to the efficient blocking effect on choline transport, hemicholinium-3 (HC-3) is considered as *the* inhibitor of the high-affinity choline carrier; radiolabeled HC-3 may be of considerable importance for a molecular identification of the choline transporter. Using this ligand, we have estimated the number of binding sites; that is, tentatively the number of choline carriers in various tissues. The number of binding sites was found to be quite low in mouse cerebellum and cortex, but very much higher in the nervous tissue from locust. Interestingly, the distribution of binding sites for HC-3 corresponds fairly well to the concentration of ACh and the rate of choline uptake in the different tissues (Table 18.2). Labeling experiments, using appropriate derivates of HC-3 will allow to identify polypeptides of the choline carrier.

18.5.2. Energetics of Choline Uptake

The basic mechanisms of choline translocation mediated by a high-affinity carrier is still poorly understood, probably because an adequate interpretation of transport data concerning aspects of energetics and regulation is complicated by metabolic activities and a complex compartmentation of in-

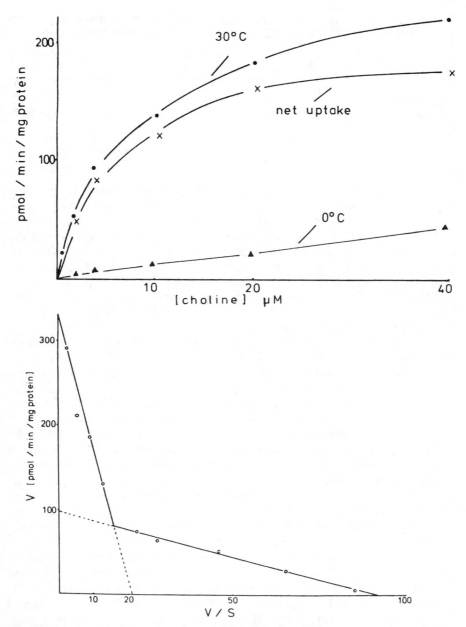

Figure 18.4. Kinetics of choline transport by nerve endings isolated from the head and thoracic ganglia of locust. Concentration-dependent experiments showed saturation at about 25 μM, indicating that the choline translocation is mediated by a specific carrier. (B) Analysis of the data revealed the existence of a high-affinity uptake system, which is supposed to be specific for cholinergic nerve endings.

TABLE 18.2. COMPARISON OF ACETYLCHOLINE (ACh), CHOLINE (Ch) UPTAKE, AND HEMICHOLINIUM (HC-3) BINDING SITES IN VARIOUS NERVOUS TISSUE

	ACh nmol/mg	Ch uptake pmol/min/mg	HC-3 binding fmol/mg
Mouse			
Cerebellum	0.17	—	4
Cortex	0.3	15	25
Locust			
Ganglia	6.8	116	120

tact nerve endings. Recently, there has been significant progress in analyzing transport activities of nerve endings by using synaptosomal ghosts (membrane vesicles) obtained by a spontaneous revesiculation of isolated presynaptic plasma membranes. Such membrane vesicles retained most of the capability of the membrane but lack the complex energetics and compartmentation of intact synaptosomes (Kanner, 1983). Unfortunately, membrane vesicles from mammalian preparations did not accumulate choline (Meyer and Cooper, 1982), probably due to the small proportion of cholinergic nerve terminals in the original synaptosomal preparation. Synaptosomal ghosts derived from the highly cholinergic insect nervous tissue, loaded with K^+ ions and resuspended in NaCl accumulated choline via high-affinity, carrier-mediated pathways; that is, choline was translocated and concentrated by membrane vesicles without ATP and with artificially imposed ion gradients as the sole energy source (Breer, 1983a).

Manipulations of the ionic conditions should elucidate whether or not the high-affinity choline transport is chemiosmotically coupled. Substitution of Na^+ ions always caused a blockade of choline uptake; sodium could not be replaced by any other monovalent cation. Vesicles without a sodium gradient (Na^+ inside and outside; or treated with appropriate ionophores) did not accumulate any choline, indicating that a sodium gradient, $[Na^+]_i/[Na^+]_o$, must be considered as the main driving force. Accordingly, the transport rate should be proportional to the sodium gradient; appropriate experiments showed a sigmoidal relationship between choline accumulation and the Na^+ gradient, indicating that more than one sodium ion interacts with the transport system. Analysis of the data results in a coefficient of about 2, suggesting a stoichiometry of 2 Na^+ per molecule of choline transported (Breer and Lueken, 1983). In similar experiments concerning the anion specificity, it was shown that a chloride gradient is important as well; however, only one Cl^- appears to be involved in choline accumulation. Thus, obviously a complex of $2 Na^+ : 1 Cl^- : 1$ choline is translocated. This points to an electrogenic transport process, that is, net charges cross the membrane. If that is the case, the membrane potential, which exists as a charge difference between inside and outside, should affect the choline transport; experiments, using compounds that modify the membrane potential, confirm this thesis: valinomycin, a potassium ionophore, which increases the potential, enhanced the uptake rate, whereas CCCP, a protonionophore, which destroys the membrane potential, reduced the choline uptake. These results clearly demonstrate that the high-affinity choline uptake is mediated by carriers, which are driven by the electrochemical ion gradient. Thus, in the intact system, the function of the Na^+, K^+-ATPase is merely to convert the energy of cellular ATP into chemiosmotic energy, which is subsequently used to drive the accumulation of choline catalyzed by tightly coupled Na^+/choline symporters.

Studies on the efflux of choline from membrane vesicles have shown that the ion dependency for influx and efflux is quite similar, suggesting that the carrier is symmetrical with respect to the membrane and that vectorial translocation under physiological conditions is caused by ion gradients and electrical potential across the membrane (Breer and Knipper, 1985b). This view is further supported by the transactivating effects of choline, showing that the carrier obviously must be returned to the outside for influx to continue. Furthermore, the data on sodium dependency of influx and efflux support the concept that the sodium-choline cotransport system involves the ordered binding of choline first and then 2 Na^+ ions on the carrier, before translocating across the membrane.

18.5.3. Regulation of Choline Transport

The close interrelationship between transport energetics and ion gradients suggests that choline translocation is probably regulated by modulating the ion gradients. This is shown in experiments with various neurotoxins: veratridine, a plant alkaloid, which activated Na^+ channels (Narahashi, 1974) significantly inhibited choline uptake; this effect was almost completely prevented by tetrodotoxin, a specific blocker of Na^+ channels, which by itself had not much of an effect (Breer and Lueken, 1983). This aspect was further extended by analyzing the effect of scorpion toxins that are known to cause a delayed inactivation of Na^+ channels. These preparations are obviously active on insect preparations as well; they potentiate the effect of veratridine on choline uptake. The combined actions of veratridine and scorpion toxins caused a drastically reduced rate of choline accumulation (Breer, 1983b; Breer and Knipper, 1985c). These results emphasize the view that choline translocation is directly related to the electrochemical gradients and can thus be regulated in multiple ways.

The choline transporter may however be regulated by other mechanisms as well. Extrasynaptosomal ATP might be a modulator, an inhibitor, of choline translocation. Extracellular ATP inhibited the choline transport into synaptosomes and membrane vesicles at submillimolar concentrations. This finding may be of particular interest, because ATP might be released together with ACh at cholinergic synapses in insects, as has already been shown for peripheral cholinergic nerve endings (Zimmermann, 1982). The effect of ATP appeared to be rather specific; neither ADP, AMP, nor adenosine showed comparable effects, and also the nonhydrolyzed ATP analogues, such as AMP-PMP, did not affect the choline transport. The specificity for ATP suggests that a kinase reaction might be involved, and that specific proteins of the synaptic plasma membrane might be phosphorylated by exokinases.

18.6. ACETYLCHOLINE RELEASE

18.6.1. Induction of Acetylcholine Release in Vitro

The release of transmitters from presynaptic nerve terminals is a fundamental step in chemical transmission, which is physiologically triggered by a nerve impulse and may be influenced by changes in the ionic environment, as well as by endogenous modulators. The complex organization of insect ganglia makes it impractical to use in vitro preparations to investigate release mechanisms; insect synaptosomes that accumulate choline and convert it to acetylcholine, are most suitable for studying transmitter release. A far reaching characterization of acetylcholine release offers the perfusion procedure originally described by Mulder et al. (1976). Synaptosomes were immobilized on small columns and perfused with isotonic media; samples of the perfusate were collected at various time intervals and analyzed (Breer and Knipper, 1985b). Perfusion with isoosmotic insect medium caused only little efflux of acetylcholine; however, the additions of depolarizing agents like veratridine induced a considerable release of acetylcholine (Fig. 18.5). Such an evoked release was shown to be essentially dependent on extracellular calcium; depolarization with calcium-free medium induced no efflux. Only the addition of millimolar concentration caused release of acetylcholine (Breer and Knipper, 1984), whereas Mg^{2+} ions antagonized the Ca^{2+} effect. These results suggest that ACh release from insect cholinergic nerve terminals is achieved in a depolarization-secretion coupled process.

Release experiments in the presence of acetylcholine and cholinesterase inhibitors always gave significantly reduced rates of release, suggesting that acetylcholine influences its own release process possibly via feedback mechanisms. To evaluate whether or not receptor-mediated reactions might be involved in this regulation, the effects of various cholinergic agents were analyzed. It became clear that muscarinic agonists, such as oxotremorine, inhibited release, whereas muscarinic antagonists enhanced the evoked acetylcholine release (Breer and Knipper, 1984). These results may be considered as support for the concept that a feedback regulation of acetylcholine release is exerted via a muscarinic autoreceptor located on the plasma membrane of cholinergic axon terminals. Furthermore, these results may represent the first evidence for a physiological role of muscarinic receptors in insect nervous tissue. Experimental evidence for the existence of muscarinic receptors in insects have emerged from binding sites, using quinuclidinyl

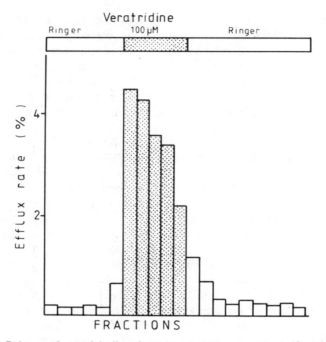

Figure 18.5. Release of acetylcholine. Insect synaptosomes were perfused with insect Ringer, only after addition of depolarizing agents like veratridine in micromolar concentration. The release of acetylcholine was induced if calcium was present in millimolar concentration.

benzilate, a potent muscarinic antagonist, as a receptor probe (Meyer and Edwards, 1980; Breer, 1981c; Lummis and Sattelle, 1985).

18.7. ACETYLCHOLINE RECEPTORS IN INSECT BRAIN

18.7.1. Characterization of Ligand Binding Sites

Receptors for acetylcholine are particularly important at cholinergic synapses, not only as recognition molecules of the transmitter but also as transducers of the molecular signals. From vertebrate neuropharmacology, it is known that there are two different types of receptors for acetylcholine, which differ in mode of action as well as in pharmacology: nicotinic and muscarinic receptors. Both receptor types can be identified and analyzed by means of specific ligands: α-bungarotoxin (BGTX), is used as a ligand for the nicotinic receptor, whereas quinuclidinic benzilate is used for muscarinic receptors.

Comparing the binding capacity for these two ligands in the central nervous tissue demonstrates that the concentration of receptors for acetylcholine (AChR) is much higher in insects than in mammals, which parallels the distribution of other cholinergic elements. However, it is most remarkable that in each case a different receptor type predominates: in mouse brain mostly QNB-binding sites (mAChR); in locust ganglia very high concentrations of α-toxin binding sites, putative nAChR. This observation, that insects as well as mammals, which are far apart on the evolutionary scale, contain nicotinic and muscarinic binding sites, demonstrates that the classical dichotomy of cholinergic receptors (nicotinic/muscarinic) developed very early in evolution. The reversed levels of α-bungarotoxin and quinuclidinyl benzilate binding sites in vertebrate and insect CNS tissue is the most striking contrast between insect and vertebrate cholinergic receptors; its significance has yet to be evaluated. Ligand-binding experiments on membrane preparations have revealed binding of both cholinergic receptor ligands in the nervous tissue of a number of insect species (Sattelle, 1985). The isotherms showed saturation in each case, indicating that there is only a limited number of high-affinity binding sites for the two ligands. Values of about 1000 fmol/mg α-BGTX and about 100 fmol/mg QNB binding sites have been reported. The pharmacological profile of the ligand-binding components obtained in competitive experiments revealed that both are distinct entities (Table 18.3), and it has been shown that the evoked EPSP of the central nerve giant fiber synapse in the abdominal ganglion of *Periplaneta americana* was depressed by α-toxin at nanomolar concentrations (Sattelle et al., 1983). These results can be considered as evidence that the toxin binding sites are associated with functional nicotinic ACh receptors on specific neurons; a conclusion, that is controversial for toxin binding in vertebrate brain.

18.7.2. Identification of Nicotinic Acetylcholine-Receptor Proteins

In the first approach to explore any molecular similarity between fish and insect receptor, a library of monoclonal antibodies (mab) against the *Torpedo* receptor was assayed for any cross-reactivity with neuronal membranes from locust ganglia in series of ELISA tests (Fels et al., 1983). It was found that 2 mab out of a collection of 20, showed considerable cross-reactivity with the insect membranes. Competition experiments, using small cholinergic ligands and α-toxins, revealed that the epitop for cross-reaction was obviously at or near the ligand-binding region of the receptor molecules. These results point to some similarities in the molecular structure of the ACh receptors in *Torpedo* electroplaques and ganglia of locust (Fels et al., 1983).

For further biochemical characterization, it was essential to solubilize the receptor protein under condition, which allows preservation of toxin binding and its pharmacology; this was achieved by using mild detergents. A detergent extract was submitted to a linear sucrose density gradient; the receptors sedimented as one major protein species indicating the homogeneity of binding proteins (Breer et al., 1985). The resulting sedimentation coefficient of approximately 10 sec suggests that the apparent molecular size of

TABLE 18.3. PROPERTIES OF CHOLINERGIC BINDING SITES IN THE NERVOUS SYSTEM OF INSECTS

A. Binding of cholinergic ligands

Tissue	B_{max} (fmol mg^{-1})	Kinetic Constants			References
		K_d (equilibrium)	K_1	K_{-1}	
I. α-Bungarotoxin					
D. melanogaster	800	1.9	2.4×10^5	1.4×10^{-4}	Dudai, 1978
L. migratoria	1175	1.1	6.7×10^5	4.2×10^{-4}	Breer, 1981
P. americana	910	4.8	2.0×10^5	1.2×10^{-4}	Lummis & Sattelle, 1985
II. Quinuclidinyl benzilate					
D. melanogaster	65	0.7	2×10^6	3.0×10^{-4}	Haim et al. 1979
L. migratoria	116	0.8	2.2×10^6	9.9×10^{-4}	Breer, 1981
P. americana	138	8.0	1.3×10^5	2.4×10^{-4}	Lummis & Sattelle, 1985

B. Pharmacology of binding sites (Inhibition constants, K_i)

Ligand	D. melanogaster[a]		L. migratoria[b]		P. americana[c]	
	BGTX	QNB	BGTX	QNB	BGTX	QNB
α-Bungarotoxin	5×10^{-10}	—	7×10^{-10}	—	9×10^{-10}	—
Nicotine	8×10^{-7}	—	4×10^{-8}	—	4×10^{-4}	—
d-Tubocurarine	2×10^{-6}	—	3×10^{-5}	—	9×10^{-7}	—
Acetylcholine	2×10^{-5}	—	4×10^{-4}	—	2×10^{-5}	—
Artropine	5×10^{-5}	4×10^{-9}	1×10^{-4}	4×10^{-7}	2×10^{-5}	2×10^{-7}
Scopolamine	—	1×10^{-9}	—	1×10^{-7}	—	6×10^{-8}
Pilocarpine	—	3×10^{-6}	—	5×10^{-6}	—	1×10^{-5}

[a] From Dudai, 1978; Haim et al., 1979.
[b] From Breer, 1981a.
[c] From Lummis and Sattelle, 1985.

IMMUNOCHEMICAL LOCALISATION OF ACH RECEPTORS

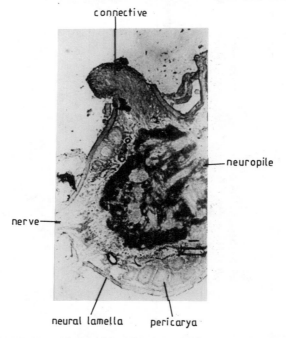

Figure 18.6. Immunocytochemical localization of nicotinic acetylcholine receptors in a thoracic ganglion of locust using monospecific anti-locust AChR antiserum and the PAP technique. Note that specific areas in the neuropil are selectively labeled.

the insect receptor is slightly higher than that reported for receptors from electric tissue (Conti-Tronconi and Raftery, 1982).

The receptor proteins enriched in the peak fraction of the sucrose gradient were further purified by affinity chromatography (Breer et al., 1984). Subsequent analysis of the receptor molecules by means of nondenaturing gradient gel electrophoresis revealed a single band, corresponding to a molecular weight of 300 Kd (Breer et al., 1985). To elucidate its possible subunit composition receptor, the protein was subjected to electrophoresis under denaturing conditions. Only one polypeptide, corresponding to a molecular weight of 65 Kd, was found, suggesting that the receptor complex of insect is probably composed of four to five identical subunits.

Further evidence for establishing the receptor nature of the identified protein, was gained by immunohistochemical approaches, investigating its location in the nervous system. Monospecific antisera against the purified receptor protein, raised in rabbits, was used for a topochemical localization. It could be demonstrated that the antigenic sites are located in very distinct areas of the neuropil (Fig. 18.6). No labeling was detected in the peripheral

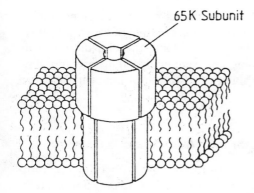
65K Subunit

Figure 18.7. Schematic drawing of a non-specific model of the ACh receptor from insect nervous system.

sheets surrounding the ganglia and in the peripheral cell layers. Thus, the appearance of the antigenic sites in immunocytochemical preparations is consistent with its location at synapses.

18.7.3. Reconstitution of Purified Receptor Proteins

The ultimate proof that the isolated toxin-binding proteins are not only ligand-binding sites but in fact the complete acetylcholine receptor can only be evaluated in reconstitution experiments. Current approaches have shown that the isolated receptor protein from locust can be reconstituted into virtually solvent-free planar lipid bilayers. There it forms a cationic channel that can be activated by cholinergic agonists. The conductance of this channel (70 pS) and its opening time (msec) is quite similar to the values found for the reconstituted *Torpedo* receptor; furthermore, the pharmacology and the substructure of the channel fluctuations resemble each other in details (Hanke and Breer, 1986). Thus, it seems reasonable to conclude that the purified protein is in fact the acetylcholine receptor itself, or an integral part of what would be a larger functional receptor complex in the membrane. It appears that the native acetylcholine receptor in the CNS of insects is a membrane protein with a molecular size of 250–300 Kd, which represents an oligomeric complex of four identical or very similar polypeptides (Fig. 18.7).

The homooligomeric organization of the insect receptor is of particular interest in view of the evolution of the nicotinic acetylcholine receptor. Based on immunological cross-reactivity and the amino acid sequence homology of the four different polypeptides forming the *Torpedo* receptor, a common genetic origin during evolution of the receptor subunits has been suggested; genes coding for the subunits of the receptor protein must have originated from a common ancestral gene via duplication (Raftery et al., 1980). Thus, the ancestral acetylcholine receptor is supposed to have been a homooligomeric complex from which the recent heterooligomeric receptors have evolved. Whether or not the homooligomeric acetylcholine receptor from insects can be considered as a representative of an ancient form of the acetylcholine receptor is unclear as yet. Only the elucidation of its complete primary structure by cloning and sequencing the corresponding nucleic acids will show if there is any significant structural similarity between the insect and vertebrate receptor.

18.7.4. Expression of Receptor mRNA in *Xenopus* Oocytes

As a first approach, the nervous system of locust was probed for receptor-specific mRNA, using *Xenopus* oocytes as an efficient expression system. It was found that after microinjecting nanogram amounts of polyA$^+$-RNA from the ganglia of young locusts, α-bungarotoxin binding protein appeared in the oocytes (Breer and Benke, 1985). Immunoprecipitation experiments showed that indeed a 65 Kd protein was synthesized by oocytes, and in ion-flux studies, it was confirmed that the toxin-binding sites coded by insect mRNA represent functional receptors, which are activated by cholinergic agonists, and increase the cation permeability of the oocyte membrane. Thus, the amphibia oocyte represents an ideal system to analyze the biosynthesis and the coding nucleic acids of the acetylcholine receptor in insects.

18.7.5. Properties of Muscarinic Receptors in Insects

Due to the low concentration of the QNB binding sites and the fact that a useful tool, such as α-BGTX, for the nicotinic receptor is not available, the putative muscarinic AChR is characterized in much less detail. Very recently, the molecular weight of a muscarinic binding site in the nervous tissue of *Periplaneta americana*, was estimated, using the radiation-inactivation procedure; values of about 78 Kd were found for cockroach (Lummis et al., 1984) and a similar figure was meanwhile received for locust by this technique. This number compared quite well with the data reported for the vertebrate muscarinic receptor. Recently, the molecular weight for the muscarinic AChR of the *Drosophila* brain was estimated at about 80 Kd, using SDS polyacrylamide electrophoresis (Venter et al., 1984). By means of immunoprecipitation, it has been shown that monoclonal antibodies raised against rat brain receptors cross-reacted with muscarinic receptors from *Drosophila*, indicating that the muscarinic receptors are obviously rather conserved over a considerable evolutionary period. Thus, biochemical characterization has confirmed that the nicotinic and muscarinic ACh receptors in the CNS of insects are distinct molecular entities and have distinct physical properties.

18.8. CONCLUSION

The application of very different research approaches to explore molecular aspects of cholinergic synaptic transmission and functional elements of cholinergic nerve contracts, has greatly increased our knowledge of the neurochemistry of cholinergic synapses in the brain of insects. A refinement and miniaturization of advanced techniques will contribute to overcome the problems encountered by the small size of the insect ganglia, will allow exploration of the synapses of the highly cholinergic insect nervous tissue, and may ultimately contribute to a more detailed understanding of cholinergic contact between nerve cells in general. Analysis of molecular elements in synaptic transmission in insects will also have considerable implications for comparative neurochemistry (Breer and Jeserich, 1984). Comparative and phylogenetic aspects of a complex process like synaptic transmission may elucidate some evolutionarily concerned processes, molecules, or parts of molecules, and may thus shed some new light on the molecular organization of functional elements and may contribute to resolve some of the main interrelated processes.

Insects have a behavior repertoire, which has been studied in great detail, and a nervous system, which is much less complex than the vertebrate brain, thus a more detailed knowledge of neurochemical events at synapses and during synaptic transmission in the insect nervous system may allow insight into molecular mechanisms, which are basic for higher functions of the nervous system. Such approaches may supplement the successful experimental analysis of these aspects on the nervous system of *Aplysia*. Neurochemical studies on insect brain may always have some implications concerning targets or mode of action of insecticides. The development of neuronal preparations (cells, organelles, membranes, molecules) from insect ganglia may contribute to develop and standardize simple assay systems from studying the targets and molecular action of neurotoxic insecticides; the evaluation of critical sites for toxin, and especially comparative aspects, will allow more rational approaches for pesticide discovery that has been typical in the past and finally lead to more selective and safer compounds.

18.9. SUMMARY

Acetylcholine appears to be a major excitatory neurotransmitter in the CNS of arthropods; a neurochemical exploration of functional processes and molecular elements of cholinergic synapses and synaptic transmission is accordingly of great importance for a better understanding of insect brain function. Analysis of presynaptic parameters of the cholinergic system have been greatly facilitated by isolating morphologically and functionally intact synaptosomes from insect ganglia. Those nerve endings released acetylcholine in a specific, calcium-dependent process, which seems to be regulated via presynaptic autoreceptors.

The high-affinity choline transport system, which is supposed to be unique for cholinergic nerve endings and to be the regulatory element for acetylcholine synthesis, is essentially sodium-dependent and energized by electrochemical ion gradients.

The nervous tissue of insects contains very high concentrations of cholinergic binding sites; mostly of the nicotinic type. The nAChR from locust brain has been identified as a macromolecular membrane glycoprotein, which appears to represent an oligomeric complex of four identical subunits. The receptor protein was reconstituted in planar lipid bilayers, where it formed agonist-activated cation channels, demonstrating that the identified protein represents a functional acetylcholine receptor. Expression experiments, using *Xenopus* oocytes, showed that mRNA coding for the receptor protein can be isolated and enriched from locust nervous tissue.

REFERENCES

Breer, H. 1981a. Comparative studies on cholinergic activities in the central nervous system of *Locusta migratoria. J. Comp. Physiol.* **141:** 271–275.

Breer, H. 1981b. Characterization of synaptosomes from the central nervous system of insects. *Neurochem. Int.* **3:** 155–163.

Breer, H. 1981c. Properties of putative nicotinic and muscarinic cholinergic receptors in the central nervous system of *Locusta migratoria. Neurochem. Int.* **3:** 43–52.

Breer, H. 1982a. Properties of synaptosomes from the central nervous system of insects, pp. 158–175. In D. Evered, M. O'Connor, J. Whelan (eds.), *Neuropharmacology of Insects*. Pitman, London.

Breer, H. 1982b. Uptake of N-Me-^{3}H choline by synaptosomes from the central nervous system of *Locusta migratoria. J. Neurobiol.* **13:** 107–117.

Breer, H. 1983a. Choline transport by synaptosomal membrane vesicles isolated from insect nervous tissue. *FEBS Lett.* **153:** 345–438.

Breer, H. 1983b. Venoms and toxins in neurochemical research of insects, pp. 115–125. In F. Hucho, Y. A. Ovchinnikov (eds.), *Toxins as Tools in Neurochemistry*. De Gruyter, Berlin.

Breer, H. 1985b. Synaptosomes systems for studying insect neurochemistry. pp. 384–413. In M. Ford, P. N. R. Usherwood, G. G. Lunt (eds.), *Neurobiology and Neuropharmacology*. Ellis Horwood, Chichester.

Breer, H. and D. Benke. 1985. Synthesis of acetylcholine receptors in Xenopus oocytes induced by polyA$^+$-mRNA from locust nervous tissue. *Naturwissenschaften* **72:** 213–214.

Breer, H. and G. Jeserich. 1980. A microscale flotation technique for the isolation of synaptosomes from the nervous tissue of *Locusta migratoria. Insect Biochem.* **10:** 457–463.

Breer, H. and G. Jeserich. 1984. Invertebrate Synaptosomes—implications for comparative neurochemistry, pp. 165–210. In D. G. Jones (ed.), *Current Topics in Research on Synapses*. A. R. Liss, New York.

Breer, H. and M. Knipper. 1984. Characterization of acetylcholine release from insect synaptosomes. *Insect Biochem.* **14:** 337–344.

Breer, H. and M. Knipper. 1985a. Synaptosomes and neuronal membranes from insects, pp. 125–154. In H. Breer and T. A. Miller (eds.), *Neurochemical Techniques in Insect Research*. Springer, Berlin.

Breer, H. and M. Knipper. 1985b. Choline fluxes in synaptosomal membrane vesicles. *Cell Mol. Neurobiol.* **5:** 285–296.

Breer, H. and M. Knipper. 1985c. Effects of neurotoxins on the high affinity translocation of choline. *Comp. Biochem. Physiol.* **81C:** 219–222.

Breer, H. and W. Leuken. 1983. Transport of choline by membrane vesicles prepared from synaptosomes of insect nervous tissue. *Neurochem. Int.* **5:** 713–720.

Breer, H., R. Kleene, and D. Benke. 1984. Isolation of a putative nicotinic acetylcholine receptor from the central nervous system of *Locusta migratoria*. *Neurosci. Lett.* **46:** 323–328.

Breer, H., R. Kleene, and G. Hinz. 1985. Molecular forms and subunit structure of the acetylcholine receptor in the central nervous system of insects. *J. Neurosci.* **5:** 3385–3392.

Conti-Tronconi, B. M. and M. Raftery. 1982. The nicotinic cholinergic receptor: Correlation of molecular structure with functional properties. *Annu. Rev. Biochem.* **51:** 491–530.

Donnelan, J. F., K. Alexander, and R. Chendeik. 1976. The isolation of cholinergic terminals from flesh fly heads. *Insect Biochem.* **6:** 419–423.

Driskel, W. L., B. H. Weber, and E. Roberts. 1978. Purification of choline acetyltransferase. *J. Neurochem.* **30:** 1135–1141.

Eldefrawi, M. E., R. K. Tripathi, and R. D. O'Brian. 1970. Acetylcholinesterase isoenzymes from the housefly brain. *Biochim. Biophys. Acta* **212:** 308–314.

Fels, G., H. Breer, and A. Maelicke. 1983. Are there nicotinic acetylcholine receptors in invertebrate ganglionic tissue? pp. 127–140. In F. Hucho, Y. A. Ovchinnikov (eds.), *Toxins as Tools in Neurochemistry*. De Gruyter, Berlin.

Florey, E. and M. J. Michelson. 1973. Occurrence, pharmacology and significance of cholinergic mechanisms in the animal kingdom, pp. 11–41. In M. J. Michelson (ed.), *Comparative Pharmacology, Section 85 of International Encyclopedia of Pharmacology and Therapeutics*. Pergamon Press, Oxford.

Gautrelet, J. 1938. Existence d'un complex d'acetylcholine dans le cerveau et divers organes. Ses caractères, sa répartition. *Bull. Acad. Nat. Med.* **120:** 285–291.

Hall, J. F. and R. J. Greenspan. 1979. Genetic analysis of *Drosophila* neurobiology. *Annu. Rev. Genet.* **13:** 127–195.

Hall, J. F., R. J. Greenspan, and D. R. Kankel. 1979. Neural defects induced by genetic manipulation of acetylcholine metabolism in *Drosophila*, pp. 1–42. In M. Cowan and J. A. Ferendelli (eds.), *Soc. Neurosci. Symposia*, vol. 4. Bethesda, MD.

Hanke, W. and H. Breer. 1986. Channel properties of a neuronal acetylcholine receptor purified from insect nervous tissue reconstituted in planar lipid bilayer. *Nature* **321:** 171–173.

Kankel, D. R. and J. F. Hall. 1976. Fate mapping of nervous system and other internal tissue in genetic mosaic of *Drosophila melanogaster*. *Dev. Biol.* **48:** 1–24.

Kanner, B. I. 1983. Bioenergetics of neurotransmitter transport. *Biochim. Biophys. Acta* **726:** 293–316.

Kerkut, G. A., G. W. Oliver, J. T. Rick, and R. J. Walker. 1970. The effects of drugs on learning in a simple preparation. *Comp. Gen. Pharmacol.* **1:** 437–484.

Kingan, T. G. and J. G. Hildebrand. 1985. Screening and assays for neurotransmitters in the insect nervous system, pp. 1–24. In H. Breer and T. A. Miller (eds.), *Neurochemical Techniques in Insect Research*. Springer, Berlin.

Kuhar, M. J. and L. C. Murrin. 1978. Sodium-dependent high affinity choline uptake. *J. Neurochem.* **30:** 15–21.

Lummis, S. C. R. and D. B. Sattelle. 1985. Binding of N-propionyl-^{3}H pronionylated α-bungarotoxin and L-benzilic-4-4'-^{3}H quinuclidinyl benzilate to CNS extracts of the cockroach *Periplaneta americana*. *Comp. Biochem. Physiol.* **80C:** 75–83.

Lummis, S. C. R., D. B. Sattelle, and J. C. Ellory. 1984. Molecular weight estimates of insect cholinergic receptors by radiation inactivation. *Neurosci. Lett.* **44:** 7–12.

Metcalf, R. L. and R. B. March. 1950. Properties of acetylcholine esterases from the bee, fly and the mouse and their relation to insecticide action. *J. Econ. Entomol.* **43:** 670–677.

Meyer, E. M. and J. R. Cooper. 1982. High-affinity choline transport in proteoliposomes derived from rat cortical synaptosomes. *Science, (Washington D. C.)* **217:** 843–845.

Meyer, M. R. and J. S. Edwards. 1980. Muscarinic cholinergic binding sites in an orthopteran central nervous system. *J. Neurobiol.* **11:** 215–219.

Mikalonis, S. J. and R. H. Brown. 1941. Acetylcholine and acetylcholinesterase in the insect central nervous system. *J. Cell. Comp. Physiol.* **18:** 401–403.

Mulder, A. H., W. B. Van den Berg, and J. C. Stoof. 1976. Calcium-dependent release of radiolabelled catecholamines and serotonin from rat brain synaptosomes in a superfusion system. *Brain. Res.* **99:** 419–424.

Narahashi, T. 1974. Chemicals as tools in the study of excitable membranes. *Physiol. Rev.* **54:** 813–889.

Pitman, R. M. 1971. Transmitter substances in insects: A review. *Comp. Gen. Pharmacol.* **2:** 347–371.

Raftery, M. A., M. W. Hunkapiller, C. D. Strader, and L. E. Hood. 1980. Acetylcholine receptor: Complex of homologous subunits. *Science (Washington D. C.)* **208:** 1454–1457.

Sattelle, D. B. 1985. Insect acetylcholine receptors, pp. 395–434. In G. A. Kerkut, L. I. Gilbert (eds.), *Comprehensive Insect Physiology, Biochemistry and Pharmacology*, vol. 11. Pergamon Press, Oxford.

Sattelle, D. B., I. D. Harrow, B. Hue, M. Pelhate, J. I. Gepner, and I. M. Hall. 1983. α-bungarotoxin blocks excitatory synaptic transmission between cercal sensory neurons and giant interneurone 2 of the cockroach *Periplaneta americana*. *J. Exp. Biol.* **107:** 473–489.

Shankland, D. L., R. M. Hollingworth, and T. Smyth Jr. (eds.). 1978. *Pesticide and Venom Neurotoxicity*. Plenum Press, New York.

Silver, L. H. and D. J. Prescott. 1982. Aggregation properties of the acetylcholinesterase from the central nervous system of *Manduca sexta*. *J. Neurochem.* **38**: 1709–1717.

Slemmon, J. R., P. M. Salvaterra, G. D. Crawford, and E. Roberts. 1982. Purification of choline acetyltransferase from *Drosophila melanogaster*. *J. Biol. Chem.* **257**: 3847–3852.

Smallman, B. N. and R. Pal. 1957. The activity and intracellular distribution of choline acetylase in insect nervous tissue. *Bull. Entomol. Soc. Am.* **3**: 25–30.

Telford, J. N. and F. Matsumura. 1970. Dieldrin binding in subcellular nerve components of cockroaches. An electron microscopic and autoradiographic study. *J. Econ. Entomol.* **63**: 795–801.

Tobias, J. M., J. J. Kollross, and J. Savit. 1946. Acetylcholine and related substances in the cockroach, fly and crayfish, and the effect of DDT. *J. Cell. Comp. Physiol.* **28**: 159–182.

Tripathi, R. K., J. N. Telford, and R. D. O'Brien. 1978. Molecular and structural characteristics of house fly brain acetylcholinesterase. *Biochim. Biophys. Acta* **525**: 103–111.

Venter, C. J., B. Eddy, L. M. Hall, and C. M. Fraser. 1984. Monoclonal antibodies detect the conservation of muscarinic cholinergic respector structure from *Drosophila* to human brain and detect possible structural homology with α-adrenergic receptors. *Proc. Natl. Acad. Sci. USA* **81**: 272–276.

Zimmermann, H. 1982. Insight into the functional role of cholinergic vesicles, pp. 305–359. In R. Klein, H. Lagercrantz, and H. Zimmermann (eds.), *Neurotransmitter Vesicles*. Academic Press, London.

CHAPTER 19

Histochemistry of Acetylcholinesterase in the Insect Brain

Gary M. Booth
Department of Zoology
Brigham Young University
Provo, Utah

Joseph R. Larsen
Department of Entomology
University of Illinois
Urbana, Illinois

19.1. INTRODUCTION

The first reported evidence that insect brains contained acetylcholinesterase (AChE; E.C.3.1.1.7) was by Gautrelet (1938) in bee heads. Since then, this enzyme has been reported in virtually every insect brain or other parts of the central nervous system (CNS) that has been investigated. Titers of AChE, particularly in the insect brain, are unusually high. For example, Mengle and Casida (1960) have shown that 75% of AChE activity from houseflies was due to the heads and 25% to thoraxes and abdomens. Other estimates show AChE to be as high as 86% (Metcalf et al., 1956) in fly brains.

This enzyme became the focus of intense investigation during the 1950s when it was found that it was the primary target site for organophosphorous (OP) insecticides (Metcalf et al., 1955; Fukuto and Metcalf, 1956) and carbamate insecticides (Metcalf and March, 1950). With the recent curtailed use of chlorinated hydrocarbons, it is expected that development of new carbamate and OP compounds will grow at an exponential rate (Metcalf, 1980, 1983), particularly in light of the fact that more than 400 species of insects have developed resistance to these chemicals (Georghiou and Mellon, 1983).

Accordingly, the more information we can obtain on the functional morphology and biochemistry of the target site, the better equipped we will be to understand the ecological role of insects. Therefore, the intent of this paper is to briefly review the histochemical and related biochemical data on AChE in the insect brain and other parts of the CNS and to hopefully provide incentives for further research.

19.2. CHOLINESTERASES IN INSECTS

19.2.1. Classification

The insect brain and other parts of the CNS contain numerous enzymes that catalyze a wide variety of esters that are inactivated by anticholinesterase (anti-ChE) agents (Chadwick, 1963). Various attempts have been made to categorize these enzymes according to effects on substrates, inhibitor inactivation, or both. Insect cholinesterases (ChE) have such widely divergent properties that a simple mammalian type of classification scheme probably is not appropriate. However, for discussion purposes, a system proposed by Mounter and Whittaker (1953) will be used throughout this paper. It distinguishes AChEs (also called specific ChE) as those inhibited by both OP and carbamate insecticides, aliesterases (AliE) inhibited by OP compounds but not by carbamates, and arylesterases (ArE's) inhibited by neither.

19.2.2. Biochemical Distribution

Acetylcholinesterase isolated from cockroach tissues, such as fat body, muscle, and testis, are generally 1–2% (Chamberlain and Hoskins, 1951) of that found in the nerve cord. In contrast, the titer of AChE in housefly brains (Metcalf, 1955) is nearly equivalent on a weight basis to that in the electric organ of *Electrophorus* or *Torpedo*. In general, the literature clearly shows that AChE is predominantly associated with the insect CNS and also occurs in mites (McEnroe, 1960).

19.3. THIOL-CONTAINING SUBSTRATES AND HISTOCHEMISTRY: USEFUL TECHNIQUES IN STUDYING INSECT ACETYLCHOLINESTERASE

19.3.1. Substrates

The development of thiol-containing substrates for localizing AChE was probably the single most important discovery (Koelle and Friendenwald, 1949) in the field of toxicology-related histochemistry. Use of acetylthiocholine (AThCh) and other thiocholine esters on insect tissues (Winton et al., 1958) has clearly been helpful in studying the action of OP and carbamate insecticides. Basically, the method involves the enzymatic hydrolysis of, for example, AThCh and precipitation of the liberated thiocholine moiety as the copper thiocholine sulfate. Yellowish-brown crystals of copper sulfide are formed at the sites of AChE activity when the copper thiocholine is treated with ammonium sulfide (Malmgren and Sylven, 1955). The study of substrate preferences of insect cholinesterases (Metcalf et al., 1955) clearly shows remarkable differences in AChEs among insects. These studies should be expanded to additional species of insects. Phenylthioacetate (PT) as an aromatic AChE substrate has also been helpful (Booth and Metcalf, 1970b) in studying AChE distribution patterns.

19.3.2. Advantages of Histochemistry

There is a distinct advantage of histochemical methods over biochemical methods in studying the mechanisms of AChE inhibition in insect toxicology. Although biochemical methods have provided quantitative information on the action of AChE inhibitors, histochemical procedures have provided insites into the exact site of action (Booth and Metcalf, 1970a; Brady, 1970) of these compounds. To understand the role of AChE in insect brains and the CNS in general, both methods must obviously be used. Using in vitro methods, the data show (Brady and Sternburg, 1967) that no one level of total AChE inhibition can be correlated with poisoning symptoms. Investigators also should be cautious in interpreting data using only mass-brei techniques, as the distribution of AChE can be obscured by endogenous inhibitors that are released during homogenization procedures (Lord and Potter, 1953).

19.4. INSECT TISSUE, SUBSTRATE, AND INHIBITOR PREFERENCES FOR ACETYLCHOLINESTERASE

19.4.1. General

The subject of this volume is the arthropod brain. We agree with Chapman (1982) that the role of the brain is not well understood, except that an overall function is probably the integration of the input from the sensory system. Because AChE is responsible in helping to mediate impulses throughout the CNS, additional information on this enzyme will hopefully increase our understanding. We are just now beginning to appreciate that not all CNS AChEs are alike; as expected, insect responses to various inhibitors and substrates are often curiously different even among closely related insects (Casida, 1954). We will now examine some of these differences using histochemical and related biochemical methods.

19.4.2. Histochemical Survey of Acetylcholinesterase

Using histochemical methods, only a few insects (a total of eight have been investigated for AChE in the brain and other parts of the CNS. These include the American cockroach (Iyatomi and Kanehisa, 1958; Winton et al., 1958; Burt et al., 1966; Frontali et al., 1971; Booth and Metcalf, 1972), assassin bug (Wigglesworth, 1958), crickets (Lee et al., 1973; Booth et al., 1975), dermestids (Ellis, 1964), dipterans including mostly houseflies (Tertyshnyi and Petrenko, 1976; Farnham et al., 1966; Lord et al., 1963; Molloy, 1961; Booth and Whitt, 1970; Brady, 1970; Booth and Metcalf, 1970a; Ramade, 1965), and the honeybee (Booth and Metcalf, 1972; Booth and Metcalf, 1970b).

In the American cockroach, AChE was found in certain neuropil areas, such as the central body, protocerebral bridge, optic lobes, and the calixes of the corpora pedunculata (Frontali et al., 1971; Booth and Metcalf, 1972). Treating the cockroach with LD99 doses of paraoxon showed a substantial amount of inhibition in the brain. A consistent feature of inhibition, regardless of whether the toxicant was placed on the head, thorax, or abdomen, was complete AChE inhibition of the corpora predunculata (Booth and Metcalf, 1972). The work by Winton et al. (1958) showed that topical treatments of cockroaches with selected OP compounds completely inhibited AChE in nerve cords, whereas treatment with a carbamate produced only partial inhibition. Effects on the brain were not examined.

The most detailed CNS AChE histochemical work has been done on crickets, honeybees, and, in particular, houseflies. Normal (control) AChE ac-

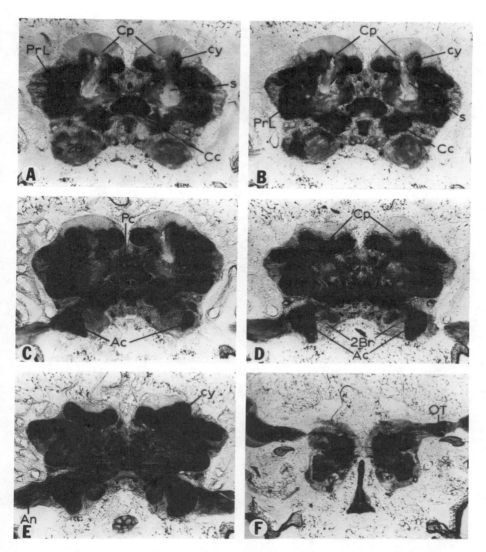

Figure 19.1. (A–F) Normal distribution of house cricket brain acetylcholinesterase. Frontal sections of the house cricket head (12 μm; 40×). Acetylcholinesterase activity appears dark in the photographs. Ac = antennal center: An = antennal nerve; 2Br = deutocerebrum; Cc = corpus centrale; Cp = corpora pendunculata; (a = α-lobe; b = β-lobe; cy = calyx; s = stalk); OT = optical tract; Pc = pons cerebralis; PrL = protocerebral lobe. (From Lee et al., 1973. Reproduced by permission of the Entomological Society of America.)

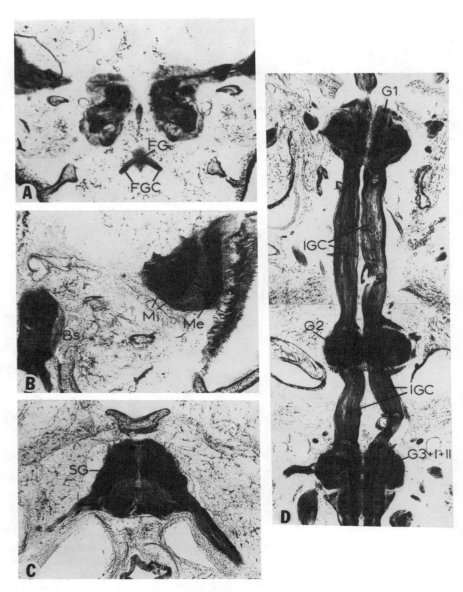

Figure 19.2. (A–D) Normal distribution of house cricket acetylcholinesterase. (A–C) Frontal sections of house cricket head (12 μm; 40×). Bs = brain stem; Fg = frontal ganglion; FGC = frontal ganglion connective; Me = medulla externa; Mi = medulla interna; SG = subesophageal ganglion. (D) Longitudinal sections of house cricket thoracic ganglia (12 μm; 20×). G1 = prothoracic ganglion; G2 = mesothoracic ganglion; G3 + I + II = compound ganglion of metathoracic, 1st and 2nd abdominal segments; IGC = interganglionic connectives. (From Lee et al., 1973. Reproduced by permission of the Entomological Society of America.)

444

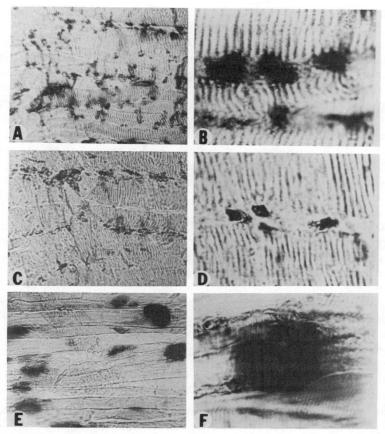

Figure 19.3. Acetylcholinesterase on the muscles. (A) House cricket thoracic muscle (250×). (B) House cricket thoracic muscle (900×). (C) Field cricket thoracic muscle (250×). (D) Field cricket thoracic muscle (900×). (E) Mouse intercostal muscle (250×) (F) Mouse intercostal muscle (900×). (From Lee et al., 1973. Reproduced by permission of the Entomological Society of America.)

tivity is shown throughout the cricket brain (Figs. 19.1A–F and 19.2A,B), subesophageal ganglion (Fig. 19.2C), and ventral nerve cord (Fig. 19.2D), and quite unexpectedly in cricket (two species) muscle (Fig. 19.3A–D). In general, the brain shows high AChE activity in neuropil areas where complex synapses are located and hence where AChE functions in neurotransmission. It is somewhat peculiar that high AChE activity is also found in apparently nonsynaptic areas, such as cell body regions and interganglionic connectives, although this finding has been confirmed in every insect that has been histochemically examined. Muscle AChE activity in the two cricket species (Fig. 19.3A–D) is very unusual because it has been observed in only one other insect, *Culiseta inornata* (Wilder, 1970). Although the details are still unknown, those activities could indicate motor endplates that are cholinergic in nature and may play an important role in integrating motor movements via the brain and other parts of the CNS. We believe this AChE activity is not an artifact because 1) the activity is inhibited by 10^{-6} M physostigmine; 2) it cannot be demonstrated with a nonspecific substrate, butyrylthiocholine; and 3) mouse motor endplates show similar distribution (Fig. 19.3E,F).

Patterns of in situ inhibition of cricket AChE by 0,0-dimethyl S-phenyl phosphorothioates show that at the hyperactive state of poisoning AChE was reduced throughout the brain (Figs. 19.4A–C; 19.5A–C; 19.6A–C; and 19.7A), in the interganglionic connectives and periphery of ganglions (Figs. 19.4B,D; 19.5B,D; 19.6B,D; and 19.7B), and completely in the muscles (figures not shown). Reactivation of cricket brain AChE and thoracic ganglia in dead insects apparently occurs (Fig. 19.7C–E). From these data, it is still not clear whether sites in the brain, ventral nerve cord, or muscle are primarily responsible for intoxication of the house cricket by 0,0-dimethyl S-aryl phosphorothioates. However, one consistent feature is the complete absence of activity in the interganglionic connectives and muscle tissue.

The majority of AChE histochemical work has been done with the housefly, probably because it is routinely available in virtually every entomology laboratory. Evidence from in vitro and in vivo data generally show that inhibition of housefly brain AChE is less important in poisoning than other sites. The remarkable finding that decapitation has little effect on the LD50 for houseflies treated with eight anticholinesterases (Mengle and Casida, 1960) makes it highly improbable that inhibition of brain AChE is important in poisoning, even though there are high concentrations (Fig. 19.8A) of AChE present.

Our laboratory was the first to show that inhibition of housefly thoracic AChE (Booth and Metcalf, 1970a) was consistently associated with poisoning symptoms, and in the same year Brady (1970) confirmed these findings independently. Figure 19.8B shows a control longitudinal section of the housefly thoracic ganglion showing the normal distribution of AChE. Note that both the synaptic and perineurium areas had AChE activity. Figure 19.8C clearly shows that an oxime phosphate (0-α-cyanobenzaldoxime phosphate) inhibits AChE peripherally at knock-down and subsequently causes complete paralysis (Fig. 19.8D) when the synaptic areas appear inhibited. Treatment with the carbamate m-isopropylphenyl methylcarbamate (Fig. 19.8E) also shows complete inhibition of AChE from the perineurium, at knock-down, without apparently affecting the synaptic areas. Furthermore, S,S-dimethyl phosphoramidodithioate was investigated, because of peculiar differences observed between the LD50 and I50 values. The housefly LD50 of this toxicant is 0.75 µg/g, which suggests that in an intact insect it is very effective in killing flies. However, the anticholinesterase (I50) activity of this chemical is lower than expected, 2×10^{-6} M. It appears (Fig. 19.8F) that when the animal is treated with this phosphoramidate that peripheral inhibition of thoracic AChE is a consistent feature of poisoning and probably

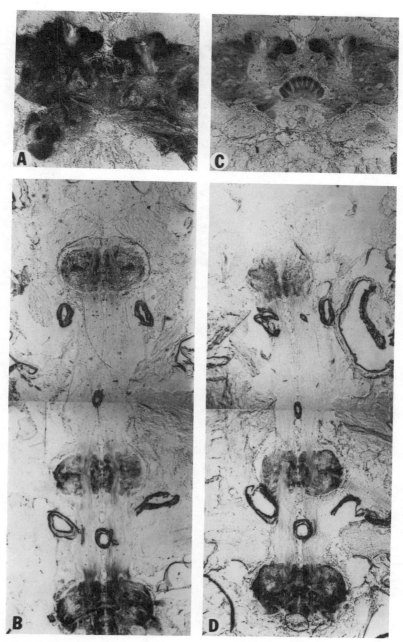

Figure 19.4. Inhibition of house cricket acetylcholinesterase in the head (A) and thoracic ganglia (B) by 0,0-dimethyl S-phenyl phosphorothioate. Inhibition of house cricket acetylcholinesterase in the head (C) and thoracic ganglia (D) by 0,0-dimethyl S-o-methylphenyl phosphorothioate. Chemicals were applied topically on the seventh abdominal tergum at 10 μg per adult female. The animals were hyperactive. (From Lee et al., 1973. Reproduced by permission of the Entomological Society of America.)

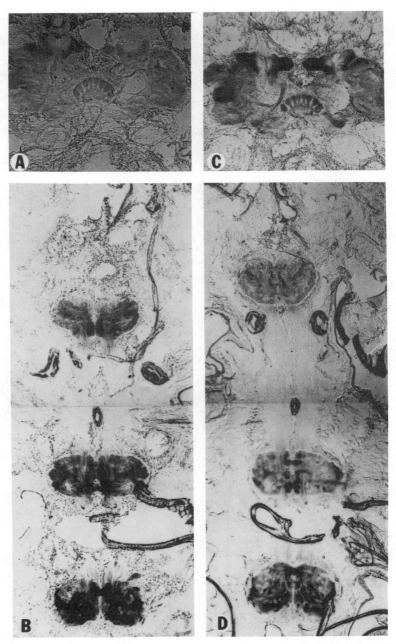

Figure 19.5. Inhibition of house cricket acetylcholinesterase in the head (A) and thoracic ganglia (B) by 0,0-dimethyl S-*m*-methylphenyl phosphorothioate. Inhibition of house cricket acetylcholinesterase in the head (C) and thoracic ganglia (D) by 0,0-dimethyl S-*p*-methylphenyl phosphorothioate. Chemicals were applied topically on the seventh abdominal tergum at 100 μg per adult female. The animals were hyperactive. (From Lee et al., 1973. Reproduced by permission of the Entomological Society of America.)

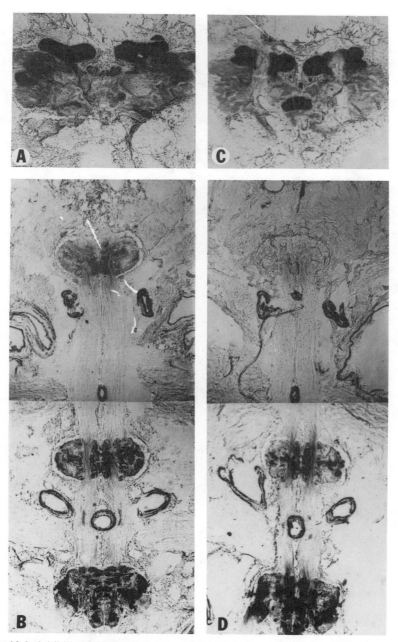

Figure 19.6. Inhibition of house cricket acetylcholinesterase in the head (A) and thoracic ganglia (B) by 0,0-dimethyl S-*o*-chlorophenyl phosphorothioate. Inhibition of house cricket acetylcholinesterase in the head (C) and thoracic ganglia (D) by 0,0-dimethyl S-*m*-chlorophenyl phosphorothioate. Chemicals were applied topically on the seventh abdominal tergum with a dosage of 100 µg per adult female. The animals were hyperactive. (From Lee et al., 1973. Reproduced by permission of the Entomological Society of America.)

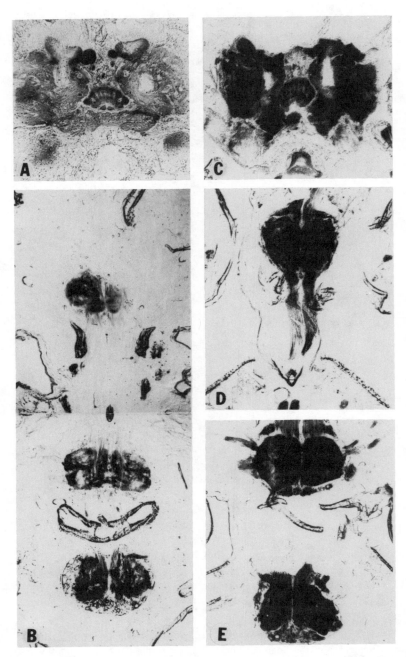

Figure 19.7. In situ acetylcholinesterase inhibition and spontaneous reactivation of the phosphorylated acetylcholinesterase. Inhibition of house cricket acetylcholinesterase in the head (A) and thoracic ganglia (B) by 0,0-dimethyl S-*p*-chlorophenyl phosphorothioate applied topically on the seventh abdominal tergum with a dosage of 10 μg per adult female.

partly explains the high toxicity of this compound. Biochemical studies (Khasawinah et al., 1978) have generally confirmed the importance of inhibition of housefly thoracic AChE by phosphoramidates in poisoning symptomology.

There is less definitive work done on honey bee brains, but Figure 19.9A shows that there is high AChE activity throughout the brain, using PT as a substrate. Pretreating brain sections with high concentrations of paraoxon (10^{-4} M) completely inhibits AChE (Fig. 19.9B), whereas pretreating brain sections with 10^{-6} M paraoxon clearly delineates a paraoxon-insensitive esterase in the deutocerebrum (Fig. 19.9B). In fact, AThCh and PT give similar AChE distribution patterns when used with paraoxon (Fig. 19.9C). Metcalf et al. (1956) biochemically confirmed the presence of an OP-insensitive esterase in bee heads.

Figure 19.9D shows electrophoretic patterns of ChE associated with brain extracts using AThCh and PT as substrates. The first two tubes (AThCh-I and PT-I) show unsolubilized enzyme patterns for the two substrates. Apparently, PT also acts as a substrate for nonspecific esterase (bands c and d). Tubes labeled AThCh-II and PT-II show solubilized enzyme patterns for the two substrates with the major activity at band b. These data show that there are multiple forms (isozymes; Booth and Metcalf, 1970b) of AChE and that most of the activity is membrane bound and associated with band b.

From this brief survey, it is evident that a great deal can be learned about the insect brain specifically and the CNS in general by studying AChE histochemistry, particularly in conjugation with OP and carbamate inhibitors. Investigators should be encouraged by the fact that only a few insects have been examined. Certainly additional work is needed, especially in areas of resistance, isozymes, and "altered AChE" (Hama, 1983). We have barely "scratched the surface" in understanding the role of AChE in the insect CNS.

19.5. SUMMARY

Histochemical analysis of CNS AChE has only been investigated in eight insects, including the American cockroach, assassin bug, crickets, dermestids, honeybees, houseflies, house mosquito, and a blood-sucking gnat. AChE is generally localized throughout the CNS, including synaptic and nonsynaptic structures. Most of the in vivo definitive work on AChE dis-

←——————————————————————————————

The animal was hyperactive. Reactivation of house cricket acetylcholinesterase in the head (C) and thoracic ganglia (D) and (E) from inhibition of 0,0-dimethyl S-p-chlorophenyl phosphorothioate applied topically on the seventh abdominal tergum with a dosage of 10 μg per adult female. The animal was dead. The sections were made 48 hours after knockdown. (From Lee et al., 1973. Reproduced by permission of the Entomological Society of America.)

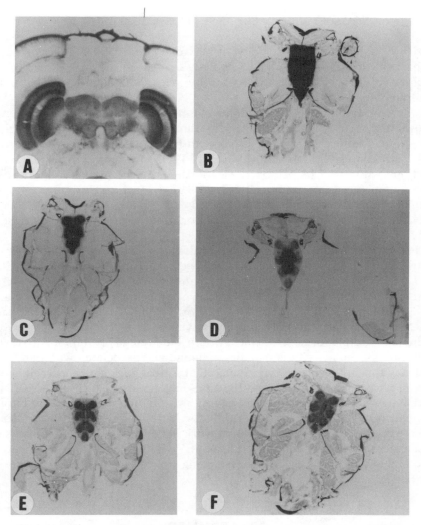

Figure 19.8. (A) Cross section of a NAIDM housefly brain showing acetylcholinesterase distribution using phenylthioacetate as substrate. (B) Control longitudinal section of the thoracic ganglion of a NAIDM housefly showing the normal distribution of acetylcholinesterase. Synaptic lobes and perineurium have an equal distribution of acetylcholinesterase. Acetylthiocholine was the substrate. (C) Longitudinal section of the thoracic ganglion of a NAIDM housefly 24 hours after treatment with 7 μg/g of an oxime phosphate. (The animal was moribund. Note the peripheral inhibition. Acetylthiocholine was the substrate.) (D) Longitudinal section of the thoracic ganglion of a NAIDM housefly 24 hours after treatment with 7 μg/g of an oxime phosphate. (The animal was dead. Note the increased peripheral inhibition. Acetylthiocholine was the substrate.) (E) Longitudinal section of the thoracic ganglion of NAIDM housefly 15 min after treatment with 90 μg/g of a carbamate. (The animal was dead. Peripheral inhibition of acetylcholinesterase is apparent. Acetylthiocholine was the substrate). (F) Longitudinal section of the thoracic ganglion of NAIDM housefly after treatment with 1.5 μg/g of a phosphoramidate. (The animal was dead. Peripheral inhibition of acetylcholinesterase is apparent. Acetylthiocholine was the substrate.) (From Booth and Metcalf, 1970a. Reproduced by permission of the Entomological Society of America.)

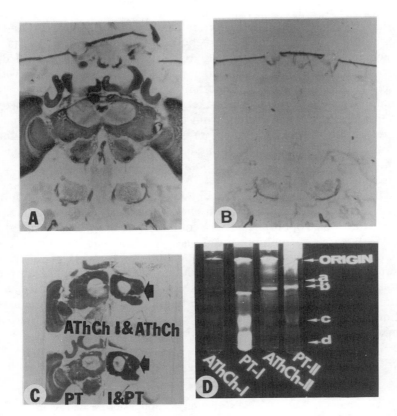

Figure 19.9. (A) Cross section of a honeybee brain showing normal distribution of acetylcholinesterase using phenylthioacetate as substrate. (B) Cross section of a honeybee brain showing complete absence of acetylcholinesterase after 30 min incubation with 10^{-4} M paraoxon and phenylthioacetate. (C) *Top left*: normal acetylcholinesterase distribution in honeybee brain using acetylthiocholine as substrate. *top right*: paraxon-insensitive esterase in deutocerebrum (*arrow*) after 30 min incubation in inhibitor (I $\times$ 10^{-6} M paraoxon) and then with acetylthiocholine. *Bottom left*: normal acetylcholinesterase distribution in honeybee brain using phenylthioacetate as substrate. *bottom right*: paraxon-insensitive esterase in deutocerebrum (*arrow*) after 30 min incubation in inhibitor (I $\times$ 10^{-6} M paraoxon) and then with phenylthioacetate. (D) AThCh-I: electrophoretic pattern of unsolubilized honeybee brain acetylcholinesterase at origin and b using acetylthiocholine as substrate; PT-I: electrophoretic pattern of unsolubilized honeybee brain esterase activity at origin, b, c, and d using phenylthioacetate as substrate; AThCh-II: electrophoretic pattern of solubilized honeybee brain acetylcholinesterase at origin, a, b, and c using acetylthiocholine as substrate; PT-II: electrophoretic pattern of solubilized honeybee esterase activity at origin, a, b, and c using phenylthioacetate as substrate. (A from Booth and Lee, 1971; C and D from Booth and Metcalf, 1970b. Copyright 1970 by the AAAS.)

tribution in conjunction with inhibitors has been done in crickets and houseflies using carbamate and OP insecticides. Published housefly data show that generalized inhibition of brain AChE is probably not critical during stages of carbamate and OP poisoning, but rather the peripheral areas of the thoracic ganglion seem to be important sites of localized inhibition. After poisoning with OP compounds, cricket brain AChE showed generalized inhibition and complete inhibition from the muscle and interganglionic connectives of the nerve cord. AChE presence in cricket muscles is unusual but could be important in integrating motor movements via the brain. Furthermore, this is evidence that a cholinergic system may exist in the neuromuscular junction of crickets.

Additionally, honeybee brains contain a paraoxon (1×10^{-6} M) insensitive esterase in the deutocerebrum but is completely inhibited at 1×10^{-4} M paraoxon. Electrophoretic patterns of honeybee brain homogenates show that AChE exists in multiple forms (isozymes) and is membrane bound. Histochemical and biochemical investigations of CNS AChE distribution patterns, substrate preferences, and inhibitor selectivity are the keys to understanding the ecological role of insects in an environment that will be continually exposed to insecticides. Certainly, histochemical techniques will continue to be useful in elucidating localized effects of OP and carbamate chemicals.

REFERENCES

Booth, G. M. and A. Lee. 1971. Distribution of cholinesterases in insects. *Bull. Wld. Hlth. Org.* **44**: 91–98.

Booth, G. M. and G. S. Whitt. 1970. Histochemical specificity of cholinesterases to phenylthioacetate in differentiated neural tissues of insects and teleosts. *Tissue Cell* **2**: 521–528.

Booth, G. M. and R. L. Metcalf. 1970a. Histochemical evidence for localized inhibition of cholinesterase in the house fly. *Ann. Entomol. Soc. Am.* **63**: 197–204.

Booth, G. M. and R. L. Metcalf. 1970b. Phenylthioacetate: A useful substrate for the histochemical and colorimetric detection of cholinesterase. *Science (Washington, D. C.)* **170**: 455–457.

Booth, G. M. and R. L. Metcalf. 1972. The histochemical fate of paraoxon in the cockroach (*Periplaneta americana*) and honeybee (*Apis mellifera*) brain. *Isr. J. Entomol.* **7**: 143–156.

Booth, G. M., C. L. Stratton, and J. R. Larsen. 1975. Localization and substrate-inhibitor specificity of insect esterase isozymes, pp. 721–738. In C. L. Markert (ed.), *Isozymes III Developmental Biology*. Academic Press, New York.

Brady, U. E. 1970. Localization of cholinesterase activity in house-fly thoraces: Inhibition of cholinesterase with organophosphate compounds. *Entomol. Exp. Appl.* **13**: 423–432.

Brady, E. U. and J. Sternburg. 1967. Studies on in vivo cholinesterase inhibition and poisoning symptoms in houseflies. *J. Insect Physiol.* **13**: 369–379.

Burt, P. E., G. E. Gregory, and F. M. Molloy. 1966. A histochemical and electro-physiological study of the action of diazoxon on cholinesterase activity and nerve conduction in ganglia of the cockroach, *Periplaneta americana. L. Ann. Appl. Biol.* **58:** 341–354.

Casida, J. E. 1954. Comparative enzymology of certain insect acetylesterases in relation to poisoning by organophosphate insecticides. *J. Physiol.* (*London*) **127:** 20–21.

Chadwick, L. E. 1963. Systemic pharmacology of the anti-cholinesterase (Anti-ChE) agents: actions on insects and other invertebrates, pp. 741–798. In G. B. Koelle (ed.), *Handbuch der Experimentellen Pharmakologie: Cholinesterases and Anticholinesterase Agents.* Springer-Verlag, Berlin.

Chamberlain, W. F. and W. M. Hoskins. 1951. The inhibition of cholinesterase in the American roach by organic insecticides and related phosphorus-containing compounds. *J. Econ. Entomol.* **44:** 177–191.

Chapman, R. F. 1982. *The Insects: Structure and Function.* Harvard University Press, Cambridge.

Ellis, V. J. 1964. The effect of some organophosphorus insecticides on *Dermestes* sp—an histochemical study. *12th Int. Congr. Entomol.* **12:** 502–503.

Farnham, A. W., G. E. Gregory, and R. M. Sawicki. 1966. Bioassay and histochemical studies of the poisoning and recovery of houseflies (*Musca domestica* L.) treated with diazinon and diazoxon. *Bull. Entomol. Res.* **57**(1): 107–117.

Frontali, N., R. Piazza, and R. Scopelliti. 1971. Localization of acetyl-cholinesterase in the brain of *Periplaneta americana. J. Insect. Physiol.* **17:** 1833–1842.

Fukuto, T. and R. Metcalf. 1956. Structure and insecticidal activity of some diethyl substituted phenyl phosphates. *J. Agric. Food. Chem.* **4:** 930–935.

Gautrelet, J. 1938. Existence d'un complexe d'acetylcholine dans le cerveau et divers organes ses caractéres, sa repartition. *Bull. Acad. Méd.* (*Paris*) **120:** 285–291.

Georghiou, G. P. and R. B. Mellon. 1983. Pesticide resistance in time and space, pp. 1–46. In G. P. Georghiou and T. Saito (eds.), *Pest Resistance to Pesticides.* Plenum Press, New York.

Hama, H. 1983. Resistance to insecticides due to reduced sensitivity of acetylcholinesterase, pp. 299–331. In G. P. Georghiou and T. Saito (ed.), *Pest Resistance to Pesticides.* Plenum Press, New York.

Iyatomi, K. and K. Kanehisa. 1958. Localization of cholinesterase in the American cockroach. *Jpn. J. Appl. Entomol. Zool.* **2:** 1–10.

Khasawinah, A. M. A., R. B. March, and T. R. Fukuto. 1978. Insecticidal properties, antiesterase activities, and metabolism of methamidophos. *Pest. Biochem. Physiol.* **9**(2): 211–221.

Koelle, G. B. and J. S. Friendenwald. 1949. A histochemical method for localizing cholinesterase activity. *Proc. Soc. Exp. Biol. Med.* **70:** 617–622.

Lee, A., R. L. Metcalf, and G. M. Booth. 1973. House cricket acetylcholinesterase: Histochemical localization and in situ inhibition by 0,0-dimethyl S-aryl phosphorothioates. *Ann. Entomol. Soc. Am.* **66**(2): 333–343.

Lord, K. A. and C. Potter. 1953. Hydrolysis of esters by extracts of insects. *Nature* (*London*) **172:** 679.

Lord, K. A., F. M. Molloy, and C. Potter. 1963. Penetration of diazoxon and acetyl choline into the thoracic ganglia in susceptible and resistant houseflies, and the effects of fixatives. *Bull. Entomol. Res.* **54:** 189–198.

Malmgren, H. and B. Sylven. 1955. On the chemistry of the thiocholine method of Koelle. *J. Histochem. Cytochem.* **3:** 441–445.

McEnroe, W. D. 1960. Cholinesterase in the two-spotted spider mite. *Bull. Entomol. Soc. Am.* **6:** 150.

Mengle, D. C. and J. E. Casida. 1960. Biochemical factors in the acquired resistance of houseflies to organophosphate insecticides. *J. Agric. Food. Chem.* **8:** 431–437.

Metcalf, R. L. 1955. *Organic Insecticides, Their Chemistry and Mode of Action.* Interscience Publishers, New York.

Metcalf, R. L. 1980. Changing role of insecticides in crop protection. *Annu. Rev. Entomol.* **25:** 219.

Metcalf, R. L. 1983. Implications and prognosis of resistance to insecticides, pp. 703–733. In G. P. Georghiou and T. Saito (eds.), *Pest Resistance to Pesticides.* Plenum Press, New York.

Metcalf, R. L. and R. B. March. 1950. Properties of acetylcholine esterases from the bee, the fly, and the mouse and their relation to insecticide action. *J. Econ. Entomol.* **43:** 670–677.

Metcalf, R. L., R. B. March, and M. G. Maxon. 1955. Substrate preferences of insect cholinesterases. *Ann. Entomol. Soc. Am.* **48**(4): 222–228.

Metcalf, R. L., M. Maxon, T. R. Fukuto, and R. B. March. 1956. Aromatic esterase in insects. *Ann. Entomol. Soc. Am.* **49**(3): 274–279.

Molloy, F. M. 1961. The histochemistry of the cholinesterase in the central nervous system of susceptible and resistant strains of the house-fly, *Musca domestica* L., in relation to diazinon poisoning. *Bull. Entomol. Res.* **52:** 667–681.

Mounter, L. A. and V. P. Whittaker. 1953. The hydrolysis of esters of phenol by cholinesterases and other esterases. *Biochem. J.* **54:** 551–559.

Ramade, F. 1965. L'action anticholinestérase de quelques insecticides organophosphores sur le systeme nerveux central de *Musca domestica. Ann. Soc. Entomol. Fr.* **1:** 549–566.

Tertyshnyi, V. N. and V. S. Petrenko. 1976. Histochemistry of esterases of the nervous system of some insects of the Diptera Order in relation to the effect of insecticides. *Fiziol. Akt. Veshchestva.* **8:** 25–29.

Wigglesworth, V. B. 1958. The distribution of esterase in the nervous system and other tissues of the insect *Rhodnius prolixus. Quart. J. Microsci. Sci.* **99:** 441–450.

Wilder, W. H. 1970. Histochemical localization of cholinesterase in larvae of *Culiseta inornata* (Diptera: Culicidae). *Ann. Entomol. Soc. Am.* **63:** 1620–1624.

Winton, M. Y., R. L. Metcalf, and T. R. Fukuto. 1958. The use of acetylthiocholine in the histochemical study of the action of organophosphorus insecticides. *Ann. Entomol. Soc. Am.* **51:** 436–441.

CHAPTER 20

Functions, Organization, and Physiology of the Olfactory Pathways in the Lepidopteran Brain

Thomas A. Christensen
John G. Hildebrand
Arizona Research Laboratories
Division of Neurobiology
University of Arizona
Tucson, Arizona

20.1. INTRODUCTION

Olfactory stimuli provide important information about the environment and can trigger profound changes in the behavior of many animals. This is particularly true for insects, many of which rely on olfaction as a principal sensory modality. Volatile chemical cues are often required to help insects orient toward and locate appropriate food sources, oviposition sites, and conspecifics (such as mating partners), whereas odors from inappropriate sources can lead to avoidance behavior. Such responsiveness to odors is most pronounced in insects that have a large proportion of the nervous system devoted to the detection and processing of olfactory information. Among the most thoroughly studied insects of this kind are the Lepidoptera. Moths and butterflies possess thousands of sensory receptors, some of which may be specialized to recognize only a single airborne chemical. Earlier studies, such as those of Schneider and co-workers, on the silk moth, *Bombyx mori*, showed that some olfactory receptors are both highly selective and extremely sensitive, whereas others exhibit broader spectra of responses and lower sensitivities (see Kaissling, 1971; Schneider, 1974, for reviews). High chemical selectivity and sensitivity are especially characteristic of receptors responsible for detecting sex pheromones (Kaissling, 1979; Priesner, 1980; Boeckh et al., 1984). The highly discriminative and sensitive detection of sex pheromones is critically important to the continued reproductive success of these insects. When an adult male moth detects the female's characteristic blend of sex pheromones, this species-specific olfactory signal triggers in him a dramatic drive to mate. To discover the physiological basis of such behavior, it is essential to explore the olfactory pathway in the central nervous system (CNS), beginning with the primary olfactory centers in the brain, the antennal lobes (ALs), to which the axons of the antennal receptor cells project. Recent research initiatives have begun to probe these centers and have revealed that the olfactory system is highly organized and physiologically complex. For example, it has become clear that the ALs are more than simple relay stations responsible for summing activity that enters by way of the myriad olfactory fibers and passing it on to higher-order centers in the brain (see, e.g., the review by Mustaparta, 1984).

In this brief review, we first describe some of the chemical substances that transmit information between these organisms. We explore how these substances, collectively called "semiochemicals" (Nordlund et al., 1981), influence the behavior of adult Lepidoptera, and discuss the anatomical, physiological, and neurochemical properties of the central neural elements that respond to and process olfactory inputs. In much of this chapter, we emphasize the olfactory system of the sphinx moth, *Manduca sexta* (Lepidoptera:Sphingidae), because most of our current knowledge of the organization and function of the olfactory centers of Lepidoptera has come from studies of *Manduca*. Moreover, because we and our co-workers focus on *M. sexta* in our own research, we can provide the most recent information available for this species. We include additional findings from other species wherever appropriate.

20.2. RECEPTION OF SEMIOCHEMICALS

Butterflies and moths rely mainly on visual and olfactory cues to communicate information to one another. These sensory systems do not play equal roles, however, in the survival of these insects. Vision is clearly the dominant modality in butterflies, which are active in daylight. Their diurnal activities are well matched to their unusual sensory capabilities, as they apparently possess the broadest visual spectrum of all animals that have been studied (Silberglied, 1977). Among other testaments to the importance of vision, are the colorful displays used by male butterflies during courtship. By contrast, in crepuscular and nocturnal Lepidoptera, which lack the benefit of unhindered visual contact with a conspecific mate, sexual rival, potential food source, or oviposition site, the olfactory system has evolved to a level of exquisite sensitivity and selectivity. In this section, we outline the various roles of olfactory signaling in the Lepidoptera, summarizing the most frequently cited functions of semiochemicals in the lives of these insects.

20.2.1. Selection of Host Plants

A female searching for a potential food source or oviposition site is guided, at least in part, by olfactory cues. Feeding behavior in butterflies can be released by visual color cues (Ilse, 1928), but certain species rely on olfactory information to identify the correct host plant. There may be great olfactory selectivity, as in the case of the tortricid moth, *Adoxophyes orana*. In this species, only the female possesses antennal receptors that detect apple odors (Den Otter et al., 1978). In other species, antennal receptors may be broadly "tuned" to respond to a wide variety of chemical stimuli, as in the female cabbage white butterfly, *Pieris brassicae* (Den Otter et al., 1980). In addition, the odor-detecting system may be very sensitive, due to the physiological characteristics of the receptors themselves and to the redundancy of receptors in the animal. Activation of only very few receptors may be required to elicit a particular behavioral response. For example, the Florida queen butterfly, *Danaus gilippus berenice,* feeds in response to honey odor even if only 5–10% of the appropriate antennal receptors are exposed (Myers and Walter, 1970). Some responsiveness to honey odor apparently persists in danaids even if the antennae have been completely covered, suggesting that extra-antennal receptors may contribute to the control of this behavior. By contrast, *Manduca sexta* appears to be completely disoriented after removal of the antennae and is unable to find its host plant (Yamamoto and Fraenkel, 1960; Yamamoto et al., 1969; I. Harrow, R. Kovelman, and J. Hildebrand, unpublished observations).

20.2.2. Discrimination of Oviposition Sites

The choice of an oviposition site may be based on the detection of host-plant odors as illustrated in Section 20.2.1 or, alternatively, on the presence of conspecific eggs or feeding larvae. In addition to visual deterrents, there is evidence that female moths and butterflies recognize pheromones emitted by the eggs or egg cement already residing on a host plant and thus avoid overloading the plant (Rothschild and Schoonhoven, 1977). At least part of this chemosensory ability is believed to be due to receptors on the tarsi (Ma and Schoonhoven, 1973; Rothschild and Schoonhoven, 1977). Feeding larvae can also be a strong deterrent, presumably due to airborne cues from the damaged plant, as in the case of *P. brassicae* (Rothschild and Schoonhoven, 1977), or to the presence of a pheromone in the larval frass, as in the case of the cabbage looper, *Trichoplusia ni* (Rendwick and Radke, 1980).

20.2.3. Location of Mates

Communication by sex pheromones has been studied in many lepidopteran species because this form of chemical communication is probably the most important mechanism for bringing the sexes together for mating. Since the first sex attractant (bombykol) was identified in the silk moth, *Bombyx mori* (Butenandt et al., 1959), scores of pheromones have been isolated from male and female Lepidoptera and subsequently identified and synthesized (see Shorey, 1976; Birch and Haynes, 1982, for references). In this section, we discuss briefly several of the best-known examples.

20.2.3.1. Detection of Male Sex Pheromones. Butterfly courtship is most often initiated by males that are visually attracted to females. Once the male is in close proximity to the female, however, he may disseminate a potent pheromone that will induce a receptive female to land and become motionless. The so-called "aphrodisiac" pheromone of the queen butterfly, *Danaus gilippus*, contains a heterocyclic ketone, which, when released from the male's brush-like hair-pencils at the tip of the abdomen, acts on the female's antennae and induces her to land and mate (Brower et al., 1965; Schneider and Seibt, 1969). Many species of butterflies and moths possess similar odor-signaling organs from which a wide variety of compounds may be released (Meinwald et al., 1966, 1969, 1974; Aplin and Birch, 1970; Bestmann et al., 1977). Unlike most male moths, the male lesser wax moth, *Achroia grisella*, releases a pheromone that can attract conspecific females over long distances (Dahm et al., 1971). Such long-range olfactory communication is more typical of female pheromonal systems.

20.2.3.2. Detection of Female Sex Pheromones.

Recent studies conducted both in the field and in wind tunnels have begun to elucidate the mechanisms of pheromone-stimulated orientation and flight behaviors in male moths (e.g., Kennedy, 1983; David et al., 1983; Willis and Baker, 1984; Baker et al., 1984; Schneiderman et al., 1986). The most obvious difference between female pheromonal communication and that mediated by male "aphrodisiac" pheromones is the range over which the chemical signals are detected. Pheromones released by calling female moths usually attract males over long distances. Thus, it has long been a source of amazement that some male moths, such as certain saturniids, can respond to calling conspecific females at a distance of 10 km or more (Bossert and Wilson, 1963). The male moths fly upwind, following a zig-zagging course that is largely restricted to the pheromone "plume" formed downwind of the pheromone source. The zig-zagging component of pheromonally influenced flight is apparently a stereotyped, preprogrammed motor act, and the frequency of counterturns in the flight depends on the concentration of pheromones in the plume (Baker and Kuenen, 1982; Kennedy, 1983; Baker et al., 1984). Optomotor anemotaxis, or visually directed orientation to the wind, appears to be an important mechanism for guiding the males upwind. Because the source of pheromone may be located in still air, however, other mechanisms must be involved (Baker and Kuenen, 1982; Kuenen and Baker, 1983). Recent evidence has shown that intermittent stimulation with pheromones is necessary to elicit or sustain the zig-zagging program (Willis and Baker, 1984). Therefore, detection of changes in pheromone concentration appears to be a crucial step in the eventual location of the female (see also Sections 20.4 and 20.5).

The antennal receptors that capture the pheromone molecules are exquisitely sensitive and specific. The most celebrated example of this extraordinary sensory capability is that of the receptors innervating the *sensilla trichodea* in the antennae of the male domestic silk moth, *B. mori*. Action potentials can be triggered in these cells by perhaps only a single molecule of the principal pheromone, bombykol [(E,Z)-10,12-hexadecadienol], whereas response thresholds for analogues or homologues of bombykol are several orders of magnitude higher (Kaissling and Priesner, 1970; Schneider, 1974; Kaissling, 1979). In several species of moths of the genus *Yponomeuta*, plant odors can also affect sexual preference via a peripheral modulatory action on antennal pheromone receptors (Van der Pers et al., 1980; Blaney et al., 1986). It is likely, however, that information from plant odors is also integrated in the CNS with that from other odor sources in the environment.

Female *Manduca sexta* call conspecific males with a blend of sex pheromones released from an abdominal lure gland. The principal pheromone in the blend is (E,Z)-10,12-hexadecadienal, or bombykal (Starratt et al., 1979; J. Tumlinson, personal communication). Also present in organic washes of the gland are several other substances, at least one of which is an essential second pheromone (J. Tumlinson, personal communication).

20.2.3.3. Chemical Deterrents to Interspecific Attraction. Years ago, when only a few sex pheromones had been identified, it was thought that moth pheromones were species specific and thus without effect on insects of other species. We now know, however, of situations in which two or more different species may be sympatric, share similar activity cycles, and even use the same pheromone substances as sex attractants, but nevertheless maintain reproductive isolation. One mechanism for achieving this isolation is the use of different ratios of pheromones in the sex-attractant blends of the different species, as in the tortricid moths, *Clepsis spectrana* and *Adoxophyes orana* (Minks et al., 1973). An alternative mechanism involves the addition of a secondary pheromone to the lure blend of one species that inhibits the attraction of males of other species. Such compounds, which have been found in the sex pheromone mixtures of many species, including members of the families Tortricidae, Noctuidae, Pyralidae, and Aegeriidae (see Shorey, 1976 for review), might exercise their inhibitory influences at the level of peripheral receptors or through central integrative mechanisms.

20.3. FUNCTIONAL ORGANIZATION OF THE CENTRAL OLFACTORY PATHWAY

Much of what has been learned so far about central olfactory mechanisms in Lepidoptera has come from studies of the subsystem for detecting and processing information about sex pheromones. This subsystem includes many sexually dimorphic neural elements that are more or less specialized for the important sensory processes underlying responses to the pheromones. These elements often can be recognized morphologically; their biologically meaningful "effective stimuli" (the pheromones) are usually known, and the functions of the pheromonal signaling system in which they participate are generally well understood. Thus, the pheromone-processing subsystem provides a sort of "keyhole" through which investigators can scrutinize the operations of at least one part of the olfactory system.

20.3.1. Antennae and Antennal Nerves

The paired antennae, which are the primary olfactory organs, carry chemoreceptors responsible for detecting pheromones and other semiochemicals. The elongated antennal flagellum, by far the largest segment of the lepidopteran antenna, is generally divided into annuli or "segments," which have branches or "side arms" in some species.

In *Manduca sexta,* for example, the male flagellum comprises about 80 annuli (Sanes and Hildebrand, 1976a). On its two extended faces, each annulus carries an orderly array of long male-specific *sensilla trichodea* as well as several other types of sensilla most of which are olfactory. Each cuticular sensillum is innervated by dendrites of a characteristic number of sensory neurons (receptor cells). In male *Manduca,* each long trichoid sensillum houses the dendrites of two receptor cells (Sanes and Hildebrand, 1976a,b). From electrophysiological recordings from single trichoid sensilla, we know that bombykal excites one of the two receptor cells innervating each sensillum. The second essential component of the female's pheromone mixture, which is different from but mimicked by (*E,Z*) 11,13-penta decadienal ("C15"), selectively stimulates the second cell in the sensillum (J. Hildebrand and K.-E. Kaissling, unpublished findings).

The axons of such male-specific sensory neurons account for approximately one third of the 2.6×10^5 fibers in the flagellar component of the antennal nerve (AN; Sanes and Hildebrand, 1976a). Of the rest of the axons in the AN of *Manduca,* most are olfactory, but some are believed to belong to mechanoreceptors, thermoreceptors, and hygroreceptors (Sanes and Hildebrand, 1976a; Camazine and Hildebrand, 1979; Hildebrand et al., 1980).

20.3.2. Antennal Lobes

The two antennal lobes (ALs, in Fig. 20.1A), which are the most prominent and anterior structures in the deutocerebrum of the lepidopteran brain, are the sites of first-order synaptic processing of olfactory information from the antennae. The general organization of the ALs in moths and butterflies is similar to that of other insects, and in many respects resembles that of the primary olfactory centers (the olfactory bulbs) of vertebrates as well (e.g., see Shepherd, 1972). Each AL comprises a central core of coarse-fibrous neuropil, which is surrounded by an array of condensed, spheroidal "knots" of neuropil called glomeruli. Bordering the glomerular neuropil are groups of somata of AL neurons. Primary-afferent olfactory axons running in fascicles in the AN project into the glomerular neuropil of the ipsilateral AL.

20.3.2.1. Glomerular Neuropil. The glomeruli are the most conspicuous and characteristic features of the ALs of most insects. In moths, such as *Manduca,* each glomerulus comprises myriad densely packed neurites surrounded by an incomplete investment of glial processes (Tolbert and Hildebrand, 1981). Each glomerulus is morphologically distinct and contains the terminals of sensory axons from the ipsilateral AN, dendrites of AL neurons (see Section 20.3.2.3.), and probably other cellular elements, such as glial processes and terminals of centrifugal fibers from higher centers in the brain. Moreover, all of the recognized chemical synapses in the AL are

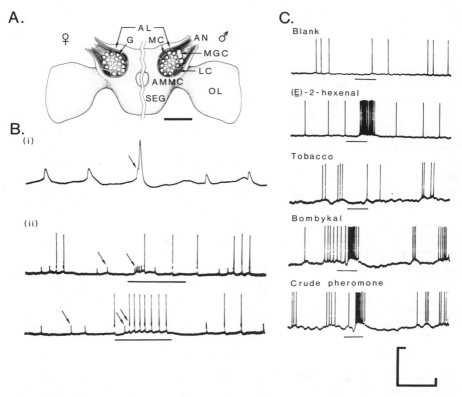

Figure 20.1. (A) Drawing of the frontal view of the brain and subesophageal ganglion (SEG) of *Manduca* showing the optic lobes (OLs) and, in cut-away diagram, a female (*left*) and a male (*right*) antennal lobe (AL). Also labeled are the antennal nerve (AN), the antennal mechanosensory and motor center (AMMC), and within the ALs, the ordinary glomeruli (G), the male-specific macroglomerular complex (MGC), and the lateral (LC) and medial (MC) groups of AL neuronal cell bodies. Scale bar = 500 μm. (B) Intracellular recordings from LNs in the AL of *Manduca* demonstrating multiple spike amplitudes. (i) Recording from an LN that showed potentials of two amplitudes. The larger spike was initiated by a smaller potential (*arrow*) and was apparently initiated closer to the site of recording. Calibrations: 40 mV, 25 msec. (ii) Another LN, in a male AL, responded differently to antennal stimulation (indicated by a bar beneath each record) with a plant odorant and a sex pheromone. The response to (*E*)-2-hexenal (*upper trace*) was characterized by a short-latency, phasic burst of smaller spikes (*arrows*), presumably actively propagated only within a part of the cell far from the site of recording. Bombykal (*lower trace*), by contrast, elicited a train of combined smaller and larger spikes, suggesting that parts of the cell closer to the site of recording were activated by this stimulus. Calibrations: 40 mV, 250 msec. (C) Intracellular recordings from a L(M2:G,MGC) cell (Table 20.1), showing that this neuron receives convergent inputs from two separate sensory "channels". one mediating pure excitation in response to antennal stimulation with (*E*)-2-hexenal, and another mediating mixed inhibition/excitation/inhibition in response to stimulation (indicated by bar) with sex pheromones. Calibration: 60 mV, 1200 msec. (B and C are reproduced, with the permission of the publisher, from Matsumoto and Hildebrand, 1981.)

464

restricted to the glomeruli (Tolbert and Hildebrand, 1981; Tolbert et al., 1983). The central, coarse neuropil is composed largely of the principal neurites of AL neurons.

In Lepidoptera as well as other insects, the glomeruli are believed to be discrete functional units responsible for processing particular sensory inputs. At present, only two glomeruli in *Manduca* ALs, the macroglomerular complex (MGC) in males and the "labial pit organ glomerulus" in both sexes (discussed in Section 20.3.3), have been identified. Because it seems that each of these glomeruli exclusively receives primary-afferent input of one characteristic type, both in a sense have defined functions. It remains to be shown in any species, however, that each and every glomerulus functions as an independent information-processing subunit of the olfactory neuropil.

The number of glomeruli in each AL varies considerably among species. Within a species, however, the number may be practically the same from AL to AL. Using computer-aided reconstructions, Rospars (1983) found that the intraindividual and intrasexual number, position, shape, and size of the glomeruli were nearly constant in the moth, *Mamestra brassicae*, whereas the smaller AL of the butterfly, *Pieris brassicae*, showed greater variability in glomerular position, shape, and size. These observations suggest that the AL may have attained greater functional specialization in the species (the moth), which relies more heavily on olfactory information for its survival and reproduction. In *Manduca*, the AL contains an average of 61 (range 57–65) ordinary glomeruli (plus the MGC in males) whose relative sizes and positions appear to be approximately constant in a series of moths (Schneiderman et al., 1983; Schneiderman, 1984).

20.3.2.2. Sexual Dimorphism. In many species that have been studied, pheromones released by one sex are detected exclusively or preferentially by insects of the opposite sex. In the case of *Manduca*, for example, only the male possesses receptors in his antennae for the female sex pheromones. It is therefore not surprising that sexually dimorphic structures have been found throughout the olfactory system in such species.

Female-Specific Glomeruli. Certain species have glomeruli that are present (or can be identified) only in the ALs of females. In *M. brassicae*, for example, the female AL possesses three glomeruli that have no recognized male homologues and two glomerular complexes (the macroglomerular and dorsoposteroexternal) that, although also present in male ALs, differ in form or location in females (Rospars, 1983). ALs of *B. mori* exhibit two identifiable lateral large glomeruli in both sexes, but these glomeruli are approximately five times larger in females than in males (M. Koontz and D. Schneider, personal communication).

Male-Specific Glomeruli. As we have outlined in Section 20.2.3.2., male moths typically rely on long-range pheromonal communication to locate a receptive female. Detection of the female's pheromones is mediated by the excitation of some of the tens of thousands of antennal sensory neurons whose axons carry the pheromonal message to the brain. One might expect that if all of these primary-afferent fibers from one antenna were to converge on a common area in the ipsilateral AL, then this area would be correspondingly large owing at least in part to the bulk of those numerous input elements. Such is the case for the MGC, one form or another of which occurs in male ALs of many species of moths.

The MGC is the most readily recognizable, sexually dimorphic structure in the ALs—indeed, in the brain—of male moths. It is absent in the butterflies that depend more on vision than on pheromonal communication to locate conspecific females (Rospars, 1983). Since a sexually dimorphic macroglomerular structure was first described in *B. mori* and *Lasiocampa* by Bretschneider (1924), similar structures have been identified in many species of moths including *Antheraea pernyi* and *A. polyphemus* (Boeckh and Boeckh, 1979), *M. sexta* (Matsumoto, 1979; Camazine and Hildebrand, 1979; Matsumoto and Hildebrand, 1981), and *Lymantria dispar* (M. Koontz and D. Schneider, personal communication). Several lines of evidence point to a specific role for the MGC as the exclusive site of first-order processing of primary-afferent information about female sex pheromones: (1) the enlarged MGC is found only in male ALs; (2) in intracellular staining studies, it has been shown that axons of receptor cells innervating the male-specific, pheromone-responsive sensilla in the antenna exclusively project to and terminate within the ipsilateral MGC (for *Manduca*: Camazine and Hildebrand, 1979; Christensen and Hildebrand, 1984; S. Camazine, T. Christensen, I. Harrow, K. Kent, B. Waldrop, and J. Hildebrand, in preparation; for *Bombyx, Antheraea,* and *Lymantria*: M. Koontz and D. Schneider, personal communication); (3) all AL neurons that respond postsynaptically to stimulation of the ipsilateral male antenna with female sex pheromones have dendritic arborizations within the MGC (for *Manduca*: Matsumoto, 1979; Matsumoto and Hildebrand, 1981; Schneiderman et al., 1982; Christensen and Hildebrand, 1987; for *Bombyx*: Light, 1982; see Section 20.4). Similar characteristics of the MGC and its associated neurons have been observed in other species of insects such as *Periplaneta americana* (see review by Boeckh et al., 1984).

The MGC exhibits considerable inter- and intraspecific variation in shape and size, but not in location within the AL. The MGC is invariably situated close to the entrance of the AN into the AL. In *Antheraea* and *Manduca*, the shape of the MGC may seem irregular or multiple-lobed, but it nevertheless appears to be a coherent structure (Boeckh and Boeckh, 1979; Matsumoto and Hildebrand, 1981; Schneiderman, 1984). In other species, such as *M. brassicae, B. mori,* and *L. dispar,* the MGC may consist of several distinct glomeruli or glomerular "levels," each of which comprises one or more glomeruli (Rospars, 1983; M. Koontz and D. Schneider, personal communication). The MGC appears to be differentiated into a central core of neurites and an outer, more condensed "cortex" in *Mamestra* and *Bombyx* but relatively homogeneous in *Manduca*. The significance of these structural differences is still unknown.

20.3.2.3. Neurons of the Antennal Lobes. Olfactory information from primary-afferent antennal axons is processed by AL neurons, some of which transmit integrated outputs of this processing to higher-order centers in the brain. The cell bodies of the AL neurons are gathered in discrete cortical groups within the AL. In *Manduca*, these are a large lateral group, a smaller medial group, and a still smaller anteroventral group. AL neurons in moths have been classified on the basis of their morphology, revealed by intracellular staining, and on the basis of their physiological responses, evoked by controlled stimulation of the ipsilateral antennal inputs (Boeckh and Boeckh, 1979; Matsumoto, 1979; Matsumoto and Hildebrand, 1981; Light, 1982; Harrow and Hildebrand, 1982; Christensen and Hildebrand, 1984; Light, 1986; Christensen and Hildebrand, 1987).

Two major classes of AL neurons have been studied in male and female moths (Fig. 20.2): *local interneurons* (LNs), confined to the AL and having elaborate dendritic arborizations in many or all glomeruli but no recognizable axon; and *projection neurons* (PNs, or output, principal, or relay neurons) with uni- or multiglomerular dendritic arborizations and an axon that projects out of the AL to one or more of several target regions in the brain. In *Manduca*, we estimate that there are about 1200 neurons in each male AL and that the ratio of PNs to LNs may be about 2.5:1 (U. Homberg, R. Montague, and J. Hildebrand, in preparation).

Since the neurons of the AL of *Manduca* were first classified (Matsumoto and Hildebrand, 1981), we have adopted a more detailed and descriptive nomenclature. This new scheme, which is presented and compared with our original terminology in Table 20.1, includes terms describing the dendritic arborizations of the cells. LNs have multiglomerular arborizations, whereas the dendrites of many PNs are confined to single glomeruli. The pattern of arborization of a LN has been described (Matsumoto and Hildebrand, 1981) as either roughly "radially symmetric" (Fig. 20.2A) or "asymmetric" (Fig. 20.2B). PNs can be further classified with respect to the projection pathways and targets of their axons. Montague et al. (1983) found in *Manduca* that both male-specific PNs (with dendrites confined to the MGC; Fig. 20.2D) and PNs associated with the ordinary glomeruli (Fig. 20.2C) have at least three distinct kinds of axonal projection patterns. The axons of the most frequently encountered type (P1) project via the inner antennocerebral tract into the protocerebrum, pass immediately rostral to the calyces of the ipsilateral mushroom body and send out branches that ramify principally in the lips of the calyces, and continue to the lateral protocerebrum, where they terminate (Fig. 20.2C and D; Montague et al., 1983; Hildebrand and Montague, 1986). The inner antennocerebral tract (nomenclature from Kenyon, 1896) is also known as the antennoglomerular tract (AGT) or the tractus

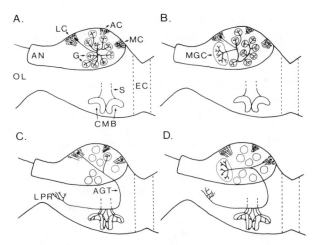

Figure 20.2. Schematic drawings of examples of major morphological types of LNs and PNs in the ALs of *Manduca* (adapted from Hildebrand and Montague, 1986). The nomenclature used to describe these examples is explained in Table 20.1. (A) LN of type L(M1:G). The most frequently encountered type of neuron in both male and female ALs, this cell has a roughly "radially symmetric" dendritic arborization that ramifies in most, if not all, of the ordinary glomeruli (G). (B) LN of type L(M2:G,MGC). Such cells, found only in males, have an "asymmetric" dendritic arborization with branches ramifying in the MGC. (C) PN of type P1(U1:G). This type of cell has its dendritic arborizations confined to a single ordinary glomerulus and sends its axon through the inner antennocerebral tract (or antennoglomerular tract, AGT) to the calyces of the ipsilateral mushroom body (CMB) and ultimately to the lateral protocerebrum (LPR). (D) PN of type P1(U2:MGC). Found only in males, this type of cell exhibits an axonal projection like that of the PN diagrammed in C, but has its dendritic arborizations exclusively within the MGC. Also shown: antennal nerve (AN); anteroventral cell group (AC); medial cell group (MC); lateral cell group (LC); esophageal canal (EC); stalk of the mushroom body (S); optic lobe (OL). Dorsal views.

olfactorio-globularis (TOG), and PNs with similar projections through this tract were described in *Antheraea* (Boeckh and Boeckh, 1979) and *Bombyx* (Kanzaki and Shibuya, 1986b), and are common in other insects as well (see review by Boeckh et al., 1984). Axons of P2 cells project directly to the lateral protocerebrum through the middle antennocerebral tract, completely bypassing the mushroom body. P3 cells send their axons through the outer antennocerebral tract into the protocerebrum, where they terminate rostral to the calyces. On the basis of the projection patterns of the AL output cells (PNs), it appears that the principal centers for "higher-order" processing of olfactory information are the calyces, the lateral protocerebrum, and a mediolateral area in the diffuse protocerebrum situated rostral to the calyces.

20.3.3. Labial-Palp Inputs to the Antennal Lobes

In *Manduca* as well as other species of Lepidoptera (e.g., *Pieris rapae*, Lee et al., 1985; *B. mori* and *A. polyphemus*, Kent et al., 1986) and many other

TABLE 20.1. SOME CHARACTERISTIC RESPONSES RECORDED INTRACELLULARLY FROM LOCAL AND PROJECTION NEURONS IN THE ANTENNAL LOBES OF _MANDUCA SEXTA_

Morphological Classification		Sex	AN Shock	Mech	Hex	Tob	Female Sex Pheromone		
New	Old						Blend	Bal	C15
Local interneurons (L)									
L(M1:G)	Ib	M,F	++/–d	0	++/–	++/–	0	0	nt
L(M2:G)	Ia	F	nt	+	++	++	0	0	nt
L(M1:G,MGC)	IIb	M	++/–d	0	++	++	++/–	++/–	nt
L(M2:G,MGC)	IIa	M	nt	0	+++	0	–/+/–	–/+/–	nt
Projection neurons (P)									
P1(U1:G)	I	M,F	–/+/–	0	+++	0	0	0	nt
P1(U1:G)	I	M,F	–/+/–	0	– –	nt	0	0	nt
P1(U2:G)	II	M,F	–/+/–	0	+	nt	0	0	nt
P1(U1:MGC)	III	M	+++	0	0	0	+++	0	+++
P1(U2:MGC)	III	M	+++	0	0	0	+++	+++	0
P1(U2:MGC)	III	M	+++	0	0	0	+++	0	+++
P1(U2:MGC)	III	M	+++	0	0	0	+++	+++	+++
P1(U2:MGC)	III	M	–/+/–	0	0	0	–/+/–	++	– –
P1(U2:MGC)	III	M	–/+/–	0	0	0	–/+/–	– –	++
P1(U2:MGC)	III	M	– –	nt	nt	nt	nt	nt	nt

Recognized types of LNs and PNs in the ALs of _Manduca sexta_ are listed according to new and previous (Matsumoto and Hildebrand, 1981) classifications. The root of the new classification scheme is a term describing the basic cellular morphology: L for LNs and P for PNs. Axons of P1 cells project through the inner antennocerebral tract (or antennoglomerular tract, AGT in Fig. 20.2) to the calyces of the ipsilateral mushroom body and further to the lateral protocerebrum, as schematized in Figure 20.2. Following this primary designation are several terms enclosed within parentheses. The first of these describes the dendritic arborization: uniglomerular (U) or multiglomerular (M). Multiglomerular neurons are further subdivided according to the pattern of the primary dendritic branching: "radially symmetric" (M1) or "asymmetric" (M2). Uniglomerular neurons are further subdivided according to the extent of their dendritic arborizations: extending through part of the glomerulus (U1) or the entire glomerulus (U2). A second term within the parentheses, preceded by a colon, describes the type of glomeruli receiving arborizations of the cell: one or more ordinary glomeruli (G) or the male-specific macroglomerular complex (MGC) or both (G, MGC).

The responsiveness of AL neurons has been tested by electrical stimulation of the antennal nerve (AN shock), mechanosensory stimulation of the antenna with puffs of clean air (Mech), and olfactory stimulation of the antenna. Plant-associated olfactory stimuli were (E)-2-hexenal (Hex) and the odor of homogenized tobacco (host plant) leaves (Tob). Pheromonal stimuli were as follows: synthetic bombykal (Bal), synthetic (E,Z)-11,13-pentadecadienal (C15), and the volatile components (principally the complete natural pheromone blend) in hexane washes of intact female lure glands (Blend). Symbols: (+, ++, +++) weak, moderate, and strong excitation; (–, – –, – – –) weak, moderate, and strong inhibition; (++/–) excitation with rebound inhibition; (–/+/–) inhibition followed by phasic excitation and subsequent rebound inhibition; (0) no response; (nt) not tested; (d) possibly direct primary-afferent excitation. (Data from Matsumoto and Hildebrand, 1981, and Christensen and Hildebrand, 1987.)

groups of insects, each of the two labial palps carries in its distalmost segment a sensory organ comprising a pit open to the environment and a number of enclosed, peg-like chemosensilla. The axons of the sensory neurons innervating these sensilla of the "labial pit organ" (LPO) join the first labial nerve and enter the subesophageal ganglion (SEG). In *Manduca*, the LPO receptors respond to a variety of plant odors and/or to a component (such as CO_2) present in all of the stimuli, and their fibers pass through the SEG and project bilaterally to a single, identified glomerulus—the "LPO glomerulus"—in each AL (Harrow et al., 1983; Kent et al., 1986). Thus. each LPO glomerulus receives primary-afferent input from both LPOs, but preliminary experiments suggest that there are few or no antennal sensory projections to the LPO glomerulus (Kent et al., 1986). On the basis of such observations, it appears that the LPO is an "accessory" olfactory organ whose inputs to the CNS are initially processed in the AL and presumably integrated with inputs from the antenna. The functions of the LPO, however, are not yet understood.

20.4. PHYSIOLOGY OF NEURONS IN THE ANTENNAL LOBES

The neurons in the ALs of *Manduca* generally show resting potentials of at least -50 mV and support overshooting action potentials. No nonspiking neurons have been found in the ALs. Some AL neurons, typically PNs, appear to be spontaneously active, and most respond postsynaptically to olfactory stimulation of the ipsilateral antenna (Fig. 20.1), whereas essentially all give postsynaptic responses to electrical stimulation of primary-afferent inputs to the AL (Matsumoto and Hildebrand, 1981).

A fundamental question about the several kinds of AL neurons is whether each receives monosynaptic, primary-afferent input from the antenna. It has been shown in a number of species that most AL neurons respond to stimulation of the ipsilateral antenna with biologically meaningful odors. The responses occur, however, only after a delay of 50–200 msec from the time of initiation of the odor delivery or 20–100 msec from the first detectable generator response in the antennal receptors (the electroantennogram response). Presumably, this delay is largely attributable to the time required for the odor molecules to reach the sites of their action on the receptor dendrites and for transduction to occur, and only in small part due to the time required for action potentials to propagate into the AL. In any case, estimates of synaptic delay based on recordings of postsynaptic responses of AL neurons to stimulation of the antenna with odors cannot by themselves serve as reliable criteria of mono- or polysynapticity. To provide more interpretable recordings for this purpose, we routinely use electrical stimulation of the AN, as well as stimulation of the antenna with odors and mechanical stimuli, when testing the responses of AL neurons. Measurements of conduction velocity and synaptic delay, along with tests of the ability of postsynaptic events to "follow" shocks delivered to the AN at high frequencies, have helped us to expose what appear to be monosynaptic connections between primary afferents and certain LNs, but not PNs, in the ALs of *Manduca* (Harrow and Hildebrand, 1982; Tolbert et al., 1983; Christensen and Hildebrand, 1987). Some LNs respond to electrical or olfactory stimulation of antennal primary-afferent inputs with excitatory postsynaptic potentials of relatively short latency (about 5 msec). This physiological evidence, combined with ultrastructural observations (Tolbert and Hildebrand, 1981) suggests that these excitatory inputs are direct. That is, antennal primary-afferent fibers, which are believed to be cholinergic (see Section 20.6), appear to connect monosynaptically to at least some LNs in the AL. This finding is reported in Table 20.1. which summarizes the physiological response profiles of various morphologically recognizable types of AL neurons in *Manduca*.

20.4.1. Local Interneurons

Four morphologically distinct types of LNs in the ALs of *Manduca* were described by Matsumoto and Hildebrand (1981) and are included in Table 20.1. Two of these morphological types are illustrated in Figure 20.2A and B. The great majority of LNs have their cell bodies in the lateral soma group in the AL (Matsumoto and Hildebrand, 1981; U. Homberg, R. Montague, and J. Hildebrand, in preparation). Two types exhibit dendritic arborizations confined to the ordinary glomeruli but differ with respect to their general patterns of branching (types L(M1:G) and L(M2:G)). Two other types, found only in males, have arborizations in the MGC as well as the ordinary glomeruli but again differ in basic branching pattern (types L(M1:G,MGC) and L(M2:G,MGC)).

Despite their morphological differences, both L(M1:G) and L(M2:G) interneurons give indistinguishable responses to leaf odors (Matsumoto and Hildebrand, 1981). Moreover, the responses to volatile extracts of leaves of the host plant (tobacco) and to the general "leaf aldehyde" (*E*)-2-hexenal are qualitatively and quantitatively the same. Whereas some L(M1:G) and L(M2:G) interneurons respond to mechanical stimulation of the antenna, these cells never respond to stimulation with female sex pheromones. It is intriguing that cells of type L(M2:G) have so far been recognized only in ALs of female *Manduca* and that they may thus occur only in females (Matsumoto and Hildebrand, 1981). Interneurons of type L(M:G,MGC) share with L(M:G) cells a general responsiveness to antennal stimulation with leaf volatiles. Both L(M1:G,MGC) and L(M2:G,MGC) interneurons respond to stimulation with (*E*)-2-hexenal, and L(M1:G,MGC) cells also respond to tobacco volatiles. The responses of these male-specific interneurons represent true "quality convergence" in that these cells also respond to stimulation of the antenna with female sex pheromones, which are detected by specialized receptors (see Section 20.3.1.) different from those responsible for detecting plant odors (J. Hildebrand and K.-E. Kaissling, unpublished observations). Furthermore, the responses of some LNs (e.g., type L(M2:G,MGC)) to plant odors and sex pheromones are very different (Fig. 20.1C), which shows that these cells are capable of discrimination between effective odor stimuli of different qualities (see Table 20.1). Because there is no evidence of primary-afferent inhibition in the AL (Matsumoto and Hildebrand, 1981; Harrow and Hildebrand, 1982; Christensen et al., 1985; Waldrop et al., 1987), the delayed synaptic inhibition at the beginning of the response of an L(M2:G,MGC) cell to sex pheromones (Fig. 20.1C; Table 20.1) must be polysynaptic and argues that these LNs participate in processing pheromonal information as higher-order interneurons. Other male-specific LNs apparently have a different role in processing pheromonal information. These cells, of type L(M1:G,MGC), receive purely excitatory and probably direct (Matsumoto and Hildebrand, 1981) synaptic input from pheromone receptors. Immunocytochemical evidence that such cells may be GABAergic (Hoskins et al., 1986) suggests that they might be responsible for mediating the pheromonally elicited inhibition recorded in L(M2:G,MGC) interneurons.

Most intracellular recordings from LNs of all types reveal action potentials of more than one amplitude (Fig. 20.1B). The smaller all-or-none potentials are thought to arise from spike-initiating zones in regions of the cell remote from the site of recording and hence are called dendritic spikes. The possibility that the dendritic spikes represent electrotonic coupling between neurons is less likely because dye coupling with Lucifer yellow has not been observed (Matsumoto and Hildebrand, 1981; Christensen and Hildebrand, 1987), nor has thin-section or freeze-fracture electron microscopy revealed any evidence of gap junctions between neurons in the AL (Tolbert and Hildebrand, 1981). Regions of conduction blockage can exist between dendrites of smaller diameter and the larger primary neurites, such that excitation of one region of the cell to firing may not be passed on to other parts of the neuron. In effect, different regions of such a neuron might function as independent local units. An example of such segregation of responses to two different odors within a single neuron is presented in Figure 20.1B.

20.4.2. Projection Neurons

The second principal class of AL neurons comprises the PNs. In *Manduca* the PNs have their cell bodies in all three cell groups in the AL, but the medial and anteroventral soma groups are particularly enriched in PNs (U. Homberg, R. Montague, and J. Hildebrand, in preparation). Each PN sends an axon out of the AL through one of the antennocerebral tracts to one or more characteristic projection areas in the brain (Hildebrand and Montague, 1986; Homberg et al., in preparation). In males, a majority of the PNs (cells of type P(U:G); Fig. 20.2C) ramify in ordinary glomeruli and appear to be indistinguishable from their female counterparts, but approximately 20–30 other, sexually dimorphic PNs (type P(U:MGC); Fig. 20.2D) have their dendritic arborizations exclusively within the MGC (Hildebrand and Montague, 1986; Homberg et al., in preparation).

Although in physiological recording experiments, all *Manduca* PNs respond postsynaptically when the ipsilateral AN is stimulated electrically (Harrow and Hildebrand, 1982; Christensen and Hildebrand, 1987), many of the P(U:G) cells tested to date have given no detectable, phase-locked response to leaf odors or female sex pheromones (Matsumoto and Hildebrand, 1981). This finding suggests that many of the P(U:G) neurons may be involved in the processing of olfactory or other sensory inputs not yet examined in our studies. By contrast, every male-specific P(U:MGC) neuron tested has responded (in a dose-dependent manner) to stimulation of the ipsilateral male antenna with the female sex pheromone mixture (see Table 20.1; Christensen and Hildebrand, 1984; Christensen and Hildebrand, 1987). Some of these cells, all of which so far have been of type P1(U2:MGC), are equally responsive to antennal stimulation with bombykal or "C15" (which mimics but is not identical with the second essential pheromone component in the female's lure blend, as described in Section 20.3.1). Other P1(U2:MGC) cells, as well as a few P1(U1:MGC) cells with restricted dendritic arborizations, are selectively responsive to either bombykal or C15, and still ot.ers show a differential response to the two pheromones. Some P1(U2:MGC) cells are inhibited by bombykal but excited by C15, whereas in others the pattern of responses to the two pheromones is reversed. In either case the response to the natural blend of the two pheromones is a mixture of excitatory and inhibitory components, suggesting that some PNs integrate inputs from the two classes of pheromone receptors. Thus, certain P(U:MGC) neurons can unambiguously encode information about the presence of either pheromone as well as the mixture of both; in a sense, such cells are "feature detectors" that signal the simultaneous detection of the essential pair of pheromones.

A similar diversity of physiological responsiveness to pheromones has also been found in AL neurons in *B. mori* (Olberg, 1983a; Kanzaki and Shibuya, 1986b). Unlike the AL neurons of *Manduca*, most pheromone-responsive neurons in the ALs of *Bombyx* apparently are multimodal, responding to mechanosensory as well as olfactory stimulation of the antenna. Feature detection nevertheless appears to be characteristic of some AL neurons in *Bombyx*. As evidence of this, several units (out of a survey of 53) were inhibited or unresponsive when tested with either of the two pheromones (bombykol and bombykal, Kaissling et al., 1978), but were strongly excited by a blend of the two (Olberg, 1983a). This indicates, as in *Manduca*, that the information derived from the pheromone blend is different from that of either pheromone alone, and this difference is reflected in the response properties of these neurons. Unfortunately not all of the neurons in *Bombyx* described in the studies above were characterized morphologically.

Another aspect of odor discrimination in moths has been revealed in the case of presumedly male-specific PNs in *Antheraea*. Two species, *A. pernyi* and *A. polyphemus*, use the same pheromones (an aldehyde and an acetate), but the male-specific PNs in *A. pernyi* exhibit a higher sensitivity for the aldehyde, whereas the corresponding PNs in *A. polyphemus* preferentially respond to the acetate (Boeckh and Boeckh, 1979). Although all of the responses to pheromones are excitatory, the responses to the aldehyde and the acetate in the two species are quantitatively different. Thus, it has been postulated that the two pheromones can be separately encoded by higher-order neurons according to the differential levels of excitation elicited by the aldehyde and the acetate. This coding scheme may not be unambiguous, however, because higher concentrations of the less effective pheromone component may elicit responses that mimic the responses to the more effective pheromone. For each species to recognize its conspecific pheromone blend, it is tempting to speculate that the ALs of *Antheraea* may have "feature-detecting" male-specific PNs that respond selectively to a specific blend of the two pheromones.

20.5. PHYSIOLOGY OF HIGHER-ORDER OLFACTORY CENTERS: MUSHROOM BODIES AND LATERAL PROTOCEREBRUM

Of the centers of multimodal integration in the insect CNS, two that are perhaps best known reside in the protocerebrum (PC). These two neuropil structures in each hemi-PC, the mushroom body (MB) and the lateral protocerebrum (LPR), receive polysynaptic inputs from visual, tactile, auditory, and gustatory relay neurons as well as olfactory PNs of the AL. Most PC neurons associated with these centers are multimodal in all insects that have been studied (e.g., honeybees: Suzuki et al., 1976; Erber, 1978; Homberg, 1984; crickets: Schildberger, 1984; silk moths: Light, 1986, and D. Light, personal communication).

As in the ALs, some cells associated with the MB are local or intrinsic interneurons (the Kenyon cells), whereas others are extrinsic neurons providing input to or output from the MB. Physiological studies in several insect species have shown that intrinsic cells and extrinsic output cells of the MB are often more complex and plastic than AL neurons (see Chapter 21 by Erber et al. in this volume). In the honeybee, *Apis*, and the cricket, *Acheta,* some of these neurons display prolonged after discharges that can alter the cells' responsiveness to subsequent inputs, and others exhibit conditional responses that depend on the cells' activity immediately before stimulation (Erber, 1978; Schildberger, 1984). These processes may participate in learning and short-term memory (Erber et al., 1980).

Multimodal interneurons with similar physiological properties as well as responsiveness to sex pheromones have been observed in the LPR of *Bombyx* by means of extracellular recording (Table 20.2; Light, 1986, and D. Light, personal communication). These neurons respond to pheromonal and

TABLE 20.2. SOME CHARACTERISTIC RESPONSES RECORDED EXTRACELLULARLY FROM NEURONS IN THE PROTOCEREBRUM OF MALE *BOMBYX MORI*

Type of Neuron	Mech	Bol	Bal	Light
Unimodal				
Olfactory	0	+ + pr	0	0
	0	0: + + pr	0	0
	0	+/− : + + + sr	0	0
	0	+ : − − pr	0	nt
	0	−	− −	0
	0	+	+ + +	0
	0	− : +	− : + + +	0
Bimodal				
Olfactory and	+	+ + +	+ + +	0
mechanosensory	+ + +	+ +	+ + +	0
	+ + +	+ +	+	0
Olfactory and	0	+ +	0	+ + +
visual	0	+	+	+ + +
	0	+ +	+	0: + + +
	0	+ +	+	+ + : + + +
	0	+ : + + + sr	− − −	+ + +
	0	+ + + sr	− − − sr	+ + +
	0	− − −	− − − sr	+ + + sr
	0	+ + sr	0	+ + + : + +
Trimodal				
Olfactory,	+ +	+ + + pr	+ +	+ + + pr
mechanosensory,	+	+ +	+ +	+ + + : + + +
and visual	+	−	+ +	+ +
	+ +	+	−	+ +
	+	+ + +	+ +	− − −

Neurons in the protocerebrum of *B. mori* were characterized by the following: stimulation of the antenna with mechanosensory (Mech) stimuli and, as olfactory stimuli, 10^{-3} µg of one of the pheromones, bombykol (Bol) or bombykal (Bal); and by visual stimulation of the entire head, including the two compound eyes, with heat-filtered white light (Light). Symbols: (pr) prolonged response lasting 2–10 sec after stimulus-off; (sr) sustained response lasting at least 10 sec or much longer after stimulus-off; other symbols, explained in the legend of Table 20.1, indicate responses to stimulus-on unless otherwise noted. Where appropriate, a response to stimulus-on is separated from the response to stimulus-off by a colon (i.e., response to stimulus-on:response to stimulus-off). Only cells that responded to olfactory stimuli are included in this table. (Data from D. M. Light, personal communication.)

visual inputs, but the responses to these two modalities are often clearly distinguishable. Moreover, these cells appear to integrate multimodal inputs, as the response to pheromone is enhanced when the odor delivery is paired with a light stimulus to the visual system. As found for AL interneurons in *Manduca*, some LPR cells in *Bombyx* also give qualitatively different responses to the two *Bombyx* pheromones (bombykol and bombykal), indicating that there must be a convergence (albeit indirect) from two separate populations of antennal receptors. Unlike the activity of male-specific P1(U2:MGC) neurons in *Manduca*, however, these responses of *Bombyx* LPR cells may outlast the presentation of the stimulus by several minutes, and may thus provide a basis for a short-term memory capability in the PC. It is conceivable that the neural circuits responsible for this sustained activity may influence the premotor and motor pathways that ultimately mediate specific behaviors.

Presumed premotor neurons with axons in the ventral nerve cord of *Bombyx* have been explored by Olberg (1983b). Designated "flip-flopping interneurons" (FFI), these elements exhibit unusual patterns of activity that alternate between high (excited) and low (inhibited) states with each stimulus presentation. A given activity state in the FFIs correlates with a particular attitude of the antennae, which in turn correlates with the orientation of walking movements as well as the occurrence of abdominal "ruddering" in the direction of a turning movement. Olberg (1983b) demonstrated that activity in FFIs is consistent with turning movements toward the antenna that receives the higher ratio of bombykol to bombykal in the antennal stimulus. The combined evidence suggests that the FFIs are premotor neurons that convey pheromone-triggered turning directions to thoracic and abdominal motor neurons as the male moth pursues his zig-zagging walk upwind in search of a calling female (Kramer, 1975). More recently, neurons that exhibit similar flip-flopping activity have been encountered in the LPR of *Bombyx* (Light, 1986, and D. Light, personal communication), suggesting a possible PC site for the "flip-flop generator." Another apparently different group of descending neurons may participate in the initiation and maintenance of wing fluttering in *Bombyx* (Kanzaki and Shibuya, 1986a).

20.6. NEUROCHEMISTRY OF THE OLFACTORY PATHWAY

Mechanistic analysis of the functioning of any part of the nervous system requires explorations of synaptic and nonsynaptic communication between neurons and ultimately an understanding of the chemical signals that transmit synaptic messages and modulate neural activities. In the case of the insect olfactory pathway, neurochemical studies are just beginning, and most of what is known has come from work on Lepidoptera and *Manduca sexta* in particular. This research arises in a context of much ignorance about synaptic mechanisms in the insect CNS, where technical obstacles and relatively late-blooming interest have delayed serious efforts. Nevertheless, despite a lack of conclusive demonstration of a neurotransmitter role for any substance in the insect CNS, abundant evidence supports the idea that many of the substances that function as transmitters and modulators in other, better-studied phyletic groups (such as mammals and crustaceans) also play similar roles in insects.

As for other primary mechanosensory and chemosensory cells in arthropods (see references in Prescott et al., 1977), several lines of evidence strongly indicate that acetylcholine (ACh) is the principal or exclusive neurotransmitter used by antennal sensory neurons in *Manduca* (Sanes and

Hildebrand, 1976c; Sanes et al., 1977; Hildebrand et al., 1979). Moreover, because putative ACh receptors in the AL glomeruli exhibit predominantly nicotinic pharmacology, it appears that the synapses between the antennal sensory axons and their targets in the ALs may be "nicotinic" (Sanes et al., 1977; Hildebrand et al., 1979). In addition to the primary-afferent inputs, some neurons in the AL also may be cholinergic (Sanes et al., 1977; Hoskins and Hildebrand, 1983).

γ-Aminobutyric acid (GABA) is also a leading neurotransmitter candidate in the ALs as well as o.her areas of the brain and SEG (Maxwell et al., 1978; Kingan and Hildebrand, 1985; Homberg et al., 1987). Recent neurophysiological and immunocytochemical findings suggest that GABA is an inhibitory neurotransmitter associated with a majority of the LNs in the lateral soma group of the AL in *Manduca* (Harrow and Hildebrand, 1982; Hoskins et al., 1984; Christensen et al., 1985; Hoskins et al., 1986). These neurons, in all probability, provide the inhibitory synaptic input to certain LNs and PNs as discussed in Sections 20.4.1 and 20.4.2. Synaptic inhibition in AL neurons can be reversibly blocked by GABA antagonists, and application of GABA can cause marked hyperpolarization in these neurons (Christensen et al., 1985; Waldrop et al., 1987). Some fibers passing between the AL and the protocerebrum, possibly belonging to the PNs that project to targets other than the calyces in the protocerebrum, also contain GABA-like immunoreactivity (Hoskins et al., 1986; Homberg et al., 1987).

The ALs of *Manduca* produce a very small amount of 5-hydroxytryptamine and more significant levels of histamine and tyramine. Only one large neuron in each AL has been found to contain 5-hydroxytryptamine-like immunoreactivity (Hoskins et al., 1985). The cellular locations of histamine and tyramine in the AL have not yet been determined. Finally, preliminary experiments have revealed that certain cells or clusters of cells in the AL contain immunoreactivities related, and possibly identical, to the neuropeptides substance P, locust adipokinetic hormone, FMRFamide, corticotropin-releasing factor, and cholecystokinin (Homberg et al., 1985; Homberg et al., 1986), but we have not yet ascertained whether these cells also contain one or more conventional neurotransmitters.

20.7. SUMMARY AND CONCLUSIONS

In Lepidoptera, as in many other groups of insects, sensitivity to odors is greatly amplified through spatial convergence of myriad olfactory receptors of several types onto far fewer central neurons in the primary olfactory centers in the brain, the ALs. Such an arrangement is exemplified by the pheromone-processing subsystem in the brain of the male moth. Some AL neurons receive inputs exclusively from a population of finely tuned, specialized receptors located in the male-specific trichoid antennal sensilla that respond selectively to a single pheromone. Through this quality convergence, these central neurons achieve a high degree of stimulus selectivity and can discriminate individual pheromones in a biologically meaningful mixture of pheromones. Other kinds of central neurons receive inputs from several populations of receptors and can, through a series of relatively complex synaptic interactions, encode the presence of a particular combination of odors ("feature detection"), such as the specific blend of pheromones liberated by a calling conspecific female moth. Many, if not all, of these central neurons exhibit a dose dependency for pheromones and can therefore code for stimulus quantity as well as quality. Inputs from other receptors, such as those responding to host-plant odors, converge on many pheromone-responsive AL neurons and undoubtedly contribute to the insect's total olfactory "picture" of the environment. Some cells also exhibit modality convergence, exemplified by responsiveness to antennal stimulation with gentle mechanosensory stimuli as well as odors. Thus, the ALs are more than mere relay stations that simply sum and transmit olfactory information to higher centers. There is considerable complexity in both the structure of the AL neurons and their physiological activities. We are currently studying the functional organization of the AL, examining in particular the synaptic relationships between LNs and PNs (Christensen and Hildebrand, 1986). Synaptic inhibition is prominent in the AL, and we are currently investigating the roles of the presumably inhibitory LNs in olfactory signal processing.

Further along in the olfactory pathway, in the higher-order centers in the protocerebrum, one encounters neurons that display still more complex properties than those of neurons in the AL. The protocerebral neurons are characterized by considerable plasticity in their responsiveness, for example to pheromones, and can show prolonged excitation as well as "conditional" responses that depend on the physiological state of the neuron before stimulation. These protocerebral cells also represent the first level in the olfactory pathway at which pheromonal information is integrated with visual inputs (and probably other inputs of other modalities as well). Such higher-order neurons presumably participate in circuits responsible for controlling the activity in descending neurons, which ultimately control or modulate mate-seeking behavior.

Relatively little is known about central processing of plant-associated odors in the brains of moths and butterflies. These important substances include odorants that influence feeding and oviposition behaviors. Studies to date in *Manduca* have shown that the nonsexually dimorphic AL neurons, and the ordinary glomeruli in which they have their dendritic arborizations, are the sites of initial processing of plant odors. Some pheromone-responsive, sexually dimorphic LNs in the AL also respond to antennal stimulation with plant odors, but because male-specific PNs are only weakly inhibited by, or unresponsive to, plant odors, the significance of the findings in LNs is unclear. Further insights into the functions of the lepidopteran olfactory system will depend upon successful unraveling of the mechanisms involved in processing and integrating information from pheromonal, plant, and other biologically important odors in the CNS. Moreover, the reproductive success of many insects depends upon the integration of processed olfactory information with input from other sensory modalities. These integrative mechanisms, which ultimately lead to the production or modulation of coordinated and stereotyped motor outputs, are among the currently intriguing problems in olfactory research.

ACKNOWLEDGMENTS

We dedicate this paper to Dietrich Schneider with appreciation and gratitude for his inspiration and friendship. We thank M. Koontz, D. M. Light, and J. H. Tumlinson for sharing with us their unpublished findings and for granting permission for the inclusion of some of them in this chapter. We are grateful to R. A. Montague for valuable assistance, U. Homberg and B. Waldrop for helpful discussions, J. Buckner and J. Svoboda of the USDA for providing *Manduca* eggs for our research, and S. Garner for excellent secretarial assistance. The recent research in our laboratory has been supported by grants AI-17711 and AI-23253 from NIH and BNS 83-12769 from NSF, and by an NIH Postdoctoral Research Fellowship to T. A. Christensen.

REFERENCES

Aplin, R. T. and M. C. Birch. 1970. Identification of odorous compounds from male Lepidoptera. *Experientia* **26**: 1193–1194.

Baker, T. C., M. A. Willis, and P. L. Phelan. 1984. Optomotor anemotaxis polarizes self-steered zigzagging in flying moths. *Physiol. Entomol.* **9**: 365–376.

Baker, T. C. and L. P. S. Kuenen. 1982. Pheromone source location by flying moths: A supplementary non-anemotactic mechanism. *Science (Washington, D. C.)* **216**: 424–427.

Bestmann, H. J., O. Vostrovsky, and H. Platz. 1977. Pheromone XII. Männchenduftstoffe von Noctuiden (Lepidoptera). *Experientia* **33**: 874–875.

Birch, M. C. and K. F. Haynes. 1982. *Insect Pheromones*. Edward Arnold Ltd., London.

Blaney, W. M., L. M. Schoonhoven, and M. S. J. Simmonds. 1986. Sensitivity variations in insect chemoreceptors; a review. *Experientia* **42**: 13–19.

Boeckh, J. and V. Boeckh. 1979. Threshold and odor specificity of pheromone-sensitive neurons in the deutocerebrum of *Antheraea pernyi* and *A. polyphemus*. *J. Comp. Physiol.* **132**: 235–242.

Boeckh, J., K. D. Ernst, H. Sass, and U. Waldow. 1984. Anatomical and physiological characteristics of individual neurones in the central antennal pathway of insects. *J. Insect Physiol.* **30**: 15–26.

Bossert, W. H. and E. O. Wilson. 1963. The analysis of olfactory communication among animals. *J. Theoret. Biol.* **5:** 443–469.

Bretschneider, F. 1924. Über die Gehirne des Eichenspinners und des Seidenspinners (*Lasiocampa quercus* und *Bombyx mori*). *Z. Wiss. Zool.* (*Abt. A*). **60:** 563–578.

Brower, L. P., J. V. Z. Brower, and F. P. Cranston. 1965. Courtship behaviour of the queen butterfly, *Danaus gilippus berenice* (Cramer). *Zoologica* **50:** 1–39.

Butenandt, A., R. Beckmann, D. Stamm, and E. Hecker. 1959. Über den Sexual-Lockstoff des Seidenspinners *Bombyx mori*. Reindarstellung und Konstitution. *Z. Naturforsch.* **14:** 283–284.

Camazine, S. M. and J. G. Hildebrand. 1979. Central projections of antennal sensory neurons in mature and developing *Manduca sexta*. *Soc. Neurosci. Abstr.* **5:** 155.

Christensen, T. A. and J. G. Hildebrand. 1984. Functional anatomy and physiology of male-specific pheromone-processing interneurons in the brain of *Manduca sexta*. *Soc. Neurosci. Abstr.* **10:** 862.

Christensen, T. A. and J. G. Hildebrand. 1986. Synaptic relationships between local and projection neurons in the antennal lobe of the sphinx moth *Manduca sexta*. *Soc. Neurosci. Abstr.* **12:** 857.

Christensen, T. A. and J. G. Hildebrand. 1987. Male-specific, sex pheromone-selective projection neurons in the antennal lobes of the moth *Manduca sexta*. *J. Comp. Physiol. A.* in press.

Christensen, T. A., B. R. Waldrop, and J. G. Hildebrand. 1985. GABA-mediated inhibition in the antennal lobes of the moth *Manduca sexta*. *Soc. Neurosci. Abstr.* **11:** 163.

Dahm, K. H., D. Mayer, W. E. Finn, V. Reinhold, and H. Röller. 1971. The olfactory and auditory mediated sex attraction in *Achroia grisella* (Fabr.). *Naturwissenschaften* **58:** 265–266.

David, C. T., J. S. Kennedy, and A. R. Ludlow. 1983. Finding of a sex pheromone source by gypsy moths released in the field. *Nature* (*London*) **303:** 804–806.

Den Otter, C. J., H. A. Scheil, and A. Sander-Van Oosten. 1978. Reception of host-plant odours and female sex pheromone in *Adoxophyes orana* (Lepidoptera:Tortricidae): Electrophysiology and morphology. *Entomol. Exp. Appl.* **24:** 570–578.

Den Otter, C. J., M. Behan, and F. W. Maes. 1980. Single cell responses in female *Pieris brassicae* (Lepidoptera:Pieridae) to plant volatiles and conspecific egg odours. *J. Insect Physiol.* **26:** 465–472.

Erber, J. 1978. Response characteristics and after-effects of multimodal neurones in the mushroom body area of the honeybee. *Physiol. Entomol.* **3:** 77–89.

Erber, J., T. Masuhr, and R. Menzel. 1980. Localization of short-term memory in the brain of the bee *Apis mellifera*. *Physiol. Entomol.* **5:** 343–358.

Harrow, I. D. and J. G. Hildebrand. 1982. Synaptic interactions in the olfactory lobe of the moth *Manduca sexta*. *Soc. Neurosci. Abstr.* **8:** 528.

Harrow, I. D., P. Quartararo, K. S. Kent, and J. G. Hildebrand. 1983. Central projections and possible chemosensory function of neurons in a sensory organ on the labial palp of *Manduca sexta*. *Soc. Neurosci. Abstr.* **9:** 216.

Hildebrand, J. G. and R. A. Montague. 1986. Functional organization of olfactory pathways in the central nervous system of *Manduca sexta*, pp. 279–285. In T.

Payne, M. C. Birch, and C. E. J. Kennedy (eds.), *Mechanisms in Insect Olfaction*. Oxford University Press.

Hildebrand, J. G., L. M. Hall, and B. C. Osmond. 1979. Distribution of binding sites for ^{125}I-labeled α-bungarotoxin in normal and deafferented antennal lobes of *Manduca sexta*. *Proc. Natl. Acad. Sci. USA* **76**: 499–503.

Hildebrand, J. G., S. G. Matsumoto, S. M. Camazine, L. P. Tolbert, S. Blank, H. Ferguson, and V. Ecker. 1980. Organization and physiology of antennal centers in the brain of the moth *Manduca sexta*, pp. 375–382. In *Insect Neurobiology and Pesticide Action (Neurotox 79)*. Soc. Chem. Ind., London.

Homberg, U. 1984. Processing of antennal information in extrinsic mushroom body neurons of the bee brain. *J. Comp. Physiol. A* **154**: 825–836.

Homberg, U., S. G. Hoskins, and J. G. Hildebrand. 1985. Immunocytochemical mapping of peptides in the brain and subesophageal ganglion of *Manduca sexta*. *Soc. Neurosci. Abstr.* **11**: 942.

Homberg, U., R. A. Montague, S. G. Hoskins, K. S. Kent, and J. G. Hildebrand. 1986. Immunocytochemical mapping of neurotransmitter candidates in the antennal lobes of *Manduca sexta*. *Soc. Neurosci. Abstr.* **12**: 949.

Homberg, U., T. G. Kingan, and J. G. Hildebrand. 1987. Immunocytochemistry of GABA in the brain and subesophageal ganglion of *Manduca sexta*. *Cell Tissue Res.* **248**: 1–24.

Hoskins, S. G. and J. G. Hildebrand. 1983. Neurotransmitter histochemistry of neurons in the antennal lobes of *Manduca sexta*. *Soc. Neurosci. Abstr.* **9**: 216.

Hoskins, S. G., U. Homberg, T. G. Kingan, and J. G. Hildebrand. 1985. Neurochemical anatomy of the brain of the sphinx moth *Manduca sexta*, pp. 84–87. In *Neuropharmacology and Pesticide Action (Neurotox 85)*. Soc. Chem. Ind., London.

Hoskins, S. G., T. G. Kingan, T. A. Christensen, and J. G. Hildebrand. 1984. Mapping GABA-like immunoreactivity in antennal lobes of the moth *Manduca sexta*. *Soc. Neurosci. Abstr.* **10**: 688.

Hoskins, S. G., U. Homberg, T. G. Kingan, T. A. Christensen, and J. G. Hildebrand. 1986. Immunocytochemistry of GABA in the antennal lobes of the sphinx moth *Manduca sexta*. *Cell Tissue Res.* **244**: 243–252.

Ilse, D. 1928. Über den Farbensinn der Tagfalter. *Z. Vgl. Physiol.* **8**: 658–692.

Kaissling, K.-E. 1971. Insect olfaction, pp. 351–431. In L. M. Beidler (ed.), *Handbook of Sensory Physiology*, vol. 4, *Chemical Senses 1. Olfaction*. Springer-Verlag, Heidelberg.

Kaissling, K.-E. 1979. Recognition of pheromones by moths, especially in saturniids and *Bombyx mori*, pp. 43–56. In F. J. Ritter (ed.), *Chemical Ecology: Odour Communication in Animals*. Elsevier, Amsterdam.

Kaissling, K.-E. and E. Priesner. 1970. Die Riechschwelle des Seidenspinners. *Naturwissenschaften*. **57**: 23–28.

Kaissling, K.-E., G. Kasang, H. J. Bestmann, W. Stransky, and O. Vostrowsky. 1978. A new pheromone of the silkworm moth *Bombyx mori*. *Naturwissenschaften*. **65**: 382–384.

Kanzaki, R. and T. Shibuya. 1986a. Descending protocerebral neurons related to the mating dance of the male silkworm moth. *Brain Res.* **377**: 378–382.

Kanzaki, R. and T. Shibuya. 1986b. Identification of the deutocerebral neurons responding to the sexual pheromone in the male silkworm moth brain. *Zool. Sci.* **3**: 409–418.

Kennedy, J. S. 1983. Zigzagging and casting as a programmed response to windborne odour: A review. *Physiol. Entomol.* **8**: 109–120.

Kent, K. S., I. D. Harrow, P. Quartararo, and J. G. Hildebrand, 1986. An accessory olfactory pathway in Lepidoptera: The labial pit organ and its central projections in *Manduca sexta* and certain other sphinx moths and silk moths. *Cell Tissue Res.* **245**: 237–245.

Kenyon, F. C. 1896. The brain of the bee. A preliminary contribution to the morphology of the nervous system of the Arthropoda. *J. Comp. Neurol.* **6**: 133–210.

Kingan, T. G. and J. G. Hildebrand. 1985. γ-Aminobutyric acid in the central nervous system of metamorphosing and mature *Manduca sexta. Insect Biochem.* **15**(6): 667–675.

Kramer, E. 1975. Orientation of the male silkmoth to the sex attractant bombykol, pp. 329–335. In D. A. Denton and J. D. Coghlan (eds.), *Olfaction and Taste,* vol. 5. Academic Press, New York.

Kuenen, L. P. S. and T. C. Baker. 1983. A non-anemotactic mechanism used in pheromone source location by flying moths. *Physiol. Entomol.* **8**: 277–289.

Lee, J.-K., R. Selzer, and H. Altner. 1985. Lamellated outer dendritic segments of a chemoreceptor within wall-pore sensilla in the labial palp-pit organ in the butterfly, *Pieris rapae* L. (Insecta, Lepidoptera). *Cell Tissue Res.* **240**: 333–342.

Light, D. M. 1982. Multimodal and pheromone sensitive interneurons in the deutocerebrum of the silk moth, *Bombyx mori,* p. 45. *Abstracts of ECRO V.* Regensburg.

Light, D. M. 1986. Central integration of sensory signals: An exploration of processing of pheromonal and multimodal information in lepidopteran brains, pp. 287–301. In T. Payne, M. C. Birch, and C. E. J. Kennedy (eds.), *Mechanisms in Insect Olfaction.* Oxford University Press, London.

Ma, W. C. and L. M. Schoonhoven. 1973. Tarsal contact chemosensory hairs of the large white butterfly, *Pieris brassicae,* and their possible role in oviposition behavior. *Entomol. Exp. Appl.* **16**: 343–357.

Matsumoto, S. G. 1979. Physiology and morphology of neurons in the antennal lobes of mature and developing *Manduca sexta. Soc. Neurosci. Abstr.* **5**: 170.

Matsumoto, S. G. and J. G. Hildebrand. 1981. Olfactory mechanisms in the moth *Manduca sexta:* Response characteristics and morphology of central neurons in the antennal lobes. *Proc. R. Soc. London B* **213**: 249–277.

Maxwell, G. D., J. F. Tait, and J. G. Hildebrand. 1978. Regional synthesis of neurotransmitter candidates in the CNS of the moth *Manduca sexta. Comp. Biochem. Physiol.* **61C**: 109–119.

Meinwald, J., Y. C. Meinwald, and P. H. Mazzochi. 1969. Sex pheromone of the queen butterfly: Chemistry. *Science (Washington, D. C.)* **164**: 1174–1175.

Meinwald, J., Y. C. Meinwald, J. W. Wheeler, T. Eisner, and L. P. Brower. 1966. Major components in the exocrine secretion of a male butterfly (*Lycorea*). *Science (Washington, D. C.)* **151**: 583–585.

Meinwald, J., C. J. Boriack, D. Schneider, M. Boppré, W. F. Wood, and T. Eisner.

1974. Volatile ketones in the hairpencil secretion of danaid butterflies (*Amauris* and *Danaus*). *Experientia* **30**: 721–722.

Minks, A. K., W. L. Roelofs, F. J. Ritter, and C. J. Persoons. 1973. Reproductive isolation of two tortricid moth species by different ratios of a two-component sex attractant. *Science* (*Washington, D. C.*) **180**: 1073–1074.

Montague, R. A., K. S. Kent, M. T. Imperato, and J. G. Hildebrand. 1983. Projections of antennal-lobe output neurons in the brain of *Manduca sexta*. *Soc. Neurosci. Abstr.* **9**: 216.

Mustaparta, H. 1984. Olfaction, pp. 37–70. In W. J. Bell and R. T. Cardé (ed.), *Chemical Ecology of Insects*. Sinauer Assoc., Sunderland, MA.

Myers, J. H. and M. Walter. 1970. Olfaction in the Florida queen butterfly: Honey odour receptors. *J. Insect Physiol.* **16**: 573–578.

Nordlund, D. A., R. L. Jones, and W. J. Lewis (eds.). 1981. *Semiochemicals. Their Role in Pest Control*. John Wiley, New York.

Olberg, R. M. 1983a. Interneurons sensitive to female pheromone in the deutocerebrum of the male silkworm moth, *Bombyx mori*. *Physiol. Entomol.* **8**: 419–428.

Olberg, R. M. 1983b. Pheromone-triggered flip-flopping interneurons in the ventral nerve cord of the silkworm moth, *Bombyx mori*. *J. Comp. Physiol.* **152**: 297–307.

Prescott, D. J., J. G. Hildebrand, J. R. Sanes, and S. Jewett. 1977. Biochemical and developmental studies of acetylcholine metabolism in the central nervous system of the moth, *Manduca sexta*. *Comp. Biochem. Physiol.* **56C**: 77–84.

Priesner, E. 1980. Sex attractant system in *Polia pisi* L. (Lepidoptera:Noctuidae). *Z. Naturforsch. C* **35**: 990–994.

Rendwick, J. A. A. and C. D. Radke. 1980. An oviposition deterrent associated with frass from feeding larvae of the cabbage looper, *Trichoplusia ni* (Lepidoptera:Noctuidae). *Env. Entomol.* **9**: 318–320.

Rospars, J. P. 1983. Invariance and sex-specific variations of the glomerular organization in the antennal lobes of a moth, *Mamestra brassicae*, and a butterfly, *Pieris brassicae*. *J. Comp. Neurol.* **220**: 80–96.

Rothschild, M. and L. M. Schoonhoven. 1977. Assessment of egg load by *Pieris brassicae* (Lepidoptera:Pieridae). *Nature* (*London*) **226**: 352–355.

Sanes, J. R. and J. G. Hildebrand. 1976a. Structure and development of antennae in a moth, *Manduca sexta*. *Dev. Biol.* **51**: 282–299.

Sanes, J. R. and J. G. Hildebrand. 1976b. Origin and morphogenesis of sensory neurons in an insect antenna. *Dev. Biol.* **51**: 300–319.

Sanes, J. R. and J. G. Hildebrand. 1976c. Acetylcholine and its metabolic enzymes in developing antennae of the moth, *Manduca sexta*. *Dev. Biol.* **52**: 105–120.

Sanes, J. R., D. J. Prescott, and J. G. Hildebrand. 1977. Cholinergic neurochemical development of normal and deafferented antennal lobes during metamorphosis of the moth, *Manduca sexta*. *Brain Res.* **119**: 389–402.

Schildberger, K. 1984. Multimodal interneurons in the cricket brain: Properties of identified extrinsic mushroom body cells. *J. Comp. Physiol. A* **154**: 71–79.

Schneider, D. 1974. The sex-attractant receptor of moths. *Sci. Am.* **23**: 28–35.

Schneider, D. and U. Seibt. 1969. Sex pheromone of the queen butterfly: Electroantennogram responses. *Science* (*Washington, D. C.*) **164**: 1173–1174.

Schneiderman, A. M. 1984. Postembryonic development of a sexually dimorphic sensory pathway in the central nervous system of the sphinx moth, *Manduca sexta*. Doctoral dissertation, Harvard University, Cambridge. MA.

Schneiderman, A. M., J. G. Hildebrand, and J. J. Jacobs. 1983. Computer-aided morphometry of developing and mature antennal lobes in the moth *Manduca sexta*. *Soc. Neurosci. Abstr.* **9**: 834.

Schneiderman, A. M., J. G. Hildebrand, M. M. Brennan, and J. H. Tumlinson. 1986. Trans-sexually grafted antennae alter pheromone-directed behaviour in a moth. *Nature (London)* **323**: 801–803.

Schneiderman, A. M., S. G. Matsumoto, and J. G. Hildebrand. 1982. Transsexually grafted antennae influence development of sexually dimorphic neurones in moth brain. *Nature (London)* **298**: 844–846.

Shepherd, G. M. 1972. Synaptic organization of the mammalian olfactory bulb. *Physiol. Rev.* **52**: 846–917.

Shorey, H. H. 1976. *Animal Communication by Pheromones*. Academic Press, New York.

Silberglied, R. 1977. Communication in the Lepidoptera, pp. 362–407. In T. A. Sebeok (ed.), *How Animals Communicate*. Indiana University Press, Bloomington, IN.

Starratt, A. M., K. H. Dahm, N. Allen, J. G. Hildebrand, T. L. Payne, and H. Röller. 1979. Bombykal, a sex pheromone of the sphinx moth, *Manduca sexta*. *Z. Naturforsch.* **34c**: 9–12.

Suzuki, H., H. Tateda, and M. Kuwabara. 1976. Activities of antennal and ocellar interneurons in the protocerebrum of the honeybee. *J. Exp. Biol.* **64**: 405–418.

Tolbert, L. P. and J. G. Hildebrand. 1981. Organization and synaptic ultrastructure of glomeruli in the antennal lobes of the moth *Manduca sexta:* A study using thin sections and freeze-fracture. *Proc. R Soc. London B* **213**: 279–301.

Tolbert, L. P., S. G. Matsumoto, and J. G. Hildebrand. 1983. Development of synapses in the antennal lobes of the moth *Manduca sexta* during metamorphosis. *J. Neurosci.* **3**: 1158–1175.

Van der Pers, J. N. C., G. Thomas, and C. J. den Otter. 1980. Interactions between plant odours and pheromone reception in small ermine moths (Lepidoptera:Yponomeutidae). *Chem. Senses* **5**: 367–371.

Waldrop, B., T. A. Christensen, and J. G. Hildebrand. 1987. GABA-mediated synaptic inhibition of projection neurons in the antennal lobes of the sphinx moth *Manduca sexta*. *J. Comp. Physiol. A.* in press.

Willis, M. A. and T. C. Baker. 1984. Effects of intermittent and continuous pheromone stimulation on the flight behaviour of the oriental fruit moth, *Grapholita molesta*. *Physiol. Entomol.* **9**: 341–358.

Yamamoto, R. T. and G. Fraenkel. 1960. The physiological basis for the selection of plants for egg-laying in the tobacco hornworm, *Protoparce sexta* (Johan.). *Proc. 11th Int. Cong. Entomol. (Wien)* **3**: 127–133.

Yamamoto, R. T., R. Y. Jenkins, and R. K. McClusky. 1969. Factors determining the selection of plants for oviposition by the tobacco hornworm *Manduca sexta*. *Entomol. Exp. App.* **12**: 504–508.

CHAPTER 21

Functional Roles
of the Mushroom Bodies
in Insects

Joachim Erber

Institute of Biology
Technical University of Berlin
Berlin, West Germany

Uwe Homberg

Division of Neurobiology
Arizona Research Laboratory
University of Arizona
Tucson, Arizona

Wulfila Gronenberg

Institute of Zoology
University of Frankfurt
Frankfurt, West Germany

Dedicated to Prof. Dr. Franz Huber on the occasion of his 60th birthday.

21.1. INTRODUCTION

The speculations about the functional roles of the mushroom bodies (MBs) in the arthropod brain started with their first anatomical description in the last century (Dujardin, 1850). Although many studies on the neuroanatomy of the MB in the different phyla exist, these investigations are in most cases insufficient to interpret the functions of these neuropils. Detailed and systematic behavioral, physiological, and neuroanatomical analyses exist only for a small number of insects: bees, cockroaches, crickets, and flies. Therefore, interpretations of the functional role of the MB have to be derived from a small sample of the arthropods. We intend to show that some common principles emerge from these studies and propose testable hypotheses, which differ in many respects from the hypothetical functions, which so far have been attributed to the MB. The neuroanatomy and pharmacological properties of the MB system are discussed in other chapters of this book.

21.2. SPECULATIONS AND HYPOTHESES

The literature on the MB in arthropods gives the impression that almost all functions that a nervous system has to perform can be attributed to the MB system. The wide spectrum of hypothetical functions of the MB ranges from purely sensory signal processing to the control of complex behavioral sequences. From neuroanatomical studies, Strausfeld (1976) concluded that the MBs serve as a sensory neuropil for the evaluation of olfactory and gustatory stimuli. A different sensory function has hypothetically been ascribed to this neuropil by Horridge (Bullock and Horridge, 1965) who assumed that they might serve as "charts" on which stimulus directions relative to the head are plotted. A variation and extension of these hypotheses assumes that the MB are association areas for olfactory and visual signals (Hanström, 1928). A hypothetical consequence of these associations could be different forms of memory (Howse, 1974).

Other hypotheses assume that the MBs are the neural substrate for complex instinctive acts (e.g., Flögel, 1878). In ? variation of these hypotheses, Vowles (1964b) proposed that the MB may serve the sensory control of motor

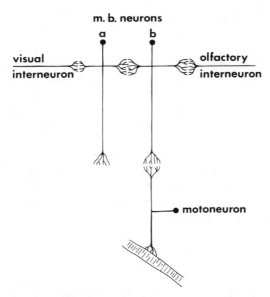

Figure 21.1. Hypothetical neuronal circuit in the mushroom body of Crustacea. (after Hanström, 1928; with permission of the publisher.)

acts, whereas Howse (1974) assumed that they generate and control behavioral sequences. Huber (1967) suggested that the MBs select and coordinate behavioral patterns. A hypothetical neural circuit proposed by Hanström (1928) is very illustrative, because it encompasses most of these hypotheses (Fig. 21.1). The neural circuit was modeled to describe visual learning experiments in hermit crabs (Spaulding, 1904). In this hypothetical circuit, olfactory interneurons impinge on MB neurons, which, in turn, are connected with motoneurons. Olfactory stimuli are assumed to elicit behavior, in this case feeding, via the MB. Visual interneurons are also hypothetically connected with MB neurons; the association between olfactory and visual stimuli occurs within this system. During learning, the efficacy of the synaptic contacts is assumed to be changed. After a number of rewards, a visual signal could then elicit feeding behavior. The evaluation of olfactory and visual signals; the association of stimuli, learning, and memory; and the elicitation of behavior were ascribed to the MB nearly 60 years ago. It is astonishing that our concepts of the MB system have only slightly changed over two generations.

21.3. LESIONS AND ELECTRICAL STIMULATION

21.3.1. Lesions

Since the last century, lesions, surgical injuries, and the extirpation of nervous tissue have been a widely used experimental approach for analyzing the functional role of brain areas in arthropods (for reviews see Bethe, 1897; Roeder, 1937; Huber, 1962; Bullock and Horridge, 1965). The interpretation of lesions in the insect brain meets major difficulties. Lesions in the calyces, for instance, will inevitably lead to damages of input neurons to the MB. Many of these cells, like those of the antennoglomerular tract (AGT), have projection areas also in the lateral protocerebrum (see Section 21.6.1). Therefore, lesions of the calyces will also have direct influence on signal processing in other brain parts, which, in turn, can influence many types of behavior. Lesions in the calyces, in many cases, will affect optical neurons bypassing the MB system. Therefore, modifications of visual responses, after calyx lesions, are not too astonishing. Taken alone, lesions are only of limited value for our understanding of the functional role of brain areas. Nevertheless, they are an important link in the chain of evidence, which includes stimulation, neuroanatomical analyses, and physiological experiments.

A summary of lesion experiments in the MB and their behavioral consequences is given in Table 21.1. The experiments can be classified in three categories: general short-term effects, general long-term effects, and specific effects on defined behaviors. Although, the degree of the lesions and the precision of localizing the focus of the injuries is extremely divergent in the different studies, some common rules emerge from these investigations.

A general short-term consequence of extensive lesions in the MB of insects is an increase in locomotor activity, and these activity changes can last some hours or days. A model in which the MB in both hemispheres mutually inhibit each other and also exert an inhibitory effect on the central body, describes this short-term general effect on locomotor behavior (Huber, 1962).

Long-term effects of MB lesions are much less pronounced (Drescher, 1960). After extirpation of both calyxes, the threshold for odor responses is increased in the cockroach. The optomotor response shows a longer latency, but male cockroaches are still responding to female pheromones and the mating behavior seems to be unaffected.

Many specific behaviors in different insect orders depend on intact MBs. There can be no doubt that olfactory responses, including conditioned reactions, are disturbed by MB lesions. Effects on acoustic behavior in Orthoptera and on the coordination of behavioral sequences in silkworms illustrate that the MB might serve different functions in different insect orders (Table 21.1).

An interesting comparison of MB function between insects and crustaceans can be drawn from large-scale lesions in the eyestalks of decapod Crustacea. Antennular functions, the elicitation of reflexes, general escape, and feeding behavior are influenced by these operations. The neural structures, subserving olfaction in the eyestalks, are located in the medulla terminalis (Hanström, 1925; Sandeman, 1982), now termed "lateral protocerebrum" (Strausfeld and Nässel, 1981). This part of the brain includes the

TABLE 21.1. THE EFFECTS OF LESIONS IN THE MUSHROOM BODIES AND HOMOLOGOUS STRUCTURES

Animal	Type of Operation	Behavioral Effect	References
Crickets (*Gryllus*)	Extirpation of one MB	Transient increase of locomotor activity	Huber, 1952, 1960, 1962
	Extirpation of both MB	Strong transient increase of locomotor activity; loss of acoustic behavior	
	Lesions in and around MB	Elicitation of mating and rivalry behavior; sometimes long lasting stridulation	
	Electrocoagulation in and around MB	Walking; turning; tilting of head	Huber, 1959
Grasshoppers (*Gomphocerus*)	Extirpation of one MB	Transient increase of locomotor activity	Huber, 1955, 1962
	Extirpation of both MB or both calyxes	Strong transient increase of locomotor activity; loss of acoustic behavior	
	Lesions around calyx	Long lasting atypical stridulation	Elsner and Huber, 1969
Bees (*Apis*)	Electrocoagulation of MB parts	Circling; antennal dysfunction	Vowles, 1954
	Lesions in MB	Disinhibition of competitive reflexes; unable to fly	Howse, 1974
	Punctures and cuts in MB	Loss of memory and various reflex responses	Voskrenskaja, 1957; Turanskaja, 1973
	Punctures in calyx and α-lobes	Interference with short-term olfactory memory	Masuhr, 1976
Ants (*Formica*)	Electrocoagulation of MB parts	Circling; antennal dysfunction	Vowles, 1954
	Extensive cuts in and around MB	Loss of olfactory memory	Vowles, 1964a
Locusts (*Schistocerca*)	Calyx lesions	Atypical movements; atypical flight reflex	Howse, 1974
Silkworms (*Cecropia*)	Electrocoagulation in one pedunculus	Spinning of aberrant structures	van der Kloot and Williams, 1954
Dragonfly larvae (*Calopteryx*)	X-ray lesions in and around MB	Change of thresholds for visual behavior	Buchholtz, 1961
Cockroaches (*Periplaneta*)	Extirpation of both calyxes	Increase of odor thresholds; no long lasting changes of other behaviors	Drescher, 1960
Various crabs and lobsters; most experiments with *Panulirus*	Ablation of one or both eyestalks, including medulla terminalis	Antennal olfactory dysfunction; disturbance of feeding behavior; change in posture and escape behavior	Maynard et al., 1963, 1968, 1970; Hazlett, 1970

MB = mushroom body.

489

"hemiellipsoid glomerulus," a structure that is homologous with the MB of insects. The large-scale lesions in Crustacea are by far more extensive than the equivalent experiments in insects. This might explain long-term alterations of behavior. Yet, in some respects, for example, olfactory dysfunction, the experimental results in Crustacea point in the same direction as those in insects.

21.3.2. Electrical Stimulation

Electrical brain stimulation in insects was introduced by Huber (1957, 1959) as an additional tool for analyzing the function of specific neural structures. A rather simplistic hypothesis would assume that stimulation of specific brain areas leads to antagonistic behavioral effects, as compared with lesions. It is astonishing that at least in some respects this simple antagonism turned out to be true (Table 21.2). Electrical stimulation in the calyx area can inhibit locomotor activity; various song types can be elicited in and around the MB, whereas the opposite effects were observed after lesions (Table 21.1).

The most careful and systematic experiments with electrical stimulation were performed by Huber and co-workers, who analyzed stridulation, locomotion, and other behaviors in Orthoptera. The stimulation foci for eliciting stridulation in crickets and grasshoppers are distributed in the median protocerbrum. The distribution of these sites shows an accumulation around the lobes and calyxes of the MB (Huber, 1962; Otto, 1971; Wadepuhl, 1983). It is remarkable that in most cases, the behavioral effect of the stimulation develops with a long latency (sometimes minutes) and shows longlasting aftereffects. It is obvious that there are no specific foci for specific songs in the MB area, because transitions from one song type to another can be elicited by varying the parameters of stimulation (Otto, 1971).

Stimulation of the AGT, a major input to the MB from the antennal neuropil, can lead to rivalry behavior and the corresponding stridulation (Huber, 1960, 1962). This finding correlates well with the behavioral observation that antennal contact between two males initiates rivalry behavior. Stimulation of the AGT probably mimics sensory inputs to the MB. In crickets, antennal, acoustic, and several types of mechanic sensory information are transmitted via the AGT to the MB (Section 21.6.1).

The size of the stimulated brain area clearly depends on the type of the electrodes and the applied currents. Wadepuhl (1983) estimated that a volume of less than 100 μm in diameter is stimulated. This appears to be a realistic estimate which is supported by stimulation experiments with visual neurons in flies (Blondeau, 1981). Nevertheless, a stimulation focus of less than 100 μm represents a significant volume in the insect brain. This might explain the multiple behavioral effects found in insects.

The stimulation sites for locomotor behavior are widely distributed in the

TABLE 21.2. EFFECTS OF ELECTRICAL STIMULATION IN THE MUSHROOM BODIES

Animal	Site of Stimulation	Behavioral Effect	References
Crickets (*Gryllus*; *Acheta*)	Calyx of MB	Movements of antennae, head, palps; walking; turning; rivalry behavior; different song-types; inhibition of locomotion and singing	Huber, 1959, 1960, 1962, 1967; Otto, 1971
	Pedunculus and lobes of MB	Movements of head, antennae, palps; walking; turning; jumping; different song-types	Huber, 1959, 1960, 1962, 1967; Otto, 1971
	AGT	Rival song; rivalry behavior; inhibition of singing	Huber, 1960, 1962
Locusts (*Schistocerca*)	In and around MB	Arousal; walking; jumping; flying; inhibition of locomotion	Rowell, 1963
Grasshoppers (*Gomphocerus*)	In and around MB	Different song-types; locomotion	Wadepuhl, 1983
Bees (*Apis*)	In and around MB	Complex coordinated patterns of movement; inhibition of motor activity	Vowles, 1964b; Turanskaja, 1972

MB = mushroom body; AGT = antennoglomerular tract.

brains of different insect orders. Therefore, locomotor behavior is not exclusively elicited by structures related to the MB.

Recent immunocytochemical findings in the bee (Bicker et al., 1985) demonstrate that feedback neurons of the MB show GABA-like immunoreactivity (see also Section 21.6.3). This finding offers a very elegant interpretation of the inhibitory effects of stimulations in the MB. Massive stimulation of the feedback neurons in the lobes or calyxes could lead to a suppression of neural activity in the MB and consequently to inhibition of ongoing behavior.

Specific behaviors can be triggered or inhibited when neurons connected with the MB system are stimulated. The stimulations can affect the inputs of the system by mimicing incoming sensory information; they can act on the outputs in the lobes and on feedback neurons. In any case, the stimulation experiments do not provide us with means of analyzing signal processing within the MB system. The distribution of stimulation foci makes it improbable that intrinsic neurons were stimulated directly. Although coordinated movements can be elicited by MB stimulation, the experimental evidence shows that the coordination of these behaviors does not occur in the MB (e.g., Otto, 1971). The long lasting aftereffects of MB stimulation point to a characteristic feature of the neuropil, which will be discussed in Section 21.6.

21.4. EXPERIENCE AND SENSORY DEPRIVATION

Sensory deprivation during sensitive phases in the development of mammals, has drastic consequences for behavior, for the physiological responses of neurons, and for the neuroarchitecture of the brain (Wiesel and Hubel, 1963; Hubel and Wiesel, 1977). Similar studies with sensory deprivation in insects revealed comparable effects for different sensory modalities. Reduced sensory experience early in life can alter the physiological responses of receptors and interneurons (Matsumoto and Murphey, 1977; Bloom and Atwood, 1980, Masson and Arnold, 1984), the phototactic behavior, and the synapse frequency in photoreceptors (Hertel, 1982, 1983). These studies demonstrate that alterations due to sensory or social deprivation can be apparent in receptors and central interneurons.

In *Drosophila,* the gross structure of the MB is dependent on age, sex, and social and sensory experience of the animals (Technau, 1984). The number of fibers in the caudal pedunculus of the MB increases by about 15% during the first week of adult life; after 3–4 weeks, the fiber number decreases again. Female flies have significantly more fibers in the pedunculus than males; different wild-type strains also differ in the absolute number of MB fibers. Technau (1984) deprived flies for 3 weeks of visual, olfactory, or social stimuli, and found that olfactory deprivation or social isolation significantly reduce the number of fibers in the pedunculus (Fig. 21.2). Visual deprivation alone did not affect the structure of the MB. This important experimental finding underlines the minor significance of the visual system for the MB in flies. A combination of olfactory, visual, and social deprivation augments the effects. The observed alteration within the MB can be attributed to a decrease in the number of intrinsic neurons after deprivation.

The analysis of the neuroanatomical structure of the MB after different experiences of the animals, seems to be fruitful approach to study the function of this brain area. At the moment, there are only a few other comparable studies that also indicate that the neuroarchitecture of the MB is dependent on age and experience. In honeybees, Coss et al. (1980) found that the structure of spines of intrinsic MB neurons is dependent on the age and experience of the animal. Foraging bees have spines with larger profile area and shorter stems, as compared with newly emerged and nurse bees. The morphology of a subgroup of spines changes rapidly after the first orientation flight of the bees (Coss and Brandon, 1982). Although these studies do not allow a

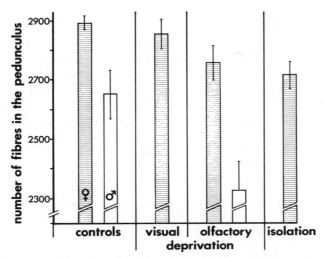

Figure 21.2. The effects of deprivation on the number of fibers in the pedunculus of *Drosophila* (wild-type Kapelle). (after Technau, 1984; with permission of the author.)

conclusion on the exact sensory experiences leading to the morphological changes, they are an important starting point for further experiments.

Sensory experience determines the structure of the MB; on the other hand, there can exist behavioral differences in animals of the same species, which differ in the structure of the MB. Studying the foraging efficiency of *Formica rufa,* Bernstein and Bernstein (1969) found that efficient foragers have larger heads, larger receptor areas, and larger calyxes of the MB. Many different genetic and environmental factors probably influence the size of the MB in different individuals. Therefore, at the moment, it remains an open question whether or not the size of this brain area really determines the differences in behavior. Differences in the number of receptors, on the other hand, might explain differences in behavior and might result in the size variations of the MB.

21.5. LEARNING AND MEMORY

Several hypotheses about the function of the MB suggest that this neuropil serves as a substrate for visual and olfactory memory (e.g., Howse, 1974). In fact, there is increasing evidence that the MBs play a decisive role in *olfactory* learning and memory formation, but, so far, there is little experimental evidence for a contribution of the MB in visual learning.

In the first experiments in bees and ants, extensive lesions were used to test various areas of MBs for their possible role in retention of a learned signal (Voskresenskaja, 1957; Vowles, 1964a; Turanskaja, 1973). Although the experimental approaches and species differed in these investigations, the results are comparable. Bilateral lesions of parts of the MB lead to significant deficits in reproducing a learned visual or olfactory response. Unilateral lesions have no or only minor effects. The extensive lesions of these first investigations probably also resulted in damages of visual interneurons and of large parts of the median protocerebrum. Therefore, it is not surprising that olfactory and visual reactions were affected. Later lesion experiments in bees were better controlled and showed that punctures in the calyces of α-lobes of the MB interrupt long-term olfactory memory formation (Masuhr, 1976; Erber et al., 1980).

21.5.1. Olfactory Short-Term Memory

Honeybees can be conditioned to odors in the laboratory. The conditioned reaction is proboscis extension, the unconditioned stimulus is sugar-water applied to one antenna. This learning paradigm, originally developed in another form by Kuwabara (1957), is very effective. A large percentage of bees responds after one reward with the conditioned reaction. Experiments with free-flying bees had shown that short-term visual memory can be interrupted by cooling the bees shortly after one-trial learning (Erber, 1976). Masuhr (1976) developed a similar paradigm for the laboratory in which small parts of the bee brain were cooled with litle metal probes after one-trial olfactory learning. Figure 21.3. shows some of these experiments.

Bilateral cooling of the antennal lobes, the α-lobes, or calyces of the MB shortly after one-trial conditioning leads to a significant reduction of the conditioned response in later tests. The susceptibility for interference lasts about 2 min in the antennal lobes, 5 min in the α-lobes, and 7 min in the calyces of the MB. The time course of memory impairment induced by cooling the whole animal after one-trial olfactory conditioning is similar to the time course found for cooling the α-lobes. Moreover, memory impairment of free-flying bees conditioned once to a visual signal also resembles these two time dependencies.

The side specificity of information processing in the MB can also be studied with this experimental approach. Bees were allowed to learn with only one antenna, and memory impairment was studied by cooling parts of the ipsi- or contralateral brain. The experimental results indicate that information processing after olfactory learning involves only the ipsilateral antennal lobes, whereas also the contralateral MB, especially the α-lobes, take part in processing the information learned with one antenna.

Obviously, neural activity in the antennal lobes and in the MB has to remain undisturbed for several minutes after one-trial learning to ensure storage of a learned olfactory signal. Whether the MB or antennal lobes are also the *sites* where the learned signals are stored, remains an open question. Although the time courses of memory impairment in the bee are similar for

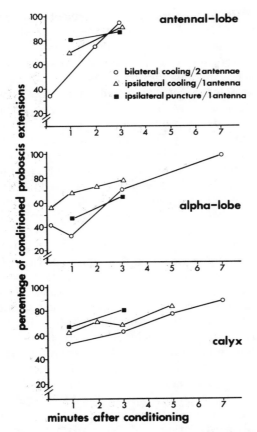

Figure 21.3. Interruption of olfactory memory formation in the honeybee. The ordinates show the percentage of conditioned proboscis extensions after cooling parts of the MB or after punctures; the abscissae show the time of treatment after one-trial learning. (after Erber et al., 1980; for details see text.)

visual and olfactory learning, there is, so far, no experimental proof that the MBs are also involved in visual learning.

21.5.2. Learning in Mushroom Body Mutants

Mutations in *Drosophila* can affect learning and memory. The mutations can be "biochemical" or "structural." The biochemical single-gene mutations in *Drosophila* affect subpopulations of enzymes in the whole animal (Dudai, 1985). Therefore, these mutants do not contribute to an understanding of the functional role of the MB. Structural mutants of the brain in *Drosophila,* on the other hand, provide a direct and very elegant access to the function of different brain areas (Heisenberg, 1980; Fischbach and Heisenberg, 1984;

Heisenberg et al., 1985). A considerable number of structural brain mutants has been isolated by Heisenberg and collaborators (1985). In our context, the MB mutants are the most interesting. Two MB mutants have been studied extensively in different types of conditioned and spontaneous behavior. In extreme forms of the mutant, *mushroom bodies deranged* (MBD), no pedunculus, α-, β-, or γ-lobes can be detected in adult flies. The antennal lobes, the AGT, the optic lobes, and the central complex appear normal. The MB defects in MBD flies develop during metamorphosis. Another MB mutant is *mushroom body miniature* (MBM), which displays a very interesting sexual dimorphism. In some females, no calyxes, peduncles, or lobes are apparent; in many others, the structural defects are less pronounced. In contrast to females, male flies have well-developed MBs. The degeneration of the MB occurs between the second and third larval instar. The sexual dimorphism of this mutant opens the possibility to test for behavioral differences in both genders.

The general behavior of the mutants does not differ from the wild-type flies. Motor coordination appears normal, and male MBD flies display normal courtship and copulation behavior. The mutant flies can distinguish odors; chemotaxis, odor-wind orientation, and osmotropotaxis appear undisturbed. The mutants were tested in several olfactory and visual learning paradigms. The MBD mutant shows no olfactory learning in an electroshock paradigm and only small learning indexes for two other olfactory paradigms. In the MBM mutant, female flies, which have structural MB defects, display small learning scores, whereas males learn much better. Different types of visual learning appear to be normal in the mutant MBD.

Additional information on the role of the MB is gained from learning experiments with larvae (Heisenberg et al., 1985). Larvae of the late third instar were tested; MBD larvae and male MBM larvae with structurally normal MBs show high olfactory learning scores, whereas female MBM larvae, which already have MB defects, do not learn to associate an odor. Olfactory learning is reduced in three other MB mutants; interestingly also four mutants with defects of the central complex are deficient in olfactory learning.

This chain of evidence allows several interpretations. The nervous system of the mutants could compensate for most of the structural defects, which would explain why the animals appear normal in most behaviors, except for olfactory learning. This would require an enormous and perhaps rather improbable degree of neural plasticity in the brain. Alternatively, the MBs or neurons associated with them are necessary sites for olfactory learning and memory. In this case, various hypotheses about the function of the MB, for example, evaluation of olfactory signals, visual memory, coordination of behavioral sequences, etc., have to be rejected. In any case, there can be no doubt that intact MBs are necessary for olfactory learning. Perhaps other behavioral deficiencies in the mutants can only be detected with more refined test methods, analyzing more complex forms of orientation behavior.

21.6. ELECTROPHYSIOLOGY OF MUSHROOM BODY NEURONS

The advance of electrophysiological techniques in the past decade, especially the combination of intracellular recording with staining of single neurons, significantly broadened our knowledge about signal processing in the MBs of different insect orders. Extrinsic neurons with arborizations in the MB have been recorded in *Acheta, Apis,* and *Periplaneta.* Some physiological information about olfactory interneurons, projecting from the antennal lobe probably to the calyxes, exists also for the moth, *Manduca* (Matsumoto and Hildebrand, 1981).

It seems reasonable to divide the extrinsic MB neurons into three groups: presumable input neurons projecting from sensory neuropils to the calyxes, presumable output neurons projecting from the lobes of the MB to other parts of the brain, and presumable feedback neurons connecting the lobes of the MB with the calyxes.

21.6.1. Input Neurons

Eletrophysiological recordings of summed potentials in the MB during electrical stimulation of the antennal nerve in the cockroach gave the first physiological indication that the calyces serve as inputs and the lobes as outputs of the system (Maynard, 1956, 1967). Neuroanatomical analyses in different insect orders confirm this finding.

Physiologically and neuroanatomically, the inputs to the calyces differ in the insect orders. The most massive input to the calyces arises from the antennoglomerular tract (AGT), which is also termed antennocerebral tract (ACT) or tractus olfacorioglobularis (TOG). We use here the term AGT, because the neuroanatomical term TOG implicates a physiological function, which is misleading as the recordings have shown.

Neurons of the AGT innervate one glomerulus in the antennal lobe. Only one exception from this rule, so far, has been found (Ernst and Boeckh, 1983). The AGT neurons innervate the calyces *and* project to the lateral protocerebral lobe (Fig. 21.4). In *Periplaneta,* two classes of these neurons respond to female pheromones. These cells in males project from the "macroglomerulus" in the antennal lobe to the calyces and the protocerebrum (Fig. 21.4). Other AGT neurons in the cockroach respond to presumable food odors. These cells innervate other glomeruli, not the macroglomerulus, in the antennal lobe (Selzer, 1979; Burrows et al., 1982; Ernst and Boeckh, 1983; Boeckh and Selsam, 1984; Boeckh et al., 1984; Schaller-Selzer, 1984). Besides these odor specific neurons, multimodal cells were found in the cockroach. Some of these neurons react to mechanical antennal inputs and odors in a differential way (Ernst and Boeckh, 1983), other cells also signal temperature shifts (Boeckh et al., 1984).

In the bee and in the cricket, most AGT neurons display multimodal characteristics (Fig. 21.5; Homberg, 1984; Schildberger, 1984), responding

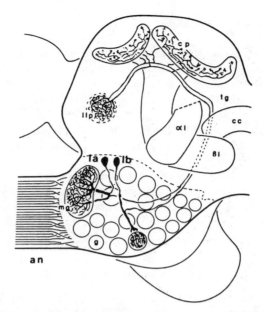

Figure 21.4. The projections of two types of neurons in *Periplaneta* from the antennal lobes to the calyces and protocerebrum (fronto-horizontal section). an = antennal nerve; αl = α-lobe; βl = β-lobe; cc = central body; cp = corpora pedunculata; g = glomerulus; llp = lateral protocerebral lobe; mg = macroglomerulus; tg = tractus olfactorioglobularis (termed "AGT" in the text). (From Ernst and Boeckh, 1983; with permission of the authors.)

to olfactory, mechanical, and chemical stimuli applied to the ipsilateral antenna. Some of these cells in bees also respond to mechanical stimuli to other parts of the body and in the cricket even to acoustic stimuli and stimulation of the cerci (Homberg, 1984; Schildberger, 1984). Apparently, the MBs receive various sensory information via the AGT. "Labeled lines" for pheromones in the cockroach and multimodal neurons in the bee and cricket represent the two ends of the input spectrum. The signaling of antennal stimuli in these cells seems to be limited to ipsilateral information.

In the honeybee, the calyces are connected by several tracts with the optic lobes (Mobbs, 1984). Neurons of the anterior-superior optic tract (ASOT) respond to light flashes (Erber, 1982); some of these cells are also movement-sensitive and respond selectively to the direction of a moving light stimulus (Gronenberg, 1984). Other neurons of the protocerebrum in bees and crickets projecting into the calyces do not belong to well-defined tracts. These cells show differing and often multimodal response characteristics.

In addition to the lesion experiments in decapod crustaceans, electrophysiological recordings of ascending neurons in the eyestalk of *Panulirus* enable a first, rather crude, comparison of information processing in insects

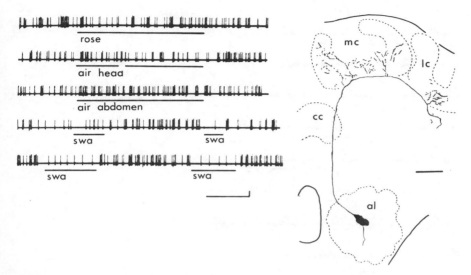

Figure 21.5. Electrophysiological recording and frontal reconstruction of AGT neuron in the honeybee. Rose = rose oil; air head = air puff applied to the head; air abdomen = air puff applied to the abdomen; swa = sugar water applied to the antenna. Calibration of recordings: horizontal bar = 1s; vertical bar = 10 mV. al = antennal lobe; cc = central body; lc = lateral calyx; mc = median calyx. Horizontal bar in reconstruction = 100 μm. (From Homberg, 1984.)

and crustaceans (Ache and Fuzessery, 1979). Extracellular recordings in the olfactory-globular tract (OGT; which is considered to be homologous to the AGT in insects) show in the spiny lobster that, similar to insects, ipsilateral olfactory information is transmitted to the neuropil of the eyestalks. Ascending multimodal neurons in this tract, which respond to visual and tactile stimulation, were only responsive to contralateral chemostimulation.

21.6.2 Output Neurons

Electrophysiological recordings and stainings of single neurons projecting from the lobes of the MB to the protocerebrum and the calyxes exist for *Acheta* and *Apis*. The distinction between "output" and "feedback" neurons is sometimes arbitrary, because feedback neurons can also project to the protocerebrum (Fig. 21.7).

In bees neurons with arborizations in the lobes of the MB often have more complex, multimodal characteristics than calycal input neurons (Fig. 21.6). These cells can respond also to light stimuli in addition to antennal or other mechanical stimuli (Homberg, 1984; Gronenberg, 1984). In the MB system, visual and other modalities are combined. This combination can either be due to the intrinsic properties of the MB system or it can be due to the integrative characteristics of the extrinsic cells.

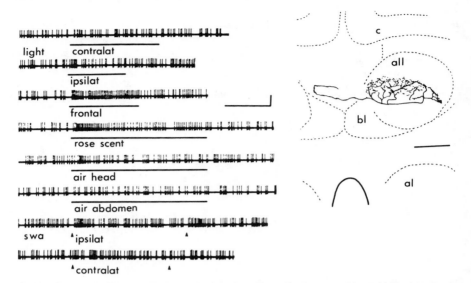

Figure 21.6. Electrophysiological recording and frontal reconstruction of MB output neuron in the honeybee. Stimuli, abbreviations, and calibrations as in Figure 21.5. all = α-lobe; bl = β-lobe; c = calyx. (From Homberg, 1984.)

In the cricket the electrophysiological results of extrinsic neurons are very similar to those in bees (Schildberger, 1981, 1984). The neurons show a high degree of multisensory responses; in some of the cells with projections at the junction between the two lobes, a significant increase of the signal-to-noise ratio, compared with calycal input neurons, can be found.

Long lasting aftereffects of extrinsic MB neurons were already found by Vowles (1964b) in extracellular recordings in the bee. These aftereffects (Figs. 21.7 and 21.8), which outlast the stimulus by many seconds, sometimes even minutes, are characteristic of many neurons with projections in the lobes of the MB (Erber, 1978, 1982; Schildberger, 1981, 1984; Gronenberg, 1984). Aftereffects are an important feature of neural integration in the MB. In the cricket, Schildberger (1984) found an extrinsic α-lobe neuron, which showed a discharge after cercal stimulation; this effect lasted up to 1 min. Only during this phase did the neuron respond to acoustic and light stimuli. Output neurons of the MB can change their response properties during the experiment, some neurons are first excited then inhibited by sensory stimuli (Gronenberg, 1984).

21.6.3. Feedback Neurons

Extrinsic neurons with projections in the lobes and the calyces probably have a feedback function within the MB by connecting the outputs and inputs of the system. Some of these cells have been characterized in *Apis* (Homberg

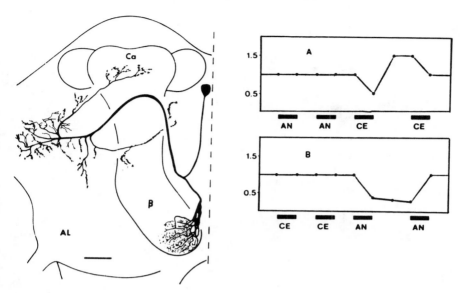

Figure 21.7. Reconstruction of extrinsic MB neuron in *Acheta* with electrophysiological responses. Left: AL = antennal lobe; β = β-lobe; Ca = calyx; calibration bar = 50 μm. Right: ordinates: quotient of actual action potential frequency and mean spontaneous frequency at the onset and offset of the stimuli. AN = mechanical stimulus to the ipsilateral antenna; CE = mechanical stimulus to the ipsilateral cercus; stimulus duration = 1 sec. (From Schildberger, 1981, with permission of the author.)

and Erber, 1979; Erber, 1984; Gronenberg, 1984) and *Acheta* (Schildberger, 1981, 1984). Similar to other extrinsic neurons, they display multimodal responses often associated with aftereffects (Fig. 21.8). An example that shows the responses and characteristics of a feedback neuron in *Acheta* is illustrated in Figure 21.7. In this cell, cercal stimulation, *after* antennal stimulation, leads to inhibition, followed by an excitatory aftereffect. Antennal stimulation, *after* cercal stimulation, leads to a lasting decrease in the frequency, which is reset by another antennal stimulus. The delay between the stimuli had to be less than 5 sec. The temporal succession of stimuli can determine the responses of MB neurons.

For the bee it has been hypothesized that the horizontal "bands" within the α-lobes represent spatially separated regions for different sensory modalities (Mobbs, 1982). Because neurons with dendrites in different bands of the α-lobe respond to multimodal stimulation, there is no support for this hypothesis in electrophysiological recordings (Gronenberg, 1984).

21.6.4. Intrinsic Neurons

Due to their small diameter, it is very difficult to record from single intrinsic MB neurons. There exists a small number of recordings from those cells in

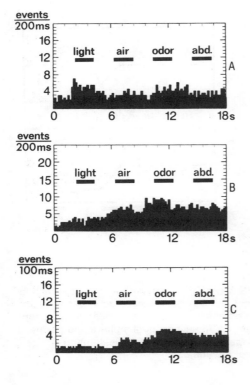

Figure 21.8. Peristimulus time histograms of actionpotentials in three feedback neurons of the honeybee. Light = light flash applied to the ipsilateral compound eye; air = mechanical stimulation of antennae; odor = olfactory stimulation with mixture of lavender, geraniol, and rose oil; abd = mechanical stimulation of abdomen. Neurons B and C show pronounced aftereffects. (After Gronenberg, 1984.)

Apis (Erber, personal observation, Gronenberg, 1984). In the calycal region intrinsic cells respond with graded potentials to antennal stimuli. Often the responses show pronounced aftereffects. Iontophoretic injection of the fluorescent dye Lucifer yellow always stains 3–40 adjacent intrinsic neurons. This dye coupling could be an indication of electrical synapses between these cells (Stewart, 1978). In the pedunculus, intrinsic cells respond with action potentials (Gronenberg, 1984).

With sophisticated extracellular recordings and a current source density analysis of evoked potentials in the bee MB system, Kaulen et al. (1984) were able to show that intrinsic neurons with dendritic fields in the lip and basal ring of the calyxes differ in their response properties from intrinsic cells from the collar. Synaptic potentials between intrinsic cells from the lip and basal ring appear to be inhibitory, whereas synaptic interactions between intrinsic cells from the collar are excitatory.

21.7. SYNTHESIS OF BEHAVIORAL AND PHYSIOLOGICAL FINDINGS

The existing behavioral and physiological experiments enable us to reject some hypotheses concerning the functional role of the MB and allow the formulation of new specific hypotheses and models, which can be tested with different experimental approaches. There is, for instance, no experimental evidence that the MBs are necessary for the generation of behavioral sequences. It is more than a semantic question whether the MBs serve "higher functions" in insects. In our view, the experimental data demonstrate that they are necessary for a number of basic operations in the nervous system.

Obviously, the MBs are involved in the processing of olfactory information. Yet, the detection of odors and other forms of olfactory behavior are also apparent in animals with defects in the MB system. The MBs represent a *parallel*, not a *serial*, system of the central olfactory pathway. Therefore, specific deficiencies of olfactory behavior can probably be detected only with more refined behavioral tests, which have to include temporal sequences of olfactory stimuli.

The differences of the physiological responses and of the neuroanatomical structure of the MB inputs in different insect orders, clearly demonstrate that the MB serves different sensory modalities in different species. Although a convergence of sensory modalities occurs in the MB and although specific stimulus combinations can be detected by the extrinsic neurons in the lobes, it does not seem to be the primary function of this structure to combine different modalities. Convergence of sensory modalities is apparent already in numerous inputs of the system.

The physiological and behavioral experiments support the hypothesis that an eminent function of the MB is the *temporal* integration and evaluation of sensory signals. Different forms of temporal integration can be distinguished. The responses of extrinsic neurons to specific successions of stimuli represent a logical "if–then" operation combined with temporal concatenation. The generation of aftereffects, following single stimuli, is a form of sensory memory, and the anatomical distribution of extrinsic cells can result in the detection of temporal stimulus changes.

The significance of the MB for signal processing in the time domain is illustrated by the olfactory learning experiments, by the lesion and stimulation experiments in crickets, by the occurrence of neuronal aftereffects, and by specific neuronal responses to temporal sequences of stimuli. In these respects, the experimental results in insects are similar to the physiological and behavioral findings in the hippocampus and prefrontal cortex of mammals (Seifert, 1983; Fuster, 1984).

The neuronal circuitry of the MB supports the hypothesis that arrays of extrinsic neurons with narrow dendritic arborizations perpendicular to the intrinsic neurons are well suited to evaluate temporal stimulus changes (Schürmann, 1974; Fig. 21.9A). In the bee, these cells are spaced along the pedunculus; their axons project to a common target area in the contralateral protocerebrum (Homberg, 1984). During stimulation these neurons are activated in succession; temporal changes of stimuli could be detected by comparing the activity of several of these cells.

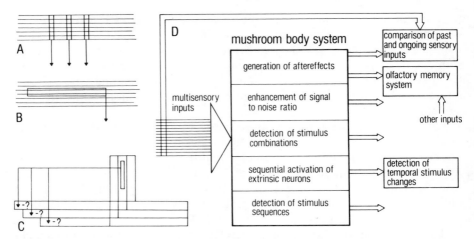

Figure 21.9. Basic neuronal circuits of extrinsic MB neurons and hypothetical scheme of MB functions; for details see text.

At present we have only estimates of the conduction velocity within the MB system. Schürmann (1974) estimated a temporal delay of 3–5 msec within the intrinsic neurons from the calyces to the lobes. The actual delays measured by comparing the latencies of calycal input and output neurons in the lobes lie in the range of 25 msec (Schildberger, 1984). Activation of intrinsic cells is a slow process (Erber, personal observation). Depending on the thresholds of extrinsic neurons, the temporal delays in the MB can be much longer than the estimated conduction delay of the intrinsic neurons.

Extrinsic neurons with dendritic projections parallel to the intrinsic cells (Mobbs, 1982), will be activated by the same intrinsic elements for a considerable time (Fig. 21.9B). Therefore, these neurons have only a very limited temporal resolution. The spacing of many synapses along the intrinsic and extrinsic cells represents a very effective amplification mechanism. One consequence of these neuronal circuits could be the observed enhancement of the signal-to-noise ratio. Another, perhaps more important consequence, could be pre- or postsynaptic potentiation, which, in turn, would explain the neuronal aftereffects observed in extrinsic cells.

Presently, we cannot determine whether the aftereffects are a property of the intrinsic or the extrinsic neurons. But as aftereffects are not observable in all extrinsic cells, they are probably a characteristic of specialized extrinsic neurons.

Recent immunological studies in bees make it probable that the feedback neurons in the MB system are inhibitory (Bicker et al., 1985). Provided this interpretation is correct, feedback neurons, innervating specific layers of intrinsic neurons in the lobes, could inhibit other intrinsic cells (Fig. 21.9C).

Electrophysiological analyses in bees have shown that the feedback neurons project to various areas in the calyces (Gronenberg, 1984) and not only to the corresponding intrinsic cells innervated in the lobes. Aftereffects in these cells are rather frequent. If inhibition and aftereffects are combined, then the activation of specific subgroups of intrinsic cells can inhibit the activity of other intrinsic groups. In this way, defined states of intrinsic activation can be preserved for a considerable time.

Several hypothetical functions of the MB system, derived from our present knowledge of this neuropil are summarized in Figure 21.9D. The mechanisms of temporal integration and aftereffects could be necessary prerequisites for olfactory learning and certain types of orientation. Parallel signal processing in the MB and the presence of aftereffects, which in many cases occur after antennal stimuli, could be the basis for a comparison of past and present sensory inputs. Such a comparison is necessary for klinotactic orientation.

In insects the orientation toward a distant odor source is controlled by nonchemical cues. The chemical stimulus, in most cases a pheromone, acts only as a trigger or conditioning stimulus, which starts anemotactic flight (Rust et al., 1976; Kennedy, 1978). In this type of orientation, remarkable aftereffects lasting many seconds are observable when the chemical stimulus is turned off (Marsh et al., 1978). In addition, chemical cues can modulate the responses to visual targets. The known mechanisms of signal processing in the MBs provide neural substrates for all of these observed behaviors.

At the moment, it is not clear whether or not the MBs are also the neural site of olfactory memory. It is possible that the MBs provide only one of the necessary conditions for olfactory memory formation in the form of lasting aftereffects. In this case, olfactory memory is located somewhere else in the insect brain, perhaps even in the antennal lobes.

The hypothetical functions of the MB system can be tested in various ways. An important task of future analyses is the opening of the multiple feedback loops of the system, which is necessary for a system analysis. Neurotoxins, pharmacological methods, and also the "old fashioned" lesions are the appropriate tools for such an analysis. A second necessary step of future analyses is the development of behavioral experiments, which show reliable responses of the animals when temporal successions or combinations of stimuli are applied. Associative learning is only one form of such experiments. In this context, it would be very illustrative to use reversible blocking of the MB during pheromone-triggered orientation.

The hypotheses about the function of the MB started with speculations about the intelligence of the insects. These vague ideas can now be reduced to a few basic functions every higher nervous system has to perform: the concatenation and comparison and integration of past and present multisensory information. We think that the MBs in arthropods solve these problems.

21.8. SUMMARY

The MB system in insects is involved in many different behaviors as revealed by lesion and stimulation experiments. Sensory deprivation and experience have consequences for the neuroarchitecture of the MB. Intact MBs are necessary for olfactory learning and memory formation. The calyces of the MB receive olfactory, mechanical, and other nonantennal inputs. Outputs and feedback neurons of the system are often multimodal; many of these neurons display pronounced aftereffects. The behavioral and physiological experimental data are summarized in a model of the MB. The hypothesis is proposed that the MBs serve the temporal integration, concatenation, and comparison of sensory information.

ACKNOWLEDGMENTS

The analysis of the MB in bees was supported by the DFG (grants Er 79/2-1; 79/2-2). We thank the following people: A. Elepfandt for the translation of two Russian publications; K. Schildberger, G. M. Technau, K. D. Ernst, J. Boeckh, and Springer-Verlag for the permission to use published data; H. Markl, R. Müller-Tautz, and J. Tautz for their hospitality and help while J. Erber was working in the library of the University of Konstanz; and W. Roloff and J. Buchholz for the production of drawings and photographs.

REFERENCES

Ache, B. and Z. Fuzessery. 1979. Chemosensory integration in the spiny lobster: Ascending activity in the olfactory-globular tract. *J. Comp. Physiol.* **130**: 63–69.

Bernstein, S. and R. A. Bernstein. 1969. Relationships between foraging efficiency and the size of the head and component brain and sensory structures in the red wood ant. *Brain Res.* **16**: 85–104.

Bethe, A. 1897. Vergleichende Untersuchungen über die Funktion des Zentralnervensystems der Arthropoden. *Pflüg. Arch.* **68**: 449–545.

Bicker, G., S. Schäfer, and T. G. Kingan. 1985. Mushroom body feedback interneurons in the honeybee show GABA-like immunoreactivity. *Brain Res.* **360**: 394–397.

Blondeau, J. 1981. Electrically evoked course control in the fly *Calliphora erythocephala. J. Exp. Biol.* **92**: 143–153.

Bloom, J. W. and H. L. Atwood. 1980. Effects of sensory experience on the responsiveness of the locust descending contralateral movement detector neuron. *J. Comp. Physiol.* **135**: 191–199.

Boeckh, J. and P. Selsam. 1984. Quantitative investigation of the odour specificity of central olfactory neurons in the American cockroach. *Chem. Sens.* **9**: 369–380.

Boeckh, J., K. D. Ernst, H. Sass, and U. Waldow. 1984. Anatomical and physiological characteristics of individual neurons in the central antennal pathway of insects. *J. Insect Physiol.* **30:** 15–26.

Buchholtz, Ch. 1961. Eine verhaltensphysiologische Analyse der Beutefanghandlung von *Calopteryx splendens* Harr. (Odonata) unter Berücksichtigung des optischen AAM nach partieller Röntgenbestrahlung des Protocerebrums. *Verhdl. Dtsch. Zool. Ges. Saarbrücken Zool. Anz. Suppl.* **25:** 401–412.

Bullock, T. H. and G. A. Horridge. 1965. *Structure and Function in the Nervous System of Invertebrates.* Freeman, San Francisco.

Burrows, M., J. Boeckh, and J. Esslen. 1982. Physiological and morphological properties of interneurons in the deutocerebrum of male cockroaches which respond to female pheromone. *J. Comp. Physiol.* **145:** 447–457.

Coss, R. G. and J. G. Brandon. 1982. Rapid changes in dendritic spine morphology during the honeybees first orientation flight, pp. 338–342. In M. D. Breed, C. D. Michener, and H. E. Evans (eds.), *The Biology of Social Insects.* Westview Press, Boulder, CO.

Coss, R. G., J. G. Brandon, and A. Globus. 1980. Change in morphology of dendritic spines on honeybee calycal interneurons associated with cummulative nursing and foraging experience. *Brain Res.* **192:** 49–59.

Drescher, W. 1960. Regenerationsversuche am Gehirn von *Periplaneta americana* unter Berücksichtigung von Verhaltensänderung und Neurosekretion. *Z. Morphol. Ökol. Tiere* **48:** 576–649.

Dudai, J. 1985. Genes, enzymes and learning in *Drosophila. Trends Neurosci.* **8:** 18–21.

Dujardin, F. 1850. Memoires sur le systeme nerveux des insectes. *Ann. Sci. Nat. (Zool.)* **14:** 195–205.

Elsner, N. and F. Huber. 1969. Die Organisation des Werbegesanges der Heuschrecke *Gomphocerippus rufus L.* in Abhängigkeit von zentralen und peripheren Bedingungen. *Z. Vgl. Physiol.* **65:** 389–423.

Erber, J. 1976. Retrograde amnesia in honeybees (*Apis mellifica carnica*). *J. Comp. Physiol. Psychol.* **90:** 41–46.

Erber, J. 1978. Response characteristics and aftereffects of multimodal neurons in the mushroom body area of the honeybee. *Physiol. Entomol.* **3:** 77–89.

Erber, J. 1982. Electrophysiological analysis of central neurons in the bee and correlations with behavior, pp. 343–346. In M. D. Breed, C. D. Michener, and H. E. Evans (eds.), *The Biology of Social Insects.* Westview Press, Boulder CO.

Erber, J. 1984. Response changes of single neurons during learning in the honeybee, pp. 275–285. In D. L. Alkon and J. Farley (eds.), *Primary Neural Substrates of Learning and Behavioral Change.* Cambridge University Press, Cambridge.

Erber, J., T. Masuhr, and R. Menzel. 1980. Localization of short-term memory in the brain of the bee, *Apis mellifica. Physiol. Entomol.* **5:** 343–358.

Ernst, K. D. and J. Boeckh. 1983. A neuroanatomical study on the organization of the central antennal pathways in insects. *Cell Tissue Res.* **229:** 1–22.

Farkas, S. R., H. H. Shorey, and L. K. Gaston. 1974. Sex pheromones of Lepidoptera. Influence of pheromone concentration and visual cues on aerial odortrail following by males of Pectinophora. *Ann. Entomol. Soc. Am.* **67:** 633–638.

Fischbach, K. F. and M. Heisenberg. 1984. Neurogenetics and behaviour in insects. *J. Exp. Biol.* **112**: 65–93.

Flögel, J. H. L. 1878. Über den einheitlichen Bau des Gehirns in den verschiedenen Insekten-Ordnungen. *Z. Wiss. Zool. Suppl.* **30**: 556–592.

Fuster, J. M. 1984. Behavioral electrophysiology of the prefrontal cortex. *Trends Neurosci.* **7**: 408–414.

Gronenberg, W. 1984. Das Protocerebrum der Honigbiene im Bereich des Pilzkörpers—eine neurophysiologisch-anatomische Charakterisierung. Doctoral dissertation, Freie Universität, Berlin.

Hanström, B. 1925. The olfactory centers in crustaceans. *J. Comp. Neurol.* **38**: 221–250.

Hanström, B. 1928. *Vergleichende Anatomie des Nervensystems der wirbellosen Tiere.* Springer-Verlag, Berlin, Heidelberg, New York.

Hazlett, B. 1971. Non-visual functions of crustacean eyestalk ganglia. *Z. Vgl. Physiol.* **71**: 1–13.

Heisenberg, M. 1980. Mutants of brain structure and function: What is the significance of the mushroom bodies for behavior, pp. 373–390. In O. Siddiqui, P. Babu, L. M. Hall, and J. C. Hall (eds.), *Development and Biology of Drosophila.* Plenum, New York, London.

Heisenberg, M. and R. Wolf. 1984. Vision in *Drosophila*. Genetics of microbehavior. Springer-Verlag, Berlin, Heidelberg, New York.

Heisenberg, M., A. Borst, S. Wagner, and D. Byers. 1985. *Drosophila* mushroom body mutants are deficient in olfactory learning. *J. Neurogenet.* **2**: 1–30.

Hertel, H. 1982. The effect of spectral light deprivation on the spectral sensitivity of the honeybee. *J. Comp. Physiol.* **147**: 365–369.

Hertel, H. 1983. Change of synapse frequency in certain photoreceptors of the honeybee after chromatic deprivation. *J. Comp. Physiol.* **151**: 477–482.

Homberg, U. 1984. Processing of antennal information in extrinsic mushroom body neurons in the bee brain. *J. Comp. Physiol.* **154**: 825–836.

Homberg, U. and J. Erber. 1979. Response characteristics and identification of extrinsic mushroom body neurons of the bee. *Z. Naturforsch.* **34c**: 612–615.

Howse, P. E. 1974. Design and function in the insect brain, pp. 180–194. In L. Barton-Browne (ed.), *Experimental Analysis of Insect Behaviour.* Springer-Verlag, Berlin, Heidelberg, New York.

Howse, P. E. 1975. Brain structure and behavior in insects. *Annu. Rev. Entomol.* **20**: 359–379.

Hubel, D. H. and T. N. Wiesel. 1977. Ferrier Lecture: Functional architecture of macaque monkey visual cortex *Proc. R. Soc. London B. Biol. Sci.* **198**: 1–59.

Huber, F. 1952. Verhaltensstudien am Männchen der Feldgrille nach Eingriffen am Zentralnervensystem. *Verh. Dtsch. Zool. Ges. Freiburg Zool. Anz. Suppl.* **17**: 138–149.

Huber, F. 1955. Über die Funktion der Pilzkörper (Corpora pedunculata) beim Gesang der Keulenheuschrecke *Gomphocerus rufus* L. (Acrididae). *Naturwissenschaften* **20**: 566–567.

Huber, F. 1957. Elektrische Reizung des Insektengehirns mit einem Impuls- und Rechteckgenerator. *Industrie-Elektronik* **2**: 17–20.

Huber, F. 1959. Auslösung von Bewegungsmustern durch elektrische Reizung des Oberschlundganglions bei Orthopteren (Saltatoria: Gryllidae, Acridiidae). *Zool. Anz. (Suppl.)* **23**: 248–269.

Huber, F. 1960. Untersuchungen über die Funktion des Zentralnervensystems und insbesondere des Gehirns bei der Fortbewegung und der Lauterzeugung der Grillen. *Z. Vgl. Physiol.* **44**: 60–132.

Huber, F. 1962. Lokalisation und Plastizität im Zentralnervensystem der Tiere. *Zool. Anz (Suppl.)* **26**: 200–267.

Huber, F. 1967. Central control of movements and behavior of invertebrates, pp. 333–351. In C. A. G. Wiersma (ed.), *Invertebrate Nervous Systems,* University of Chicago Press, Chicago, London.

Kaulen, P., J. Erber, and P. Mobbs. 1984. Current source-density analysis in the mushroom bodies of the honeybee (*Apis mellifera carnica*). *J. Comp. Physiol. A* **154**: 569–582.

Kennedy, J. S. 1978. The concepts of olfactory "arrestment" and "attraction." *Physiol. Entomol.* **3**: 91–98.

van der Klott, W. G. and C. M. Williams. 1954. Cocoon construction by the *Cecropia* silkworm. III. The alteration of spinning behavior by chemical and surgical techniques. *Behaviour* **6**: 233–255.

Kuwabara, M. 1957. Bildung des bedingten Reflexes von Pavlovs Typus bei der Honigbiene (*Apis mellifica*). *J. Fac. Sci. Hokkaido Univ. Ser. VI Zool.* **13**: 458–464.

Marsh, D., I. S. Kennedy, and A. R. Ludlow. 1978. An analysis of anemotactic zigzagging flight in male moths stimulated by pheromone. *Physiol. Entomol.* **3**: 221–240.

Masson, C. and G. Arnold. 1984. Ontogeny, maturation and plasticity of the olfactory system in the workerbee. *J. Insect Physiol.* **30**: 7–14.

Masuhr, T. 1976. Lokalisation und Funktion des Kurzzeitgedächtnisses der Honigbiene (*Apis mellifica L.*). Doctoral dissertation, Fachbereich Biologie, TH Darmstadt.

Matsumoto, S. G. and R. K. Murphey. 1977. Sensory deprivation during development decreases the responsiveness of cricket giant interneurons. *J. Physiol.* **268**: 533–548.

Matsumoto, S. G. and J. G. Hildebrand. 1981. Olfactory mechanisms in the moth *Manduca sexta*: Response characteristics and morphology of central neurons in the antennal lobes. *Proc. R. Soc. London B* **213**: 249–277.

Maynard, D. M. 1956. Electrical activity in the cockroach cerebrum. *Nature (London)* **177**: 529–530.

Maynard, D. M. 1967. Organization of central ganglia, pp. 231–255. In C. A. G. Wiersma (ed.), *Invertebrate Nervous Systems*. University of Chicago Press, Chicago, London.

Maynard, D. M. and H. Dingle. 1963. An effect of eyestalk ablation on antennular function in the spiny lobster *Panulirus argus*. *Z. Vgl. Physiol.* **46**: 515–540.

Maynard, D. M. and J. G. Yager. 1968. Function of an eyestalk ganglion, the medulla terminalis, in olfactory integration in the lobster, *Panulirus argus*. *Z. Vgl. Physiol.* **59**: 241–249.

Maynard, D. M. and A. Sallee. 1970. Disturbance of feeding behavior in the spiny lobster, *Panulirus argus,* following bilateral ablation of the medulla terminalis. *Z. Vgl. Physiol.* **66:** 123–140.

Mobbs, P. G. 1982. The brain of the honeybee *Apis mellifera.* I. The connections and spatial organization of the mushroom bodies. *Philos. Trans. R. Soc. London Ser. B* **298:** 309–354.

Mobbs, P. G. 1984. Neural networks in the mushroom bodies of the honeybee. *J. Insect Physiol.* **30:** 43–58.

Otto, D. 1971. Untersuchungen zur zentralnervösen Kontrolle der Lauterzeugung von Grillen. *Z. Vgl. Physiol.* **74:** 227–271.

Roeder, K. D. 1937. The control of tonus and locomotor activity in the praying mantis *Mantis religiosa L. J. Exp. Zool.* **76:** 353–374.

Rowell, C. H. F. 1963. A method for chronically implanting stimulating electrodes into the brains of locusts, and some results of stimulation. *J. Exp. Biol.* **40:** 271–284.

Rust, M. K., T. Burk, and W. Bell. 1976. Pheromone stimulated locomotory and orientation responses in the American cockroach. *Anim. Behav.* **24:** 52–67.

Sandeman, D. C. 1982. Organization of the central nervous system, pp. 1–61. In H. L. Atwood, D. C. Sandeman (eds.), *The Biology of Crustacea,* vol. 3. Academic Press, New York, London.

Schaller-Selzer, L. 1984. Physiology and morphology of the larval sexual pheromone-sensitive neurones in the olfactory lobe of the cockroach *Periplaneta americana J. Insect Physiol.* **30:** 537–546.

Schildberger, K. 1981. Some physiological features of mushroom body linked fibers in the house cricket brain. *Naturwissenschaften* **67:** 623.

Schildberger, K. 1984. Multimodal interneurons in the cricket brain: Properties of identified extrinsic mushroom body cells. *J. Comp. Physiol.* **154:** 71–79.

Schürmann, F. W. 1974. Bemerkungen zur Funktion der Corpora pedunculata im Gehirn der Insekten aus morphologischer Sicht. *Exp. Brain. Res.* **19:** 406–432.

Seifert, W. (ed.). 1983. *Neurobiology of the Hippocampus.* Academic Press, London, New York.

Selzer, R. 1979. Morphological and physiological identification of food specific neurones in the deutocerebrum of *Periplaneta americana. J. Comp. Physiol.* **134:** 159–163.

Spaulding, E. G. 1904. An establishment of association in hermit crabs, *Eupagurus longicarpus. J. Comp. Neurol.* **14:** 49–61.

Stewart, W. W. 1978. Functional connections between cells as revealed by dye-coupling with a highly fluorescent naphthalimide tracer. *Cell* **14:** 741–759.

Strausfeld, N. J. 1976. *Atlas of an Insect Brain.* Springer-Verlag, Berlin, Heidelberg, New York.

Strausfeld, N. J. and D. R. Nässel. 1981. Neuroarchitectures serving compound eyes of crustacea and insects, pp. 1–132. In H. Autrum (ed.), *Handbook of Sensory Physiology,* vol. 7/6B. Springer-Verlag, Berlin, Heidelberg, New York.

Technau, G. M. 1984. Fiber number in the mushroom bodies of adult *Drosophila melanogaster* depends on age, sex and experience. *J. Neurogenet.* **1:** 113–126.

Turanskaja, V. M. 1972. Motor activity in the honeybee *Apis mellifera* during electrical stimulation of the supraoesophageal ganglion. *Zh. Evol. Biokhim. Fiziol.* **8:** 452–453.

Turanskaja, V. M. 1973. The brain and behavior of the bee. *Pcelovodstvo* **9:** 17–19.

Voskresenskaja, A. K. 1957. On the role played by mushroom bodies (corpora pedunculata) on the supraesophageal ganglion in the conditioned reflexes of the honey bee. *Dokl. Akad. Nauk. SSSR* **112:** 964–967.

Vowles, D. M. 1954. The function of the corpora pedunculata in bees and ants. *Br. J. Anim. Behav.* **2:** 116.

Vowles, D. M. 1964a. Olfactory learning and brain lesions in the wood ant (*Formica rufa*). *J. Comp. Physiol. Psychol.* **58:** 105–111.

Vowles, D. M. 1964b. Models and the insect brain, pp. 377–399. In R. F. Reiss (ed.), *Neural Theory and Modelling*. Stanford University Press, Stanford.

Wadepuhl, M. 1983. Control of grasshopper singing behavior by the brain: Responses to electrical stimulation. *Z. Tierpsychol.* **63:** 173–200.

Wiesel, T. N. and D. H. Hubel. 1963. Single-cell responses in striate cortex of kittens deprived of vision in one eye. *J. Neurophysiol.* **26:** 1003–1017.

CHAPTER 22

Mucosubstances of the Central Nervous System
Their Distribution and Histochemical Characterization

Doreen E. Ashhurst

Department of Anatomy
St. George's Hospital Medical School
University of London
London, England

22.1. EXTRACELLULAR MATRICES OF THE CENTRAL NERVOUS SYSTEM

The first extracellular matrices of the Arthropoda to be studied were those of the insect nervous system. It has been recognized since Michels's (1880) and Schneider's (1902) work that the entire insect nervous system, both

513

central and peripheral, is surrounded by an acellular sheath and an underlying layer of cells, which Schneider named the neural lamella and the perineurium, respectively. Much later, Wigglesworth (1960) described the glial lacunar system in cockroach ganglia, and since that time it has been realized that channels of varying widths and containing an extracellular material permeate throughout the neuropil. Similar extracellular matrices have been described in the nervous systems of other arthropods (Martoja and Cantacuzène, 1966, 1968; Baccetti and Lazzeroni, 1969; Lemire and Deloince, 1970; Abbott, 1971; Binnington and Lane, 1980).

The original descriptions of these extracellular matrices were based on observations of the thoracic and abdominal ganglia, and the same is true of the subsequent investigations aimed to determine the nature of the molecules of which they are composed. The first histochemical investigations suggested that collagen, together with neutral, and possibly acidic, mucosubstances (mucopolysaccharides, glycosaminoglycans), are present in the neural lamellae of the thoracic ganglia of cockroaches, grasshoppers, and locusts (Baccetti, 1955, 1956; Ashhurst, 1959, 1961a), and that hyaluronate is present in the glial lacunar system of cockroaches (Ashhurst, 1961b). Later with the newly introduced alcian blue techniques, it was shown that the neural lamellae of these insects contain chondroitin and keratan sulfates and that the glial lacunar systems of many insects contain hyaluronate (Ashhurst and Costin, 1971a,b).

There are few published histochemical studies of the cerebral ganglia. Those of the neural lamella merely report the presence of acidic and neutral mucosubstances (Mustafa and Kamat, 1973; Dybowska and Dutkowski, 1977, 1979; Malhotra and Taneja, 1978; Choudhuri et al., 1981). Martoja and Cantacuzène (1966, 1968) identified hyaluronate in the glial lacunar system of several species of arachnids, chilopods, crustaceans, and insects, and in some instances they examined the cerebral ganglion. Other similar work on Crustacea has concentrated on the thoracic ganglia (Elston and Abbott, personal communication).

The small amount of material available for study within the arthropods necessitates the use of histochemical methods in the characterization of the extracellular matrices of the nervous system. This chapter is devoted to an assessment of the histochemical methods presently available for some of these molecules.

22.2. COMPOSITION OF THE EXTRACELLULAR MATRICES

The molecules, of which the matrices associated with the nervous system are composed, are the same as those found in matrices elsewhere in the animal. The neural lamella consists of a dense network of collagen fibrils, with glycosaminoglycans (GAGs) and glycoproteins, whereas the lacunae of the glial lacunar system appear to contain only GAGs.

The collagen fibrils of the neural lamella may be of large diameter and clearly banded, as in many insects such as cockroaches and locusts (Ashhurst, 1985), pseudoscorpions, scorpions (Baccetti and Lazzeroni, 1969), ticks (Binnington and Lane, 1980), and crustaceans (Abbott, 1971), or thin and indistinctly banded, as in the Lepidoptera and several other insect orders (Ashhurst, 1985) and in Crustacea (Abbott, 1971). Biochemical studies have revealed that the collagen molecules of which the fibrils are composed are similar to the collagen molecules in other animals from coelenterates to mammals (Ashhurst and Bailey, 1980; François et al., 1980). Basement membrane collagens are also present in insects (Ashhurst and Bailey, 1980; Fessler et al., 1984), but basement membranes are not found in the nervous system. Although it is often considered that some trichrome stains demonstrate the extent of the connective tissues, and hence the presence of collagen, there is no stringent chemical basis to these staining methods. The most straightforward way to determine the presence of fibrous collagens is by electron microscopy; in this context it is important to note that the banding pattern, and not the actual measurement of the periodicity of the banding, is the criterion on which to identify collagen fibrils (see Ashhurst and Bailey, 1980).

The glycosaminoglycans (GAGs) are another major component of the connective tissue matrices, and these have been identified in many arthropod connective tissues (Ashhurst and Costin, 1971a,b; Mustafa and Kamat, 1973; Dybowska and Dutkowski, 1977, 1979). They are long-chain polymers of repeating disaccharide units, each of which contains an O-sulfated N-acetylhexosamine and a uronic acid residue (Heinegård and Paulsson, 1984). There are six GAGs: hyaluronate, chondroitin-4-sulfate, chondroitin-6-sulfate, dermatan sulfate, heparin, and heparan sulfate; keratan sulfate is a related disaccharide that has no uronic acid. The diversity of these disaccharides derives from small variations in the constituent residues, for example, whether the uronic acid is glucuronic acid or iduronic acid, and the position of the sulfate group. In addition, the degree of sulfation is very variable and leads to undersulfated and highly sulfated forms of the chondroitin, heparan, and keratan sulfates.

Each GAG chain, or polymer, consists of only one type of disaccharide unit; of these only hyaluronate and heparin can exist independently. All the others are attached to a core protein via a linkage region to form a proteoglycan. The structure of arthropod proteoglycans is unknown, although by analogy with observations of the other matrix molecules, it seems very probable that arthropod GAGs do form proteoglycans. The proteoglycans of mammals are very diverse and are tissue-specific (Heinegård and Paulsson, 1984). The diversity arises from the length of the polypeptide core protein, the length of the polymer chains, and their arrangement on the core protein. Thus, an infinite variety of proteoglycans is possible. In cartilage the proteoglycans are attached to a single chain of hyaluronate to form a large branching aggregate.

Because of their large size, the proteoglycans occupy an appreciable volume of the matrix and the carboxyl and sulphate groups along the polymers both contribute to their high negative charge. Thus, they form a negatively charged network through which other molecules must pass to get, for example, from the hemolymph to the cells. Many studies of the effects of GAGs on the movement of molecules through the extracellular matrices have been made by Laurent and his group (Comper and Laurent, 1978). More recently, the GAG in the endothelial basement membrane has been implicated in the control of capillary permeability (Kanwar and Farquhar, 1979).

Other macromolecules have been isolated from the matrices. These include fibronectin, laminin, and several tissue-specific molecules (Hakomori et al., 1984). A laminin-like molecule has recently been isolated from *Drosophila* (Fessler et al., 1984), and no doubt in the future others will be isolated from the Arthropoda.

22.3. HISTOCHEMICAL METHODS FOR THE IDENTIFICATION OF CONNECTIVE TISSUE MUCOSUBSTANCES

The details of the methods, described below, are not included in this chapter; they are given in the papers referred to in the text, or by Ashhurst (1979) or Pearse (1985).

22.3.1. Methods for Glycosaminoglycans

The histochemical methods available for the identification of the different types of GAGs depend primarily on the electrostatic binding of cationic dyes by the polyanions of the GAG. There is no specificity inherent in this binding, and it must be pointed out that carboxyl and sulfate ester groups are also present on many glycoprotein molecules, such as those of mucins. In addition, the phosphate groups of nucleic acids and phosphoproteins also bind cationic dyes.

The basic copper phthalocyanin dye, alcian blue, is most frequently used to determine the presence of GAGs as it has a high affinity for these molecules. It is a large dye molecule with a central chromophore and four positive charges around the periphery (Scott, 1970). For steric reasons, it is bound more readily by the carboxyl and sulfate groups of GAGs than by the phosphate groups of nucleic acids (Scott and Willett, 1967); the opposite is true of the smaller basic dye, pyronin. Using alcian blue at different pHs and in the presence of other cations, it is possible to distinguish molecules that possess carboxyl, phosphate, or sulfate groups.

Alcian blue is frequently used at pH 2.5, when it is bound by all the anionic groups, that is, by the phosphates of nucleic acids and proteins and by the carboxyl and sulfate ester groups of glycoproteins and GAGs. At pH 1.0 (in 0.1 N HCl), only sulfate groups bind the dye because the carboxyl and

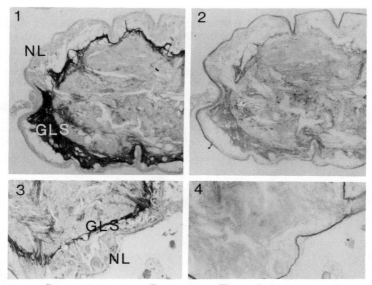

Figures 22.1. to 22.4. Photomicrographs of the thoracic ganglia of *L. migratoria* and *P. americana* to illustrate the alcian blue binding properties of the glial lacunar system (GLS) and the neural lamella (NL) at pH 2.5 and 1.0. The glial lacunar system binds strongly at pH 2.5, but not at pH 1.0, which suggests that only carboxyl groups are present. The neural lamella binds weakly at pH 2.5, but more strongly at pH 1.0, which indicates that sulfate groups are present in addition to some carboxyl groups. (Fig. 22.1. *P. americana*, pH 2.5; Fig. 22.2. *P. americana*, pH 1.0; Fig. 22.3. *L. migratoria*, pH 2.5; Fig. 22.4. *L. migratoria*, pH 1.0; magnification of all figures 90×.) (From Ashhurst and Costin, 1971a; reproduced by courtesy of *Histochemical Journal.*)

phosphate groups are not dissociated at this pH. The dye binding by sulfate-containing molecules may be enhanced compared with that at pH 2.5, because the complexes formed between the polyanions of the proteins are split by the chloride ions in the dye bath, which releases more sulfate groups to bind the dye (Quintarelli et al., 1964a,b). These characteristics of alcian blue binding in insect ganglia are illustrated in Figures 22.1. through 22.4.

A further refinement of the use of alcian blue is the stepwise addition of salt (Scott and Dorling, 1965), usually magnesium or sodium chloride, to the dye solution. The added cations compete with the dye cations for the anionic sites, and the amount of dye bound depends on the competitive affinity of the dye and added cations. The concentration of added electrolyte necessary to prevent dye binding is characteristic for each GAG (Scott, 1980a). Carboxyl and phosphate groups only bind alcian blue in the presence of less than 0.2 M magnesium chloride. Thus, hyaluronate and sialic acid-containing glycoproteins bind alcian blue when in the presence of less than 0.2 M magnesium chloride. As the other GAGs are sulfated, the concentrations of magnesium chloride are higher. Chondroitin and dermatan sulfates will bind

TABLE 22.1. THE REACTIONS OF THE PERIPHERAL NERVOUS SYSTEMS OF
***PERIPLANETA AMERICANA* AND *GALLERIA MELLONELLA* TO A BATTERY OF**
HISTOCHEMICAL TESTS[a]

Procedure	*P. americana* Thoracic Ganglia		*G. mellonella* Adult Abdominal Nerve Cord	
	Glial Lacunar System	Neural Lamella	Dorsal Mass of Connective Tissue	Neural Lamella
Alcian blue, pH 2.5	4+	0–1+	3–4+	1+
Alcian blue, pH 1.0	0–1+	0–3+	0–3+	0–1+
Alcian blue, pH 5.7				
with 0.05 M MgCl$_2$	4+	3+	3–4+	4+
with 0.1 M MgCl$_2$	4+	3+	3+	3+
with 0.2 M MgCl$_2$	0	3+	1–2+	2+
with 0.4 M MgCl$_2$	0	1–2+	0–±	1+
with 0.6 M MgCl$_2$	0	1+	0	±
with 0.8 M MgCl$_2$	0	1+	0	0
with 1.0 M MgCl$_2$	0	±	0	0
Enzymatic digestion/alcian blue				
Hyaluronidase-AB,pH 2.5	1–2+	±	2+	0–1+
Control-AB, pH 2.5	4+	±	3–4+	1+
Hyaluronidase-AB, pH 1.0	0	1+	0–1+	±
Control-AB, pH 1.0	0	1+	0–3+	1+
Neuraminidase-AB, pH 2.5	4+	±	3–4+	1+
Control-AB, pH 2.5	4+	±	3–4+	1+
Periodic acid/Schiff	0	3+	3–4+	4+
PAS (diastase control)	0	3+	3–4+	4+
High Iron Diamine	0	2+	2+	1–2+

[a] The intensity of the reactions or staining is indicated as follows: no reaction or staining, 0; increasing intensity of reaction or staining, 1–4 + (1 = weak, 4 = very strong); ±, very weakly positive reaction or staining.

alcian blue in the presence of less than 0.6 M concentrations of magnesium chloride, heparan sulfate, and heparin in less than 0.9 M concentrations of magnesium chloride. Keratan sulfate, which is the most highly sulfated, still binds alcian blue in the presence of 1.0 M magnesium chloride. Thus, using alcian blue under controlled conditions, it is possible to distinguuish the different GAGs. Typical results obtained by using these methods to determine the types of GAGs in the neural lamella and glial lacunar system of the thoracic ganglia of *Periplaneta americana* and the dorsal connective tissue mass of the abdominal nerve cord of *Galleria mellonella* are given in Table 22.1. The results indicate that in *P. americana* the glial lacunar system contains only hyaluronate, whereas the neural lamella contains chondroitin and keratan sulfates, and in *G. mellonella* the dorsal connective tissue mass contains chrondroitin, but no keratan, sulfate (Ashhurst and Costin, 1971a,b). It must be stressed that the *full staining conditions* must be given when such studies are reported.

Other tests, which can be used to distinguish between molecules with carboxyl and sulfate groups, are based on the use of diamine dyes (Spicer, 1965). The dye bath contains a mixture of meta- and para-dimethylphenelene diamines to which ferric chloride is added in low or higher concentrations to provide the Low Iron Diamine (LID) or High Iron Diamine (HID) tests. The cationic oxidation product in the mixture is bound by the polyanions. With the lower concentration of iron, both carboxyl and sulfate groups bind the dye, but with the higher concentration, only sulfate groups bind the dye. Thus, sulfated molecules can be distinguished from those with only carboxyl groups.

Colloidal iron is another reagent that has been used to detect the presence of acidic substances in tissue sections. It is, however, more capricious than the methods outlined above. Metachromatic dyes, such as toluidine blue or azure A, have also been widely used in the past to detect the presence of acidic substances. The metachromatic color change occurs when the dye molecules are bound in large numbers by substances that provide a high density of negative charges. These methods did not prove as sensitive as those using alcian blue for the characterization of the GAGs of neural lamellae (see Ashhurst, 1959, 1961; Ashhurst and Costin, 1971a).

22.3.2. Methods for Neutral Mucosubstances

One of the most frequently used tests for carbohydrates is the Periodic Acid/Schiff (PAS) test (McManus, 1946). This test is specific for molecules that possess vic-glycol groups. These are oxidized by periodic acid which breaks the C-C bonds where they occur as 1-2 glycol groups. The resulting two aldehydes are detected by Schiff's reagent, which attaches to them and becomes a magenta color as a result of molecular reorganization. Many molecules, such as glycoproteins and glycolipids, are PAS-positive because their carbohydrate side chains have vic-glycol groups. Thus, in connective tissue matrices, collagens and other glycoproteins are PAS-positive. GAGs are not PAS-positive with the standard PAS test in which the initial periodic acid oxidation lasts between 10 and 15 min. The uronic acid moiety of the GAG chains contains vic-glycol groups, but these are not oxidized unless the periodic acid treatment is extended for 18 to 24 hours. It is thought that the periodic acid is prevented from reaching the vic-glycol groups for steric reasons (Scott and Dorling, 1969; Scott and Harbinson, 1969). The various types of mucus, both acidic and neutral, are PAS-positive under the standard conditions.

22.3.3. Enzymatic Digestions

Enzymatic digestions are a useful confirmatory tool in investigations of the GAGs. The most readily available enzyme is testicular hyaluronidase, which hydrolyzes uronic acid and so breaks the polymer chains of hyaluronate, and the chondroitin sulfates. *Streptomyces* hyaluronidase, which is much more expensive, breaks the glucosamine ring and so it removes only hyaluronate. A series of chondroitinases, one which reacts with all chondroitins and others, chondroitinases A, B, and C, which are specific for chondroitin-4-sulfate, dermatan sulfate, and chondroitin-6-sulfate, respectively, are available (Yamagata et al., 1968). A specific keratanase is also available commercially, but it is not useful for the digestion of tissue sections.

Some sialic acid residues are labile to neuraminidase. This enzyme is not directly useful in characterizing the GAGs, but it may be used to confirm that sialic acid-containing mucosubstances are not responsible for the alcian blue binding.

The enzyme digestions are carried out on sections and are followed by the appropriate test (Table 22.1.). Many commercially available enzyme preparations are impure and contain proteolytic enzymes; this must be taken into account in the interpretation of the results of enzymatic digestions.

Only testicular hyaluronidase has been used so far in studies of arthropod connective tissues, and examples of the results obtained are shown in Table 22.1.

22.3.4. Electron Microscopical Methods

The methods for locating GAGs with the electron microscope depend on the use of stains that have intrinsic electron density. The most commonly used are ruthenium red and colloidal iron, which are electrostatically bound by the GAGs (Pearse, 1985). Because these are usually added to the fixative solution, it is impossible to control the staining conditions in the same way as is possible for sections. Schofield et al. (1975) tried to rectify this problem when he used alcian blue by adding magnesium chloride to the staining solutions to try to exploit the light microscopical technique. Scott has pursued the same ideas using the newly synthesized phthalcyanin dyes, cuprolinic and cupromeronic blue (Scott, 1980b; Scott and Orford, 1981). Sannes et al. (1979) used a modification of the HID technique to detect sulfated complex carbohydrates.

Few of these methods have been used so far on arthropod tissues and those that have been used have yielded little information about the GAGs and their distribution. There are two major problems: firstly, these methods do not allow the individual GAGs to be identified; and, secondly, there are problems of stain penetration. The methods are usually executed on small tissue blocks or thick sections that are then processed and sectioned for electron microscopy. The stain molecules may only penetrate a short distance, or remain on the surface. Penetration was a major problem when Schofield's methods were applied to thoracic ganglia of the cockroach (Ashhurst, unpublished observations) and this is also illustrated in Dybrowska and Dutkowski's (1977) study of the neural lamella of the cerebral ganglion of *Galleria mellonella*, in which the ruthenium red was confined to the surface layers of the neural lamella.

22.3.5. Immunohistochemical Methods

Immunohistochemical methods are increasingly being used to locate connective tissue macromolecules. Their application to arthropod tissues depends on the availability of antibodies that will react with the arthropod antigens. Although some antibodies have been made to arthropod connective tissue macromolecules (Fessler et al., 1984), this is a laborious process. Some monoclonal antibodies have been made to mammalian proteoglycans (Caterson et al., 1983; Couchman et al., 1984). These antibodies detect antigens in the linkage regions of the GAG chains. Antibodies against chondroitin-4- and chondoitin-6-sulfate and keratan sulfate proteoglycans have been used on several insect tissues using the standard immunohistochemical methods with promising results (Ashhurst, unpublished observations). In this way, it will be possible to distinguish more precisely which GAGs are present in any tissue.

22.4. CONCLUDING REMARKS

The foregoing is a brief account of the application of readily available histochemical methods to an analysis of the complex connective tissue carbohydrates in the central nervous systems of arthropods. These tests, with some further modifications and pretreatments, are also applicable to the characterization of the glycoproteins in the neurosecretory cells in the nervous system. For a discussion of the application of these and other techniques to intracellular granules, the reader is referred to papers by Ashhurst (1979) and Costin (1975).

It can be seen from the work mentioned in the Introduction that there have been few studies of the mucosubstances or GAGs of the connective tissues of arthropod brains and these only go so far as to show whether or not acidic molecules are present. Unfortunately, many authors do not state the conditions under which the tests were conducted so that it is not possible to evaluate their results fully.

In conclusion, methods are available for the characterization of the complex carbohydrates in the extracellular matrices of arthropod brains. The results published so far indicate that the two extracellular matrices, that is the neural lamella and glial lacunar system, are very similar both in the different parts of the central nervous system and in the different arthropods examined so far; more work however, is needed to confirm this.

22.5. SUMMARY

The extracellular matrices of the nervous system of arthropods are composed of collagen, glycosaminoglycans, and glycoproteins. The presence of collagen fibrils in a matrix can be clearly demonstrated electron microscopically. The banding pattern, and *not* the periodicity of the banding, is the criterion on which to identify collagen fibrils.

The GAGs are long polymers of repeating disaccharide units that consist of an O-sulfated N-acetylhexosamine and a uronic acid residue. They have anionic charges, carboxyls, and sulfates along their length. Methods for their identification use cationic dyes, such as alcian blue. By using dye baths at different pHs and containing other cations, such as magnesium, it is possible to distinguish the different GAGs. Thus, it has been demonstrated that hyaluronate is present in the glial lacunar system, whereas chondroitin and keratan sulfates are found in the neural lamella.

The presence of neutral mucosubstances can be detected using the PAS test. GAGs are not PAS-positive under the standard conditions, but glycoproteins are.

Confirmation of the presence of GAGs may be obtained by digestion of tissue sections with specific enzymes, such as *Streptomyces* hyaluronidase and chondroitinases A, B, or C; keratanase is not so useful.

Electron histochemical methods are briefly reviewed, but problems remain in their application to arthropod tissues. Immunohistochemical methods using monoclonal antibodies to GAGs seem to offer a promising approach to the characterization of GAGs in the future.

REFERENCES

Abbott, N. J. 1971. The organization of the cerebral ganglion in the shore crab, *Carcinus maenas*. I. Morphology. *Z. Zellforsch.* **120:** 386–400.

Ashhurst, D. E. 1959. The connective tissue sheath of the locust nervous system: a histochemical study. *Q. J. Microsc. Sci.* **100:** 401–412.

Ashhurst, D. E. 1961a. A histochemical study of the connective tissue sheath of the nervous system of *Periplaneta americana*. *Q. J. Microsc. Sci.* **102:** 455–461.

Ashhurst, D. E. 1961b. An acid mucopolysaccharide in cockroach ganglia. *Nature (London)* **191:** 1224–1225.

Ashhurst, D. E. 1979. Histochemical methods for hemocytes, pp. 581–599. In A. P. Gupta (ed.), *Insect Hemocytes*. Cambridge University Press, Cambridge.

Ashhurst, D. E. 1985. Connective tissues, pp. 249–297. In G. A. Kerkut and L. I. Gilbert (eds.), *Comprehensive Insect Physiology, Biochemistry and Pharmacology,* vol. 3. Pergamon Press, Oxford.

Ashhurst, D. E. and A. J. Bailey. 1980. Locust collagen: Morphological and biochemical characterization. *Eur. J. Biochem.* **103:** 75–83.

Ashhurst, D. E. and N. M. Costin. 1971a. Insect mucosubstances. II. The mucosubstances of the central nervous system. *Histochem. J.* **3:** 297–310.

Ashhurst, D. E. and N. M. Costin. 1971b. Insect mucosubstances. III. Some mucosubstances of the nervous systems of the wax-moth (*Galleria mellonella*) and the stick insect (*Carausius morosus*). *Histochem. J.* **3:** 379–387.

Baccetti, B. 1955. Ricerche sulla fine struttura del perilemma nel sisterna nervoso degli insetti. *Redia* **40:** 197–212.

Baccetti, B. 1956. Lo stroma di sostegna di organi degli insetti esaminato a luce polarizzata. *Redia* **41:** 259–276.

Baccetti, B. and G. Lazzeroni. 1969. The envelopes of the nervous system of pseudoscorpions and scorpions. *Tissue Cell* **1:** 417–424.

Binnington, K. C. and N. J. Lane. 1980. Perineurial and glial cells in the tick *Boophilus microplus* (Acarina:Ixodidae): freeze-fracture and tracer studies. *J. Neurocytol.* **9:** 343–362.

Caterson, B., J. E. Christner, and J. R. Baker. 1983. Identification of a monoclonal antibody that specifically recognizes corneal and skeletal keratan sulfate. *J. Biol. Chem.* **258:** 8848–8854.

Choudhuri, D. K., S. Mandal, and B. Ghosh. 1981. Histochemical and biochemical studies on the brain of *Schizodactylus monstrosus (Orthoptera: Schizodactylidae)*. *Nucleus* **24:** 10–15.

Comper, W. D. and T. C. Laurent. 1978. Physiological function of connective tissue polysaccharides. *Physiol. Rev.* **8:** 255–315.

Costin, N. M. 1975. Histochemical observations of the haemocytes of *Locusta migratoria*. *Histochem. J.* **7:** 21–43.

Couchman, J. R., B. Caterson, J. E. Christner, and J. R. Baker. 1984. Mapping by monoclonal antibody detection of glycosaminoglycans in connective tissues. *Nature (London)* **307:** 650–652.

Dybowska, H. E. and A. B. Dutkowski. 1977. Ruthenium red staining of the neural lamella of the brain of *Galleria mellonella*. *Cell Tissue Res.* **176:** 275–284.

Dybowska, H. E. and A. B. Dutkowski. 1979. Developmental changes in the fine structure and some histochemical properties of the neural lamella of *Galleria mellonella* (L.) brain. *J. Submicrosc. Cytol.* **11:** 25–37.

Fessler, J. H., G. Lunstrum, K. G. Duncan, A. G. Campbell, R. Sterne, H. P. Bächinger, and L. I. Fessler. 1984. Evolutionary constancy of basement membrane components, pp. 207–219. In R. L. Trelstad (ed.), *Role of Extracellular Matrix in Development*. Alan R. Liss, New York.

François, J., D. Herbage, and S. Junqua. 1980. Cockroach collagen: isolation, biochemical and biophysical characterization. *Eur. J. Biochem.* **112:** 389–396.

Hakomori, S., M. Fukuda, K. Sekiguchi, and W. G. Carter. 1984. Fibronectin, laminin, and other extracellular glycoproteins, pp. 229–275. In K. A. Piez and A. H. Reddi (eds.), *Extracellular Matrix Biochemistry*. Elsevier, New York.

Heinegård, D. and M. Paulsson. 1984. Structure and metabolism of proteoglycans, pp. 277–328. In K. A. Piez and A. H. Reddi (eds.), *Extracellular Matrix Biochemistry*. Elsevier, New York.

Kanwar, Y. S. and M. G. Farquhar. 1979. Presence of heparan sulfate in the glomerular basement membrane. *Proc. Natl. Acad. Sci. USA* **76:** 1303–1307.

524 DOREEN E. ASHHURST

Lemire, M. and R. Deloince. 1970. Études des mucopolysaccharides du système nerveux prosomien du Scorpion saharien *Androctonus australis* L. *C.R. Acad. Sci. Paris* **271D:** 1630–1633.

Malhotra, R. K. and S. Taneja. 1978. A histochemical profile of the cockroach brain. *Zool. Anz.* **200:** 114–118.

Martoja, R. and A. M. Cantacuzène. 1966. Données histologiques sur les mucoprotéines du système glial des Orthoptéroïdes et des Blattotéroïdes. *C.R. Acad. Sci. Paris* **263D:** 152–155.

Martoja, A. and A. M. Cantacuzène. 1968. Sur les mucopolysaccharides acides des centres nerveux des Arthropodes. *C.R. Acad. Sci. Paris* **267D:** 1607–1610.

Michels, H. 1880. Beschreibung des Nervensystems von *Oryctes nasicornis* im Larven-, Puppen- und Köferzustande. *Z. Wiss. Zool.* **34:** 641–702.

Mustafa, M. and D. N. Kamat. 1973. Mucopolysaccharide histochemistry of *Musca domestica*: VII. The brain. *Acta Histochem.* **45:** 254–269.

McManus, J. F. A. 1946. Histological demonstration of mucin after periodic acid. *Nature (London)* **158:** 202.

Pearse, A. G. E. 1985. *Histochemistry, Theoretical and Applied. Vol. 2. Analytical Technology.* Fourth edition. Churchill Livingston, Edinburgh.

Quintarelli, G., J. E. Scott, and M. C. Dellovo. 1964a. The chemical and histochemical properties of alcian blue. II. Dye binding of tissue polyanions. *Histochemie.* **4:** 86–98.

Quintarelli, G., J. E. Scott, and M. C. Dellovo. 1964b. The chemical and histochemical properties of alcian blue. III. Chemical blocking and unblocking. *Histochemie.* **4:** 99–112.

Sannes, P. L., S. S. Spicer, and T. Kutsuyama. 1979. Ultrastructural localization of sulfated complex carbohydrates with a modified iron diamine procedure. *J. Histochem. Cytochem.* **27:** 1108–1111.

Schneider, K. C. 1902. *Lehrbuch der Vergleichenden Histologie der Tiere.* G. Fischer, Jena.

Schofield, B. H., B. R. Williams, and S. B. Doty. 1975. Alcian blue staining of cartilage for electron microscopy. Application of the critical electrolyte principle. *Histochem. J.* **7:** 139–149.

Scott, J. E. 1970. Histochemistry of alcian blue. I. Metachromasia of alcian blue, astrablau and other cationic phthalocyanin dyes. *Histochemie* **21:** 277–285.

Scott, J. E. 1980a. The molecular biology of histochemical staining by cationic phthalocyanin dyes: The design of replacements for alcian blue. *J. Microsc.* **119:** 373–381.

Scott, J. E. 1980b. Collagen-proteoglycan interactions. *Biochem. J.* **187:** 887–91.

Scott, J. E. and J. Dorling. 1965. Differential staining of acid glycosaminoglycans (mucopolysaccharides) by alcian blue in salt solutions. *Histochemie.* **5:** 221–233.

Scott, J. E. and J. Dorling. 1969. Periodate oxidation of acid polysaccharides. III. A PAS method for chondroitin sulphates and other glycosamino-glycuronans. *Histochemie.* **19:** 295–301.

Scott, J. E. and R. J. Harbinson. 1969. Periodate oxidation of acid polysaccharides. II. Rates of oxidation of uronic acids in polyuronides and acid mucopolysaccharides. *Histochemie.* **19:** 155–161.

Scott, J. E. and C. R. Orford. 1981. Dermatan sulphate rich proteoglycan associates with rat tail-tendon collagen at the d band in the gap region. *Biochem. J.* **197:** 312–16.

Scott, J. E. and I. H. Willett. 1967. Binding of cationic dyes to nucleic acids and other biological polyanions. *Nature (London)* **209:** 985–987.

Spicer, S. S. 1965. Diamine methods for differentiating mucosubstances histochemically. *J. Histochem. Cytochem.* **13:** 211–234.

Wigglesworth, V. B. 1960. The nutrition of the central nervous system of the cockroach *Periplaneta americana* L. *J. Exp. Biol.* **37:** 500–512.

Yamagata, T., H. Saito, O. Habuchi, and S. Suzuki. 1968. Purification and properties of bacterial chondroitinases and chondrosulfatases. *J. Biol. Chem.* **243:** 1523–1535.

PART III
TECHNIQUES

CHAPTER 23

Identification of Monoamine-Containing Neurons in Arthropods

Nikolai Klemm
Department of Zoology
University of Lund
Lund, Sweden

23.1. INTRODUCTION

Monoamines (MAs) are neuroactive substances that occur in the central nervous system of all animals so far studied. The most frequently occurring and best studied MAs are dopamine (DA), noradrenaline (NA), octopamine, and 5-hydroxytryptamine (5-HT or serotonin). Adrenaline has not been detected in arthropod nervous tissue. MAs are synthesized within the neuron (Maxwell et al., 1978; Wallace, 1976). When released after stimulation, they react with receptors (Evans, 1980) and are subsequently inactivated by reuptake into its presynaptical site (Klemm, 1976; Evans, 1980) and by metabolism (Maxwell et al., 1978; Mir and Vaugham, 1981; Omar and Murdock, 1982). There is little information on other MAs such as tyramine, tryptamine, and histamine in the arthropod nervous system. Because no specific histochemical method is presently available for octopamine and tryptamine, this chapter will focus on identification of primary catecholamines (CAs) (DA, NA) and 5-HT.

23.2. METHODS

Several methods are available to demonstrate MAs intraneuronally. Chemical analyses of single neurons were successively applied to large neurons of gastropodes. Contamination from adjacent neurons could, however, not always be avoided (Osborne, 1979). As the neurons of arthropods are smaller than the giant neurons in molluscs, this method is difficult to apply to arthropod preparations. Many methods were used to "histochemically" identify intraneuronal MAs. Thus, this chapter will deal only with those methods, for which specificity has been established and whose histochemical reactions are understood.

Generally, the identification can be performed either by direct methods or indirect ones. The direct methods are highly specific. They imply techniques that demonstrate the molecule of interest directly or after a slight and well-controlled and understood modulation. Indirect methods imply techniques that either make the wanted compound visible after coupling it to a labeled compound, or that demonstrate enzymes or their activity, that are related to the presence of MAs, and from which the presence or absence of the MAs of interest can be deduced. The application of an indirect method (e.g., autoradiography, immunohistochemistry) alone is not sufficient to prove or disprove the existence of MAs. The value of these methods lies rather in their application in addition to other cytochemical analyses. The only direct method for the demonstration of certain MAs so far is the aldehyde-induced fluorescence method, which involves condensation of MAs by aldehydes into a specific fluorophore (Björklund et al., 1975).

23.2.1. Aldehyde Fluorescence Methods

23.2.1.1. Theoretical Considerations. CAs and 5-HT react in the presence of aldehydes and form fluorophores that can be visualized in the fluorescence microscope with a proper filter setting (Fig. 23.1.). The chemical reactions of the fluorophore-forming process are well understood. The initial reaction requires a high electron density at the point of ring closure, for example, in

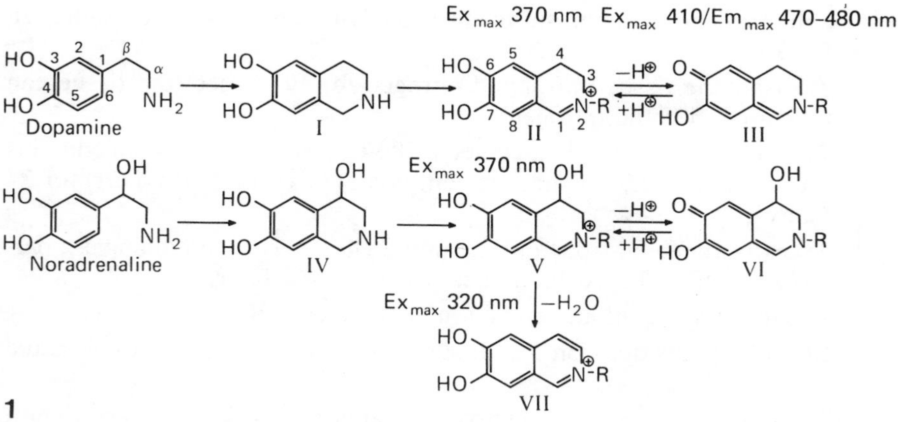

Ex$_{max}$ 370 nm   Ex$_{max}$ 410/Em$_{max}$ 470–480 nm

1

Figure 23.1. Formation of fluorophores from CAs by aldehyde. I = dihydroxytetrahydroisoquinoline; IV = trihydroxytetradoisoquinoline; II and V = their dihydro derivates. At neural pH, the fluorophores are in their tautomeric quinoidal forms (III, VI). After short exposure to low pH, the fluorophores convert to nonquinoidal forms (II, V). After prolonged acidification, the NA-fluorophore is further converted into a fully aromatic quinoline derivate (VII). (After Björklund et al., 1975; from Klemm, 1980.)

position 6 in the case of phenylethylamines (Figs. 23.1. and 23.2.) and position 2 in the case of indolylalkylamines (Fig. 23.2.). Tyramine and octopamine lack such substitutes in appropriate positions and are thus nonfluorogenic (Björklund et al., 1975). For fluorophore formation either gaseous formaldehyde (Falck-Hillarp method) or gaseous or liquid glyoxylic acid, or both, can be used.

23.2.1.2. Falck-Hillarp Method. The following procedure is adapted from Klemm (1980, 1983):

1. Dip the tissue in a mixture of propane:propylene (commercially avail-

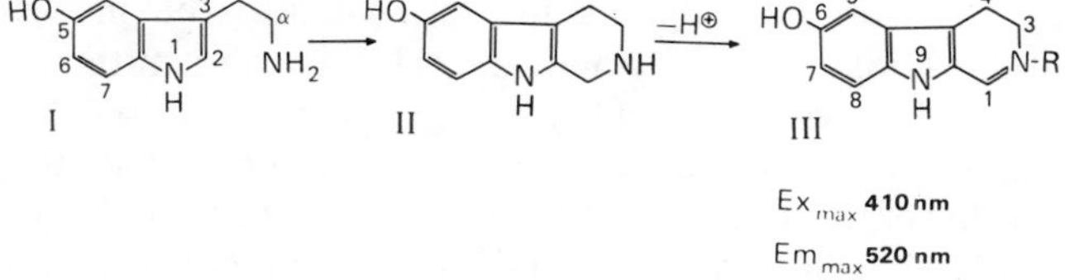

Ex$_{max}$ **410 nm**

Em$_{max}$ **520 nm**

2

Figure 23.2. The formation of fluorophores from 5-HT by aldehyde. II = 6-hydroxy-tetrahydro-β-carboline; III = its dihydroderivate. (After Björklund et al., 1975.)

able propane contains sufficient propylene) cooled in liquid nitrogen for 5–10 sec.

2. Transfer the tissue into liquid nitrogen for 10–15 sec (the tissue can be stored in liquid nitrogen).

3. Freeze-dry the tissue in a cooled ($-80°C$) freeze dryer which contains a water trap (i.e., phosporous pentoxyde) in vacuo (10^{-2} Torr) for 3–5 days.

4. Transfer the deep-frozen tissue into a glass vessel containing 5 g paraformaldehyde/l l, seal, and heat to 80°C for 1 hour.

5. Let the tissue cool and embed in vacuo in paraffin and section.

6. Stretch the sections on a slide and warm gently. The paraffin should not melt.

7. View in a fluorescence microscope equipped with a primary Schott BG 12 or BG 3 and a Schott OG 4 as secondary filter.

Results. CAs emit green and 5-HT yellow fluorescence. A high concentration of the green CA-fluorophore appears yellow to the eye (Betzold-Brücke effect) and may be misinterpreted as 5-HT (Fig. 23.3.).

Specificity Tests. Specificity tests must be done to avoid misinterpretation.

1. Compare the aldehyde-treated specimens with aldehyde-untreated ones but otherwise proceed similarly.

2. The specific fluorophore should be photolabile when exposed to excitation light.

3. The specific fluorophore diffuses in a humid environment.

4. Reserpin treatment should drastically decrease fluorescence.

5. Record emission and excitation spectra (most specific test).

The MAs are not equally distributed throughout the MA-containing neuron. Usually, their concentration is very high in the terminals, less so in the perikarya, and the concentration in its preterminal part can be too low to be visualized in the fluorescence microscope. To achieve a higher fluorescence yield, the MA-containing neuron can be loaded with externally applied fluorogenic MAs using the uptake mechanism into the presynaptic site, which usually terminates the action of the released transmitter. This uptake mechanism requires energy and the presence of certain cations (Na^+, Ca^{2+}). Loading of MA-containing neurons can be achieved by incubating the tissue with MAs (Myhrberg et al., 1979; Flanagan, 1983) or with fluorogenic non-metabolizable MAs ("false transmitters") for example, 6-HT or methyl-noradrenaline (Klemm and Schneider, 1975; Klemm and Falck, 1978; Myhrberg, 1979; Klemm, 1980). The concentration should not exceed 10^{-6} M for 6-HT and 10^{-7} M for methyl-noradrenaline. When a higher concentration is used,

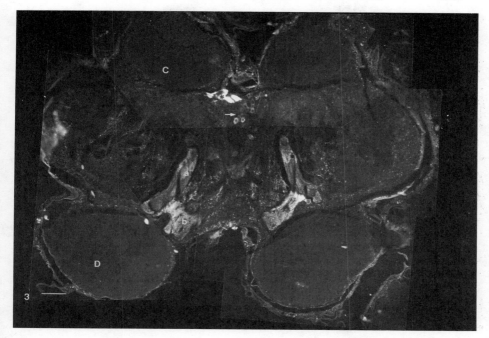

Figure 23.3. Frontal section of the brain of adult *Schistocerca gregaria*, showing CA-fluorescence (Falck-Hillarp method). α-lobe (traversed by non-CA-containing fiber bundels, b = β-lobe, C = calyx, D = olfactory lobe; CA-containing cell bodies in the *pars intercerebralis* (*arrow*), asterisk = nonspecific fluorescence. Scale: 100 μm.

more neurons get filled, which originally may not contain 5-HT or CAs. This uptake process is not very selective. The majority of MA-containing neurons can take up almost all these amines. In some neurons of the olfactory lobe of the locust, *Schistocerca gregaria*, however, the uptake is highly selective (Klemm and Schneider, 1975). For differentiation between nonspecific accumulation (which can be selective, as for instance into the ocellar nerve of insects; Klemm, 1976) and specific uptake, control tissue should be incubated at 0°C or in a medium lacking Na^+ or Ca^{2+}. As the neuronal uptake is not very selective, the filling of neurons with a certain amine does not necessarily mean that the neuron contains this or a related amine; for example, DA can be taken up into 5-HT-containing neurons, and similarly 5-HT can be taken up into DA-containing neurons, as can be demonstrated in certain strata in the optic lobe of insects (Klemm, 1980; Nässel and Klemm, 1983). Differentiation between, for instance, DA- and 5-HT-containing neurons, can be achieved by blocking the uptake into one or both neurons by specific uptake blockers. Uptake of 6-HT into 5-HT-containing neurons of the olfactory lobe of *Schistocerca gregaria* has been prevented by preincubation with chlorimipramine (10^{-4} M), a blocker into 5-HT-con-

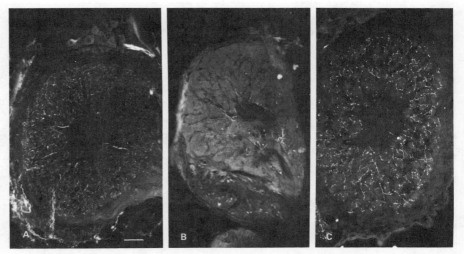

Figure 23.4. *Schistocerca gregaria.* Olfactory lobe. (A) After incubation in 6-HT. (B) After preincubation in chlorimipramine and subsequent incubation in 6-HT (Falck-Hillarp method). (C) 5-HT immunoreactivity (cf. Fig. 23.3.). Scale: 50 μm.

taining neurons (Klemm and Schneider, 1975). The presence of 5-HT in these neurons was later verified by immunocytochemical methods (Klemm and Sundler, 1983; Fig. 23.4.). It must be stressed that the uptake mechanism and the function of blockers in arthropods are poorly understood. Besides amines, the precursor amino acids L-DOPA (Klemm and Schneider, 1975; Klemm, 1976; Myhrberg et al., 1979; Flanagan, 1984) and 5-HTP (Klemm, unpublished observation) can also be taken up into MA-containing neurons. The uptake mechanism of amino acids remains to be understood.

The Falck-Hillarp method was successfully applied to brains of *Xiphosura* (O'Connor et al., 1982), *Chelicerata* (Meyer and Jehnen, 1980), *Crustacea*, and *Insecta* (Klemm, 1976, 1980, 1983). Its advantages are high specificity; high sensitivity for CAs (a concentration of 10^{-6} pmol DA or NA can readily be detected; Björklund et al., 1975); and the fluorophore-forming process can be manipulated and adjusted to the specimen's requirements. Its disadvantages are as follows: the method is very laborious; freeze-drying can be a problem, especially in marine animals; the photoinstability of the fluorophore (especially of 5-HT); the lower sensibility for 5-HT; and the technique does not allow differentiation between the MAs and their amino acid precursors (L-DOPA, 5-HTP). In adult insects, both amino acids occur in very low concentration (Clarke and Donnellan, 1982; Klemm, 1985) and can be neglected. High concentration of L-DOPA was reported in larval flies (Nässel and Laxmyr, 1982) and in crustaceans after stressful handling (Elofsson et al., 1982; Laxmyr, 1984). Chemical analysis showed that the concentration of DOPA in the brain of crayfish is resistant to reserpin (Elofsson

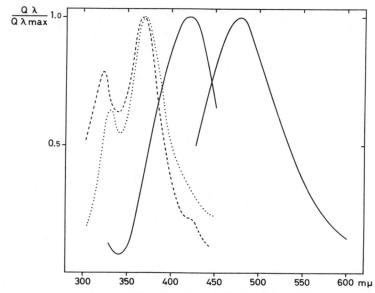

5

Figure 23.5. Microspectrofluorometric recordings of a DA fluorophore in the β-lobe of the caddis fly *Anabolia nervosa*. Left side excitation, right side emission (*solid line*); after short exposure to HCl (*dotted line*); after prolonged exposure to HCl (*dashed line*; Fig. 23.1.). (Adapted from Klemm and Björklund, 1971.)

et al., 1982). In histochemical studies, certain neurons in the brain and optic lobes in *Pacifastacus leniusculus* maintained their fluorophore even after prolonged treatment with the highest sublethal doses of reserpin (Myhrberg et al., 1979) and thus may contain DOPA. This makes quantitative analysis of MAs and their precursor amino acids mandatory for the correct interpretation of the fluorophore. The histochemical results vary in different species due to different intraneuronal concentration of fluorogenic MAs and/or because of different conditions in the tissue for fluorophore formation. For example, contrary to the findings in the olfactory lobe of the locust, *S. gregaria*, in the related species, *Locusta migratoria*, 5-HT fibers have been demonstrated with the Falck-Hillarp method (Klemm, unpublished results). The aldehyde fluorescence techniques cannot be applied on an ultrastructural level.

23.2.1.3. Microspectrofluorometry. Microspectrofluorometrical recording of the emission and excitation spectra is the only exact way to specify the fluorophore. For comparison of results from different laboratories, the spectra should be corrected and expressed in quanta per unit of wavelength. In models, the excitation maximum for the CA fluorophore is approximately at 410 nm and the peak of emission at 470 nm (Ex max 410/Em max 470). The data for 5-HT are Ex max 410/Em max 520 nm (see Björklund et al.,

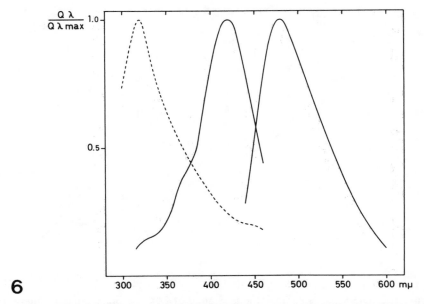

6

Figure 23.6. Microspectrofluorometric recordings from a NA fluorophore in the *stratum caudale* of *A. nervosa*. Excitation curve after prolonged acid treatment (*dashed line*). (Adapted from Klemm and Björklund, 1971.)

1975; Fig. 23.1.). In arthropod tissue, the spectra are 410 nm/470–480 nm for CAs and 390 nm/510–530 nm for 5-HT (Klemm and Björklund, 1971; Klemm and Axelsson, 1973; Klemm, 1976; Klemm and Falck, 1978; Figs. 23.5.–23.7.). The yellow fluorophore of 5-HT and that of a high concentration of CA-fluorophore can most accurately be differentiated by their emission spectrum. After aldehyde treatment DA and NA emit green light of the same wavelength. After acidification the Ex max of both compounds shift to 370 nm (the Em max is not influenced; Fig. 23.1., III, V). After prolonged acidificiation, the Ex max of the NA fluorophore shifts further to 310–320 nm, whereas the Ex max of the DA fluorophore remains at 370 nm (Fig. 23.1.).

23.2.1.4. Glyoxylic Acid Techniques. CAs and 5-HT can be condensed into fluorophore by other aldehydes than formaldehyde, for example, glyoxylic acid (Lindvall et al., 1974). Two methods should be mentioned here that successfully were applied on arthropod (cockroach, A, and bee, B) tissue.

A. According to Klemm, 1982.

1. Mount and freeze the brain on a cryostat chuck (20–25°C).
2. Pick up sections on chrome alum-gelatine pretreated slide.
3. Place slide with sections into an ice-cold sodium-buffered saline so-

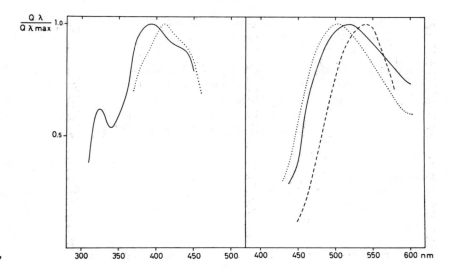

7

Figure 23.7. Microspectrofluorometric recordings of an indolylalkylamine fluorophore in the pars intercerebralis of the locust *Schistocerca gregaria*. Excitation (*dotted line*) emission (*dashed line*) spectrum from a typical 5-HT fluorphore. Other curves show atypical spectra. (Altered from Klemm and Axelsson, 1973.)

lution containing glyoxylic acid (SPG-solution: sucrose 0.2–0.6 M, KH_2PO_4 0.236 M, glyoxylic acid 1% (w/v), titrate with 10 N NaOH to pH 7.4 (de la Torre and Surgeon, 1976) for 5–10 min.

4. Remove from solution and blot excess liquid.

5. Dry in warm air stream for 1–3 hours.

6. View under fluorescence microscope.

7. For intensification of the fluorescence, heat the preparation in a 1 L vessel containing 5 g paraformaldehyde at 80°C for 1 hour.

Comments. The method is very fast, and the fluorophore is stable and does not diffuse. The sensitivity is lower than in the Falck-Hillarp method. Only CA-fluorescence was seen.

B. According to Mercer et al., 1983. Another method, using glyoxylic acid, was described by Mercer et al. (1983). The tissue was quickly frozen for cryostat sectioning. A drop of SPG solution (see above, step 3) was placed over the section and frozen on a cold glass slide. The section was thawed, the excess of glyoxylic acid was quickly removed by filter paper, and the section was dried under a warm air stream and subsequently heated at 100°C for 10 min. This method proved to be very sensitive as 5-HT-containing fibers were demonstrated in the optic lobe of the bee, which could not be seen after applying the Falck-Hillarp method. The findings have been verified by immunohistochemical methods (Schürmann and Klemm, 1984).

Comments. The advantage of the glyoxylic method is the relatively easy procedure, its sensibility, and the relatively stable fluorophore. The disadvantages are mainly due to the high temperature used for fluorophore formation (up to 100°C), which results in low tissue preservation, high background fluorescence, and unsatisfactory reproducibility.

23.2.1.5. Wholemount Techniques. Flanagan (1983) used the glyoxylic acid technique for wholemount preparation of larval thoracic ganglia. The specimens were incubated in a Ca^{2+}-free, and Mg^{2+}-rich (25 mM) saline solution containing 0.2 M sucrose and 5% glyoxylic acid for 1 hour at room temperature. The specimens were dried over calcium carbonate for 3–5 days, covered with a drop of mineral oil, and quickly heated on a copper block to 85°C for 2 min.

Comments. Fluorescent wholemount preparations of the central nervous system suffer from high background fluorescence and light scattering due to thickness of the preparation and due to the neurolemma covering it. The reproducibility is low and only the most peripheral neurons can be seen. Aldehyde-induced fluorescence wholemount techniques, however, gave good results in thin tissue preparation and may successfully be used for tissue of early developmental stages.

23.2.2. Ultrastructural Identification of Biogenic Monoamines

No direct identification of MAs on the ultrastructural level is yet possible. Correlation between MA and vesicle size, form, or electron density in conventionally fixed tissue is generally not possible (Schürmann and Klemm, 1973; Klemm, 1976). Identification of MA-containing neurons on the submicroscopical level may, however, be possible on identified MA-containing neurons (Lafon-Cazal and Arluison, 1976; Rutschke et al., 1976). Its validity should be proven on different species, tissues, and MAs.

23.2.3. Immunohistochemical Methods

23.2.3.1. Theoretical Considerations. Immunohistochemical methods are increasingly used to identify MA-containing neurons. The antibodies are raised in mammals and are usually directed against mammalian enzymes involved in synthesis of MA or the MA itself. In tissue preparation, the antibody couples to the antigen of the tissue. In the commonly used "indirect methods" (Sternberger, 1979; Vandesande, 1979; Klemm, 1983), the antigen-antibody complex is subsequently made visible by a labeled second or third layer of antibody, which couples to the antigen-antibody complex.

Antibodies against MA-synthesizing enzymes are routinely used in mammalian nervous tissue. Our knowledge of enzymes involved in MA synthesis in arthropods is fragmentary. Enzymes involved in MA synthesis in mammals often are species specific. They may share the same function as the enzymes in the MA synthesis of invertebrates (Klemm, 1976, 1985; Evans, 1980). The enzymes may not, however, be chemically identical in different animals and thus may not share in immunogenic properties.

Immunohistochemical labeling should be interpreted with caution. One should be aware that the antibodies are directed against antigenic sites of the immunogen and not against the whole molecule (Pool et al., 1983). Related, and even nonrelated molecules, which share the antigenic site of the immunogen, react with the antibody and may be sources of nonspecific labeling. For specificity testing, antisera absorbed with (uncoupled) "antigens" are commonly used, and specificity is defined when no more labeling occurs. The validity of such an absorption test is overestimated, because the "antigen" is not identical with the immunogenic compound used to induce the antigenic reaction in the donor. Thus, for absorption tests coupled antibodies (identical to the immunogenic compound) should be preferred. even when the antiserum is absorbed with coupled antigen, the absorption test will rather characterize the reactivity of the antibody than provide information about the reactive site of the tissue (Pool et al., 1983; Klemm et al., 1986). The chemical identification of the immunoreactive site in the tissue provides some important contribution to the specificity of antigen binding. However, only the identification of the antigen in the immunoreactive cell definitely determines the specificity of the antibody labeling. Unless such identification test has been made, all "specific" immunoreactions should be characterized as, for example, DβH- or 5-HT-like immunoreactivity (DβHi, 5-HTi). On the other hand, negative results may not exclude the presence of an antigen. Fixation procedures may denaturate or mask the antigen. Improper fixation results in translocation of the antigen. In the PAP method, negativity may result from high concentration of antigen-antibody complex, which may cause binding to both antigenic sites of the second antibody and thus fails to bind to the PAP complex (Pool et al., 1983). All the above factors reduce the specificity of immunohistochemistry unless other additional techniques have been used.

Antibodies raised in rabbits against bovine dopamine-β-hydrolase (DβH) were recently successfully applied to insect nervous tissue (Klemm et al., 1986), but failed to label neurons in Crustacea (Klemm, unpublished observation). DβH converts DA into NA and may also convert tyramine into octopamine. The distribution of DβHi was found to be partly identical with CA-containing neurons and is obviously not related to octopamine, and some DβHi is not correlated to CA. There is evidence that enzymes, which are normally involved in the synthesis of CAs, do not necessarily indicate the presence of CAs (Grzanna and Coyle, 1978; Klemm et al., 1981).

Small molecules, such as those of MAs, are identical in all animals; thus, one can expect that antibodies to MAs can be used regardless of the species studied. Several antibodies to different MAs were tested on arthropod tissue. So far, only the antibodies against 5-HT (Steinbusch et al., 1983) had been proved to be specific. Because serotonin is not immunogenic itself, it was coupled to bovin serum albumen (BSA) by formaldehyde as a coupling reagent. The antibodies against the 5-HT-protein conjugate were raised in rabbits. Antibodies with BSA affinity were absorbed with the carrier protein. According to Schipper and Tilders (1983) the affinity of the antibodies is higher to 6-hydroxytetrahydro-β-carboline than to 5-HT. As can be seen in Figure 23.2., 6-hydroxytetrahydro-β-carboline is formed when 5-HT reacts with aldehydes, for example, during formaldehyde coupling between 5-HT and BSA and during the fixation of tissue with formaldehyde. Thus, fixation of the tissue contributes to the specificity of antibody coupling.

Antibodies against 5-HT were successfully used to identify 5-HT in species of many systematic groups (see Steinbusch et al., 1983; Klemm, 1985). The specificity of anti-5-HT coupling in neurons of insects was verified by chemical identification of 5-HT in the nervous system by comparison with results obtained by aldehyde fluorescence methods (Klemm and Sundler, 1983; Klemm et al., 1984; Schürmann and Klemm 1984). Not only neurons emitting typical 5-HT fluorescence (Ex max 410/Em max 520–530 nm) after aldehyde treatment react with anti-5-HT, but also those with an atypical fluorophore (Ex max 390–410/Em max 510 nm; Klemm and Axelsson, 1973; Klemm, 1976). 5-HTi was found in the olfactory lobe in the locust *Schistocerca gregaria* in which no 5-HT fluorescence has been seen with the Falck-Hillarp method, but in which neurons revealed uptake characteristics of indolylalkylamine-containing neurons (Klemm and Schneider, 1975; Klemm and Sundler, 1983; Figs. 23.2. and 23.4.). In the olfactory lobe of the related locust, *Locusta migratoria* 5-HT fluorophore (see Section 23.2.1.2.) and 5-HTi were detected (Tyrer et al., 1984). Due to the high sensitivity (Steinbusch et al., 1983) of the antibodies, more 5-HTi sites were found than 5-HT fluorescence. Cross-reaction with other MAs is neglectable after formaldehyde fixation (Steinbusch et al., 1983; Klemm et al., 1984).

23.2.3.2. Procedures for Light Microscopical Identification of 5-HTi (and DβHi) Neurons (see Klemm, 1983).

Procedures for Sections. A. Fluorescein-isothiocyanate (FITC) technique:

1. Dissect out the nervous tissue and fix at 4°C in freshly prepared physiological solution containing 4% (v/v) paraformaldehyde (pH 7.2–7.6) for 2–12 hours.

2. Rinse the tissue in phosphate-buffered saline (PBS) several times for 24–48 hours. Add 15–20% (w/v) sucrose to the last change.

3. Freeze the specimens on a cryostat chuck with Tissue Teck II (Lab-

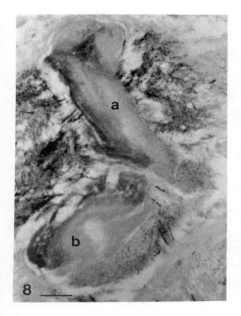

Figure 23.8. Frontal section of the brain of adult *Schistocerca gregaria*, showing 5-HTi in the α- (a), β- (b) lobe and in the surrounding neuropil. Compare with the distribution of CAs in Figure 23.3. Scale: 50 μm.

Tech Products, Naperville IL, section (8–12 μm) in a cryostat (knife temperature: −20−−25°C), and adhere to cold microscopic slides pretreated with chrome alum-gelatine.

4. Incubate sections in antiserum against 5-HT (or DβH) dissolved 1:1000 in PBS containing 0.25% (v/v) Triton X-100, 0.25% (w/v) human (HSA), or bovine (BSA) serum albumen, and 0.1% (w/v) sodium azid for 12–16 hours at 4°C.

5. Rinse solution at room temperature in PBS twice for 15 min, and blot sections.

6. Incubate with diluted FITC-labeled sheep anti-rabbit (secondary) serum (SBL, Stockholm, Sweden), diluted 1:50 in PBS containing Triton X-100, HSA and sodium azide (see step 4) at room temperature for 1 hour.

7. Same as step 5.

8. Mount in PBS-glycerol (1:1–1:3 v/v).

9. View in fluorescence microscope equipped with proper filter setting (e.g., Zeiss filter BP 485, FT 510, LP 520).

B. PAP technique (Sternberger, 1979; Fig. 23.8.):

1. Fix tissue either according to step 1 in FITC technique or in ice-cold 0.01 M PBS containing 4% paraformaldehyde and 0.2% (v/v) saturated picric acid (Stefanini et al., 1967) for 5–12 hours.

2. Follow step 2 in FITC technique.

3. Dehydrate in alcohol, clear in xylene, and embed in paraffin.

4. Section (8–12 μm) and adhere the sections on albumen-glycerol pretreated microscopic slides.

5. Deparaffinize and rehydrate via absolute methanol containing 0.03% H_2O_2 for 30 min.

6. Wash in PBS 5–12 hours.

7. Follow step 4 and 5 in FITC technique.

8. Incubate with (secondary) unlabeled goat anti-rabbit serum (SBL, Stockholm, Sweden or Nordic, Tilburg, The Netherlands) diluted in 1:50 (see step 6 in FITC technique) for 1 hour at room temperature.

9. Blot and incubate the sections in rabbit peroxidase-antiperoxidase (PAP) complex diluted 1:50 (see step 8) for 1 hour at room temperature.

10. Rinse in 0.05 M Tris-buffer (pH 7.6) for 5 min.

11. React with 40–60 mg DAB (3,3'-diaminobenzidine tetrahydrochloride, Sigma) diluted in 100 mL 0.05 M Tris-buffer (pH 7.6), including 0.01% (v/v) H_2O_2 for 10–30 min at room temperature.

12. Rinse, dehydrate in alcohol and xylene, and mount in Permount (Fisher, USA).

Procedures for Wholemounts. Because many MA-containing neurons are wide-field neurons (Beltz and Kravitz, 1983; Bishop and O'Shea, 1983; Klemm and Sundler, 1983; Nässel and Klemm, 1983; Klemm, 1983a; Klemm et al., 1984; Schürmann and Klemm, 1984; Tyrer et al., 1984) wholemount techniques can be advantageous to identify entire neurons and their projections. The wholemount preparation of insect brain suffers from little penetration of the antibody, partly due to the neurilemma. The results on embryonic brains are thus more satisfactory than on adult brains (Nässel and Cantera, 1985). The technique described below is adapted from Nässel and Elekes (1984) and Nässel and Klemm (unpublished information). It can also be used for electron microscopical preparations after slide variations (Fig. 23.9.). (For wholemount FITC technique see Beltz and Kravitz, 1983; Klemm et al., submitted):

1. Fix tissue in 0.01 M phosphate buffer containing 4% paraformaldehyde for 4–12 hours at 4°C.

2. Wash in buffer for 48 hours.

3. Dehydrate in alcohol, clear in xylene (5 min), and rehydrate.

4. Incubate in anti-5-HT (1:2000) for 48 hours at 4°C.

5. Wash in buffer five times, 10 min.

6. Incubate blotted tissue in secondary (unlabeled) antibody (1:50) at room temperature for 2–3 hours (see PAP technique step 8).

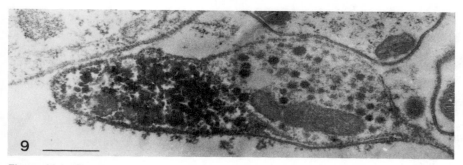

Figure 23.9. Electron micrograph of two nerve fibers distal to the lamina synaptic layer. One process with 5-HTi. Scale: 0.5 μm. (From Nässel et al., 1985; with permission.)

7. Same as step 5.

8. Blot and incubate tissue in PAP complex (1:50) (see PAP technique, step 9).

9. Rinse in 0.05 M Tris-HCl buffer (pH 7.6) for 20 min.

10. React with DAB for 20–30 min (see PAP technique, step 11).

11. Wash in sodium cacodylate buffer (three times, 10 min).

12. Mount and view under microscope.

23.2.3.3. Procedure for Electron Microscopical Identification (Fig. 23.9.).

1. Fix specimens in 0.01 M phosphate buffer containing 4% paraformaldehyde and 0.25% glutaraldehyde (pH 7.2) for 4–12 hours.

2. Wash in buffer five times in 15 min.

3. Follow wholemount technique from step 4 to 11.

4. Postfix in 2% (w/v) OsO_4 in 0.16 M sodium cacodylate buffer for 2 hours at 4°C.

5. Wash in cacodylate buffer.

6. Dehydrate in alcohol and propylene oxide and embed in Durcupan (Fluka).

7. Cut thick (25 μm) sections.

8. Remount sections in Durcupan (Nässel, 1983).

9. Cut ultrathin sections.

10. Contrast sections with uranyl acetate (30 min) and lead citrate (4 min).

Comments. The DAB precipitate, indicative for 5-HTi, is seen in profiles with large (80–120 nm) granular vesicles along the inner surface of the plasmalemma, on membranes of mitochondria, and large granular vesicles (Nässel and Elekes, 1984; Nässel et al., 1985). As the DAB reaction product often prevents the observation of details, especially at sites of membrane contacts, additional conventional electromicroscopic preparations are helpful. The intraneuronal distribution of 5-HTi does not necessarily reflect the original localization of antigen. One should also be aware that in wholemount preparations as well as in electron microscopical preparations, the activity of the endogenous peroxidase was not prevented. Thus, knowledge about the specificity of the labeling is necessary before the results can be interpreted.

23.3. SUMMARY

The most specific methods presently known for intraneuronal identification of monoamines are described and discussed. Generally, the methods can be devided into direct and indirect methods. The direct method (aldehyde-induced fluorescence) is highly specific for certain (fluorogenic) monoamines. It is very sensitive to primary catecholamines. Indirect methods either demonstrate an enzyme or a specific reaction (uptake), which is suggested to be involved with the presence of monoamines, or immunohistochemically label the monoamines in the tissue. Especially for the indirect methods, their specificity has to be verified by other methods. The antibodies against monoamine-synthesizing enzymes are mostly species specific. Because they were raised against mammalian enzymes, they often fail to react in arthropod tissue. So far the only reaction achieved was with bovine antidopamine-β-hydrolase in insect tissue. Its immunoreactivity is only partly correlated to the presence of catecholamines (noradrenaline) but not to octopamine. Immunohistochemical techniques can be used for wholemounts and on the ultrastructural level.

ACKNOWLEDGMENTS

This work was performed during my stay at the Department of Zoology in Lund and was sponsored by grants from the Swedish Natural Science Research Council to Professor R. Elofsson. For comments and technical help I thank professor R. Elofsson, Dr. D. Nässel, Ylwa Andersson, and Inger Norling.

REFERENCES

Beltz, B. S. and E. A. Kravitz. 1983. Mapping of serotonin-like immunoreactivity in the lobster nervous system. *J. Neurosci.* **3**: 585–602.

Bishop, C. A. and M. O'Shea. 1983. Serotonin immunoreactive neurons in the central nervous system of an insect (*Periplaneta americana*). *J. Neurobiol.* **14**: 251–269.

Björklund, A., B. Falck, and O. Lindvall. 1975. Microspectrofluorometric analysis of cellular monoamines after formaldehyde or glyoxylic acid condensation, pp. 249–294. In P. B. Bradley (ed.), *Methods in Brain Research*. John Wiley, London.

Clarke, B. S. and J. F. Donnellan. 1982. Concentrations of some putative neurotransmitters in the CNS of quick-frozen insects. *J. Insect Physiol.* **12**: 623–638.

Elofsson, R., L. Laxmyr, E. Rosengren, and Ch. Hansson. 1982. Identification and quantitative measurements of biogenic amines and DOPA in the central nervous system and haemolymph of the crayfish *Pacifastacus leniusculus* (Crustacea). *Comp. Biochem. Physiol.* **71C**: 195–201.

Evans, P. 1980. Biogenic amines in the insect nervous system. *Adv. Insect Physiol.* **15**: 317–473.

Flanagan, T. R. J. 1983. Monoaminergic innervation in a hemipteran nervous system: A wholemount histofluorescence survey, pp. 317–338. In N. J. Strausfeld (ed.), *Functional Neuroanatomy*. Springer-Verlag Heidelberg.

Grzanna, R. and J. T. Coyle. 1978. Dopamine-β-hydrolase in rat submandibular ganglion cells which lack norepinephrine. *Brain Res.* **151**: 206–214.

Klemm, N. 1976. Histochemistry of putative transmitter substances in the insect brain. *Prog. Neurobiol.* **7**: 99–169.

Klemm, N. 1980. Histochemical demonstration of biogenic monoamines (Falck-Hillarp method) in the insect nervous system, pp. 521–73. In N. J. Strausfeld, T. A. Miller (eds.), *Neuroanatomical Techniques*. Springer-Verlag, New York.

Klemm, N. 1982. A modified glyoxylic acid-formaldehyde technique for histofluorescence of catecholamine-containing neurons in cryostat sections of the insect brain. *J. Histochem. Cytochem.* **30**: 398–400.

Klemm, N. 1983a. Monoamine-containing neurons and their projections in the brain (supraoesophageal ganglion) of cockroaches. An aldehyde fluorescence study. *Cell Tissue Res.* **229**: 379–402.

Klemm, N. 1983b. Detection of serotonin-containing neurons in the insect nervous system by antibodies to 5-HT, pp. 303–316. In N. J. Strausfeld (ed.), *Functional Neuroanatomy*. Springer-Verlag, Heidelberg.

Klemm, N. 1985. The distribution of biogenic monoamines in invertebrates, pp. 280–296. In R. Gilles and J. Balthazart (eds.), *Neurobiology*. Springer-Verlag, Berlin.

Klemm, N. and S. Axelsson. 1973. Detection of dopamine, noradrenaline and 5-hydroxytryptamine in the cerebral ganglion of the desert locust, *Schistocerca gregaria* Forsk. (Insecta, Orthoptera). *Brain Res.* **57**: 289–298.

Klemm, N. and A. Björklund. 1971. Identification of dopamine and noradrenaline in nervous structures of the insect brain. *Brain Res.* **26**: 459–464.

Klemm, N. and B. Falck. 1978. Monoamines in the pars intercerebralis-corpus carciacum complex of locusts. *Gen. Comp. Endocrinol.* **34**: 180–192.

Klemm, N. and L. Schneider. 1975. Selective uptake of indolamine into nervous fibers in the brain of the desert locust, *Schistocerca gregaria* Forskal (Insecta). A fluorescence and electron microscopical investigation. *Comp. Biochem. Physiol.* **50C**: 177–182.

Klemm, N. and F. Sundler. 1983. Organization of catecholamine and serotonin-immunoreactive neurons in the *corpora pedunculata* of the desert locust, *Schistoerca gregaria* Forsk. *Neurosci. Lett.* **36**: 13–17.

Klemm, N., H. W. M. Steinbusch, and F. Sundler. 1984. Distribution of serotonin-containing neurons and their pathways in the suboesophageal ganglion in the cockroach *Periplaneta americana* (L.) as revealed by immunocytochemistry. *J. Comp. Neurol.* **225**: 387–395.

Klemm, N., D. R. Nässel, and N. N. Osborne. 1985. Dopamine-β-hydrolase-like immunoreactive neurons in two insect species, *Calliphora erytrocephala* and *Periplaneta americana*. *Histochemistry* **83**: 159–164.

Lafon-Cazal, M. and M. Arluison. 1976. Localization of monoamines in the *corpora cardiaca* and the hypercerebral ganglion of locusts. *Cell Tissue Res.* **172**: 517–527.

Laxmyr, L. 1984. Biogenic amines and DOPA in the central nervous system of decapod crustaceans. *Comp. Biochem. Physiol.* **77C**: 139–143.

Lindvall, O., A. Björklund, and L.-A. Svensson, 19744. Fluorophore formation from catecholamines and related compounds in the glyoxylic acid fluorescence histochemical method. *Histochemistry* **39**: 197–227.

Maxwell, G. D., J. F. Tait, and J. G. Hildebrand, 1978. Regional synthesis of neurotransmitter candidates in the CNS of the moth *Manduca sexta*. *Comp. Biochem. Physiol.* **61C**: 109–119.

Mercer, A. R., P. G. Mobbs, A. P. Davenport, and P. D. Evans. 1983. Biogenic amines in the brain of the honeybee, *Apis mellifera*. *Cell Tissue Res.* **234**: 655–677.

Meyer, W. and Jehnen, R. 1980. The distribution of monoamine oxidase and biogenic monoamines in the central nervous system of spiders (Arachnida: Araneida). *J. Morphol.* **164**: 69–91.

Mir, A. K. and P. F. T. Vaugham. 1981. Biosynthesis of N-acetyldopamine and N-acetyloctopamine by *Schistocerca gregaria* nervous tissue. *J. Neurochem.* **36**: 441–446.

Myhrberg, H., R. Elofsson, R. Aramant, N. Klemm, and L. Laxmyr. 1979. Selective uptake of exogenous catecholamines into nerve fibers in crustaceans. A fluorescence histochemical investigation. *Comp. Biochem. Physiol.* **62C**: 141–150.

Nässel, D. R. 1983. Horseradish peroxidase and other heme proteins as neuronal markers, pp. 44–91. In N. J. Strausfeld (ed.), *Functional Anatomy*. Springer-Verlag, Heidelberg.

Nässel, D. R. and R. Cantera. 1985. Mapping of serotonin-immunoreactive neurons in the larval nervous system of the flies *Calliphora erythrocephala* and *Sarcophaga bullata*. A comparison with ventral ganglia in adult animals. *Cell Tissue Res.* **239**: 423–434.

Nässel, D. R. and K. Y. Elekes. 1984. Ultrastructural demonstration of serotonin-immunoreactivity in the nervous system of an insect (*Calliphora erytrocephala*). *Neurosci. Lett.* **48**: 203–210.

Nässel, D. R. and N. Klemm. 1983. Serotonin-like immunoreactivity in the optic lobes of three insect species. *Cell Tissue Res.* **232**: 129–140.

Nässel, D. R. and L. Laxmyr. 1983. Quantitative determination of biogenic amines and DOPA in the CNS of adult and larval blowflies, *Calliphora erythrocephala*. *Comp. Biochem. Physiol.* **75C**: 259–265.

Nässel, D. R., E. P. Meyer, and N. Klemm. 1985. Mapping and ultrastructure of serotonin-immunoreactive neurons in the optic lobes of three insect species. *J. Comp. Neurol.* **323**: 190–204.

O'Connor, E. F., W. H. Watson, and G. A. Wyse. 1982. Identification and localization of catecholamines in the nervous system of *Limulus polyphemus*. *J. Neurobiol.* **13**: 49–60.

Omar, D., and L. L. Murdock. 1982. Determination of N-acetyldopamine by liquid chromatography with electrochemical detection. *J. Chromatogr.* **224**: 310–314.

Osborne, N. N. 1979. Is Dale's principle valid? *TINS* **2/3**: 73–75.

Pool, Ch. W., R. M. Buijs, D. F. Swaab, G. J. Boer, and F. W. van Leeuwen. 1983. On the way to a specific immunocytochemical localization, pp. 1–46. In A. C. Cuello (ed.), *Immunocytochemistry. IBRO, Methods in the Neurosciences*, vol. 3. Wiley, Chichester.

Rutschke, E., D. Richter, and H. Thomas. 1976. Autoradiographische Untersuchungen zum Einbau von 3H-Dopamin und 3H-Hydroxytryptamin in das Gehirn von *Periplaneta americana* L. *Zool. Jahrb. Anat.* **95**: 439–447.

Schipper, J. and F. J. H. Tilders. 1983. A new technique for studying specificity of immunocytochemical procedures: Specificity of serotonin immunostaining. *J. Histochem. Cytochem.* **31**: 12–18.

Schürmann, F. W. and N. Klemm. 1973. Zur Monoaminverteilung in den *Corpora pedunculata* des Gehirns von *Acheta domesticus* L. (Orthoptera, Insecta). Histochemische Untersuchungen, mit Vergleich zur Struktur und Ultrastruktur. *Z. Zellforsch.* **136**: 393–414.

Schürmann, F. W. and N. Klemm. 1984. Serotonin-immunoreactive neurons in the brain of the honeybee. *J. Comp. Neurol.* **225**: 570–580.

Stefanini, M., C. de Martino, and L. Zamboni. 1967. Fixation of ejaculated spermatozoa for electron microscopy. *Nature (London)* **216**: 173–174.

Steinbusch, H. W. M., A. A. J. Verhoefstad, and H. W. J. Joosten. 1983. Antibodies to serotonin for neuroimmunocytochemical studies: Methodological aspects and applications, pp. 193–214. In C. A. Cuello (ed.), *Immunocytochemistry, IBRO, Methods in Neurosciences,* vol. 3. Wiley, Chichester.

Sternberger, L. A. 1979. *Immunocytochemistry*. Second edition. John Wiley, New York.

de la Torre, J. C. and J. W. Surgeon. 1976. A methodological approach for rapid and sensitive monoamine histofluorescence using a modified glyoxylic acid technique. *Histochemistry* **49**: 81–93.

Tyrer, N. M., J. D. Turner, and J. S. Altman. 1984. Identifiable neurons in the locust central nervous system that react with antibodies to serotonin. *J. Comp. Neurol.* **227**: 313–330.

Vandesande, F. 1979. A critical review of immunocytochemical methods of light microscopy. *J. Neurosci. Meth.* **1**: 3–23.

Wallace, B. G. 1976. The biosynthesis of octopamine. Characterization of lobster tyramine-β-hydrolase. *J. Neurochem.* **2**: 761–770.

CHAPTER 24

Strategies for Neuronal Marking in Arthropod Brains

Dick R. Nässel

Department of Zoology
University of Lund
Lund, Sweden

24.1. INTRODUCTION

Many techniques exist for displaying the morphology of nerve cells and the architectonics of neuron populations in the brain. These can be used for qualitative and quantitative studies to acquire a catalogue of neuron types and an idea of the packing and numbers of neurons in different parts of the central nervous system (CNS). A more functional approach to neuroanatomy requires selective marking of neurons, so that one can resolve, for instance, (1) sets of connected neurons, (2) neurons with common origin or destination, (3) chemically defined neurons, (4) active neurons, and (5) neurons whose synaptic connections and/or responses to various stimuli have been monitored electrophysiologically. Neuronal marking is also useful in conjunction with experimental work, such as developmental studies and surgical and pharmacological lesioning.

Blackstad et al. (1981) divided the neuroanatomical techniques into four main categories: (1) Golgi methods and intracellular dye injections, which display the morphology of normal adult and developing neurons, (2) methods to display degeneration or reactive changes in various parts of a neuron as a consequence of mechanical or chemical lesions; (3) tract-tracing methods, which rely on transport or diffusion of a marker throughout the length of a neuron; and (4) histochemical methods including monoamine histofluorescence methods, immunohistochemistry, activity staining (e.g., radiolabeled 2-deoxyglucose method), lectin binding, and so on. Most techniques in these four categories can be brought to the electron microscope level for fine structural analysis of neuron morphology and synaptic contacts.

This chapter does not give complete schedules for all marking techniques, rather as the title suggests, it lists approaches for analyzing the functional anatomy of the arthropod brain. Where schedules are not complete, references are given to original descriptions of methods.

24.2. PROBLEMS THAT MAY BE SOLVED WITH VARIOUS MARKING TECHNIQUES

In this section examples will be given of different types of neurobiological problems that can be approached with neural marking techniques. To many researchers, these approaches are obvious, but still much work remains to be done even in the most well-studied arthropod model species. Although there are exquisite data on electrophysiologically monitored and dye-injected single cells from the CNS of many arthropods, rarely is the neuronal matrix, in which these cells are embedded, known in any detail. The functional organization of the largest portion of the arthropod brain is unfortunately beyond our grasp with methods such as Golgi impregnation and simple dye infusion. Rather, one should use multiple combinations of techniques.

To reveal the morphology of groups of sensory, motor, and interneurons and their projections as well as neuropil organization, some of the following strategies may be used: (1) Backfill the severed peripheral portions of the sensory or motorneurons. This mass filling technique is useful also for developmental studies of these neuron types as well as for interspecies comparisons. (2) Fill intersegmentally running neurons via the severed connectives between ganglia. This way ascending and descending neurons of the brain may be labeled. (3) Apply label into localized lesions or use focal injection of dye into tracts or termination areas of the brain. By this method, the morphologies and projections of relay neurons connecting different centers in the CNS can be resolved. (4) More specifically, focal injections of, for example, cobalt chloride may result in labeling of assemblies of neurons forming a portion of the matrix of a neuropil region (Strausfeld and Hausen, 1978). This technique when properly executed may label entire populations of various neuron types and can be used for quantitative studies as well as for revealing architectures and local variations in these in neuropil areas of the brain. (5) Transneuronal labeling and dye coupling may provide useful information on functional neuronal pathways and synaptically coupled sets of neurons (Stewart, 1978; Nässel, 1981; Strausfeld and Bassemir, 1983). At least, when using cobalt chloride this phenomenon can also be used for quantitative and architectural studies. (6) Histochemical (and immunocytochemical) identification of neurons provides information about contents of, for example, metabolic enzymes or putative neuroactive substances. Besides, it can also be used as a neuroanatomical tool. These techniques often display the neurons in their entirety, and the architectonics of whole populations of chemically defined neurons can be displayed in adult and developing brains. As will be described later, such displays may be very revealing and help in understanding of brain organization. Into this category can also be included studies using uptake of transmitters, their ligands, or different neurotoxins. These compounds can be used as they are or radiolabeled. The uptake can then be demonstrated autoradiographically, histochemically, or by the possible toxic effects on cell cytology. (7) Activity staining gives information on groups of active cells (see Buchner and Buchner, 1983). (8) Finally, a useful approach is to combine several of the above techniques for double or multiple marking. By this means, besides untangling complex architectonics (superimposed neuronal patterns), one can analyze convergent and divergent pathways, synaptic contacts between identified neurons, and the presence of one or more neuroactive substances in an identified neuron. Most of the above techniques may further be used at the electron microscopic level for synaptic studies.

Used singly, no technique is sufficient for a functional analysis of the nervous system. Adams et al. (1983) elegantly describe a multidisciplinary approach for neuron identification that allows studies of the action of certain transmitters released by identified insect neurons at specific targets. They combine intracellular dye injection and recording with vital staining, biochemical analysis after single-cell isolation, bioassay, and immunocytochemistry. This approach applies well to neurons in which the target of the released neuroactive substance is accessible. The actions of neurotransmitters and modulators is less easy to assay in the brain; hence, the present chapter will be limited to the morphological and histochemical aspects of neuronal analysis. All the described methods basically rely on mass labeling of neurons with some common denominator (chemical identity, destination of processes, innervation field, etc). They can, however, be combined with intracellular techniques, as described by Adams et al. (1983).

24.3. MARKERS USED FOR UPTAKE BY OR FILLING OF NEURONS

24.3.1. Markers and Their Visualization

A number of fluorescent dyes have been used for marking arthropod neurons either by backfilling (iontophoretically or by passive diffusion), extra- or intracellular injection, or tissue flooding, resulting in selective uptake. The most common markers are Procion yellow, Lucifer yellow CH, and propidium iodide (Stewart, 1978; DeOlmos and Heimer, 1980). These dyes are normally dissolved in water at concentrations of 2–5%, but they can also be used in their crystalline or powder form or as pellets.

Uptake time varies, depending on the distance the dye must migrate. Intracellularly injected Lucifer yellow spreads very fast throughout a neuron in less than 1 min (Stewart, 1978). The other dyes probably spread as fast (unless they combine with cytoplasmic macromolecules), but due to lower fluorescence yield, more dye needs to pass into the neuron for visibility and hence longer uptake time is needed. For filling cut axons with Lucifer yellow, filling time should be at least 20 min for each millimeter of dye passage.

After filling with a fluorescent dye, the tissue is fixed in 4% paraformaldehyde in phosphate buffer (for Lucifer yellow, the aldehyde can be dissolved in absolute methanol) for approximately 4 hours. The filled neurons can be viewed in cleared wholemounts, cryostat sections, paraffin sections, or semithin sections (using Spurr's embedding medium). Lucifer yellow-filled tissue can also be postfixed in 2% osmium tetroxide, and the fluorescent structures seen in the light microscope can be correlated with profiles in the electron microscope (Stewart, 1981; Heinrichs, personal communication). Another strategy is to expose Lucifer yellow-filled neurons to diaminobenzidine and intense blue light to produce an electron-dense precipitate inside the marked neurons (Maranto, 1982).

Cobalt and nickel chloride have been used extensively for axonal filling. Solutions of 1–6% of either chloride alone or in various mixtures (see Sakai and Yamaguchi, 1983) in distilled water are commonly used. Also, the crystals can be applied to the cut axons. Uptake time for cobalt and nickel varies, but approximately 20 min/mm should be allowed. This is followed by removing the source of marker and subsequently leaving the tissue for further diffusion of the marker. This diffusion time can be as long as the uptake time. The cobalt or nickel is precipitated by immersing the exposed nervous tissue in ammonium sulphide (4 drops of $(NH_4)_2S$ in 10 ml 0.1 M sodium cacodylate buffer) for 2–4 min (on a rotator). Thereafter, the tissue may be fixed in either a coagulant fixative, such as Carnoy, acetic acid-alcohol-formalin, or alcoholic Bouin, or in aldehyde fixatives, such as 4% paraformaldehyde with 0.25% glutaraldehyde in 0.1 M sodium cacodylate buffer. Fixation time should be 4–12 hours.

An alternative method for precipitation of cobalt and nickel chloride is to use rubeanic acid (Quicke and Brace, 1979). With this compound, the two chlorides produce different colors and can thus be used for double-labeling experiments.

It is often necessary to intensify the precipitated cobalt or nickel to achieve detailed labeling of finer processes or spines. The most reliable method (Bacon and Altman, 1977) uses silver intensification. After intensification the filled neurons can be viewed as cleared wholemounts or in thick or semithin sections of resin-embedded tissue. It is also possible to perform the silver intensification on cryostat or paraffin sections or directly on semithin or ultrathin sections (Tyrer et al., 1980).

Heme peptides constitute a third group of neuronal markers. A number of them have been used for neuronal marking or axonal transport studies: catalase, lactoperoxidase, horseradish peroxidase (HRP), hemoglobin, myoglobin, cytochrome c, and microperoxidase (Malmgren et al. 1978; Nässel, 1983a,b). These peptides can be transported actively in axons or diffuse passively in intact or injured neurons (Malmgren et al., 1978; Nässel, 1983b). HRP has been extensively used to trace pathways and neuronal connections in the CNS of vertebrates. In studies of arthropods, cobalt and Lucifer yellow have in most cases been preferred to HRP labeling. As will be shown in this chapter, however, heme peptides can be quite useful alone or in combination with other markers for the analysis of the arthropod brain.

The heme peptides are used in solution at concentrations of 2–4% (in 0.1 M KCl or an appropriate saline) or as dry pellets. Pellets are made either by air-drying a concentrated peptide solution and then fragmenting needle-shaped pieces (Nässel, 1983a) or by ejecting a small amount of solution out of the tip of a fine glass microcapillary, which after air-drying, can be advanced into the nervous tissue (Meyer, 1984). The uptake time varies for the different heme peptides. For HRP, the time is about twice that for cobalt uptake. For the other heme peptides, at least twice the time for HRP is required. This is probably because of the lower peroxidase activity of these peptides. The heme peptides useful for neuronal marking are limited to HRP, cytochrome c, and myoglobin (Nässel, 1983b). The preparations from Sigma have been most commonly used, and others are listed in Nässel (1983a). Fixation must be executed in aldehydes, for example, 2.5% glutaraldehyde in sodium cacodylate buffer, 4% paraformaldehyde in phosphate buffer, or a mixture of 4% paraformaldehyde and 0.25% glutaraldehyde in phosphate buffer. Glutaraldehyde in cacodylate buffer gives the best results for light and electron microscopy. Formaldehyde partially inactivates the peroxidase, but in low concentration and with thorough washes in buffer this fixation gives the best results when combining HRP labeling with Lucifer yellow or antibody labeling.

After fixation and thorough washes in buffer, the heme peptide is visualized by means of a chromagen and H_2O_2. The most commonly used chromagens are 3,3'-diaminobenzidine tetrahydrochloride (DAB), tetramethyl benzidine, or p-phenylenediamine-pyrocatechol (PPD-PC) (Hanker-Yates agent). The chemistry and histochemistry of these compounds and original references to their use are given in Nässel (1983a). The tissue is treated histochemically either as whole brains, or as cryostat-, vibratome-, or tissue chopper sections. For whole brains (or ganglia) a preincubation in 0.04% DAB without H_2O_2 for 1–3 hours is recommended, followed by a 1–3-hour incubation in 0.04% DAB, containing 0.1% H_2O_2 (all dissolved in, for example, 0.1 M sodium cacodylate buffer). Tissue sections require no preincubation and should have shorter reaction time with chromagen (about 10–20 min). After washes in buffer, the tissue may be postfixed in 1–2% osmium tetroxide in 0.1 M cacodylate buffer. This enhances the contrast of labeling due to the osmophilic nature of the DAB reaction product, and also makes the tissue useful for electron microscopical analysis. The whole brains that have been treated histochemically are embedded in an appropriate resin (e.g., Spurr's medium or Araldite), and thick sections (20–40 μm) are made for light microscopical analysis. Thick sections containing interesting neuronal profiles can be reembedded and cut for electron microscopy (see Nässel, 1983a). As will be described later, marking with HRP and other heme peptides can be combined with a large number of other marking techniques.

Different lectins have been successfully used as axonally transported markers: wheat germ agglutinin tagged with HRP is the most commonly used (Schwab et al., 1978). When using HRP-tagged lectins, the histochemistry is as after normal HRP tracing. An alternative approach is to use unlabeled lectins and then to display them immunocytochemically with specific antibodies (see Section 24.5.). The lectins appear to be more sensitive and detailed tracers than the heme peptides. They can also be used for a variety of multiple labeling techniques.

Several compounds can be used as neuronal tracers when radiolabeled or tagged with fluorescent compounds: tetanus toxin, leucin, fucose, bovine serum albumen, etc. (for original references see Nässel, 1983a).

24.3.2. Application of Markers for Filling of or Uptake by Neurons

24.3.2.1. Filling Cut Axons in Peripheral Nerves and Connectives. Sensory hairs and bristles can simply be cut off and their stumps inserted into a microcapillary filled with marker, or when axons from a sensory field are to be filled, a trough of wax or vaseline can be built up around the field, the bristles cut, and a drop of marker applied to the trough. Isolated peripheral nerves or connectives can similarily be inserted into dye-filled capillaries or troughs. When central portions of sensory or motorneurons are to be filled, the appendage can be cut at the desired level and inserted into a marker-filled capillary (or tube). Photoreceptors in the retina or ocelli may be filled by lesioning them with a razor blade or needle and applying the dye to the lesion. In all cases, it is crucial to seal off around the cut axons to avoid spreading of the dye along the outside of the nerves or connectives. Such spreading often leads to extracellular flooding of the nervous tissue which may result in labeling of undesired neurons unrelated to the peripheral cut.

24.3.2.2. Intracellular Injection of Markers. Details on intracellular injection of markers are given in Kater and Nicholson (1973) and Nässel (1983a).

24.3.2.3. Injection Into Tracts or Termination Areas. With this type of focal or stereotactic dye injection, information can be achieved on origin and/or destination of populations of neurons, on architectures of processes from classes of neurons, and distribution of sensory neurons in the periphery having common central termination areas. Dye-filled electrodes are inserted stereotactically and filling achieved by passive diffusion into damaged neurons or by means of active uptake into intact neurons near the electrode tip. another strategy is to implant small pellets of HRP or Lucifer yellow (e.g., with the tip of a microcapillary) into a desired focus in the brain. The dye will dissolve in extracellular fluid, resulting in a local concentrated pool from which marker enters the neurons. In both techniques, the uptake time should be kept short to avoid too extensive extracellular spread. When using electrodes, these can be withdrawn after 10–30 min (depending on the aim of study and area to be flooded). The timing thus has to be tried out for each desired filling type.

24.3.2.4. Marker Flooding of Living Neural Tissue. Intact neurons become labeled after flooding neuronal tissue with HRP. The labeling is specific to sensory- and motorneurons, which pass through peripheral nerves. Hence, for example, retinal photoreceptors take up marker after applying a drop of HRP inside the head capsule of an insect; leg sensory- and motorneurons get marked after flooding thoracic ganglia. After uptake times of 1–2 hours, such flooding may be a reproducible marking technique for certain populations of neurons (see Nässel, 1983a). Other heme peptides are not taken up by intact neurons (Nässel, 1983b), and flooding with cobalt chloride causes nonspecific darkening of brain tissue.

Lucifer yellow will be taken up selectively by certain populations of cells after tissue (Detwiler and Sarthy, 1981). Wilcox and Franceschini (1984) showed that Lucifer yellow can be incorporated by light-stimulated photoreceptor cells of house flies. Hence, this dye may be used as an activity stain. Tissue exposure to and subsequent selective uptake of neurotoxins and transmitter analogues may also be used for neuronal marking after histochemical processing or by means of toxic effects on the neural cytology (see Klemm, 1976; Nässel and Elekes, 1985). Finally, the vital stain neutral red may, after tissue flooding, selectively mark certain classes of monoamine-containing neurons (Stuart et al., 1974; Evans, 1980).

24.3.2.5. Transneuronal Labeling. Several compounds used for neuronal marking have been shown to pass from the neurons originally exposed to contiguous sets of neurons. These compounds are, for example, tritiated sugars, amino acids and nucleotides, radiolabeled tetanus toxin, acetyl choline, wheat germ agglutinin, cobalt ions, fluorescent low-molecular-weight compounds, and HRP (for references see Nässel, 1983a). The mechanisms by which these markers pass between neurons probably varies. Some of the compounds are metabolic intermediates that may pass selectively through membranes, others are low-molecular-weight compounds that may pass through gap junctions. Larger molecules, such as HRP, may pass through membranes by means of exoendocytosis (Triller and Korn, 1981).

In flies cobalt reproducibly labels entire sets of specific neurons that are presynaptic to the primarily filled neurons (Strausfeld and Bassemir, 1984). HRP transneuronally labels the same sets of neurons as cobalt, but it appears as if the labeling is less complete (Nässel, 1981, 1982). Hence, transneuronal labeling with cobalt may be more useful for quantitative studies, whereas HRP is useful for EM analysis. Lucifer yellow passes transneuronally mainly after intracellular injection (Stewart, 1978).

24.4. IMMUNOCYTOCHEMISTRY FOR NEURONAL MARKING

Immunocytochemical methods for displaying neural antigens, as applied to insects, have been described in several recent reviews (Adams et al., 1983; Duve and Thorpe, 1983; Eckert and Ude, 1983; Klemm, 1983; Nässel, 1987) and in other chapters in this book. These contributions stress the use of immunocytochemistry as a tool to map certain antigens, such as neuroactive substances or their metabolic enzymes. Immunocytochemistry may also be used as a purely anatomical tool to display special populations of neurons with their processes and connections within different brain regions. Hence, this technique can be an aid in studies of adult brain architecture and the development of neural connections. Polyclonal antisera or monoclonal antibodies can be raised against a variety of antigens associated with nerve cells for these purposes.

The use of antibodies to serotonin (5-HT) will be shown here as an example. These antibodies are very specific (see Klemm, 1983) and seem to display serotonin-containing neurons in their entirety in a large number of invertebrate and vertebrate species. When applied to insect brains, constellations of 5-HT-immunoreactive (5-HTi) neurons can be resolved in great detail (Figs. 24.1. and 24.2.A,B). The patterns of immunoreactive processes reveal brain architectures not seen with other techniques, and some new neuron types could be resolved that had been refractory to other labeling techniques (Nässel et al., 1985). An analysis of the segmented nervous system is facilitated by the mapping of 5-HTi neurons both in larval and adult insects (Nässel and Cantera, 1985). Furthermore, the 5-HT-immunoreactivity seems to remain in the same neurons throughout metamorphosis and appears early in newly differentiating cells during postembryonic development. Hence, the morphogenesis of 5-HTi neurons and their formation of connections can be studied immunocytochemically (Nässel et al., 1987; Nässel, 1987; see Fig. 24.1.). Similar studies may be performed, using other antisera, provided that the immunoreactivity appears early and that the labeling occurs also in the finest processes of the neurons.

An interesting finding is that antibodies to HRP are specific markers for insect nerve cells (Jan and Jan, 1982). These antibodies have been extensively used to study the development and axonal navigation of pioneer cells and sensory cells in appendages of insects (see Keshishian and Bentley, 1983; see also Chapter 3 in this volume).

A different approach is to produce monoclonal antibodies to antigens in brain homogenates (Zipser and McKay, 1981; Fujita et al., 1982; Zipursky et al., 1984). Without initially knowing the antigens one can screen adult or developing brain tissue for specific antibody binding to different sets of neurons. As many antigens may be surface molecules in neuronal membranes, monoclonal antibodies can be probes for the molecular basis of cell interactions during axonal pathfinding and synaptogenesis (see Goodman et al., 1984). Monoclonal antibodies can also be used to identify molecules associated with particular patterns in the CNS, providing links between observable structures and the genes (Zipursky et al., 1984).

The antibody binding sites are normally resolved by indirect marking either using a fluorochrome- or HRP-labeled, secondary antibody directed against the primary antibody or with a sandwich technique (more economical and sensitive) using an unlabeled secondary antiserum and then a peroxidase antiperoxidase complex (Sternberger, 1979). For most immunocytochemistry the brain tissue has to be fixed, usually in paraformaldehyde (see papers listed above). The antibody can then be spread onto paraffin, crysostat, or vibratome sections. The best tissue preservation is achieved when using resin-embedded semithin sections (see Meyer et al., 1986). Whole brains can also be incubated in antibodies for production of cleared wholemounts, or after osmium fixation and embedding in resin, thick sections can be cut. These can also be resectioned for electron microscopy (see Nässel and Elekes, 1984, 1985). This latter technique, which uses osmium postfixation and resin embedding, produces good tissue preservation and increased light microscopical resolution of very fine processes due to the osmophilic nature of the reaction product of the peroxidase-antiperoxidase complex (Fig. 24.2.A,B). Immunocytochemical marking can be combined with other mark-

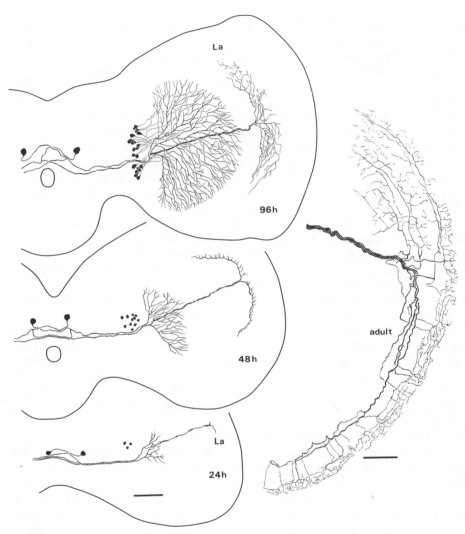

Figure 24.1. Tracings of optic lobe neurons immunoreactive to antibodies raised against serotonin (5-HT) during postembryonic development of the blowfly, *Calliphora*. Drawn from immuno-PAP preparations that were osmium postfixed and cut at 25 μm. Four stages are shown; 24, 48, and 96 hours after pupariation and adult stage (only the lamina portion of cells is shown in adult). The smaller cell bodies increase in number during development. Note the growth of immunoreactive processes in the medulla and lamina (La). Scale bars = 100 μm. (Tracing provided by L. Ohlsson.)

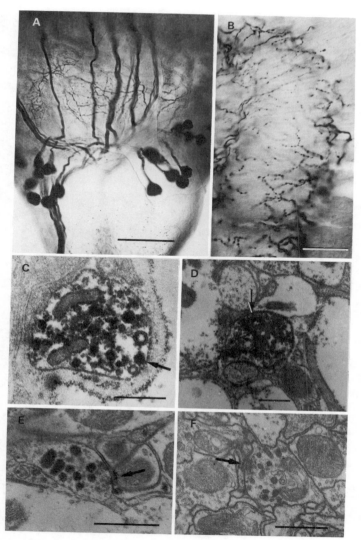

Figure 24.2. 5-HT-immunoreactive neurons in insect brains, immuno-PAP method with osmium postfixation (in A, no osmium was used). (A) Part of optic lobes in the beetle, *Thermophilum*. Wholemount showing cell bodies and processes of immunoreactive neurons. (B) Fine varicose immunoreactive fibers in portion of the lamina from the same species. 25-μm araldite section. (C) Ultrastructure of 5-HT-immunoreactive (5-HTi) terminal in neuronal sheath of cranial nerve of *Calliphora*. *Arrow* indicates large granular vesicle characteristic of 5-HTi terminals. (D) 5-HTi terminal in the medulla of the *Calliphora* optic lobe. *Arrow* points at large granular vesicle. (E) Weakly immunoreactive profile in the antennal lobe of *Calliphora*. This profile forms synapse with other element at *arrow*. (F) In tissue fixed for conventional EM (medulla of *Calliphora*) profiles with large granular vesicles are forming synaptic contacts with other elements. Correlations between immuno-PAP preparations and conventional EM can therefore be made. Scales: A, 100 μm; B, 50 μm; C–F, 0.5 μm.

559

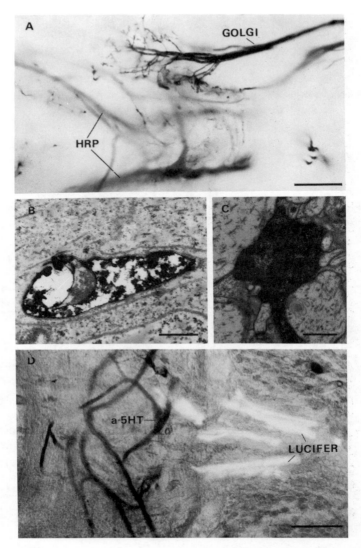

Figure 24.3. Examples of differential marking in the *Calliphora* brain. (A) Golgi-HRP combination marking. These markers can be distinguished easily at the EM level, as shown in the electron micrographs B and C, where B is a Golgi-EM and C a HRP-EM. The highly electron-dense "Golgi precipitate" often becomes displaced, leaving white spaces. (D) Lucifer yellow-filled ocellar interneurons and 5-HT-immunoreactive brain neurons. Paraffin section, using immuno-PAP method. Combined epifluorescence and transmitted light. This method can be used, for example, to search for interactions between the ocellar system and 5-HTi neurons. Scales: A, 50 μm; B and C, 0.5 μm; D, 50 μm.

ers (Fig. 24.3.D). Furthermore, it is possible to produce different primary antisera from two different species (e.g., anti-5-HT and anti-dopamin in rabbit and goat, respectively) for simultaneous double labeling, using differently tagged secondary antisera (directed to the two different animals). Electron microscopical immunocytochemistry is described later.

24.5. HISTOCHEMICAL METHODS

Neuronal monoamines, such as dopamine, noradrenaline, and serotonin, and some related compounds can be visualized after condensation with formaldehyde vapor, glyoxylic acid, or elaborations of these basic procedures (see Klemm, 1980 and Chapter 23 in this volume). The former method uses freeze-dried tissue and subsequent paraffin sectioning. The latter can be performed on cryostat or vibratome sections or intact brains for cleared wholemounts.

Lectins are sugar-binding proteins of nonimmuno origin. They can be used as reagents to recognize complex carbohydrate structures in glycoproteins and glycopeptides of cellular membranes. Hundreds of lectins have been purified and some of these are commercially available (listed in Leathem and Atkins, 1983; Ponder, 1983). Carbohydrate components of cell membrane molecules are important in defining cell types and changes in these components are associated with cellular differentiation and maturation and in cell recognition during axonal pathfinding and synaptogenesis (Edelman, 1976). Lectins can thus be used as a complement to immunocytochemical studies to analyze neuronal development.

Two simple methods for detection of lectin binding will be described briefly. For detailed descriptions see Ponder (1983) and Leathem and Atkins (1983). One is a direct method using lectins conjugated to HRP or a fluorescent tag, the other uses immunocytochemical detection. Fixation can be in buffered paraformaldehyde or in coagulant fixatives (must be tested for each lectin). Cryostat or paraffin sections can be used. The direct method is as follows: 1) dewax sections, 2) incubate in appropriate dilution of lectin conjugate (e.g., 100 µg/ml) for about 30 min; 3) wash three changes in buffer, 4) examine in fluorescence microscope or process for HRP reaction. The two-step antibody technique: (1) dewax, (2) block endogenous peroxidases with 3% hydrogen peroxide in methanol for 10 min, (3) wash in buffer, (4) incubate in lectin (e.g., 10 µg/ml) for 30 min, (5) wash in buffer, (6) incubate in primary unlabeled antilectin serum 1:100 for 30 min, (7) wash in buffer, (8) incubate in label-conjugated secondary antiserum 1:50–1:100 for 30 min, or use the PAP method, and (9) process and examine. Specificity tests are made by using the specific sugars to inhibit or compete with lectin binding to tissues (on alternate sections). Phosphate-buffered saline, with 0.5% bovine serum albumen added, can be used throughout as buffer (Ponder, 1983).

24.6. COMBINATIONS OF TECHNIQUES: DOUBLE LABELING AND DIFFERENTIAL MULTIPLE LABELING

As pointed out in the previous sections, many of the listed markers can be used simultaneously to depict various populations of neurons. Three categories of such labeling can be distinguished: (1) double labeling may be used when resolving, for example, whether a population of neurons project to several areas of the CNS or through several peripheral nerves. This technique (mainly used in the vertebrate brain) involves introduction of two different markers in two different suspected termination areas or axonal paths. If, after processing, each cell body in a given population contains both markers, the assumption was right. (2) Double labeling of identified neurons can be used, for instance, to determine the contents of a putative neuroactive substance within a neuron or neuron population that has been marked by intracellular injection or axonal filling; immunocytochemistry can be performed on dye or tracer-filled neurons. When immunocytochemistry or histochemistry does not label the entire neuron with all its processes or if the axonal projection of such cells is too long to be unambiguously followed, neuronal tracing technique can be used as a complement; dye injection into termination area or axonal tract can be followed by immuno- or histochemistry. A further application is to search for possible co-occurrence of two or more substances within one neuron: different antibodies distinguishable after processing can be used. (3) Differential double or multiple labeling can be used to resolve convergence or divergence of neurons, to determine synaptic contacs between specific neurons, and to analyze superimposed patterns of neuronal architectures. The different markers can be introduced in different centers, termination areas, axonal tracts, or nerves. Another way is to introduce one or more markers as above and then subject the tissue to histo- or immunocytochemical treatment. Finally, various combinations of histo- and immunocytochemical methods can be used.

Many types of markers can be combined for the types of analysis listed above (for further details see Steward, 1983; Nässel, 1983a). The combinations are simply listed here and original references are given: (1) HRP labeling can be combined with Golgi silver impregnation (Fig. 24.3.A; Somogyi et al., 1979; Nässel, 1983a), fluorescent markers (Macagno et al., 1981; Heinrichs, 1985), degeneration (Somogyi et al., 1979), enzyme histochemistry (Lewis and Henderson, 1980), catecholamine histofluorescence (Berger et al., 1978), and immunocytochemistry (Priestly et al., 1981). (2) Cobalt marking can be combined with Golgi impregnation (Mobbs, 1976), with nickel labeling, using rubeanic acid development (Quicke and Brace, 1979), and immunocytochemistry (Eckert and Ude, 1983). (3) Fluorescent markers can be combined with other fluorescent markers (DeOlmos and Heimer, 1980), histochemistry (Björklund and Skagerberg, 1979), immunocytochemistry (Fig. 24.3.D; Adams et al., 1983). (4) Multiple marking with antibodies (Lechago et al., 1979; Gu et al., 1981). (5) Immunocytochemistry with histochemistry (NcNeill and Sladek, 1980). Many of the marking techniques can also be used for electron microscopy (see Section 24.7.). When combining markers, it is important that the tissue processing is compatible for the combined markers. Often compromises must be made.

24.7. MARKERS FOR ELECTRON MICROSCOPY

A problem often facing the electron microscopist studying neuroanatomy, is how to unravel the mass of small profiles in an ultrathin section, and to be certain which processes belong to a given cell. Constructions of wiring diagrams require tedious serial sectioning unless one can selectively identify certain cells with electron-dense markers. Single cells can be marked with Golgi impregnation (Fig. 24.3.B) or intracellular HRP injection, populations of cells with HRP or cobalt mass-filling, or degeneration. Chemically identified populations of neurons can be marked histo- or immunocytochemically or by uptake of tritiated compounds. Here, we shall consider only marking with heme peptides and immunocytochemistry and techniques that can be combined with these.

Heme peptide-labeled neurons are relatively easy to resolve in the electron microscope (Figs. 24.3.C and 24.4.A–C). The osmophilic peroxidase reaction product is distributed throughout the neuronal processes after heavy labeling (Figs. 24.3.C and 24.4.C) (after intracellular injection and massive filling of lesioned neurons). In more weakly labeled neurons (transneuronally filled, or after passage of marker over long distances or after insufficient histochemical reation), reaction product is mainly seen along plasmalemma, neurotubules, postsynaptic membranes, and mitochondria (Fig. 24.4.A,B). The cytological features of the neuronal processes are normally well preserved.

Electron microscopical immunocytochemistry can be done in various ways, either using preembedding labeling or postembedding labeling. In both cases, the tissue must be chemically fixed with an aldehyde fixative. For the preembedding technique, whole brains can be incubated in antisera if they are small, but preferably fragments of brains (cryostat or vibratome sections) should be used to improve penetration of antibodies. This penetration can be further improved with the use of freezing-thawing and/or by use of detergents such as Triton X-100 or saponin. This treatment decreases the tissue preservation slightly. Methods for preembedding incubation with antibodies and subsequent processing for EM are given for peptides by Eckert and Ude (1983) and for serotonin by Nässel and Elekes (1984, 1985) and in the appendix (Fig. 24.2.C–F). These methods use unlabeled secondary antisera and the peroxidase-antiperoxidase complex (Sternberger, 1979).

The postembedding technique relies on embedding the fixed tissue in a resin (e.g., Araldite) before exposure to antibodies. It is crucial that the antigen survives the dehydration and embedding in resin. The detailed pro-

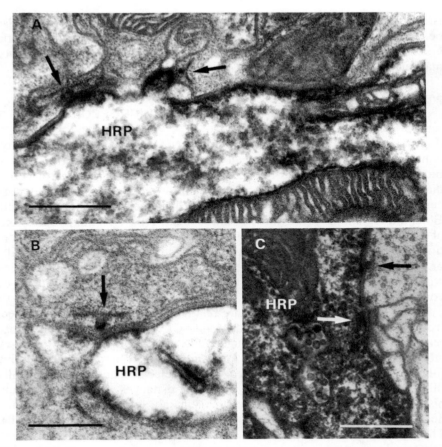

Figure 24.4. Electron microscopy of HRP-labeled visual interneurons (HRP) in *Calliphora*. In A and B transneuronally filled neurons in the lobula plate are postsynaptic (at *arrows*) to other elements. (C) A primarily filled (hence, more reaction product) profile, which is both pre- (*white arrow*) and post- (*black arrow*) synaptic. Scales: A and C, 0.5 μm; B, 0.2 μm.

cedures for semithin-ultrathin sections are given by Eckert and Ude (1983). The sections are first treated with a drop of sodium ethoxide to remove the resin (see also Meyer et al., 1986). Then they are treated with antisera, and the PAP method is performed as for cryostat or paraffin sections, followed by fixation in osmium tetroxide to enhance contrast. The semithin sections can be analyzed by light microscope and then reembedded and sectioned for EM.

The protein A colloidal gold method (Roth et al., 1978) is also described in detail by Eckert and Ude (1983). This procedure can be performed directly on ultrathin sections. Protein A from *Staphyllococcus aureus* interacts with

IgG immunoglobulins from mammals and can be coupled to colloidal gold (as an electron-dense marker). Specific antibodies applied to the tissue sections will bind to antigenic sites, then the protein A-gold complex is applied and bound to the FC region of the γ-globulin of the tissue-bound antibodies. This technique allows marking of subcellular distribution of antigen, for example, a neuroactive substance within vesicles (Eckert and Ude, 1983).

Several markers can be used in combination for electron microscopic multiple marking: HRP and the Golgi method (see Fig. 24.3.B,C), HRP and degeneration, HRP and cobalt labeling, HRP and tritiated compounds, HRP and immuno-PAP method, HRP and immunoprotein A-gold method, immuno-PAP and immunoprotein A-gold method (for details see Stewart, 1981; Nässel, 1983a).

24.8. APPENDIX

HRP Labeling for Electron Microscopy.

1. Fix filled tissue in 2.5% glutaraldehyde in 0.1 M sodiumcacodylate buffer.
2. Wash thoroughly in cacodylate buffer.
3. Preincubate in 0.04% DAB in cacodylate buffer, 1–3 hours in the dark.
4. Incubate in 0.04% DAB and 0.1% H_2O_2 in same buffer, 1–3 hours in the dark.
5. Wash in buffer and fix in 1–2% OsO_4 for 2 hours at 4°C.
6. Wash, dehydrate, and embed in resin. Cut thick or ultrathin sections.

Immuno-PAP Method for Wholemounts, Thick Sections, and EM.

1. Fix in ice-cold 4% paraformaldehyde in 0.1% phosphate buffer. For better ultrastructural preservation add 0.25% glutaraldehyde.
2. Wash in buffer, dissect brains, cut slices, or use whole brain.
3. Incubate in primary antiserum at appropriate dilution 24–48 hours for light microscopy (LM) or 12–24 hours for EM. Serum diluted in phosphate buffered saline (PBS) with 0.25% Triton X-100, 0.25% human serum albumin, and 0.1% sodium azide.
4. Wash in PBS and then incubate in unlabeled secondary antiserum (diluted about 1:50), 12–24 hours for LM and 2 hours for EM. Use same PBS mixture as in step 3.
5. Wash in PBS and then incubate in PAP complex (1:50 in PBS as in step 3) 12–24 hours for LM or 2 hours for EM.
6. Wash in PBS and incubate in 0.06% DAB and 0.1% H_2O_2 in 0.05 Tris-HCl buffer, 1–2 hours.

7. Wash in PBS and then cacodylate buffer. From here tissues can be taken for wholemounts (dehydrate, clear in xylene and methyl salicylate, and embed in Permount or Canada balsam). For EM, continue instead to step 8.
8. Fix in 1–2% OsO_4 for 2 hours- at 4°C. Dehydrate, embed in resin, and cut thick or ultrathin sections.

Lucifer Yellow Combined With Immuno-PAP Labeling.

1. After dye filling, fix in 4% paraformaldehyde in 0.1 M phosphate buffer.
2. Wash in buffer, dehydrate, and embed in paraffin or infiltrate with 25% sucrose and freeze. Cut paraffin or cryostat sections (about 10 μm).
3. Incubate in primary antiserum 24–48 hours at 4°C.
4. Run PAP method as above. When using sections, incubation times can be shortened to about 1 hour for the two subsequent antisera and 20 min for the DAB reaction. Dehydrate and mount with Fluoromount or Entellan.

HRP Filling Combined With Immuno-FITC Labeling.

1. After filling, fix in paraformaldehyde and wash as above. Dissect out brain.
2. Infiltrate with 25% sucrose in buffer. Freeze and cut cryostat sections.
3. Apply primary antiserum for 12–24 hours at 4°C.
4. Wash in PBS and incubate in FITC-labeled secondary antiserum (1:30 in PBS) for 1 hour.
5. Wash in buffer and incubate in 0.06% DAB and 0.1% H_2O_2 in Tris buffer for 10–20 min. Mount with coverslips in PBS-glycerol (1:1).

24.9. SUMMARY

This chapter summarizes some strategies for neuronal marking, useful for a functional analysis of adult and developing brains of arthropods. All marking techniques described display groups of neurons with a common denominator, such as chemical identity, common destination or origin of processes, being part of a pathway, and so on. The methods include tract-tracing techniques, immunocytochemical and histochemical methods, and combinations of these for light and electron microscopy. The emphasis is on how to approach an analysis of the brain by combining different neuronal markers. A catalogue of relevant techniques and their uses is given, and where not all details of the methodology are given, references to original contributions are made. In an appendix, the complete schedules are given for HRP-EM labeling, immuno-PAP EM, and combinations of HRP filling and immuno-FITC as well as Lucifer yellow-immuno-PAP labeling. The conclusion one can draw after attempting an anatomical-functional analysis of an arthropod brain is that no single technique suffices; rather, by using many methods separately or in combination, the relations and roles of single neurons within the dense matrix of nerve cells can slowly be understood.

REFERENCES

Adams, M. E., C. A. Bishop, and M. O'Shea. 1983. Strategies for the identification of amine- and peptide-containing neurons, pp. 239–249. In N. J. Strausfeld (ed.), *Functional Neuroanatomy*. Springer-Verlag, Berlin, Heidelberg, New York.

Bacon, J. P. and J. S. Altman. 1977. A silver intensification method for cobalt-filled neurons in wholemount preparations. *Brain Res.* **138**: 359–363.

Berger, B., J. Nguyen-Legros, and A. M. Thierry. 1978. Demonstration of horseradish peroxidase and fluorescent catecholamines in the same neuron. *Neurosci. Lett.* **9**: 297–302.

Björklund, A. and G. Skagerberg. 1979. Simultaneous use of retrograde fluorescent tracers and fluorescence histochemistry for convenient precise mapping of monoaminergic projections and collateral arrangements in the CNS. *J. Neurosci. Methods* **1**: 261–277.

Blackstad, T. W., L. Heimer and E. Mugnaini. 1981. Experimental neuroanatomy. General approaches and laboratory procedures, pp. 1–53. In L. Heimer and M. J. Robards (eds.), *Neuroanatomical Tract-Tracing Methods*. Plenum, New York.

Buchner, E. and S. Buchner. 1983. Neuroanatomical mapping of visually induced nervous activity in insects by H-deoxyglucose, pp. 623–634. In M. A. Ali (ed.), *Photoreception and Vision in Invertebrates*. Plenum, New York.

DeOlmos, J. and L. Heimer. 1980. Double and triple labeling of neurons with fluorescent substances; the study of collateral pathways in the ascending Raphe system. *Neurosci. Lett.* **19**: 7–12.

Detwiler, P. B. and P. V. Sarthy. 1981. Selective uptake of Lucifer yellow by bipolar cells in turtle retina. *Neurosci. Lett.* **22**: 227–232.

Duve, H. and A. Thorpe. 1983. Immunochemical identification of vertebrate-type brain-gut peptides in insect nerve cells, pp. 250–266. In N. J. Stausfeld (ed.), *Functional Neuroanatomy*. Springer-Verlag, Berlin, Heidelberg, New York.

Eckert, M. and J. Ude. 1983. Immunocytochemical techniques for the identification of peptidergic neurons, pp. 267–301. In N. J. Strausfeld (ed.), *Functional Neuroanatomy*. Springer-Verlag, Berlin, Heidelberg, New York.

Edelman, G. M. 1976. Surface modulation in cell recognition and cell growth. *Science (Washington, D.C.)* **192**: 218–226.

Evans, P. D. 1980. Biogenic amines in the insect nervous system. *Adv. Insect Physiol.* **15**: 317–473.

Fujita, S. C., S. L. Zipursky, S. Benzer, A. Ferrus, and S. L. Shotwell. 1982. Monoclonal antibodies against the *Drosophila* nervous system. *Proc. Natl. Acad. Sci. USA* **79**: 7929–7933.

Goodman, C. S., M. J. Bastiani, C. Q. Doe, S. du Lac, S. L. Helfand, J. Y. Kuwada, and J. B. Thomas. 1984. Cell recognition during neuronal development. *Science (Washington, D.C.)* **225**: 1271–1279.

Gu, J., D. DeMey, M. Moeremans, and J. M. Polak. 1981. Sequential use of the PAP and immunogold staining methods for the light microscopical double staining of tissue antigens. Its application to the study of regulatory peptides in the gut. *Regul. Pept.* **1**: 365.

Heinrichs, S. 1985. Differential retrograde labeling with horseradish peroxidase (HRP) and Lucifer yellow (LY) in an invertebrate nervous system—HRP fluorescence and LY preservation limit choice of fixative. *J. Neurosci. Methods* **15**: 85–93.

Jan, L. Y. and Y. N. Jan. 1982. Antibodies to horseradish peroxidase as specific neuronal markers in *Drosophila* and in grasshopper embryos. *Proc. Natl. Acad. Sci. USA* **79**: 2700–2704.

Kater, S. and C. Nicholson. 1973. *Intracellular Staining in Neurobiology.* Springer-Verlag, Berlin, Heidelberg, New York.

Keshishian, H. and D. Bentley. 1983. Embryogenesis of peripheral nerve pathways in grasshopper legs. I. The initial nerve pathway to the CNS. *Dev. Biol.* **96**: 89–102.

Klemm, N. 1976. Histochemistry of putative transmittersubstances in the insect brain. *Progr. Neurobiol.* **7**: 99–169.

Klemm, N. 1980. Histochemical demonstration of biogenic monoamines (Falck-Hillarp method) in the insect nervous system, pp. 51–73. In N. J. Strausfeld and T. A. Miller (eds.), *Neuroanatomical Techniques: Insect Nervous System.* Springer-Verlag, New York, Heidelberg, Berlin.

Klemm, N. 1983. Detection of serotonin-containing neurons in the insect nervous system by antibodies to 5-HT, pp. 302–316. In N. J. Strausfeld (ed.), *Functional Neuroanatomy.* Springer-Verlag, Berlin, Heidelberg, New York.

Leathem, A. J. C. and N. J. Atkins. 1983. Lectin binding to paraffin sections, pp. 39–70. In G. R. Bullock and P. Petrusz (eds.), *Techniques in Immunocytochemistry*, vol. 2. Academic Press, New York.

Lechago, J. N., C. J. Sun, and W. M. Weinstein. 1979. Simultaneous visualization of two antigens in the same tissue section by combining immunoperoxidase with immunofluorescence techniques. *J. Histochem. Cytochem.* **27**: 1221–1225.

Lewis, P. R. and Z. Henderson. 1980. Tracing putative cholinergic pathways by a dual cytochemical technique. *Brain Res.* **196**: 489–493.

Macagno, E. R., K. J. Muller, B. W. Kristan, S. A. DeRiemer, R. Stewart, and B. Granzow. 1981. Mapping of neuronal contacts with intracellular injection of horseradish peroxidase and Lucifer yellow in combination. *Brain Res.* **217**: 143–149.

Malmgren, L. T., Y. Olsson, T. Olsson, .and K. Kristensson. 1978. Uptake and retrograde axonal transport of various exogenous macromolecules in normal and crushed hypoglossal nerves. *Brain Res.* **153**: 477–493.

Maranto, A. R. 1982. Neuronal mapping: A photooxidation reaction makes Lucifer yellow useful for electron microscopy. *Science (Washington D.C.)* **217**: 953–955.

McNeill, T. H. and J. R. Sladek Jr. 1980. Simultaneous monoamine histofluorescence and neuropeptide immunocytochemistry: V. A methodology for examining correlative monoamine-neuropeptide neuroanatomy. *Brain Res. Bull.* **5**: 599–608.

Meyer, E. P. 1984. Retrograde labeling of photoreceptors in different regions of the compound eyes of bees and ants. *J. Neurocytol.* **13:** 825–836.

Meyer, E. P., C. Matute, P. Streit, and D. R. Nässel. 1986. Insect optic lobe neurons identifiable with monoclonal antibodies to GABA. *Histochemistry* **84:** 207–216.

Mobbs, P. G. 1976. Golgi staining of material containing cobalt-filled profiles in the insect CNS. *Brain Res.* **105:** 563–566.

Nässel, D. R. 1981. Transneuronal labeling with horseradish peroxidase in the visual system of the house fly. *Brain Res.* **206:** 431–438.

Nässel, D. R. 1982. Transneuronal uptake of horseradish peroxidase in the central nervous system of dipterous insects. *Cell Tissue Res.* **225:** 639–662.

Nässel, D. R. 1983a. Horseradish peroxidase and other heme proteins as neuronal markers, pp. 44–91. In N. J. Strausfeld (ed.), *Functional Neuroanatomy*. Springer-Verlag, Berlin, Heidelberg, New York.

Nässel, D. R. 1983b. Extensive labeling of injured neuron with seven different heme peptides. *Histochemistry* **179:** 95–104.

Nässel, D. R. 1987. Serotonin and serotonin-immunoreactive neurons in the insect nervous system. *Progr. Neurobiol.* in press.

Nässel, D. R. and R. Cantera. 1985. Mapping of serotonin-immunoreactive neurons in the larval nervous system of the flies *Calliphora erythrocephala* and *Sarcophaga bullata*. A comparison with ventral ganglia in adult animals. *Cell Tissue Res.* **239:** 423–434.

Nässel, D. R. and K. Elekes. 1984. Ultrastructural demonstration of serotonin-immunoreactivity in the nervous system of an insect (*Calliphora erythrocephala*). *Neurosci. Lett.* **48:** 203–210.

Nässel, D. R. and K. Elekes. 1985. Serotonergic terminals in the neural sheath of the blowfly nervous system: electron microscopical immunocytochemistry and 5,7-dihydroxytryptamine labeling. *Neuroscience* **15:** 293–387.

Nässel, D. R., E. P. Meyer, and N. Klemm. 1985. Mapping and ultrastructure of serotonin-immunoreactive neurons in the optic lobes of three insect species. *J. Comp. Neurol.* **232:** 190–204.

Nässel, D. R., L. Ohlsson, and P. Sivasubramanian. 1987. Differentiation of serotonin-immunoreactive neurons in fly optic lobes developing in situ or cultured in vitro without eye discs. *J. Comp. Neurol.* **255:** 327–340.

Ponder, B. A. J. 1983. Lectin histochemistry, pp. 129–142. In J. M. Polak and S. van Noorden (eds.), *Immunocytochemistry*. Wright PSG, Bristol, London, Boston.

Priestly, J. V., P. Somogyi, and A. C. Cuello. 1981. Neurotransmitter specific projection neurons revealed by combining PAP immunohistochemistry with retrograde transport of HRP. *Brain Res.* **220:** 231–240.

Quicke, D. L. J. and R. C. Brace. 1979. Differential staining of cobalt- and nickel-filled neurons using rubeanic acid. *J. Microsc.* **115:** 1–4.

Roth, J., M. Bendayan, and L. Orci. 1978. Ultrastructural localization of intracellular antigens by the use of protein A-gold complex. *J. Histochem. Cytochem.* **26:** 1074–1081.

Sakai, M. and T. Yamaguchi. 1983. Differential staining of insect neurons with nickel and cobalt. *J. Insect Physiol.* **29:** 393–397.

Schwab, M. E., F. Javory-Agid, and Y. Asid. 1978. Labeled wheat germ agglutinin (WGA) as a new highly sensitive retrograde tracer in the rat hippocampal system. *Brain Res*. **152**: 145–153.

Somogyi, P., A. Hodgson, and A. Smith. 1979. An approach to tracing neuron networks in the cerebral cortex and basal ganglia. Combination of Golgi staining, retrograde transport of horseradish peroxidase and anterograde dengeration of synaptic boutons in the same material. *Neuroscience* **4**: 1805–1852.

Sternberger, L. A. 1979. *Immunocytochemistry*. Second edition. John Wiley, New York.

Steward, O. 1983. Horseradish peroxidase and fluorescent substances and their combination with other techniques, pp. 279–310. In L. Heimer and M. J. Robards (eds.), *Neuroanatomical Tract-Tracing Methods*. Plenum, New York.

Stewart, W. W. 1978. Functional connections between cells, as revealed by dye-coupling with a highly fluorescent naphthalimide tracer. *Cell* **14**: 741–759.

Stewart, W. W. 1981. Lucifer dyes—highly fluorescent dyes for biological tracing. *Nature (London)* **292**: 17–21.

Strausfeld, N. J. and U. K. Bassemir. 1983. Cobalt-coupled neurons of a giant fibre system in Diptera. *J. Neurocytol*. **12**: 971–991.

Strausfeld, N. J. and K. Hausen. 1977. The resolution of neuronal assemblies after cobalt injection into neuropil. *Proc. R. Soc. Lond. B*. **199**: 463–476.

Stuart, A. E., A. J. Hudspeth, and Z. W. Hall. 1974. Vital staining of specific monoamine-containing cells in the leech nervous system. *Cell Tissue Res*. **153**: 55–61.

Triller, A. and H. Korn. 1981. Interneuronal transfer of horseradish peroxidase associated with exo/endocytic activity in adjacent membranes. *Exp. Brain Res*. **43**: 233–236.

Tyrer, N. M., M. K. Shaw, and J. S. Altman. 1980. Intensification of cobalt-filled neurons in sections (light and electron microscopy), pp. 426–446. In N. J. Strausfeld and T. A. Miller (eds.), *Neuroanatomical Techniques: Insect Nervous System*. Springer-Verlag, New York, Heidelberg, Berlin.

Wilcox, M. and N. Franceschini. 1984. Illumination induces dye incorporation in photoreceptor cells. *Science (Washington D.C.)* **225**: 851–853.

Zipser, B. and R. McKay. 1981. Monoclonal antibodies distinguish identifiable neurons in the leech. *Nature (London)* **289**: 549–554.

Zipursky, S. L., T. R. Venkatesh, D. B. Teplow, and S. Benzer. 1984. Neuronal development in the Drosophila retina: Monoclonal antibodies ans molecular probes. *Cell* **36**: 15–16.

Taxonomic Index

Subject Index

575